OXFORD STUDIES IN NUCLEAR PHYSICS

General editor: P. E. Hodgson

Nucleon–Hadron Many-Body Systems

From Hadron–Meson
to Quark–Lepton
Nuclear Physics

Edited by

HIROYASU EJIRI
and
HIROSHI TOKI

Research Center for Nuclear Physics
Osaka University

UNIVERSITY PRESS

OXFORD
UNIVERSITY PRESS

Great Clarendon Street, Oxford OX2 6DP

Oxford University Press is a department of the University of Oxford.
It furthers the University's objective of excellence in research, scholarship,
and education by publishing worldwide in

Oxford New York

Athens Auckland Bangkok Bogota Buenos Aires Calcutta
Cape Town Chennai Dar es Salaam Delhi Florence Hong Kong Istanbul
Karachi Kuala Lumpur Madrid Melbourne Mexico City Mumbai
Nairobi Paris São Paolo Singapore Taipei Tokyo Toronto Warsaw

with associated companies in Berlin Ibadan

Oxford is a trade mark of Oxford University Press
in the UK and in certain other countries

Published in the United States
by Oxford University Press, Inc., New York

© Oxford University Press, 1999

The moral rights of the author have been asserted

Database right Oxford University Press (maker)

First published 1999

All rights reserved. No part of this publication may be reproduced,
stored in a retrieval system, or transmitted, in any form or by any means,
without the prior permission in writing of Oxford University Press,
or as expressly permitted by law, or under terms agreed with the appropriate
reprographics rights organisation. Enquiries concerning reproduction
outside the scope of the above should be sent to the Rights Department,
Oxford Unviersity Press, at the address above

You must not circulate this book in any other binding or cover
and you must impose this same condition on any acquiror

British Library Cataloguing in Publication Data
Data available

Library of Congress Cataloguing in Publication Data
Data available

ISBN 0 19 851900 1

Typeset by
Newgen Imaging Systems (P) Ltd., Chennai, India
Printed in Great Britain by
Biddles Ltd, Guildford and King's Lynn

Preface

Nuclear physics is changing very rapidly. The nucleon–hadron many-body systems, usually described in terms of the nucleon–meson degrees of freedom, are now described more microscopically in terms of the quark–lepton degrees of freedom. The Research Center for Nuclear Physics (RCNP) was founded in the middle of this development at Osaka University in Japan and has witnessed some glorious activities in nuclear physics for nearly 25 years. This was made possible by the existing medium-to-high energy ($K = 60$–400 MeV) cyclotron at the Center. An additional accelerator has been planned to be established in the near future, which will help the Center to continue to participate in these developments. At RCNP, an active programme of theoretical studies has been pursued, which utilizes the exciting experimental data produced with these world-class frontier machines. Given this situation, we felt that it is very important to document the significance of various subjects and achievements in these fields.

We asked several active researchers in Japan who have made and continue to make great contributions to the development of nuclear physics associated with RCNP to write the various chapters of this book. We have targetted advanced graduate students and young researchers in physics as the readers of this book. Hence, each chapter has an introduction of appropriate size, with a brief history to the subject and enough materials, e.g. the theoretical background and the experimental techniques, to understand and appreciate the subject. This is followed by a discussion of the achievements in the subject leading to the current status, so that those who are interested in pursuing the subject know clearly what could be done at the next stage.

We start with an introductory description of developments in individual areas in Chapter 1. The details of various current activities are followed in Chapters 2–10. As for the future directions, we discuss nucleon–hadron many-body systems in terms of quarks and leptons in Chapters 11 and 12. The contributing authors are all active and leading scientists in the field and have written on subjects they are most familiar with. We would like to thank all the authors for making time for writing the respective chapters in spite of their busy schedules. In addition, we acknowledge Professor P. Hodgson for suggesting various improvements in the present book and also for reading the manuscripts.

Osaka
April 1998
H. E.
H. T.

Contents

List of contributors   ix

Figure acknowledgements   xii

1. Introduction   1
   H. Ejiri (RCNP Osaka) and H. Toki (RCNP Osaka)

2. Elastic and inelastic scattering   18
   H. Sakaguchi (Kyoto)

3. Spin–isospin responses and charge-exchange reactions   50
   M. Fujiwara (RCNP Osaka)

4. Giant resonances and highly excited states   77
   Y. Fujita (Osaka)

5. Breakup processes and cluster structures of light composite projectiles   126
   Y. Sakuragi (Osaka City)

6. Nuclear clusters   150
   S. Ohkubo (Kochi W.U.), T. Yamaya (Tohoku), and P. E. Hodgson (Oxford)

7. Physics of snowballs—impurity ions in superfluid helium   190
   N. Takahashi (Osaka) and T. Shimoda (Osaka)

8. Deeply bound pionic atoms   210
   *H. Toki (RCNP Osaka), S. Hirenzaki (Nara W.U.), and
   N. Matsuoka (RCNP Osaka)*

9. Atomic physics with cyclotron beams   226
   I. Katayama (IPNS-KEK)

10. Nuclear physics with strange flavour   250
    T. Kishimoto (Osaka) and T. Motoba (Osaka E.C.)

11. Quark nuclear physics   292
    H. Toki (RCNP Osaka) and H. Suganuma (RCNP Osaka)

12. Lepton nuclear physics   313
    H. Ejiri (RCNP Osaka)

Index   337

Contributors

H. Ejiri

Research Center for Nuclear Physics (RCNP), Osaka University,
Mihogaoka 10-1, Ibaraki, Osaka 567-0047, Japan
e-mail: ejiri@rcnp.osaka-u.ac.jp

Y. Fujita

Department of Physics, Osaka University,
Machikaneyama 1-1, Toyonaka, Osaka 560-0043, Japan
e-mail: fujita@rcnp.osaka-u.ac.jp

M. Fujiwara

Research Center for Nuclear Physics (RCNP), Osaka University,
Mihogaoka 10-1, Ibaraki, Osaka 567-0047, Japan
e-mail: fujiwara@rcnp.osaka-u.ac.jp

S. Hirenzaki

Department of Physics, Nara Women's University,
Nishimachi, Kitauoya, Nara 630-8506, Japan
e-mail: zaki@phys.nara-wu.ac.jp

P. E. Hodgson

Department of Physics, Nuclear Physics Laboratory,
Oxford OX1 3RH, UK

I. Katayama

Center for Nuclear Study (CNS), University of Tokyo, Midorichou 3-2-1,
Tanashi, Tokyo 188-0002, Japan
e-mail: ktymichi@tanashi.kek.jp

T. Kishimoto

Department of Physics, Osaka University,
Toyonaka, Osaka 560-0043, Japan
e-mail: kisimoto@phys.sci.osaka-u.ac.jp

N. Matsuoka

Research Center for Nuclear Physics (RCNP), Osaka University,
Mihogaoka 10-1, Ibaraki, Osaka 567-0047, Japan
e-mail: matsuoka@rcnp.osaka-u.ac.jp

T. Motoba

Department of Physics, Osaka Electrocommunication University,
Hatsumachi 18-8, Neyagawa, Osaka 572-8530, Japan
e-mail: motoba@isc.osakac.ac.jp

S. Ohkubo

Department of Applied Science, Kochi Women's University,
Eikokuji 5-15, Kochi 780-8515, Japan
e-mail: ohkubo@yukawa.kyoto-u.ac.jp

H. Sakaguchi

Department of Physics, Kyoto University,
Kitashirakawa, Sakyoku, Kyoto 606-8502, Japan
e-mail: sakaguchi@ne.scphys.kyoto-u.ac.jp

Y. Sakuragi

Department of Physics, Osaka City University,
Sugimoto 3-3-138, Sumiyoshi, Osaka 558-8585, Japan
e-mail: sakuragi@ocunp.hep.osaka-cu.ac.jp

T. Shimoda

Department of Physics, Osaka University,
Machikaneyoma 1-16, Toyonaka, Osaka 560-0043, Japan
e-mail: shimoda@rcnp.osaka-u.ac.jp

H. Suganuma

Research Center for Nuclear Physics (RCNP), Osaka University,
Mihogaoka 10-1, Ibaraki, Osaka 567-0047, Japan
e-mail: suganuma@rcnp.osaka-u.ac.jp

N. Takahashi

Department of Physics, Osaka University,
Machikaneyama 1-16, Toyonaka, Osaka 560-0043, Japan
e-mail: ntakahas@rcnp.osaka-u.ac.jp

H. Toki

Research Center for Nuclear Physics (RCNP), Osaka University,
Mihogaoka 10-1, Ibaraki, Osaka 567-0047, Japan
e-mail: toki@rcnp.osaka-u.ac.jp

T. Yamaya

Department of Physics, Tohoku University,
Aramaki, Aoba, Sendai 980-8578, Japan

Figure acknowledgements

We are grateful to the following authors for permission to reproduce figures from their papers or books: Prof. H. Sakaguchi (Figs 2.1, 2.6, 2.7, 2.8, 2.9, 2.10, 2.11, 2.12, 2.14, 2.15, 2.16, 2.17, 2.19, 2.20, 2.22, 2.23, and 2.24), Prof. N. Olsson (Fig. 2.4), Prof. N. Yamaguchi (Fig. 2.5), Prof. M. Fujiwara (Figs 3.1, 3.6, 4.10, 4.16, and 12.9), Prof. B. L. Berman (Fig. 4.4), Prof. F. E. Bertrand (Fig. 4.5), Prof. D. H. Youngblood (Fig. 4.6), Prof. A. van der Woude (Fig. 4.7), Prof. J. M. Moss (Fig. 4.8), Prof. A. M. Rozsa (Fig. 4.9), Prof. Y. Fujita (Figs 4.11, 4.12, 4.13, 4.14, 4.15, 4.17, 4.18, 4.19, 4.29, 4.30, 4.32, and 4.33), Prof. T. Yamagata (Figs 4.20 and 4.21), Prof. S. Nakayama (Figs 4.22 and 4.23), Prof. H. Ejiri (Figs 4.24, 4.26, 4.27, 4.28, 12.3, 12.5, 12.10, 12.11, 12.12, and 12.15), Prof. M. Yahiro (Fig. 5.4), Prof. H. Nishioka (Fig. 5.4), Prof. Y. Sakuragi (Figs 5.5, 5.4, 5.5, 5.6, 5.7, and 5.8), Prof. E. Uegaki (Fig. 6.1), Prof. Y. Fujiwara (Figs 6.2 and 6.3), Prof. S. Ohkubo (Figs 6.4, 6.5, 6.6, 6.7, 6.9, 6.10, 6.11, 6.12, 6.13, 6.14, 6.15, 6.16, 6.17, 6.18, 6.19, and 6.20), Prof. A. C. Merchant (Fig. 6.8), Prof. F. London (Fig. 7.1), Prof. K. R. Atkins (Fig. 7.2), Prof. P. V. E. McClintock (Figs 7.3 and 7.4), Prof. T. Shimoda (Figs 7.7, 7.8, and 7.10), Prof. N. Takahashi (Figs 7.9 and 7.11), Prof. H. Toki (Figs 8.1, 8.2, 8.3, 8.4, 8.5, 8.9, 8.10, 11.1, 11.2, 11.3, 11.4, 11.5, and 11.6), Prof. N. Matsuoka (Figs 8.6 and 8.8), Prof. S. Ito (Figs 9.1 and 9.2), Prof. I. Katayama (Figs 9.6, 9.10, 9.11, 9.12, 9.13, 9.14, 9.15, 9.16, 9.17, 9.18, and 9.19), Prof. W. Schorsch (Fig. 10.5).

Figs 2.1, 2.8, 2.9, 2.11, and 2.12 are reprinted with permission from the American Physical Society; *Phys. Rev.* **C26** (1982) 944. Fig. 2.4 is reprinted with permission from Elsevier Science Publications B.V.; *Nucl. Phys.* **A472** (1987) 237. Fig. 2.5 is reprinted with permission from the Editorial Office of *Progress of Theoretical Physics*; *Prog. Theor. Phys.* **76** (1986) 459. Fig. 2.6 is reprinted with permission from the Physical Society of Japan; *J. Phys. Soc. Japan* **55** (Suppl.) (1986) 61. Fig. 2.7 is reprinted with permission from Elsevier Science Publications B.V.; *Nucl. Phys.* **A366** (1981) 189. Figs 2.10, 2.14, 2.15, and 2.16 are reprinted with permission from the American Physical Society; *Phys. Rev.* **C29** (1984) 1228. Figs 2.17, 2.19, and 2.20 are reprinted with permission from the American Physical Society; *Phys. Rev.* **C36** (1987) 1754. Figs 2.22, 2.23, and 2.24 are reprinted with permission from the American Physical Society; *Phys. Rev.* **C57** (4) (1998) 1749. Fig. 3.1 is reprinted with permission from the American Physical Society; *Phys. Rev.* **C52** (1995) 604. Fig. 3.6 is reprinted with permission from Elsevier Science Publications B.V.; *Nucl. Phys.* **A599** (1996) 223c.

Fig. 4.4 is reprinted with permission from the American Physical Society; *Rev. Mod. Phys.* **47** (1975) 713. Fig. 4.5 is reprinted with permission from Elsevier Science Publications B.V.; *Nucl. Phys.* **A354** (1981) 129c. Fig. 4.6 is reprinted with permission from the American Physical Society; *Phys. Rev.* **C13** (1976) 994. Fig. 4.7 is reprinted with permission from World Scientific Publishing Co. Pte. Ltd; *World Scientific* (1991), p. 99. Fig. 4.8 is reprinted with permission from the American Physical Society; *Phys. Rev.* **C18** (1978) 741. Fig. 4.9 is reprinted with permission from the American Physical Society; *Phys. Rev.* **C21** (1980) 1252. Fig. 4.10 is reprinted with permission from Elsevier Science Publications B.V.; *Nucl. Instrum. and Meth.* **175** (1980) 335. Figs 4.11, 4.13, and 4.14 are reprinted with permission from the American Physical Society; *Phys. Rev.* **C37** (1988) 45. Fig. 4.12 is reprinted with permission from the American Physical Society; *Phys. Rev.* **C32** (1985) 425. Fig. 4.15 is reprinted with permission from the American Physical Society; *Phys. Rev.* **C45** (1992) 993. Fig. 4.16 is reprinted with permission from the American Physical Society; *Phys. Rev.* **C39** (1989) 1286. Figs 4.17, 4.18, and 4.19 are reprinted with permission from the American Physical Society; *Phys. Rev.* **C40** (1989) 1595. Fig. 4.20 is reprinted with permission from the American Physical Society; *Phys. Rev.* **C23** (1981) 937. Fig. 4.21 is reprinted with permission from Elsevier Science Publications B.V.; *Nucl. Phys.* **A381** (1982) 277. Fig. 4.22 is reprinted with permission from Elsevier Science Publications B.V.; *Nucl. Instrum. and Meth.* **A302** (1991) 472. Fig. 4.23 is reprinted with permission from the American Physical Society; *Phys. Rev.* **C46** (1992) 1667. Figs 4.24, 4.26, and 4.28 are reprinted with permission from IOP Publishing Limitted and Elsevier Science Publications B.V.; *J. Phys.* **C4** (1984) 135; *Nucl. Phys.* **A482** (1988) 471c. Fig. 4.27 is reprinted with permission from the American Physical Society; *Phys. Rev. Lett.* **48** (1982) 1382. Figs 4.29 and 4.30 are reprinted with permission from the American Physical Society; *Phys. Rev.* **C55** (1997) 1137. Figs 4.32 and 4.33 are reprinted with permission from Elsevier Science; *Phys. Lett.* **365B** (1996) 29. Fig. 5.4 is reprinted with permission from the Editorial Office of *Progress of Theoretical Physics* and Elsevier Science Publications B.V.; *Prog. Theor. Phys.* (Suppl.) **89** (1986) 32, *Nucl. Phys.* **A415** (1984) 230, and *Nucl. Phys.* **A415** (1984) 271. Fig. 5.5 is reprinted with permission from the Editorial Office of *Progress of Theoretical Physics* and Elsevier Science Publications B.V.; *Prog. Theor. Phys.* **70** (1983) 1047 and *Nucl. Phys.* **A480** (1988) 361. Fig. 5.6 is reprinted with permission from Elsevier Science; *Phys. Lett.* **B220** (1989) 22. Fig. 5.7 is reprinted with permission from the American Physical Society; *Phys. Rev. Lett.* **69** (1992) 1892.

Fig. 5.8 is reprinted with permission from Elsevier Science; *Phys. Lett.* **B205** (1988) 204. Fig. 6.1 is reprinted with permission from the Editorial Office of *Progress of Theoretical Physics*; *Prog. Theor. Phys.* (Kyoto) **57**, 1262 (1977); *Prog. Theor. Phys.* (Kyoto) **59**, 1031 (1978); **62**, 1621 (1979). Fig. 6.2 is reprinted with permission from the Editorial Office of *Progress of Theoretical Physics*; *Prog. Theor. Phys.* (Kyoto) (Suppl.) **68** (1980). Fig. 6.3 is reprinted with permission from the Editorial Office of *Progress of Theoretical Physics*; *Prog. Theor. Phys.* (Kyoto) **62**, 122 (1979). Fig. 6.5 is reprinted with permission from the American Physical Society; *Phys. Rev.* **C34**, 1248 (1986). Fig. 6.6 is reprinted with permission from the American Physical Society; *Phys. Rev. Lett.* **57**, 1215 (1986); *Phys. Rev.* **C37**, 292 (1988). Fig. 6.7 is reprinted with permission from the American Physical Society; *Phys. Rev.* **C38**, 2377 (1988). Fig. 6.8 is reprinted with permission from IOP Publishing Limited; *J. Phys.* **G15**, 601 (1989). Figs 6.9, 6.10, and 6.11 are reprinted with permission from the American Physical Society;

Phys. Rev. **C42**, 1935 (1990). Fig. 6.12 is reprinted with permission from Springer-Verlag GmbH & Co. KG; *Z. Phys.* **A349**, 363 (1994). Figs 6.13 and 6.14 are reprinted with permission from Elsevier Science; *Phys. Lett.* **B306**, 1 (1993). Fig. 6.15 is reprinted with permission from the American Physical Society; *Phys. Rev.* **C49**, 149 (1994). Figs 6.16 and 6.17 are reprinted with permission from the American Physical Society; *Phys. Rev.* **C51**, 586 (1995). Fig. 7.9 is reprinted with permission from Baltzer Science Publishers, Amesterm, The Netherlands; *Hyperfine Interactions* **97/98**, 469. Fig. 7.10 is reprinted with permission from Elsevier Science Publications B.V.; *Nucl. Phys.* **A588**, 235c. Fig. 7.11 is reprinted with permission from Springer-Verlag GmbH & Co. KG; *Z. Phys.* **B98**, 347. Fig. 8.1 is reprinted with permission from Elsevier Science; *Phys. Lett.* **B213** (1988) 129. Fig. 8.2 is reprinted with permission from Elsevier Science Publications B.V.; *Nucl. Phys.* **A501** (1989) 653. Fig. 8.3 is reprinted with permission from the American Physical Society; *Phys. Rev.* **C43** (1991) 1099. Fig. 8.4 is reprinted with permission from Elsevier Science Publications B.V.; *Nucl. Phys.* **A530** (1991) 679. Fig. 8.5 is reprinted with permission from Elsevier Science Publications B.V.; *Nucl. Phys.* **A553** (1993) 221c. Figs 8.6 and 8.8 are reprinted with permission from Elsevier Science; *Phys. Lett.* **B359** (1995) 39. Figs 8.9 and 8.10 are reprinted with permission from Springer-Verlag GmbH & Co. KG; *Z. Phys.* **A355** (1996) 219. Figs 9.1 and 9.2 are reprinted with permission from IOP Publishing Limited; *J. Phys.* **B20** (1987) L597. Fig. 9.6 is reprinted with permission from Elsevier Science; *Phys. Lett.* **A144** (1990) 357. Figs 9.6 and 9.7 are is reprinted with permission from Elsevier Science Publications B.V.; *Nucl. Instr. Meth.* **B56/57** (1991) 180. Fig. 9.10 is reprinted with permission from Elsevier Science Publications B.V.; *Nucl. Instr. Meth.* **A262** (1987) 23. Figs 9.10, 9.11, 9.12, and 9.13 are reprinted with permission from the American Physical Society; *Phys. Rev.* **A52** (1996) 23. Fig. 9.14 is reprinted with permission from Elsevier Science Publications B.V.; *Nucl. Instr. Meth.* **171** (1980) 195. Figs 9.15, 9.16, and 9.17 are reprinted with permission from the American Physical Society; *Phys. Rev.* **A43** (1991) 11 370. Figs 9.18 and 9.19 are reprinted with permission from the American Physical Society; *Phys. Rev.* **A50** (1994) 3533 and *Phys. Rev.* **A51** (1995) 3868. Fig. 10.5 is reprinted with permission from Elsevier Science Publications B.V.; *Nucl. Phys.* **B25** (1970) 179. Figs 11.1, 11.2, and 11.3 are reprinted with permission from Elsevier Science Publications B.V; *Nucl. Phys.* **B435** 207; *Nucl. Phys.* **A557** (1994) 353c; *Prog. Theor. Phys.* **120** (Suppl.) (1995). Figs 11.4 and 11.5 are reprinted with permission from the American Physical Society; *Phys. Rev.* **D52** (1995) 2994. Fig. 11.6 is reprinted with permission from Elsevier Science; *Phys. Lett.* **B339** (1997) 141. Fig. 12.3 is reprinted with permission from Elsevier Science Publications B.V.; *Nucl. Phys.* **A522** (1991) 305c; *Int. J. Mod. Phys.* **E6** (1) (1997) 1. Fig. 12.5 is reprinted with permission from Elsevier Science Publications B.V., Elsevier Science, IOP Publishing Limited, and the American Physical Society; *Nucl Instr. Meth.* **A302** (1991) 304; *Phys. Lett.* **258B** (1991) 17; *J. Phys. G. Nucl. Phys.* **17** (1991) S155; *Phys. Rev.* **D48** (1993) 5412; *Phys. Rev.* **C46** (1992) R2132. Fig. 12.9 is reprinted with permission from Elsevier Science; *Phys. Lett.* **B394** (1997) 23. Figs 12.10, 12.11, and 12.12 are reprinted with permission from Elsevier Science; *Phys. Lett.* **B317** (1993) 14; private communication (1995). Fig. 12.15 is reprinted with permission from Elsevier Science; *Phys. Lett.* **B228** (1989) 24.

Every effort has been made to obtain permission to reproduce copyright figures and tables, but we regret that there are instances in which we have been unable to trace or elicit a reply from the copyright holder. If notified, the publisher will be pleased to rectify any errors or omissions at the earliest opportunity.

1

Introduction

H. Ejiri and H. Toki

1.1 Nucleon–hadron many-body systems

The nuclear physics of nucleon–hadron many-body systems is concerned with fundamental structures of interacting nuclear particle systems. Here, nuclear particles are particles involved in nucleon/hadron systems and interactions therein. The particles to be considered are primarily quarks in the case of hadrons, and are mostly hadrons in the case of nuclei. Nuclear interactions associated with them are strong, weak, and electromagnetic ones. They are given by gauge fields with relevant gauge bosons. Then the gauge bosons to be considered in quark–lepton systems are gluons, weak bosons, and photons. Leptons are necessarily involved in nuclear weak processes. Gravitational interactions are significant only in special cases in astrophysics, such as neutron stars and supernovae. In the case of nucleon–hadron systems, meson exchange interactions and effective nucleon interactions are used in place of the QCD gluon field due to confinement of quarks and spontaneous chiral symmetry breaking.

Nuclear systems are cold systems, where symmetries are spontaneously broken. The nuclear particles (quarks and hadrons) have different masses, and nuclear interactions (gauge bosons and mesons) are quite different from each other in strength (mass). They are mostly in low-energy quantum states with good quantum numbers of E and J^π, reflecting symmetries in the time transformation and the space rotation.

In nuclear systems, however, partial symmetries are realized so that low-lying states have nearly good quantum numbers such as baryon and lepton numbers, parities, flavours, isospins, and so on. These quantum numbers are associated with large asymmetries (gaps) between quarks and leptons, strong and weak interactions, individual generations, and between electromagnetic and strong interactions. Thus nuclear systems are good microlaboratories for studying fundamental particles and interactions. Precision studies of symmetries and asymmetries in nuclear microlaboratories give stringent verifications of the standard theory and may lead to unified theories beyond the standard theory.

Nuclear many-body systems are interacting finite-body systems with particle numbers $A = 2$–10^3. Consequently, there exist interesting interplays of individual particle motions in effective (one-body) mean fields and coherent motions due to effective (two-body) interactions. Here, various types of the coherent motions are described as relevant vibrations in Random Phase Approximation (RPA). Interplays and mixings of different phases, which are characteristic of finite-body systems, are studied in nuclear systems.

There are various ways of studying nuclear many-body systems, depending on subjects to be investigated and on methods to be used. Experimental studies of nuclear many body systems emphasize the following in the sensitivity/quality frontier.

1. High-quality medium low energy probes of light GeV/c particles such as e, γ, and light ions. Here, nuclear systems to be studied consist mostly of such light quarks and leptons in the first and second generations as u, d, s quarks, e, ν_e and μ, ν_μ leptons and nucleons, hyperons, mesons, and others in the nucleon–hadron systems.
2. High-sensitivity detectors, high-resolution spectrometers, high-speed computers, and low-background (underground) measurements to detect small and important features in nuclear many-body systems.
3. Multiple methods with hadron, lepton, and photon probes through strong, weak, and electromagnetic interactions.

These points are complementary to large-scale laboratories with high-energy/high-intensity accelerators. There are many medium-energy laboratories in the world, which have unique accelerators, powerful spectrometer–detector systems, and strong research groups. Consequently, collaborations between these medium-size laboratories are of vital importance.

The Research Center for Nuclear Physics (RCNP) at Osaka University is a medium-energy laboratory. Extensive studies on many-body nucleon–hadron systems have been made at RCNP. Some of them have been carried out in collaboration with other medium-energy laboratories. The recent RCNP activities are summarized briefly in books [1,2] and in proceedings of recent symposia [3].

1.2   RCNP laboratory complex

The present RCNP is a laboratory complex. It consists of a cyclotron laboratory, the Oto Cosmo Observatory, and a laser electron photon laboratory. The cyclotron laboratory is a major laboratory located at Osaka University campus. The accelerator consists of the AVF cyclotron with $K = 0.14\,\text{GeV}$ and the ring cyclotron with $K = 0.4\,\text{GeV}$. The AVF cyclotron is the injector cyclotron for the ring cyclotron. The ring cyclotron provides polarized protons up to $0.42\,\text{GeV}$, ^3He up to $0.5\,\text{GeV}$, and other light ions up to $0.1\,\text{GeV/nucleon}$. Polarized and unpolarized ions are produced externally, and are introduced into the injector cyclotron through the axial injection device [4].

The accelerated beams are distributed into five beam lines. The major spectrometer–detector system is set at the WS beam line. It is a two-arm spectrometer coupled with multi-detector arrays [5]. This system is used extensively for studies of nucleon–meson nuclear physics. The two-arm spectrograph consists of the high-resolution spectrograph GRAND RAIDEN with a momentum resolution $\Delta p/p \sim 3 \cdot 10^{-5}$ and the Large-Acceptance Spectrograph (LAS) with an acceptance $\Delta\Omega \sim 20\,\text{m\,sr}$. GRAND RAIDEN has a polarimeter at the focal plane to measure polarizations of the scattered particles.

Three kinds of multi-detector arrays are used at the target spot of the spectrometer. The multi-Si(Li) detector array, the neutron detector array, and the γ-ray detector array are used to measure decay particles from excited states produced by nuclear reactions.

The long-baseline TOF analysis, with flight length up to 100 m for fast neutrons, is made at the WN beam line [6].

The secondary particle beam line EN provides spin-polarized unstable projectiles used to study reaction mechanisms of light and heavy nuclei. Hyperon production via flavour-changing weak processes is employed for studies of the strangeness-related nuclear physics at the beam line ES.

The Oto Cosmo Observatory is a low-background underground laboratory [7]. It is located at the middle of the Tentsuji tunnel, 100 km south of Osaka. High-sensitivity detectors ELEGANT (ELEctron GAmma-ray Neutrino Telescope) V and VI are set in the two observation rooms of the underground laboratory to study neutrinos (ν's) and Dark Matter (DM).

The laser electron photon laboratory is built to study low-energy quark nuclear physics. Multi-GeV polarized photons are obtained by the Compton backscattering of ultraviolet laser photons from 8 GeV electrons of SPring-8, which is located 120 km west of Osaka [8]. SPring-8 is the 8 GeV electron ring used primarily for synchrotron radiation sources.

The Compton scattered photon energy E_γ at angle θ is given by

$$E_\gamma = E_{\rm o} \cdot 4f/(1 + f\theta^2), \tag{1.1}$$

$$f = E_{\rm e}^2/(1 + 4E_{\rm o}E_{\rm e}), \tag{1.2}$$

where the electron energy $E_{\rm e}$ and the laser photon energy $E_{\rm o}$ are given in units of the electron mass. Since SPring-8 has the highest-energy electron of $E_{\rm e} = 8$ GeV, one can get the highest-energy photon with energy up to 3.5 GeV.

Polarized GeV photons obtained by using polarized laser photons are very useful for spectroscopic studies of hadron structures and non-perturbative QCD phenomena.

1.3 Nucleon many-body systems

Protons and light projectiles with medium energies of 0.1–0.4 GeV are adequate for studying the effects of the nuclear medium and spin–isospin responses therein. Since nucleon inelastic (Δ-production) cross sections are small in this energy region, these medium-energy probes access directly the nuclear interior without many distortion effects.

1.3.1 *Nucleon and nuclear interactions in nuclear medium*

Nucleon–Nucleon (NN) interactions in nuclear medium are studied by measuring quasi-elastic NN scatterings [9],

$$p_{\rm i} + p_0 = p_1 + p_2, \tag{1.3}$$

where $p_{\rm i}$ and p_0 are incident and target nucleons, and p_1 and p_2 are scattered protons measured by the two-arm spectrograph. Selecting p_0 in various s-orbits of nuclei, one can study the NN interactions as a function of the nucleon density in the nuclear medium. Differential cross sections, analysing powers, and polarization transfers have been studied for quasi-free scatterings at several nuclei.

The observed values for the analysing powers show clearly that effective nucleon masses are reduced in the nuclear medium, depending on the nucleon density. They are interpreted partly as the relativistic effects of Dirac particles in the nuclear medium. Further studies of quasi-free scatterings are of interest in the observation of nuclear medium effects of mesons mediating nuclear interactions, as the low-energy QCD suggests.

Precision measurements of proton elastic scatterings from nuclei provide the nuclear optical potentials and, accordingly, allow the effective NN interactions to be folded to build the optical potential. Nuclear elastic scatterings of polarized protons were measured to study nucleon–nucleus interactions [10]. The observed values show clearly deviations from values calculated by using the optical potential based on the free NN interactions. This fact suggests finite effects of the nuclear medium. Elastic and inelastic scatterings of medium-energy nucleons from nuclei are described in Chapter 2.

Effective nucleon interactions in breakup processes of light projectiles are studied by measuring $^6\mathrm{Li} + \mathrm{A} \to \mathrm{d} + \alpha + \mathrm{A}'$ at $E/A = 0.1\,\mathrm{GeV}$ [11]. Breakup processes and cluster structures of light composite projectiles are discussed in Chapter 5. The present focus of the cluster structure studies is the α-cluster structure in the $A = 40$ mass region as discussed in detail in Chapter 6.

1.3.2   *Spin–isospin waves and interactions in nuclei*

Spin–isospin $(\sigma\tau)$ modes of nucleons are important modes of nucleon motions in the medium-energy region below the threshold energy of Δ isobar productions, which are spin–isospin modes of quarks in a nucleon. The nucleon motions associated with spins and isospins are interesting from several viewpoints. Spin–isospin interactions give rise to spin–isospin giant resonances. These interactions are associated with π and ρ meson exchange interactions. Nuclear spin–isospin responses are relevant to axial-vector weak responses in nuclei. Actually, they are crucial for studies of neutrinos in nuclei.

The isospin interaction involves the Coulomb interaction, which gives rise to finite isospin mixing (proton–neutron asymmetries) in nuclei. Since the Coulomb interaction is long-range and weak in strength, isospin becomes a rather good quantum number. Then, the Isobaric Analogue State (IAS), $|\,\mathrm{IAS}\rangle = T^-\,|\,0\rangle$, becomes a good eigenstate with a narrow width Γ_τ. Thus, most isospin strengths are included in the sharp IAS resonance region.

Nucleon spin interactions, however, are quite different from nucleon isospin interactions because of the strong short-range nature of the former. The $\sigma\tau$ supermultiplet is not valid in nuclei, and σ itself is not a good quantum number. Spin–isospin (Gamow–Teller) giant resonances $|\mathrm{GT}\rangle$ are not sharp resonances but rather broad ones. Consequently, they include only a fraction of the sum-rule limit of the total nucleon $\sigma\tau$ strength in their resonance regions because of strong couplings with other nuclear states and nucleon resonance states such as the Δ isobars. The couplings lead to finite $\sigma\tau$ responses for low-lying states as well, which are relevant to low-energy axial weak β, ν, and $\beta\beta$ processes.

The nucleon $\sigma\tau$ modes have extensively been studied by means of charge exchange nuclear reactions at RCNP since the ring cyclotron energy is just adequate for preferential excitation of the $\sigma\tau$ modes. Systematic studies of (p, n) reactions show considerable strengths spread well above the GT ($\sigma\tau\ J^\pi = 1^+$) resonance due to coupling with other excitation modes there [12].

Extensive studies of charge exchange (^{3}He, t) reactions have been made to study $\sigma\tau$ structures in finite-body nucleon systems. It is shown that (^{3}He, t) reactions proceed essentially as (p, n) reactions with the singlet pn pair acting as a spectator in both ^{3}He projectiles and t-ejectiles. Here the charged spectator makes their high energy-resolution studies possible using the magnetic spectrograph. The GT resonances with $J^\pi = 1^+$ and spin–isospin multipole resonances with $J^\pi = 0^-, 1^-$, and 2^- have been studied in a wide range of nuclear mass. Particle and γ-decays of $\sigma\tau$ resonances have shown microscopic structures of nucleon motions involved in the resonances. The particle hole configuration of the GT resonance was studied by measuring the (^{3}He, tp) reaction on ^{208}Pb [13].

Spin–isospin modes with the isospin raising operator τ^+ have been studied by means of (d, ^{2}He) and (^{7}Li, ^{7}Be) reactions [14]. These reactions proceed as a (n, p) reaction with charged spectators of p and ^{6}Li acting as respective spectators. Medium-energy d and ^{7}Li projectiles with $E/A \sim 0.1\,\text{GeV}$ are very useful for studies of $\tau^+\sigma$ modes in comparison with the radioactive neutral projectile n used in (n, p) reactions.

^{7}Li has the spin-parity of $J^\pi = (3/2)^-$, while ^{7}Be has the ground state with $J^\pi = (3/2)^-$ and the first excited state with $J^\pi = (1/2)^-$. Thus, the (^{7}Li, ^{7}Be) reaction feeding the ground state of ^{7}Be involves spin-flip and non-spin-flip processes, while the reaction feeding the excited state of ^{7}Be involves only the spin-flip process. These are identified by observing the γ-rays from the first excited state. The (d, ^{2}He) reaction has been used to study $\sigma\tau^+$ resonances and the neutron orbits involved in the resonances. The (^{7}Li, ^{7}Be) reaction has been used to study τ^+ and $\sigma\tau^+$ mode excitations in light nuclei.

The (t, ^{3}He) reaction proceeds as a (n, p) reaction with the singlet np pair being the spectator. The reaction was studied at MSU by the RCNP–MSU collaboration [15]. Here the $0.6\,\text{GeV}$ α-beam from the MSU cyclotron was used to obtain the $0.4\,\text{GeV}$ t-beam through the α-projectile fragmentation. The triton is just suitable for $\sigma\tau$ excitation.

Spin–isospin giant resonances have extensively been studied by investigating charge-exchange spin-flip reactions, as described in Chapter 3.

The high-resolution studies of the (t, ^{3}He), (^{3}He, t) and (p, p′) reactions were used to study isospin multiplets in light nuclei [16]. Here the GT resonance shows isospin-dependent gross and microscopic structures. The former reflects the symmetry energy, while the latter reflects the coupling with nearby particle–hole states. These features are characteristic of the finite-body nuclear system. Microscopic structures of giant resonances are described in Chapter 4.

1.3.3 *Neutrino nuclear responses*

Neutrinos are key particles in view of fundamental weak interactions and symmetries. They are well studied in nuclear microlaboratories for studies of neutrinos and weak interactions. Neutrino nuclear responses are crucial for neutrino studies in nuclei and

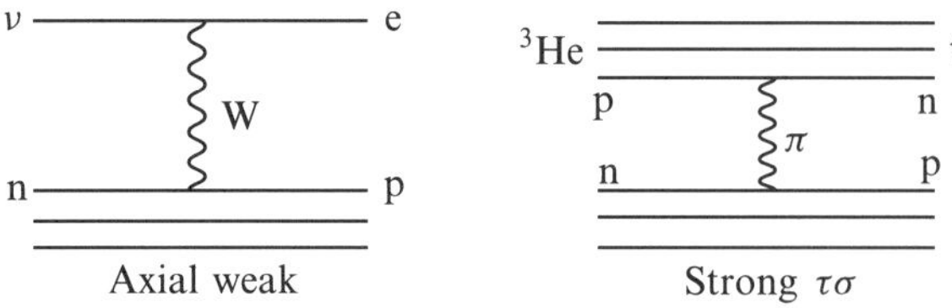

Fig. 1.1 Nuclear spin–isospin responses for axial charged weak processes and those for spin-flip charge-exchange strong processes.

involve vector and axial-vector weak responses. They are effectively studied by charge exchange spin-non-flip and spin-flip nuclear reactions at the cyclotron laboratory, as shown in Fig. 1.1.

Solar neutrinos are studied by measuring inverse β-decays by solar-ν absorptions into nuclei:

$$\nu + \mathrm{A} \to \mathrm{B} + \beta^+. \tag{1.4}$$

^{71}Ga, ^{37}Cl, ^{127}I, and other nuclei are used because of low $Q(\beta^+)$ values. Solar-ν responses for ^{71}Ga were studied by measuring the charge exchange ^{71}Ga(^{3}He, t) ^{71}Ge and ^{71}Ga(^{3}He, tγ) ^{71}Ge reactions [17]. The ^{3}He beam with $E(^3$He$) = 0.45\,$GeV from the ring cyclotron was used for studying spin–isospin responses. The (^{3}He, t) reaction was momentum analysed by the high-resolution spectrometer. The solar-ν flux is measured by measuring the production rate of the residual ^{71}Ge nuclei. The solar-ν absorption rates for individual ν components were obtained.

Contributions from particle unbound states through γ-transitions to low-lying bound states were obtained by measuring the ^{71}Ga(^{3}He, tγ) reaction. It is interesting to note that the $(5/2)^+$ states with energies up to $1\,$MeV above the neutron threshold energy contribute to the production of ^{71}Ge through γ-decays to lower states.

Fundamental properties of ν and weak interactions are studied by investigating double beta ($\beta\beta$) decays, as discussed in Chapter 12. Nuclear responses (matrix elements) for $\beta\beta$ decays are studied by measuring charge exchange (^{3}He, t) and (d, ^{2}He) reactions [18]. The nuclear matrix element $M_s(\beta)$ for the β^--decay from the 0^+ ground state of ^{100}Mo to the Single Particle–hole 1^+ state (SP) of ^{100}Tc was obtained by using (^{3}He, t) reactions.

The nuclear matrix element $M_s(\beta')$ for the β^--transition from SP 1^+ state of ^{100}Tc to the 0^+ ground state of ^{100}Ru was obtained from the β^- half-life. Using the $M_s(\beta)$ and $M_s(\beta')$, the $\beta\beta$ matrix element $M_s(\beta\beta) = M_s(\beta)M_s(\beta')/\Delta_s$ for the successive single β-decays through the SP 1^+ state in the intermediate nucleus was derived. The value obtained for $M_s(\beta\beta)$ is found to reproduce the observed two-neutrino $\beta\beta$ matrix element $M^{2\nu}(\beta\beta)$. Thus one gets

$$M^{2\nu}(\beta\beta) \sim \frac{M_s(\beta) \cdot M_s(\beta')}{\Delta_s}. \tag{1.5}$$

The spin–isospin Giant Resonance (GR) in the intermediate nucleus, which absorbs most of the single-β strength, does not contribute to $2\nu\beta\beta$ process. This supports the theoretical analysis in terms of the coupling of the SP and GR modes.

Studies of nuclear responses for $\beta\beta$ and double spin–isospin resonances are made by the double charge-exchange reaction of (^{11}B, ^{11}Li).

1.3.4 *Exotic nuclear beams and spin polarizations*

Polarizations of nuclear fragments are produced by projectile fragmentation of light nuclear beams from the ring cyclotron. Radioactive ^{12}B beams are produced by fragmentation of medium to low energy ^{14}N beams with $E/A \sim 0.04\,\mathrm{GeV}$. Here, the reaction used is the light ion reaction on the light ^{9}Be target,

$$^{14}\mathrm{N} + {}^9\mathrm{Be} \rightarrow {}^{12}\mathrm{B} + \mathrm{X}. \tag{1.6}$$

Polarizations of the ^{12}B fragment are measured by observing asymmetric β-decays from ^{12}B.

High-intensity ^{12}B beams with large polarization ($P = -0.3$ to -0.4) are obtained [19]. This is due to the cooperative process of the nucleon transfer and the projectile fragmentation.

High-intensity large-polarization radioactive beams are used to carry out symmetry studies of β-decays and exotic nuclear reactions relevant to astronuclear physics. These radioactive nuclei are used as spin-polarized nuclear probes for condensed matter. Actually, polarizations of ^{12}B nuclei implanted into snowballs are found to be retained well. Then, studies of nuclear moments of exotic nuclei can be carried out therein. Physics of snowballs is discussed in Chapter 7.

1.4 Hadron–nucleon systems

1.4.1 *Nucleons, mesons, and isobars in NN channels*

Fundamental NN interactions in the GeV/c region are studied by investigating NN $\rightarrow$ NNγ and NN $\rightarrow$ NNπ processes. The interactions involve NN off-shell interactions with momentum transfers $q = 0.2$–$0.5\,\mathrm{GeV}/c$, meson currents and Nπ couplings, and the Δ isobar currents and NΔ couplings. These studies link nucleon–meson nuclear physics to quark nuclear physics.

The pp bremsstrahlung is a simple inelastic NN process given by

$$\mathrm{p} + \mathrm{p} \rightarrow \mathrm{p} + \mathrm{p} + \gamma. \tag{1.7}$$

The interaction, involved explicitly in the ppγ process is only the electromagnetic one, which is weak in strength and simple in its process. Since mesons in the pp channel are neutral, meson exchange currents have no direct couplings with photons. On the other hand, isobar currents are involved implicitly in the energy region of the ring cyclotron.

The two-arm spectrograph is used to measure the two scattered protons in the pp bremsstrahlung [20]. The missing-mass spectrum shows a clean peak at $\Delta m = 0$, indicating the photon channel. The γ-ray detector is used together with the spectrograph to check the photon channel.

Pion productions in pn and pp channels are studied via [21]

$$p + n \rightarrow {}^2\mathrm{He} + \pi^-, \tag{1.8}$$

$$p + p \rightarrow {}^2\mathrm{He} + \pi^0. \tag{1.9}$$

Protons with $E_\mathrm{p} = 0.3$–$0.4\,\mathrm{GeV}$ are used to bombard neutrons and protons in CD_2 and CH_2 targets. ${}^2\mathrm{He}$ at $\theta = 0°$ is analysed by measuring the two protons from ${}^2\mathrm{He}$, using the spectrograph LAS. The missing-mass spectrum shows a peak at $\Delta m = m_\pi c^2$, indicating the π channel. Since the pn channel includes both isospin $T = 0$ and 1 channels, s- and p-wave pions contribute to the process $p + n \rightarrow {}^2\mathrm{He} + \pi^-$. The observed cross section at $\theta_\pi = 180°$ is reproduced by the $\pi\mathrm{N}$ model including off-shell NN interaction.

Pionic atoms were observed by measuring ${}^{208}\mathrm{Pb}(p, {}^2\mathrm{He})$ reactions [22]. The deeply bound pionic states, discussed in Chapter 8, provide a strong constraint on the π-nucleus optical potential.

1.4.2 *Hyperon interactions and flavour nuclear physics*

Hyperons (Y) with strangeness $S = -1$ include one s quark. Hyperon–nucleon interactions are useful in observing the effects of s quark, which is much heavier than the u and d quarks, in strong and weak interactions. Hyperons can also be used to probe deep interior regions of hypernuclei without being Pauli-blocked.

Hyperons in hypernuclei give unique opportunities for studies of weak interactions with $\Delta S \neq 0$ in nuclei. Non-mesonic weak decays $\Lambda\mathrm{N} \rightarrow \mathrm{NN}$ are used for selective studies of parity-violating non-leptonic weak processes since strong processes with $\Delta S \neq 0$ are forbidden. Recent progresses in strange-flavour nuclear physics are discussed in Chapter 10.

So far, hyperon–nucleon interactions have been studied by using strangeness-conserving strong processes $\mathrm{N} + \mathrm{K} \rightarrow \pi + \mathrm{Y}$ and $\mathrm{N} + \pi \rightarrow \mathrm{K} + \mathrm{Y}$ [23]. Here K and π, in the GeV/c region, are produced as secondary beams from high-energy (10–$30\,\mathrm{GeV}$) protons. In contrast to these methods, RCNP provides two other ways for studying hyperon interactions; one is to use strangeness-non-conserving weak processes and other is to use strangeness-conserving electromagnetic processes.

The lambda hyperon (Λ) is produced by the weak process with $\Delta S = 1$ at the ring cyclotron laboratory:

$$p + n \rightarrow \Lambda + p. \tag{1.10}$$

This is an inverse process of the non-mesonic weak decay of Λ in a hypernucleus. Protons with $E_\mathrm{p} = 0.41\,\mathrm{GeV}$ from the ring cyclotron are used for producing Λ by the weak process of neutrons in deuteron target nuclei. Polarized Λ particles are produced by using polarized protons. Then parity-violating weak processes with the flavour change of $\Delta S = 1$ are studied by measuring the P-odd term of $J_P \cdot k_P$. Actually, the non-mesonic weak decay $\Lambda\mathrm{N} \rightarrow \mathrm{NN}$ for polarized Λ-hypernuclei shows for the first time the asymmetry with respect to the Λ-spin polarization, indicating the P-odd term in the process with $\Delta S \neq 0$ [23]. Since the momentum transfer in the weak process is as large as $\Delta q \sim 0.4\,\mathrm{GeV}/c$, heavy meson currents and quark currents contribute to some extent.

It is interesting to study the T-odd term of $(J_P \times k_P) \cdot J_\Lambda$. The Λ spin (J_Λ) is measured by observing the asymmetric weak decay $\Lambda \rightarrow p + \pi^-$. An experimental programme to study the $NN \rightarrow N\Lambda$ process is under progress at the ES beam line of the cyclotron laboratory [24].

Hyperons are also produced by strangeness-conserving electromagnetic processes

$$\gamma + N \rightarrow K^+ + Y. \tag{1.11}$$

Multi-GeV polarized photons at the laser electron photon laboratory in SPring-8 are used to produce polarized hyperons. Although the hyperon momentum is as large as $P_Y \sim 1\,\mathrm{GeV}/c$, it is expected to be trapped in the nucleus with a probability of 10–15% to produce a hypernucleus. Using deuteron targets, one can study hyperon–nucleon interactions.

1.5 Quark nuclear physics

Quark nuclear physics is concerned with studies of hadron structures and hadron interactions in terms of the quarks and gluons, which involve the low-energy (multi-GeV) region. Here, the colour interaction (charge) is strong, and necessarily non-perturbative QCD must be used. The strong colour interaction leads to very interesting hadron properties and structures, which are characteristic of interacting quark–gluon many-body systems. In this sense, the quark–gluon system is essentially similar to the nucleon–meson system, although they look very different from each other in their explicit properties.

Quark nuclear physics programmes, both theoretical and experimental, are currently under progress at RCNP.

1.5.1 *DGL theory and colour confinement*

The non-perturbative QCD is studied in two ways with the emphasis on the dual Higgs mechanism for colour confinement. One way is to use the lattice QCD theory and the other is to employ an effective infrared theory, the Dual Ginzburg–Landau (DGL) theory. In the dual Higgs model, the quark confinement and the chiral symmetry breaking are shown to be linked to the QCD monopole condensation.

The quark and colour field confinement in hadrons is considered by analogy with the Meissner effect in superconductors, where the magnetic field is confined into the vortex-like configuration. The colour electric field between a quark–antiquark pair can be confined in a vortex-like configuration in a dual QCD superconductor. The dual QCD superconductor requires magnetic QCD monopoles and their condensations.

The DGL theory is developed so as to incorporate the dual QCD superconductor picture [25,26]. Here the QCD Lagrangian, which is non-abelian in nature, is reduced to the abelian gauge theory with the QCD monopole. The DGL Lagrangian is constructed following the dual superconductor picture, where the monopole and its condensation are the main ingredients. The DGL Lagrangian is written as

$$L = L_\mathrm{g} + \bar{q}\left(i\gamma_\mu \partial^\mu - e\gamma_\mu A^\mu \vec{H}\right)q$$
$$+ \left|\left(i\gamma_\mu \partial^\mu - g\vec{\varepsilon}_a \partial_\mu \vec{B}^\mu\right)\chi_a\right|^2 - \lambda\left(|\chi_a|^2 - v^2\right)^2, \tag{1.12}$$

where A_μ and B_μ are, respectively, the gauge and dual-gauge fields, L_g represents the Lagrangian for gluons, q the quark, and χ the QCD monopole. The last term provides the monopole condensation, giving a mass to the dual-gluon field by the Higgs mechanism.

The static q–$\bar{q}$ potential derived by the DGL theory shows a linear behaviour at large distance, which is characteristic of the quark confinement. The quark acquires finite mass of 0.3–0.4 GeV due to QCD monopole condensation. Thus the chiral symmetry breaking takes place there. This symmetry restores at high temperature.

The QCD monopole is shown to be related to the topological object of the instanton, which is the classical solution of the non-abelian gauge theory [27,28]. At high density of the instanton, long QCD monopole loops become conspicuous. This indicates a strong correlation in the four-dimensional space, reflecting the monopole condensation. The lattice QCD calculation shows increase of long QCD monopole loops at low temperature, supporting the correspondence of the QCD monopole condensation with the appearance of the long-loop monopoles. This fact indicates that instantons are responsible for the quark confinement. The details are described in Chapter 11.

The lattice QCD is used to extract the effective degree of freedom in non-perturbative QCD phenomena. The key aspect of the lattice QCD study is the use of the Maximally Abelian (MA) gauge, where the abelian gluons acquire much of the gluon dynamics through a gauge transformation. We see clearly the appearance of colour monopoles and also the monopole condensation at low energy. Hence, the colour monopoles are demonstrated to be the effective degrees of freedom in non-perturbative phenomena such as colour confinement and chiral symmetry breaking.

1.5.2 *Baryon structure*

Energy spectra of baryons, which are confined three-quark systems, are expected to show features characteristic of the colour field confinement. Quarks in a nucleon are excited by flipping spins, isospins, and flavours to form isobars, hyperons, and other baryons. This can be done without any colour field changes in space according to the SU(3) symmetry of u, d and s quarks. On the other hand, orbital excitations of quarks in space are restricted by the colour field confinement. The baryon energy spectra seem to show that the excited baryons are deformed in shape.

Spin-independent parts of the hadron spectra are shown to be reproduced by introducing deformed potential with volume conservation [29]:

$$H = \sum_i \left[p^2/2m + m\left(\omega_x^2 x^2 + \omega_y^2 y^2 + \omega_z^2 z^2\right)/2 \right]_i, \qquad (1.13)$$

where $\omega_1 \cdot \omega_2 \cdot \omega_3 = \omega^3 = \text{const}$. The $N = 0$ state is found to have the spherical shape with $\omega_1 = \omega_2 = \omega_3$ and forms the spherical ground state. On the other hand, the $N = 1$ and $N = 2$ states are found to be deformed with $\omega_1 = \omega_2 = 2\omega_3$ $(N = 1)$ and $\omega_1 = \omega_2 = 3\omega_3$ $(N = 2)$. They may be identified as the first negative-parity state and the roper, respectively. The rotational states are built on those deformed states and are clearly seen in the experimental baryon spectra.

These interesting characteristics are investigated by measuring intraband and inter-band γ-decays in the rotational band spectra.

1.5.3 *Quark nuclear physics with multi-GeV laser electron photons*

Quark nuclear physics programmes at the laser electron photon laboratory are intensively being pushed by both experimental and theoretical groups. Multi-GeV polarized photons are unique and powerful probes for studies of hadron structures and non-perturbative QCD. Physics programmes under considerations are as follows.

1. Glueballs and pomerons studied by photoproduction of vector mesons as functions of momentum transfer, as shown in Fig. 1.2. Glueballs, which are predicted by QCD, are not established experimentally. Pomerons introduced to explain the photoproduction of vector mesons are considered to be not quark configurations, but rather likely gluon configurations. Identification of glue-like states is crucial for establishing non-perturbative QCD.
2. Search for QCD monopole. The QCD monopole is predicted as a key particle of the Higgs mechanism for quark confinement by the DGL theory.
3. Nucleon spin structures and GDH sum rule.
4. Quark effects in photo-disintegrations of deuterons.
5. Baryon structures by baryon spectroscopy.

Detailed discussions on quark nuclear physics programmes at the laser electron photon laboratory are discussed in Ref. [8].

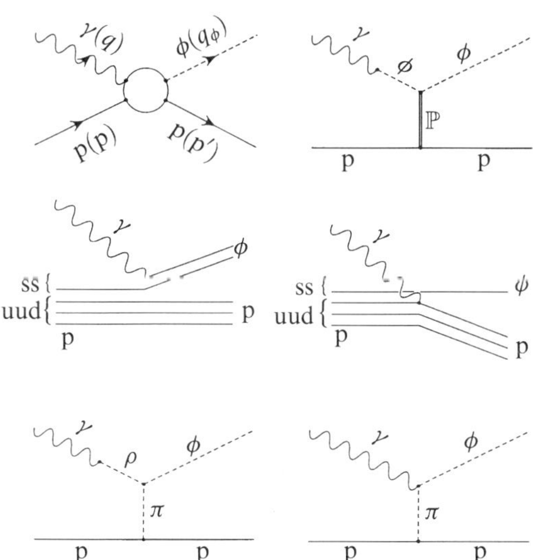

Fig. 1.2 Photoproduction of vector mesons and mediating processes [30].

1.6 Lepton nuclear physics and symmetry studies in nuclei

Nuclei, which are cold broken-symmetry quantum systems of interacting particles, are used as excellent microlaboratories for studying fundamental interactions and symmetries, as discussed in Section 1.1. Leptons, weak interactions, and symmetries have been extensively studied by investigating nuclear weak processes in nuclear microlaboratories, as discussed in Chapter 12. Neutrinos, being the simplest elementary particles, are very sensitive probes for symmetry studies associated with weak interactions. Neutrinos have only the left-handed weak charge. They have no colour charge, no mass, no flavour mixing and no right-handed helicity within the present experimental limits and in the present standard theory.

1.6.1 *Neutrino nuclear physics and double beta decay*

Symmetries are completely broken for massless neutrinos with pure left-handed weak charges within the framework of the standard $SU(2)_L \times U(1)$ theory. Search for small deviations from completely broken symmetries are very interesting for finding limits on the standard theory and evidence for unified theories beyond the standard theory.

Fundamental weak interactions and symmetry problems associated with ν are studied effectively by investigating the $\beta\beta$ decays, viz. the two-neutrino $\beta\beta$ decays ($2\nu\beta\beta$) and the neutrinoless $\beta\beta$ decays ($0\nu\beta\beta$):

$$2\nu\beta\beta: \quad A \rightarrow B + \beta + \beta + \nu + \nu, \tag{1.14}$$

$$0\nu\beta\beta: \quad A \rightarrow B + \beta + \beta. \tag{1.15}$$

In eqn (1.15), ν is antineutrino in the case of β^--decays and a neutrino in the case of β^+-decays. The $2\nu\beta\beta$ process, which conserves the lepton number L, occurs within the scope of standard theory, whereas the $0\nu\beta\beta$ process, which violates the lepton number conservation law by $\Delta L = 2$, occurs beyond the standard theory. Actually, $0\nu\beta\beta$ is very sensitive to the Majorana-ν mass, the left–right mixing of weak currents, the majoron–ν coupling, the SUSY coupling, and so on, which are beyond the scope of standard theory, are shown in Fig. 1.3. The $2\nu\beta\beta$ and $0\nu\beta\beta$ transition rates are written as

$$T^{2\nu} = G^{2\nu} \, |M^{2\nu}|^2 \tag{1.16}$$

and

$$T^{0\nu} = G^{0\nu} \, |M^{0\nu}|^2 \, (\langle m_\nu \rangle^2 + \langle \lambda \rangle^2 + \langle \eta \rangle^2 + \langle m_\nu \rangle \langle \lambda \rangle + \langle m_\nu \rangle \langle \eta \rangle + \langle \lambda \rangle \langle \eta \rangle)^2, \tag{1.17}$$

respectively, where $G^{2\nu}(G^{0\nu})$ and $M^{2\nu}(M^{0\nu})$ are the phase space factor and the nuclear matrix element for $2\nu\beta\beta$ ($0\nu\beta\beta$) decay, respectively. Equation (1.17) gives the rate for $0\nu\beta\beta$ due to the Majorana-ν mass term $\langle m_\nu \rangle$ and the right-handed current terms of $\langle \lambda \rangle$ and $\langle \eta \rangle$. Here λ is the square of the mass ratio for left- and right-handed weak bosons, and η is their mixing amplitude.

The nuclear responses are given by $|M^{2\nu}|^2$ and $|M^{0\nu}|^2$ for $2\nu\beta\beta$ and $0\nu\beta\beta$ processes, respectively. The nuclear response for the $2\nu\beta\beta$ process is obtained from the measured half-life. It is mainly the double GT ($J^\pi = 1^+$) response since the Fermi-type response with $J^\pi = 0^+$ is concentrated only in the double IAS. The nuclear response for the $0\nu\beta\beta$

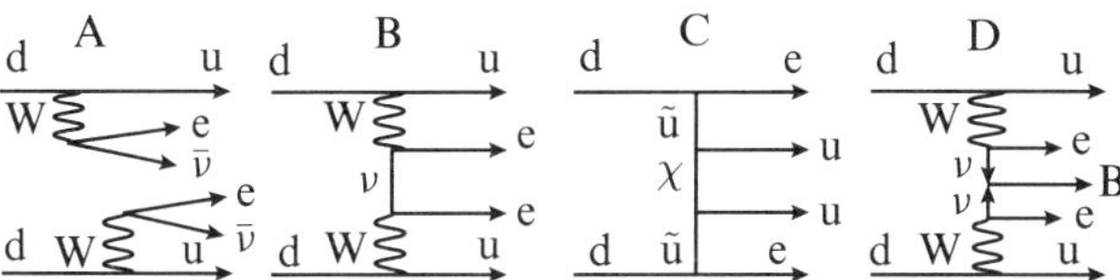

Fig. 1.3 Schematic diagrams of $\beta\beta$ decay processes. A: two-neutrino process, B: neutrinoless process with the Majorana-ν exchange; C: neutrinoless process with the SUSY particles, and D: neutrinoless process followed by the Goldstone boson (majoron). d and u represent the down and up quarks, respectively [31].

process is crucial for obtaining particle physics quantities or the (limits on them) such as $\langle m_\nu \rangle$, $\langle \lambda \rangle$, and $\langle \eta \rangle$, from the observed value of (or the limit on) the $0\nu\beta\beta$ half-life. The $0\nu\beta\beta$ process given in eqn (1.18) is associated with a virtual ν exchange between two nucleons in a nucleus. Thus it involves large ν momenta of $q_\nu = 0.2$–$0.4\,\text{GeV}/c$, and necessarily large angular momenta up to $J^\pi \sim 6$.

The $2\nu\beta\beta$ and $0\nu\beta\beta$ processes were studied by means of high-sensitivity detectors ELEGANT (ELEctron GAmma-ray Neutrino Telescope) III, IV, and V at the Kamioka underground laboratory [31–33]. EL III is a high-purity Ge detector surrounded by 4π NaI detector array. It was used to study the $0\nu\beta\beta$ decay of ^{76}Ge. EL IV is a multi-layer Si detector array surrounded by 4π NaI detector array. It was used to study the $2\nu\beta\beta$ decay of ^{100}Mo. EL V is a detector complex which consists of multilayer drift chambers for $\beta\beta$ trajectories, plastic scintillator arrays for $\beta\beta$ energies and times, and NaI scintillator arrays for X- and γ-rays [33]. It is used to study the $0\nu\beta\beta$ and $2\nu\beta\beta$ decays of ^{100}Mo and ^{116}Cd. Here ^{100}Mo and ^{116}Cd are chosen because of the large phase space factors $G^{2\nu}$ and $G^{0\nu}$.

Finite half-lives for the $2\nu\beta\beta$ decay of ^{100}Mo and ^{116}Cd were obtained for the first time [33]. The observation of the $2\nu\beta\beta$ decay for ^{100}Mo is the first such observation by a direct-counting method alone. The $2\nu\beta\beta$ matrix elements for ^{100}Mo and ^{116}Cd are found to be smaller, by one order of magnitude, than the single-particle values. This is considered to be due to the destructive interference with the spin–isospin giant resonances

The observed values for the $2\nu\beta\beta$ matrix elements are analysed in terms of single particle–hole states, spin–isospin giant resonances and their couplings. The $2\nu\beta\beta$ processes between the 0^+ ground states in even–even nuclei are found to proceed through low-lying single particle–hole states in intermediate nuclei [34]. High-lying GT and IAS giant resonances do not contribute to the $2\nu\beta\beta$ process.

No finite $0\nu\beta\beta$ yields were observed for ^{100}Mo. Half-life limits on individual $0\nu\beta\beta$ modes were derived. Limits on $\langle m_\nu \rangle$, $\langle \lambda \rangle$, $\langle \eta \rangle$, and other quantities are deduced by using theoretical nuclear matrix elements [33]. Presently, several groups are studying $\beta\beta$ decays of several nuclei. So far, no finite half-lives for $0\nu\beta\beta$ decay have been observed. The limits obtained on $\langle m_\nu \rangle$ are in the range of 1–2 eV, depending on the nuclear matrix elements used. High-precision studies of $\beta\beta$ decays of several nuclear species by various methods are very important because of large ambiguities in the $0\nu\beta\beta$ nuclear responses and very low counting rates.

The $\beta\beta$ decays are studied with EL V and EL VI at Oto Cosmo Observatory, which is a new underground laboratory with small Rn contents. EL V, being improved in sensitivity, has now been set up at the new laboratory.

EL VI is a newly developed detector system to study the $\beta\beta$ decay of ^{48}Ca. It consists of CaF$_2$ scintillator arrays surrounded by 4π CsI detector arrays [35]. They aim at studying $\langle m_\nu \rangle$ in the sub-eV region and right-handed currents in the 10^{-8}–10^{-7} region.

1.6.2 *Nuclear instabilities, symmetries, and solar neutrinos*

Nucleons in nuclei decay to leptons, photons, and/or mesons in unified theories where the baryon number is not conserved. So far, nucleon decays N $\to$ X + Y, ... have been studied by measuring decay particles of X, Y, ... Thus limits on decay rates depend on decay modes and the measured decay particles.

Nucleon decays in nuclei leave nucleon holes in the parent nuclei. Thus they are studied by measuring deexcitations of the nucleon holes [36]. This is an inclusive measurement, which is independent of decay modes. This is very effective for studying modes with invisible decay particles such as N$\to$ 3ν.

Nucleon decays in nuclei were studied by measuring deexcitations of nucleon holes in a large NaI detector array and in a large water Cerenkov counter. Stringent limits on inclusive (mode-independent) nucleon decays were obtained as Γ(n $\to$ 3ν, x, ...; 2n $\to$ Y, ...) $< (10^{28}$y$)^{-1}$.

In nuclei nucleons decay to lighter exotic hadrons, e.g. the dihyperon H. It is the singlet six-quark state uuddss with $J^\pi = 0^+$ and strangeness $S = -2$. Then, two nucleons in a nucleus decay to H by double weak processes with $\Delta S = 2$ as nn $\to$ H and np $\to$ He$^+\nu$. High-sensitivity studies of the double weak processes exclude the light H with $M_{\rm H} \leq 2M_{\rm N}$ [37].

Nuclear deexcitation processes of nucleon holes are well studied by investigating deexciting particles and γ-rays from nucleon holes produced by (p, 2p) and (p, d) reactions. Quasi-free scatterings for nucleons in the nuclear medium leave nucleon holes in the parent nuclei. Nucleon correlations of deep nucleon holes are studied by investigating deexcitations of the holes. Proton holes in light nuclei are found to decay partly by emitting tritons, suggesting that proton holes are produced in α-clusters in the nuclear medium.

Nucleon holes are produced in various ways such as nucleon decays in nuclei, flavour changes in the $N\pi \to \Lambda K$ process in nuclei, non-mesonic weak decays $\Lambda N \to NN$, and so on. These decays are studied by measuring deexcitations of nucleon holes.

Solar neutrinos are of current interest in particle physics, astrophysics, and nuclear physics. Particle physics aspects include possible ν oscillations due to ν-mass differences and ν-flavour mixing; astrophysics aspects are concerned with the solar modes; and the nuclear physics aspects include nuclear reaction rates associated with ν-productions in the sun.

Nuclear reaction rates in the sun are very slow because of the very low temperature there. Their determination requires high-sensitivity low-background studies. The solar nuclear reaction ^{3}He $+$ ^{3}He $\rightarrow$ 2p $+$ α is being studied in the energy region $E \leq$ 100 keV. The reaction is concerned with the ^{8}B-ν and ^{7}Be-ν production rates.

The solar neutrinos are detected by inverse β-decays of ^{71}Ga, ^{37}Cl, and other nuclei. The nuclear responses for the solar neutrinos are studied by means of ^{71}Ga (^{3}He, t) and ^{71}Ga (^{3}He, tγ) reactions at the cyclotron laboratory, as discussed in Section 1.3.3 [38].

1.6.3 *Dark matter and weakly interacting massive particles*

Dark matter is of great interest in particle physics, astrophysics, and nuclear physics. A major component of the mass of the universe is considered to be the invisible DM. Weakly Interacting Massive Particles (WIMPs) are likely to be candidates for cold DM. Lightest SUSY Particle (LSP) such as the neutralino, which is beyond the scope of standard theory, is one such candidate for WIMPs. Direct measurements of WIMPs are carried out at nuclear microlaboratories by studying elastic and inelastic scatterings of DM from nuclei.

Interactions of WIMPs with nucleons (quarks) are very weak, DM densities are extremely small, and DM signals are very low. Thus, DM studies have to be made by using high-sensitivity detectors at low-background underground laboratories.

The high-sensitivity detectors, ELEGANTs, which are used for neutrino studies of $\beta\beta$ decays at the Oto Cosmo Observatory with low Rn and cosmic-ray backgrounds are just adequate for WIMPs studies. Furthermore, medium-energy beams from the RCNP ring cyclotron are used to investigate nuclear responses for DM.

Vector-coupled DM with spin-independent interactions are studied by measuring coherent elastic scatterings from all nucleons in a nucleus. Axial-vector (spin-) coupled DM with spin-dependent interactions are studied by measuring elastic and inelastic scatterings from a valence nucleon in an odd-A (spin $J \neq 0$) nucleus. Thus scattering cross sections for spin-coupled DM are much smaller than those for the vector-coupled DM.

The vector- and axial-vector-coupled WIMPs have been studied by using the large-volume NaI detector array of EL V and the CaF$_2$ scintillator array of EL VI. The NaI detector array has a very large volume of 0.76 tons and a low noise level of around 4 keV. It is composed of odd-A isotopes of ^{23}Na ($J^\pi - (3/2)^+$) and ^{127}I ($J^\pi = (5/2)^+$). Thus the NaI detector array is useful for DM studies, particularly for the spin-coupled WIMPs [39]. Inelastic scatterings from the $(5/2)^+$ ground state of ^{127}I to the $(5/2)^+$ first excited state at $E = 58$ keV have two merits; one is the large energy addition of the deexciting 58 keV γ-rays and other is the nuclear spin response known from the M1 γ-decay rate. The large-volume NaI is also useful for studying annular modulations of the DM spectrum. Studies of WIMPs by using the NaI detectors are illustrated in Fig. 1.4.

The CaF$_2$ detector array of EL VI consists of 25 modules of CaF$_2$(En) crystals surrounded by active shields of pure CaF$_2$ crystals and CsI crystals [35]. It contains 3.5 kg ^{19}F isotopes with a large nuclear spin matrix element. Studies with the NaI and CaF$_2$ scintillators are now under progress to search for vector- and axial-vector-coupled WIMPs in the critical density region of $\Omega \approx 0.3$ GeV cm^{-3}.

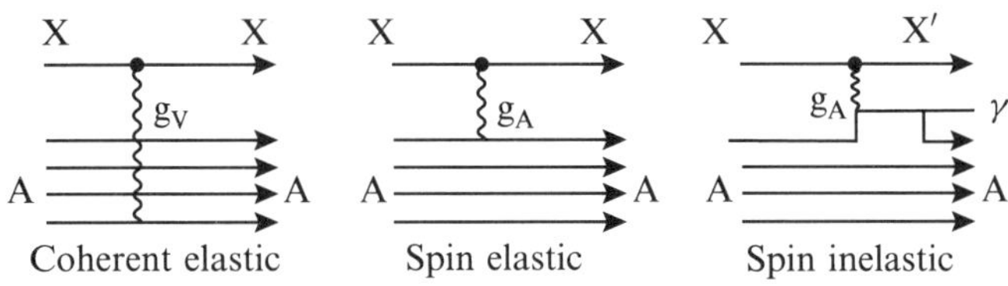

Fig. 1.4   Dark matter nuclear elastic and inelastic scattering processes.

References

1. H. Ejiri and H. Toki, *Nuclear reaction dynamics*, Oxford (1998).
2. T. Yamazaki, Y. Mizuno, and H. Ejiri, eds, *Nuclear physics at RCNP, from nucleon meson nuclear physics to quark lepton nuclear physics*, RCNP, Osaka University (1996).
3. H. Ejiri, T. Noro, K. Takahisa, and H. Toki, eds, *Proc. 14th RCNP Osaka int. symp. nuclear reaction dynamics of nucleon hadron many body systems*, December 1995, Osaka, World Scientific (1996); H. Ejiri, T. Kishimoto, and T. Sato, eds, *Proc. int. symp. weak and electromagnetic interactions in nuclei*, June 1995, Osaka, World Scientific (1995).
4. K. Hatanaka *et al.*, *Nucl. Instr. Methods* **A384** (1997) 575.
5. M. Fujiwara *et al.*, *Proc. int. symp. heavy ion research with magnetic spectrographs*, MSU, January 1989, p. 283; A. Tamii *et al.*, *IEEE Transactions Nucl. Science* **43** (1996) 2488; N. Matsuoka *et al.*, *RCNP annual report* (1991), p. 186.
6. H. Sakai *et al.*, *Nucl. Instr. Methods* **A320** (1992) 479; **A369** (1996) 120.
7. H. Ohsumi, H. Ejiri *et al.*, *Proc. XV RCNP symp. nucl. phys. frontiers with electroweak probes*, H. Toki *et al.*, eds, March 1996, Osaka, World Scientific (1996), p. 274.
8. H. Ejiri, M. Fujiwara, T. Kinashi, and H. Toki, *Proc. int. symp. high energy spin physics*, C. W. de Jager *et al.*, eds, September 1996, Amsterdam, World Scientific, p. 300; T. Nakano, *Proc. int. conf. quark lepton nuclear physics*, H. Ejiri *et al.*, eds, May 1997, Osaka, *Nucl. Phys.* (1997).
9. K. Hatanaka, T. Noro *et al.*, *Phys. Rev. Lett.* **78** (1997) 1014.
10. H. Sakuragi *et al.*, *Phys. Rev. C* (1978), to be published.
11. C. Samanta *et al.*, private communication.
12. T. Wakasa *et al.*, *Phys. Rev.* **C55** (1997) 2909.
13. H. Akimune *et al.*, *Phys. Rev.* **C52** (1995) 604.
14. S. Nakayama *et al.*, *Nucl. Instr. Methods* (1997); *Phys. Rev. Lett.* **67** (1991) 1082; *Phys. Rev.* **C46** (1992) 1667.
15. I. Daito *et al.*, *Nucl. Instr. Methods* **A396** (1997) 465.
16. Y. Fujita *et al.*, *Phys. Lett.* **B365** (1996) 29.
17. M. Fujiwara *et al.*, *Nucl. Phys.* **A569** (1994) 245c; **A577** (1994) 43c; H. Ejiri, K. Ishibashi *et al.*, *Nucl. Phys.* (1992).
18. H. Akimune, H. Ejiri *et al.*, *Phys. Lett.* **B394** (1997) 23.

19. N. Takahashi *et al.*, *Hyperfine Interactions* **97/98** (1996) 469; T. Shimoda *et al.*, *Phys. Lett.* (1998), to be published.

20. M. Nomachi *et al.*, *Proc. int. conf. quark letpon nuclear physics*, H. Ejiri *et al.*, eds, May 1997, Osaka, *Nucl. Phys.* (1997).

21. K. Tamura *et al.*, to be published; Y. Maeda, K. Tamura *et al.*, *Proc. int. symp. meson nuclear physics and the structure of the nucleon* (*MENU97*), TRIUMF, July 1997, πN *News Letter*.

22. N. Matsuoka *et al.*, *Hyperfine Interactions* **103** (1996) 299.

23. S. Ajimura, H. Ejiri, T. Kishimoto *et al.*, *Phys. Lett.* **B282** (1993) 293.

24. T. Kishimoto, *Proc. weak electromagnetic interactions in nuclei*, H. Ejiri *et al.*, eds, June 1995, Osaka, World Scientific (1995), p. 514.

25. S. Suganuma, S. Sasaki, and H. Toki, *Nucl. Phys.* **435** (1995) 207.

26. S. Sasaki, H. Suganuma, and H. Toki, *Prog. Theor. Phys.* **94** (1995) 373.

27. M. Fukushima, H. Suganuma, and H. Toki, *Phys. Lett.* **B399** (1997) 141.

28. M. Fukushima, H. Suganuma, and H. Toki, *Nuclear Phys.* **B53** (1997) 494.

29. H. Fujimura, H. Toki, and H. Ejiri, *Proc. XV RCNP int. symp. nucl. phys. frontiers with electro-weak probes*, March 1996, Osaka, World Scientific (1996), p. 110.

30. A. I. Titov *et al.*, *Phy. Rev. Lett.* **79** (1997) 1634.

31. H. Ejiri, *Int. J. Modern Phys.* **E6** (1997) 1.

32. H. Ejiri *et al.*, *Nucl. Phys.* **A448** (1986) 271; *J. Phys. G. Nucl. Phys.* **13** (1987) 839; K. Kamikubota *et al.*, *Nucl. Instr. Methods* **A245** (1986) 379; T. Watanabe, Ph.D. Thesis, Osaka University (1986).

33. H. Ejiri *et al.*, *Phys. Lett.* **B258** (1991) 17; H. Ejiri *et al.*, *Nucl. Instr. Methods* **A302** (1991) 304; H. Ejiri *et al.*, *Nucl. Phys.* **A611** (1996) 85.

34. H. Ejiri and H. Toki, *J. Phys. Soc. Japan Lett.* **65** (1996) 7.

35. R. Hazama *et al.*, *Proc. int. symp. weak electromagnetic interactions in nuclei*, H. Ejiri *et al.*, eds, World Scientific (1995) p. 635.

36. H. Ejiri, *Phys. Rev.* **C48** (1993) 1442; H. Ejiri and H. Toki, *Phys. Lett.* **B306** (1993) 218; R. Hazama *et al.*, *Phys. Rev.* **49** (1994) 2407.

37. H. Ejiri *et al.*, *Phys. Lett.* **B228** (1989) 24.

38. T. Rizawa *et al.*, *Proc. 11th symp. acc. science technology*, Harima, October 1997; K. Hatanaka *et al.*, *Proc. 16th RCNP int. symp. multi-GeV high-performance accelerators and related technology*, March 1997, Osaka, World Scientific (1998); K. Sato, *Proc. 16th RCNP int. symp. multi-GeV high-performance accelerators and related technology*, March 1997, Osaka, World Scientific (1998).

39. K. Fushimi *et al.*, *Phys. Rev.* **C47** (1993) R425; H. Ejiri *et al.*, *Phys. Lett.* **B317** (1993) 14.

2

Elastic and inelastic scattering

H. Sakaguchi

2.1 Introduction

When we bombard a nucleus with a nucleon or with light ions like deuterons, ^{3}He, α-particles, etc., various nuclear phenomena occur, including elastic scattering, inelastic scattering, nucleon transfer reactions, and projectile fragmentation, depending on the projectile species and the bombarding energy. The simplest among these phenomena—and one with the smallest amount of internal freedom—is the elastic scattering of a nucleon by a nucleus which we treat in this chapter.

In elastic scattering the target nucleus is not excited, so its initial and final states are the same. Furthermore, the initial states in such experiments are the target ground states, the charge distributions of which can be measured precisely by electron scattering. The wave function of the proton part of the target ground state is restricted and almost defined by the experiment. Thus elastic scattering is an appropriate type of reaction to use as a probe for the interactions and the dynamics which occur inside the nucleus. It is quite understandable that the first successful attempt to explain nuclear reactions microscopically began with elastic scattering.

Historically, elastic scattering of light ions has been studied since the 1950s. Since that time, measured target isotopes and bombarding energies used by the elastic scattering have been increasing with the increased energy resolution of the detector system and improved accelerator techniques. Polarized ion sources were in use in the mid-1960s and were established technically in the late 1960s mainly in Saclay [1] and Wisconsin [2]. Today, a polarized ion source is standard equipment employed in nuclear physics. For the analysis of experimental data from elastic scattering, a mean field acting on the projectile, namely a potential with real and imaginary parts—optical potential—has been proposed. Initially, the relation between the optical potential used in the nuclear reaction and the harmonic oscillator potential used for the calculation of the nuclear structure was vague and indirect. Now, however, they both are thought to represent mean fields acting on nucleons or ions inside the nuclei.

In the 1970s there was great progress made in the microscopic explanation of the optical potential. By applying the nuclear matter theory [3] developed in the 1950s to nuclear reactions, it became possible [4] to construct an optical potential from the Nucleon–Nucleon (NN) interaction and the target ground state wave function. At the international symposium on the microscopic optical potential held in Hamburg in 1978 the form of the microscopic optical potential was established. Using the G-matrix deduced from the NN interaction in nuclear matter and folding [5] with the density distribution of the target ground state, the optical potential was calculated and compared with the experimental data accumulated during the previous 20 years.

During this period, progress made in experimental techniques for accelerators, polarized ion sources, and spectrometers was quite significant and as a result the precision of the experimental data improved a great deal. These data were thus able to provide a stringent test of these microscopic theories. Applying these modern experimental techniques, the RCNP cyclotron [6] began to provide experimental data in 1977.

2.2 Proton elastic scattering around 65 MeV

At RCNP the elastic scattering of polarized protons has been measured for 55 nuclei from ^{12}C to ^{238}U at 65 MeV [7–13]. Differential cross sections and analysing powers are shown in Fig. 2.1. The polarization axis of the spin-polarized beam is usually selected to be parallel to the magnetic field direction of dipole magnets in the beam transport system, and is usually chosen to be in the direction of $k_i \times k_f$, where $\hbar k_i$ is the incident momentum and $\hbar k_f$ is the momentum of the outgoing particle. This is done because the scattering is usually measured horizontally. If we denote the yield of the left detector by $\sigma_L(\theta)$ and that of the right detector, placed at the same scattering angle but on opposite sides of the beam direction, by $\sigma_R(\theta)$, the cross section and the analysing power are defined as

$$\frac{d\sigma}{d\Omega}(\theta) = \frac{1}{2}(\sigma_L(\theta) + \sigma_R(\theta)), \tag{2.1}$$

$$A_y(\theta) = \frac{1}{P_y} \frac{\sigma_L(\theta) - \sigma_R(\theta)}{\sigma_L(\theta) + \sigma_R(\theta)}, \tag{2.2}$$

where P_y is the incident beam polarization. According to the Madison convention [14] shown in Fig. 2.2, we define the y-axis to be parallel to $k_i \times k_f$, while the z-axis is along the direction of the momentum of the particle. We use primes to denote the axis after scattering. In this case the y-axis is equivalent to the y'-axis.

Beam polarization is monitored continuously during the measurement downstream from the scattering target as shown in Fig. 2.3. In the energy region below 100 MeV, elastic scattering from carbon is normally used to monitor the beam polarization, because its analysing power is already known by the double scattering method [15] or by other calibration runs [16]. Compared with the analysing power data, there exist some ambiguities in the absolute values of the cross section due to inhomogeneities in the target foil, difficulties in measuring the integrated beam current and target thickness, and so on. Special care is neccesary in measuring the cross sections at extreme forward angles because of difficulties in current integration at very low beam intensity.

When the incident energy per nucleon is below 100 MeV, experimental data are first analysed using the phenomenological optical potential. This result is then compared with the microscopic theory. The optical potential used in the phenomenological analysis has

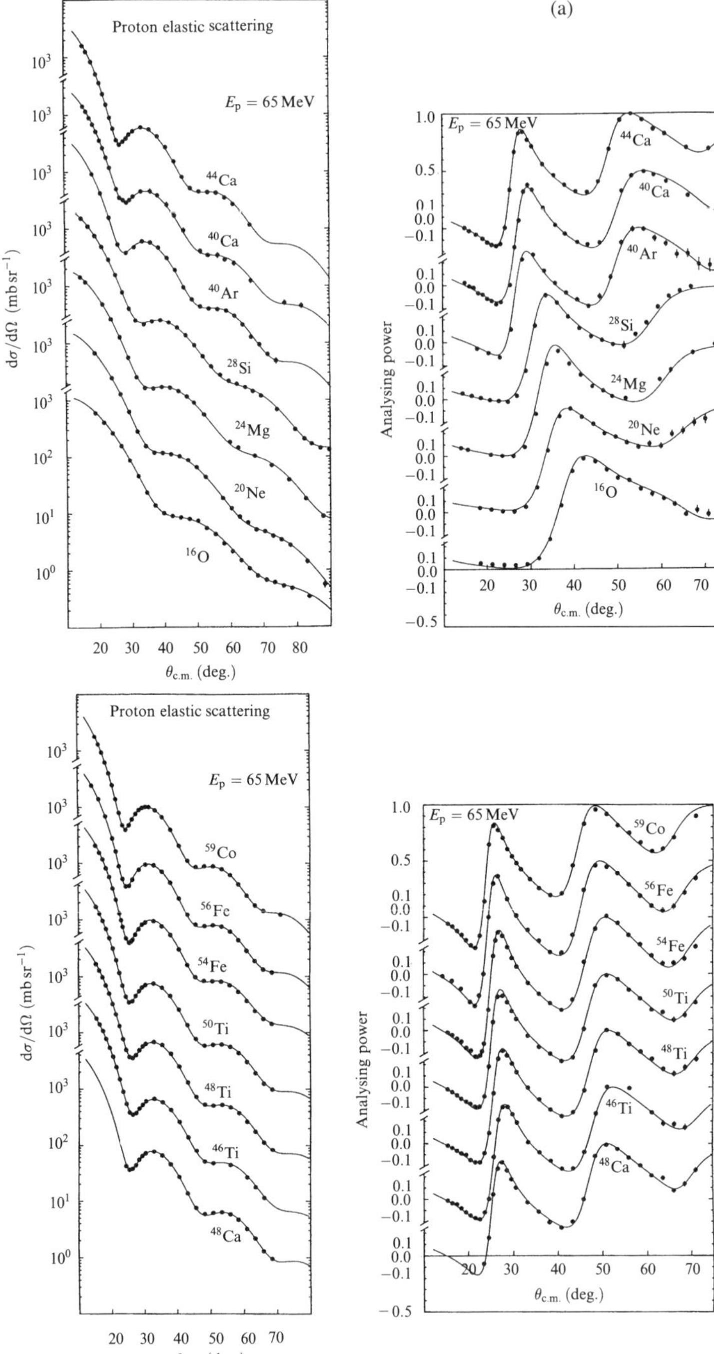
(a)
Proton elastic scattering
$E_p = 65\,\mathrm{MeV}$
^{44}Ca
^{40}Ca
^{40}Ar
^{28}Si
^{24}Mg
^{20}Ne
^{16}O
$\mathrm{d}\sigma/\mathrm{d}\Omega\ (\mathrm{mb\,sr}^{-1})$
$\theta_{\mathrm{c.m.}}$ (deg.)
Analysing power
$E_p = 65\,\mathrm{MeV}$
Proton elastic scattering
^{59}Co
^{56}Fe
^{54}Fe
^{50}Ti
^{48}Ti
^{46}Ti
^{48}Ca

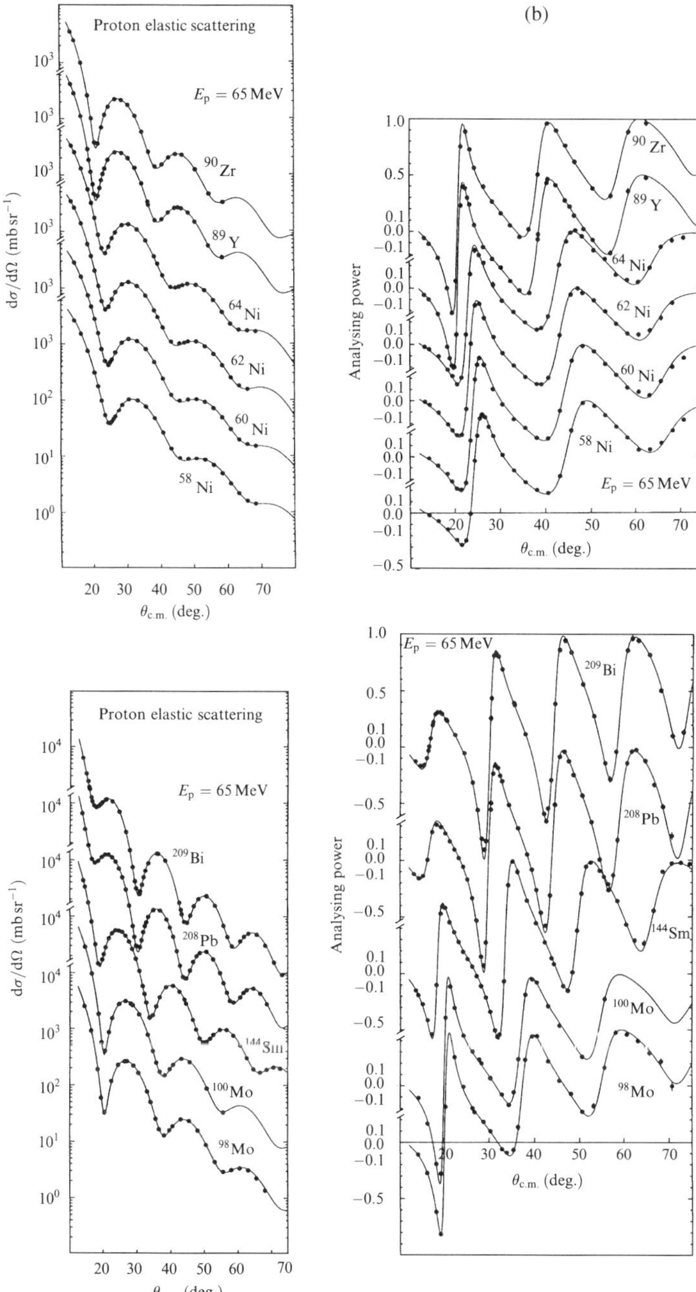

Fig. 2.1 Measured cross sections and analysing powers for proton elastic scattering from various nuclei from (a) ^{16}O to ^{59}Co and (b) ^{58}Ni to ^{209}Bi at 65 MeV. The solid curves represent phenomenological optical potential calculations. Values of 12 potential parameters were adjusted to fit the experimental data. (Taken from Sakaguchi *et al.* [8].)

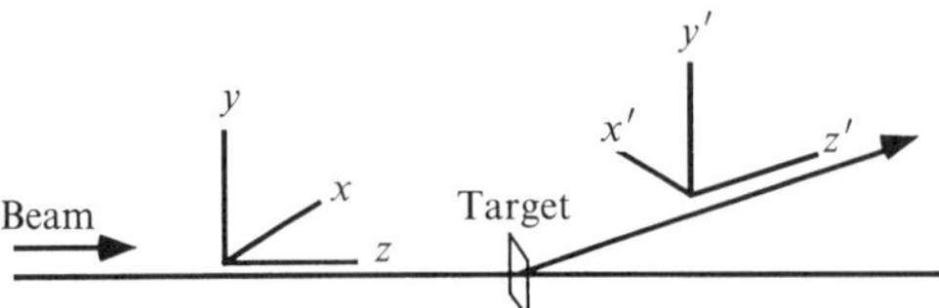

Fig. 2.2 The Madison conventions. For the notation of the polarization observables and axes, the Madison conventions defined in the figure are used. The z-axis coincides with the particle propagation direction and the y-axis is selected to be parallel to $\mathbf{k}_i \times \mathbf{k}_f$. Primes are used to denote the axes after scattering.

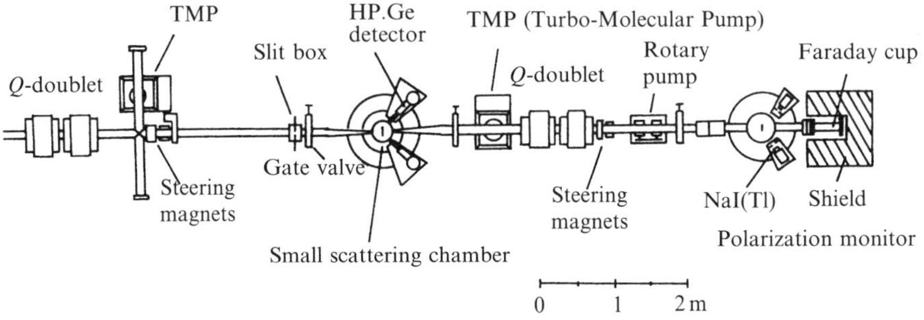

Fig. 2.3 Typical experimental arrangement for measuring cross sections and analysing powers. The asymmetry of scattering is measured by the solid state detectors placed to the left and right of the beam axis at the same scattering angle. The beam polarization during the measurement is monitored continuously by the beam line polarimeter installed downstream of the target chamber.

the Woods–Saxon form:

$$U(r) = V_{\text{Coulomb}} - V_R f(r; r_R, a_R) - iW_V f(r; r_{WV}, a_{WV})$$

$$+ 4ia_{WS} W_S \frac{\mathrm{d}}{\mathrm{d}r} f(r; r_{WS}, a_{WS}) + V_{ls} \left(\frac{\hbar}{m_\pi c}\right)^2 \frac{1}{r} \left(\frac{\mathrm{d}}{\mathrm{d}r} f(r; r_{ls}, a_{ls})\right) (\mathbf{l} \cdot \mathbf{s})$$

$$+ iW_{ls} \left(\frac{\hbar}{m_\pi c}\right)^2 \frac{1}{r} \left(\frac{\mathrm{d}}{\mathrm{d}r} f(r; r_{wls}, a_{wls})\right) (\mathbf{l} \cdot \mathbf{s}), \tag{2.3}$$

where

$$f(r; r_j, a_j) = \frac{1}{1 + \exp((r - r_j A^{1/3})/a_j)}. \tag{2.4}$$

The form factor $f(r; r_j, a_j)$ is usually taken as a Woods–Saxon type, since density distributions of the medium and heavy nuclei are assumed to be approximately Woods–Saxon type in shape, due to the saturation property of the nucleus, and since the potential depth is thought to be proportional to the density in the first-order approximation. The imaginary part of the central potential has both surface and volume components. The surface term represents collective excitations of the nucleon orbiting in the surface region of the nucleus, and the volume term reflects the quasi-free knock-out which is the main process at higher energy.

As for the spin–orbit part, the target centre could be known, in order to define the angular momentum l, only from the variation of the density, because of the short-range nature of the spin-dependent NN interaction. Thus, the form factor of the spin–orbit part for spherical nuclei is a derivative form of the density distribution, namely a Thomas term. The imaginary part of the spin–orbit part of the optical potential is very small, as was found in spin-rotation parameter measurements [17]. Thus, we have set $W_{ls} = 0$ and searched for a total of 12 potential parameters.

The solid curves in Fig. 2.1 correspond to calculations using the best-fit optical potentials. The fits to the experimental data are remarkably good. The χ^2 per $(N - F - 1)$ (N: number of experimental points, F: number of parameters used in the search) values are nearly 1.0. Thus, we believe that we can represent the experimental data by the phenomenological optical potential. The next step is to explain the phenomenological optical potential microscopically.

2.3 Microscopic optical potential

When a nucleon enters the nucleus, it seems to feel a mean field, because the phenomenological optical potential discussed in the previous section explains the scattering very precisely. In this section we briefly review the microscopic optical potential for energies below 100 MeV. When we use the word *microscopic*, it means that we explain the nucleon–nucleus elastic scattering from only the superposition of the NN interaction. If we superimpose the bare NN interaction, it overestimates the depth of the resultant potentials. If we estimate the imaginary potential from the mean free path derived from the free NN cross section, we obtain an absorptive potential which is too deep. We need to consider the effects of Pauli-blocking inside the nucleus, and also various correlation effects. Because of such effects, the NN interaction inside the nucleus differs from that in free space.

The nuclear matter theory developed at the end of the 1950s [3] gives us a guideline to construct an effective NN interaction, namely the G-matrix theory, taking into account the Pauli-blocking and correlation effects in an approximate way. We have

$$\langle kj|g(\omega)|kj\rangle = \langle kj|v|kj\rangle + \sum_{a,b>k_{\mathrm{F}}} \langle kj|v|ab\rangle \frac{1}{\omega - e_a - e_b - \mathrm{i}\eta} \langle ab|g(\omega)|kj\rangle, \qquad (2.5)$$

$$e_a = \frac{a^2}{2m} + V(a, e_a), \qquad (2.6)$$

$$E = \frac{k^2}{2m} + V(k, E), \qquad (2.7)$$

$$U(k, E) = \sum_{j<k_{\mathrm{F}}} \langle kj|g(\omega = E + e_j)|kj\rangle_{\text{Antisymmetry}}$$

$$= V(k, E) + \mathrm{i}W(k, E). \qquad (2.8)$$

Here, $\hbar \mathbf{k}$ is the momentum of the incident nucleon, $\hbar \mathbf{j}$ the momentum of the nucleon inside the nucleus ($j \leq k_F$), and $\hbar \mathbf{a}$ and $\hbar \mathbf{b}$ the momenta of two nucleons in the intermediate state. We have $|\mathbf{a}|, |\mathbf{b}| > k_F$ if we take into account the Pauli-blocking effect approximately. This G-matrix includes the parameters, such as the Fermi momentum $\hbar k_F$, which shows the Pauli-blocking effect. The above equations can be solved numerically by the Brueckner theory developed to solve the nuclear structure problem, i.e. the problem of calculating the binding energy. For application to nuclear reactions such as elastic scattering we need the G-matrix in the r-representation (coordinate space representation). We usually express the G-matrix in the form of a sum of Yukawa potentials of different ranges [18–20]. The coefficient of each term in this sum depends on the incident energy and the nucleon density in the target. The G-matrix is then folded [5] with the density distributions of protons and neutrons in the nuclear ground state to give a microscopic optical potential. Proton density distributions are estimated from charge distributions, but neutron density distributions are inferred only from model calculations, such as Hartree–Fock calculations with the simplified interactions or from various assumptions.

There exist some problems in using the microscopic optical potential. In particular, there are problems with the imaginary part of the mean field, which do not appear in the binding energy case. If the incident energy of the nucleon is so high that quasi-free scattering is the main source of the imaginary part, the above formula for the mean field will be a good approximation. However, if the incident energy of the projectile is sufficiently low and the collective excitations or the process of the compound nucleus formation occupy the larger part of the reaction, the approximation given above becomes less accurate. Thus, for comparison of the microscopic theory with experimental data, calculations employing a renormalized imaginary part are often used.

In Fig. 2.4 we display experimental data for $E_n = 21.3\,\text{MeV}$ [21] along with the microscopic calculation [19,20,22]. For the solid curves in Fig. 2.4, the depths of both the real and imaginary parts of the optical potential are renormalized. However, at higher energy, for example at $E_p = 65\,\text{MeV}$ shown in Fig. 2.5, they are not renormalized [20]. In Fig. 2.5 we give an example of the complete experimental data, namely differential cross sections, analysing powers, and spin-rotation parameters, measured over a wide angular range [17]. The microscopic optical potential seems to explain the main features of scattering. The differences at backward angles suggest a different density distribution of neutrons in the nucleus, incompleteness of the G-matrix, and/or the presence of effects of different nuclear reaction processes. In Fig. 2.6 we compare microscopic optical potentials with phenomenological potentials for ^{40}Ca at $E_p = 65\,\text{MeV}$. Although the Woods–Saxon-type phenomenological optical potential, which can explain scattering in the region of $\theta \leq 75°$, agrees with the microscopic potential only in the surface region of the nucleus, the one with adjustable shape (indicated in the figure as a model-independent potential), which can reproduce the scattering over the entire angular range, agrees very well with the microscopic potentials throughout, except in the central region of the imaginary part, where the ambiguity in the phenomenological potential is large [17].

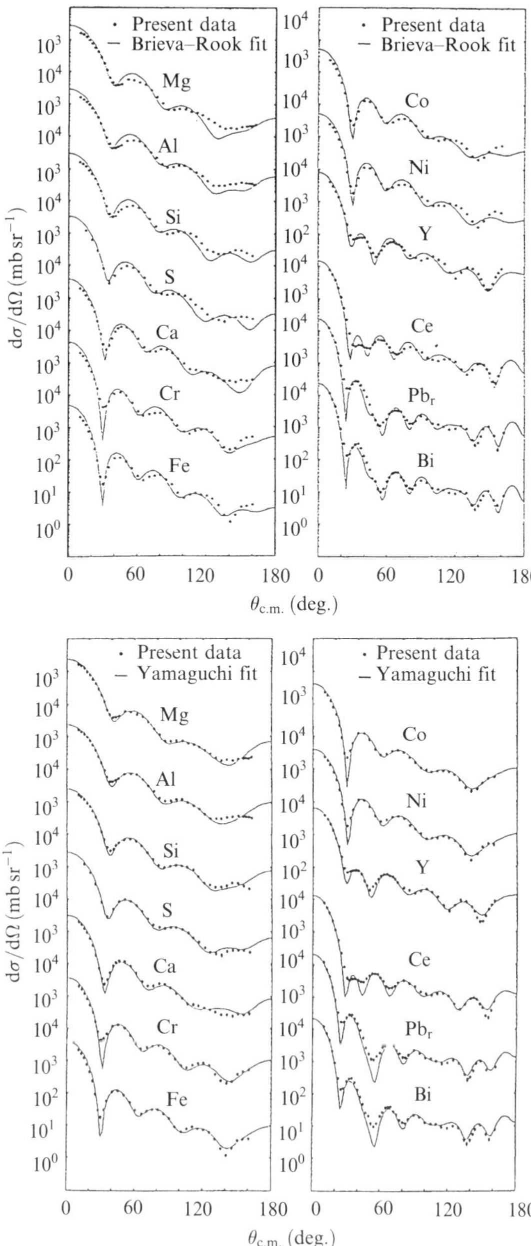

Fig. 2.4 Experimental data for neutron elastic scattering [21] from various nuclei compared with microscopic optical potential calculations by two groups. At low energies, the depths of both the real and imaginary central parts of the microscopic optical potential are renormalized (real part $\approx -10\%$, imaginary part $\approx -35\%$) to give an optimum fit to the experimental data. (Taken from Olsson *et al.* [21].)

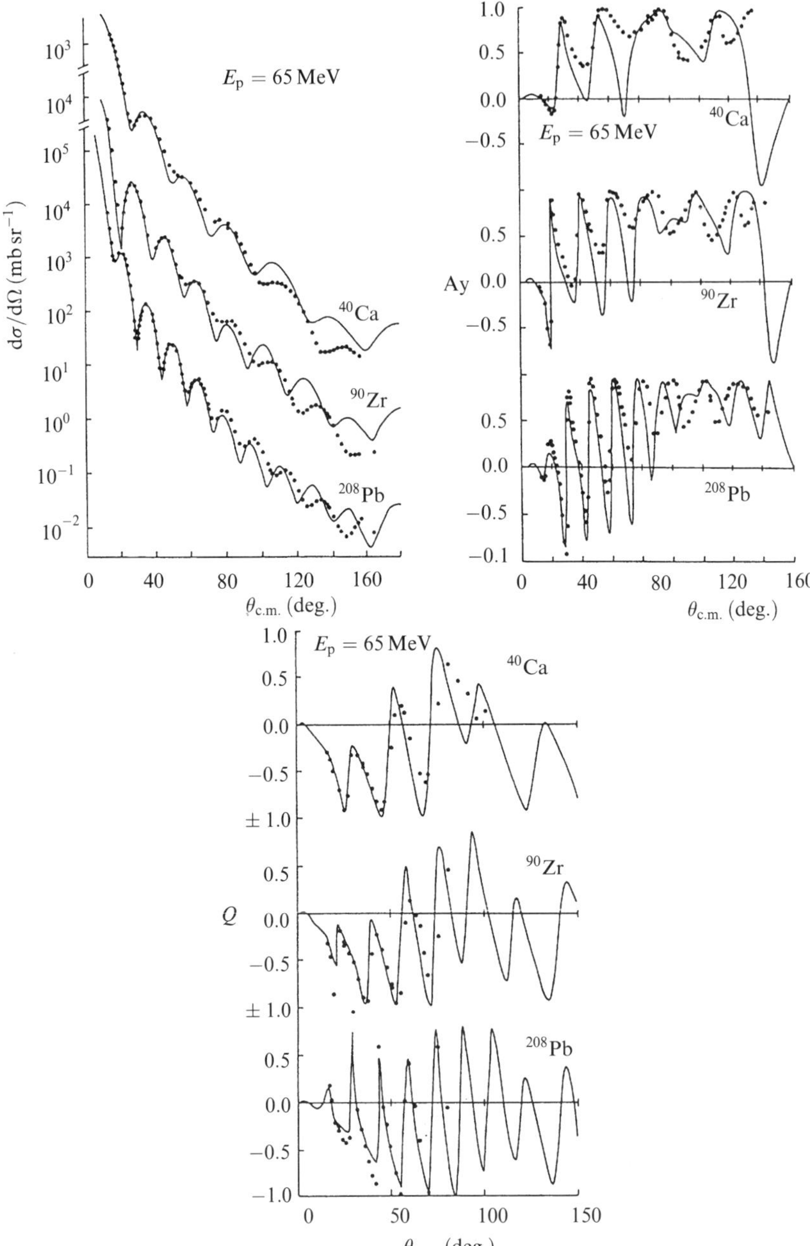

Fig. 2.5 Proton elastic scattering data including spin observables compared with one of the microscopic optical potential calculations [20] at $E_p = 65$ MeV. At this energy, renormalization of the microscopic optical potential is not necessary. (Taken from Yamaguchi *et al.* [20].)

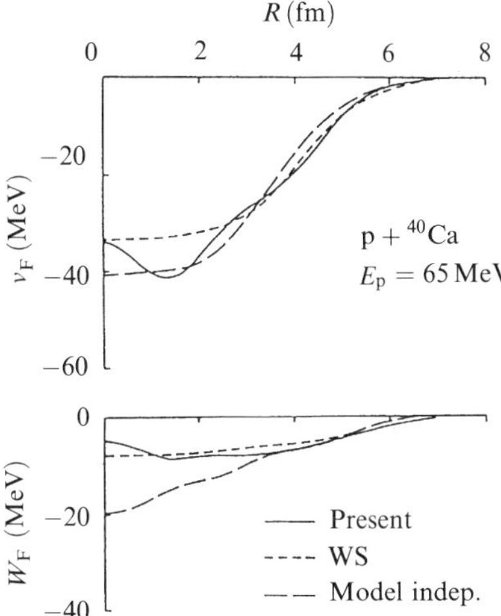

Fig. 2.6   The real and imaginary parts of the microscopic optical potential for ^{40}Ca at $E_p =$ 65 MeV. The solid curves indicate microscopic potentials, the short-dashed curves represent the phenomenological Woods–Saxon-type potentials which can reproduce the angular distribution ($\theta \leq 75°$), and the long-dashed curves correspond to phenomenological potentials (indicated in the figure as model-independent potentials), which can reproduce [17] the angular distributions, including the spin observables, at backward angles. Agreement between the phenomenological and the microscopic optical potential is good except for small differences in the central region of the nucleus. (Taken from Sakaguchi *et al.* [17].)

2.4   Comparison with microscopic theory

Since the microscopic optical potential at low energy has some ambiguities in the imaginary part, it is not appropriate to compare the theory with experimental scattering observables directly. Instead, the real parts of the phenomenological and microscopic optical potentials are compared. In addition, since the microscopic potentials are not of the Woods–Saxon shape, we compare them in the form of integrated quantities, such as the volume integral J_d of the potential and the Mean Square Radius (MSR) $\langle r^2 \rangle_{\text{pot}}$ of the potential, which are defined as follows:

$$J_d = \left| \int U(r) \, dr \right|, \qquad \langle r^2 \rangle_{\text{pot}} = \frac{\int U(r) r^2 \, dr}{\int U(r) \, dr}. \tag{2.9}$$

There exist some ambiguities in the phenomenological potential parameters, for example the ambiguities of the type $VR^n = $ constant for the case of a Woods–Saxon-type potential,

$$U(r) = \frac{V}{1 + e^{((r-R)/a)}}. \tag{2.10}$$

This means that if we change the depth of the potential V, the χ^2-value of the fit (or the quality of the fit) remains almost unchanged when varying R so as to keep $VR^n = $ constant.

In Fig. 2.7 we display a χ^2-map [7] which was generated by searching around the χ^2-minimum, changing the depth of the potential and the radius parameters. A region around the χ^2-minimum was found to exhibit the ambiguities of the type $VR^n = $ constant. For a Woods–Saxon-type potential (2.10), the volume integral of the potential $J_{\rm d}$ and the MSR of the potential, $\langle r^2 \rangle_{\rm pot}$, are given to a good approximation [8,23] by

$$J_{\rm d} = \frac{4\pi V R^3}{3}\left(1 + \frac{\pi^2 a^2}{R^2}\right), \tag{2.11}$$

$$\langle r^2 \rangle_{\rm pot} = \tfrac{3}{5}R^2 + \tfrac{7}{5}\pi^2 a^2. \tag{2.12}$$

The quantity $J_{\rm d}$ is already in the VR^n form and thus relatively free from such ambiguities. On the other hand, $\langle r^2 \rangle$ is sensitive to the positions of the peak and the valley in the diffractive angular distribution and is defined by the scattering, as shown in Fig. 2.7(b).

In Fig. 2.8 we exhibit the MSR of the real central part of the optical potential as a function of $A^{2/3}$. The solid line in the figure represents the least-squares fit to the data. We find

$$\langle r^2 \rangle = (0.937 \pm 0.012)A^{2/3} + (6.42 \pm 0.21)\,{\rm fm}^2.$$

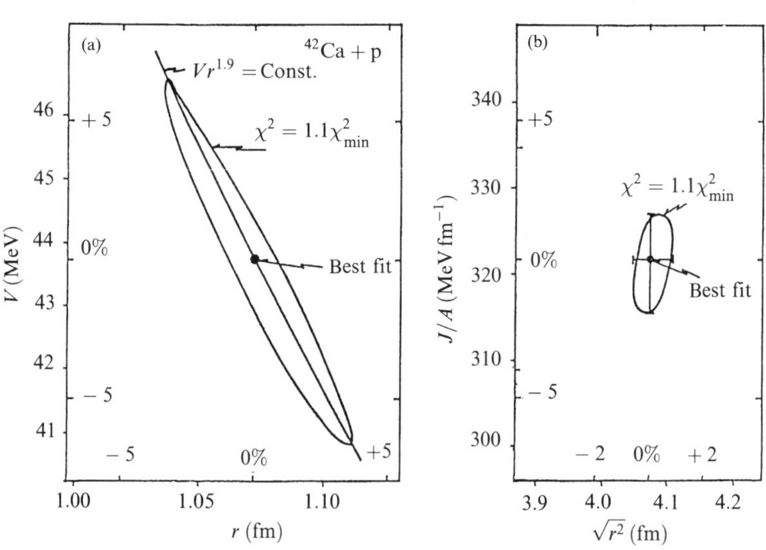

Fig. 2.7 Contours of constant χ^2 values (a) in the V_R–r_R plane and (b) in the $J_{\rm d}/A - \sqrt{\langle r^2 \rangle}$-plane. The curve for constant $Vr^{1.9}$ is also shown in (a). (Taken from Noro *et al.* [7].)

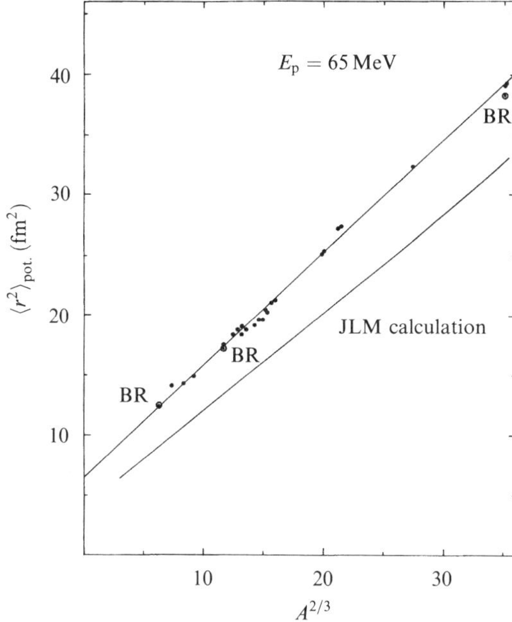

Fig. 2.8 The MSR values of the real central part of the optical potential shown as a function of some power of the target mass number A. The solid line was obtained by a least-squares linear fit to the data. The solid line marked JLM is obtained from the optical potential with the microscopically derived parameters of JLM [24] and it shows the zero-range nature of the JLM model. (Taken from Sakaguchi *et al.* [8].)

The doubly circled points denoted BR and the line denoted JLM in the figure correspond to the values calculated from the microscopic optical potentials. The Brieva and Rook [5] (BR) calculations give a better explanation of the data, while the points of Jeukenne, Lejeune, and Mahaux (JLM) [24–26] are near the MSR of the point-nucleon distribution. Since JLM constructed an optical potential directly from the G-matrix using the local energy approximation and the folding procedure, the calculation is almost a zero-range approximation for the effective NN interaction and thus is almost the same as the MSR of the density distribution of the nucleus.

For application to nuclear reactions a further 'smearing' procedure is neccesary for the JLM potential, considering the finite range of the nuclear force. The Oxford group (BR) [5] and the Miyazaki group (Yamaguchi and co-workers) [19,20,22] used the G-matrix in the r-representation and took into account the finite-range nature of the NN interaction, which is very important to explain nuclear reactions.

In Fig. 2.9 the MSR of the charge distribution obtained from the electron scattering is shown as a function of $A^{2/3}$. The solid line in the figure represents the least-squares

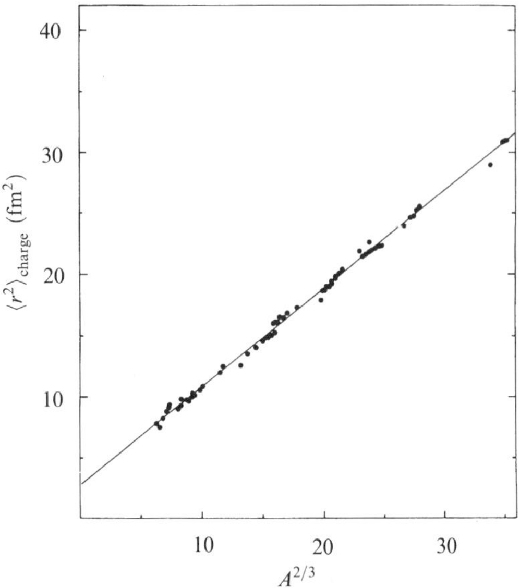

Fig. 2.9   The MSR values of charge distributions obtained from electron scattering plotted as a function of some power of the target mass number A. The solid line was obtained by a least-squares linear fit to the data. (Taken from Sakaguchi *et al.* [8].)

fit to the charge radius obtained from electron scattering:

$$\langle r^2 \rangle_{\text{charge}} = (0.799 \pm 0.006) A^{2/3} + (2.50 \pm 0.12)\,\text{fm}^2.$$

Due to nuclear saturation, the nuclear density inside a nucleus and the diffuseness of the nucleus are almost constant for stable nuclei. Thus, by assuming the Fermi shape for the density distribution, we can derive the above $A^{2/3}$ dependence for the charge distribution [8].

If we assume a simplified folding model, the real central part of the optical potential can be written as

$$U(\mathbf{r}) = \int \rho(\mathbf{r}) g(|\mathbf{r} - \mathbf{r}_0|)\,\mathrm{d}\mathbf{r}_0, \tag{2.13}$$

where $\rho(\mathbf{r})$ is the density distribution of point nucleons and $g(\mathbf{r})$ is an effective two-body interaction, which is a simplified G-matrix without k_{F}- and E-dependence. For a nucleus possessing a symmetry axis, the MSR of the potential deduced from eqn (2.13) is

$$\langle r^2 \rangle_{\text{pot}} = \langle r^2 \rangle_{\text{matt}} + \langle r^2 \rangle_{\text{int}}, \tag{2.14}$$

where $\langle r^2 \rangle_{\text{pot}}$, $\langle r^2 \rangle_{\text{matt}}$, and $\langle r^2 \rangle_{\text{int}}$ are MSR values of the potential $U(\mathbf{r})$, the density distribution of point nucleons $\rho(\mathbf{r})$, and the effective interaction $g(\mathbf{r})$, respectively. For the MSR of the charge distribution we also obtain

$$\langle r^2 \rangle_{\text{charge}} = \langle r^2 \rangle_{\text{matt}} + \langle r^2 \rangle_{\text{nucleon}}, \tag{2.15}$$

where $\langle r^2 \rangle_{\text{nucleon}}$ represents the MSR of a nucleon, which is $(0.82)^2$ fm^2. Thus, the mass number dependence of the MSR of the charge distribution and the optical potential is the same if we assume a simplified folding model.

However, as shown in Figs 2.8 and 2.9, the dependence on the target mass number is different for the charge distribution and the optical potential. This means that the shape of the optical potential is not the same as the density distribution of the nucleons, but is somewhat deeper at the nuclear surface. This can be explained by the density dependence (DD) of the effective NN interaction. In Fig. 2.10 we display an example [10] of the model calculation of the optical potential. Here, a density dependence is introduced into the M3Y effective interaction by multiplying by a density-dependent factor which simulates the microscopic calculation. The dotted line corresponds to a folding model calculation using the simple M3Y [27] effective interaction without any density dependence. When we introduce an exchange contribution δ, the MSR becomes smaller, as shown by the dot-dashed line. The MSR increases up to the observed values when we introduce density dependence. Green's [28] and JLM's [24] density dependence formulae give similar results.

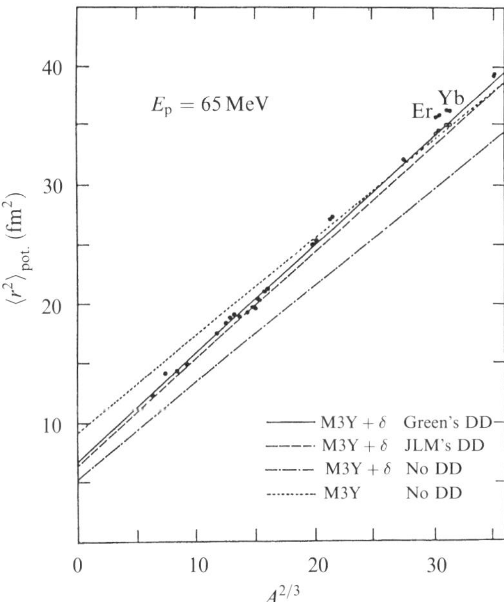

Fig. 2.10 The MSR of the potential. The closed circles represent the MSR of the optical potentials at $E_{\text{p}} = 65\,\text{MeV}$, and the open circles represent the MSR for deformed nuclei, where all the deformation parameters are set to zero to exclude the coupling effect [9]. (Taken from Ichihara *et al.* [10].)

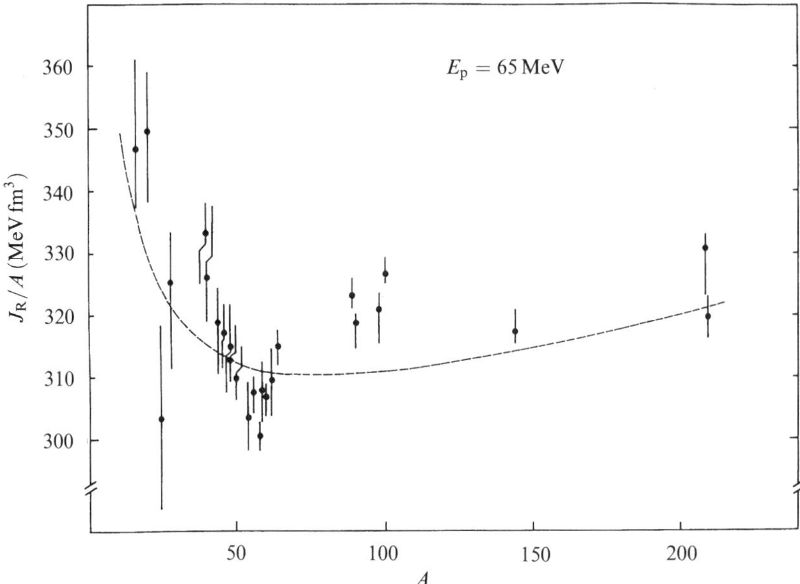

Fig. 2.11 Volume integrals per nucleon of the real central part of the optical potential as a function of the target mass number A [8]. The dashed curve represents the JLM model calculation for $E_\mathrm{p} = 65\,\mathrm{MeV}$. The low point of $^{24}\mathrm{Mg}$ was found to correspond to the effect of the level coupling to the inelastic channel [9] in deformed nuclei. (Taken from Sakaguchi *et al.* [8].)

In Fig. 2.11 the J_d/A values deduced from the phenomenological fit using eqn (2.11) are plotted as a function of the target mass number. Although the observed J_d/A values are considerably scattered, we notice global behaviour around the average value of $J_\mathrm{d}/A = 318\,\mathrm{MeV\,fm}^3$. As the mass number increases, the J_d/A value decreases rapidly down to a minimum in the Fe–Ni region and then increases gradually towards the Pb–Bi region. This global trend is reproduced remarkably well by the JLM calculation represented by the dashed curve in Fig. 2.11.

According to the JLM model, the effective interaction is density dependent. In the lower-density region, the effective interaction is stronger, due to the smaller Pauli-blocking effect. The surface-to-volume ratio decreases as $A^{-1/3}$, and the value of J_d/A decreases rapidly. The rapid change of J_d/A for lighter nuclei is an indication of the density-dependent interaction. The second gradual increase is explained in the JLM model by the isospin-dependent interaction and by the velocity dependence of the effective interaction. As the mass number increases, the velocity of the projectile inside the nucleus decreases due to the repulsive Coulomb potential.

In Fig. 2.12 the energy dependence of J_d/A is displayed for $^{40}\mathrm{Ca}$ nuclei, where the values of the volume integral per nucleon of the real central part of the optical potential are plotted as a function of the incident proton energy. The open circles represent the Milan data for $^{40}\mathrm{Ca}$ [29], and the double circle represents a mean value over 25

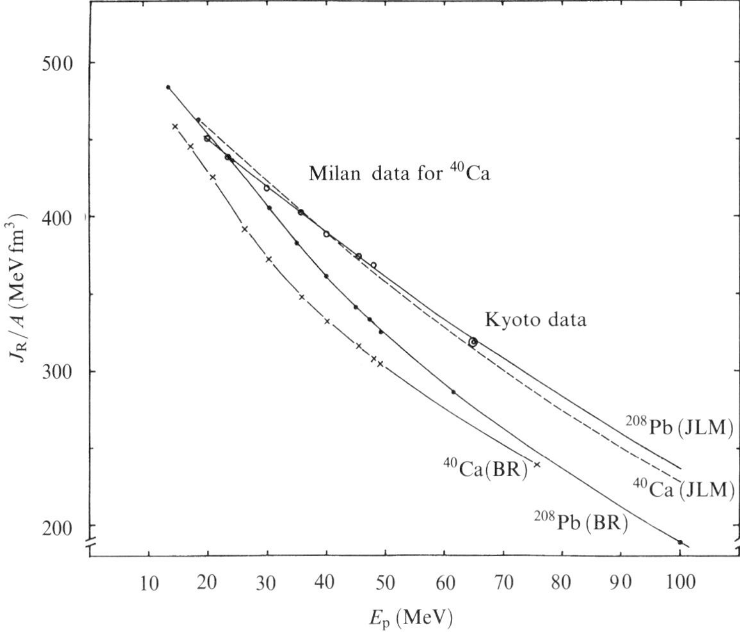

Fig. 2.12 Volume integral values per nucleon of the optical potential plotted as a function of the incident proton energy. Open circles correspond to the Milan data [29] for ^{40}Ca, and the double circle marked Kyoto data represents a mean value over 25 targets at $E_p = 65$ MeV [8]. The solid dots and the crosses represent values calculated microscopically by Brieva and Rook [5]. The curve connecting these dots is meant to serve only as a guide to the eye. The solid and broken curves labelled JLM represent values calculated using the procedure of JLM [24]. (Taken from Sakaguchi *et al.* [8].)

targets at $E_p = 65$ MeV. The solid dots and crosses correspond to values calculated microscopically by BR [5]. These are connected by curves, which serve only as a guide to the eye. The solid and broken curves labelled JLM are values calculated using the method of JLM. The energy dependence of J_d/A is reproduced well by their calculation [24].

2.5 Inelastic scattering from deformed nuclei

Stable nuclei with mass numbers between 150 and 200 are permanently deformed. The energy levels for these nuclei show typical rotational bands with level spacing proportional to $I(I+1)$ (where I represents the spin of the nuclear state). Since the level spacing between the ground and the first 2^+ states is typically 80 keV, it is experimentally difficult to separate elastic and inelastic scattering when the incident energy of the beam increases. At $E_p = 65$ MeV, measurements were performed using the high-resolution magnetic spectrograph RAIDEN with momentum resolution $p/\Delta p = 35\,000$. In Fig. 2.13 an example of the momentum spectrum is shown. We can see a clear separation of the

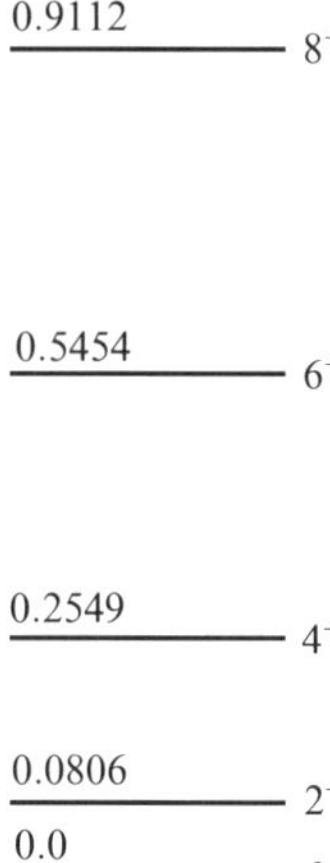

Fig. 2.13 The level scheme of the ground rotational band of ^{166}Er.

ground and the first excited states. For monitoring the beam polarization, a sampling-type beam polarimeter placed upstream from the target position was used. In order to obtain a clean spectrum, the target foil of the sampling polarimeter was inserted into the beam line during about one-tenth of the run time, during which the measurement of scattering was inhibited. Figure 2.14 shows the low-lying energy levels of the ^{166}Er nucleus, which form a typical ground state rotational band. Each state in a given rotational band possesses rotational motion corresponding to its energy, but all states within such a band have the same intrinsic state. Inelastic scattering to the level in the ground rotational band is thought to be similar to the elastic scattering for spherical nuclei, since there is no intrinsic nuclear excitation. The scattering is treated by using a deformed mean field, namely a deformed optical potential. To obtain this potential, we usually deform the half-radius of the potential ($r_j A^{1/3}$) as follows:

$$U(r, \theta) = V_{\text{Coulomb}} - V_R f(r; r_R(\theta), a_R) - iW_{WV} f(r; r_{WV}(\theta), a_R)$$

$$+ 4ia_{WS} W_s \frac{\mathrm{d}}{\mathrm{d}r} f(r; r_{WS}(\theta), a_{WS})$$

$$+ V_{ls} \left(\frac{\hbar}{m_\pi c} \right)^2 \frac{1}{r} \left(\frac{\mathrm{d}}{\mathrm{d}r} f(r; r_{ls}(\theta), a_{ls}) \right) (\mathbf{l} \cdot \mathbf{s})$$

$$+ iW_{ls} \left(\frac{\hbar}{m_\pi c} \right)^2 \frac{1}{r} \left(\frac{\mathrm{d}}{\mathrm{d}r} f(r; r_{wls}(\theta), a_{wls}) \right) (\mathbf{l} \cdot \mathbf{s}), \qquad (2.16)$$

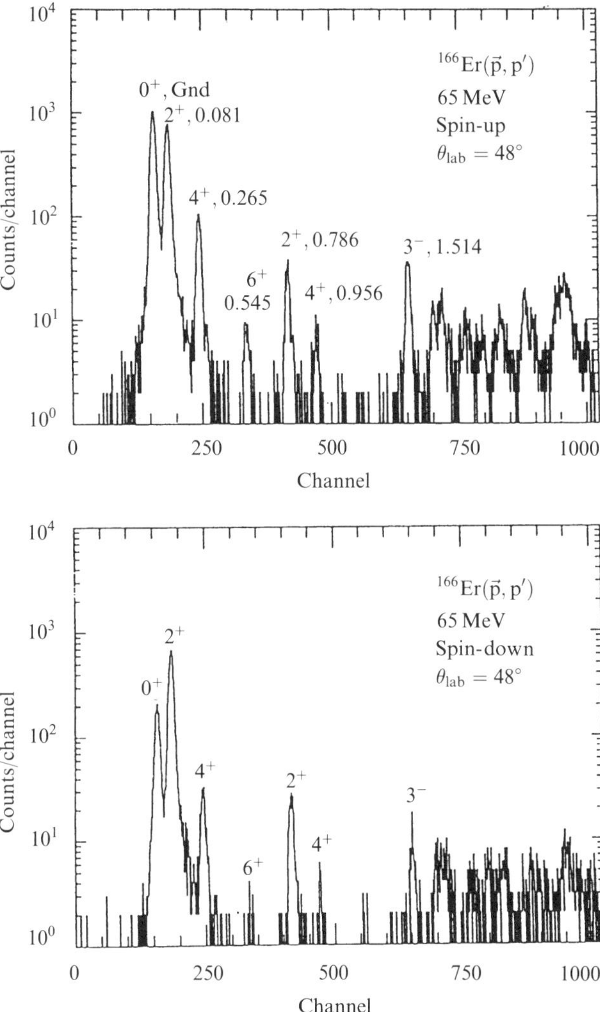

Fig. 2.14 Typical momentum spectra for ^{166}Er(p, p') scattering at $E_p = 65$ MeV. Beam spin-up and spin-down spectra are shown at $\theta_{\text{lab}} = 48°$. Asymmetries in the scattering are noticed by the relative heights of the peaks. (Taken from Ichihara *et al.* [10].)

where

$$f(r; r_j(\theta), a_j) = \frac{1}{1 + \exp((r - r_j(\theta)A^{1/3})/a_j)},$$

$$r_j(\theta) = r_0^j \left(1 + \sum_\lambda \beta_\lambda^j Y_{\lambda 0}(\theta)\right), \tag{2.17}$$

where β_λ^j is a static deformation parameter for the j part of the optical potential, and the index j corresponds to each part of the optical potential such as the real central part, the imaginary part, and so on.

Scattering by this deformed potential which does not excite the intrinsic nuclear state is described by the the coupled-channel formalism using the Schrödinger equation as follows:

$$H\Psi = E\Psi, \tag{2.18}$$

$$H = H_A - \left(\frac{\hbar^2}{2\mu_\alpha}\right)\frac{d^2}{dr^2_\alpha} + U(r_\alpha), \tag{2.19}$$

$$\Psi = \sum_{A'}\chi_{A'}(r_\alpha)\psi_{A'}, \qquad H_A\psi_A = \epsilon_A\psi, \tag{2.20}$$

$$\left(\frac{d^2}{dr_\alpha{}^2} - U_{A,A}(r_\alpha) + k_A^2\right)\chi_A(r_\alpha) = \sum_{A'\neq A}\chi_{A'}(r_\alpha)U_{A,A'}(r_\alpha), \tag{2.21}$$

$$k_A^2 = \frac{2\mu_\alpha(E - \epsilon_A)}{\hbar^2}, \tag{2.22}$$

$$U_{A,A'} = \frac{2\mu_\alpha}{\hbar^2}\langle\psi_A|U|\psi_{A'}\rangle. \tag{2.23}$$

The deformed components of the potential connect different rotational states in the ground rotational band. In Fig. 2.15 we show a typical example of inelastic scattering from deformed nuclei. The solid curves in the figure represent the best-fit coupled-channel calculations. From the best-fit potential we can deduce the multipole moments of the various parts of the potential. The λth multipole moment is defined as

$$Q_\lambda^j = Ze\frac{\int U_j(r)Y_{\lambda 0}(\theta)r^\lambda\,dr}{\int U(r)\,dr}, \tag{2.24}$$

where j indicates the various components of the optical potential such as the real or imaginary central part, and the real or imaginary spin–orbit part. Here, we have multiplied the formula by Ze in order to compare it with the electron scattering data or the values obtained with electromagnetic probes. As shown in Fig. 2.16, the quadrupole moments (Q_2) obtained from the proton data are about 5% larger than those obtained using the electromagnetic probes. As discussed in Ref. [10], Satchler's theorem [30] plays an important role in understanding the differences among multipole moments found using measurements with different probes. For a folding model potential in which the underlying effective NN interaction is assumed to depend only on the magnitude of the distance between the projectile nucleon and the nucleons in the target, the multipole moments of the potential are equal to those of the nuclear density, as shown by Satchler's theorem [30]. That is, if the Deformed Optical Potential (DOP) is given by a simplified

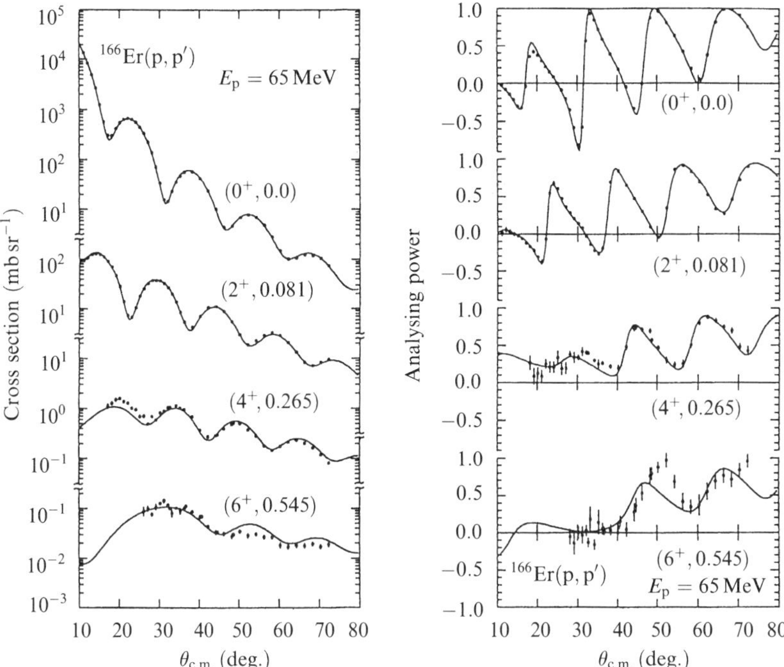

Fig. 2.15 Angular distributions of cross sections and analysing powers for ^{166}Er(p, p$'$) scattering at 65 MeV. The solid curves represent the results of coupled-channel calculations with the deformed optical potential. (Taken from Ichihara *et al.* [10].)

folding model,

$$U(r) = \int \rho(r_1) t(|r - r_1|) \, dr_1, \tag{2.25}$$

then the multipole moments of the potential (Q_λ^{DOP}) are equal to multipole moments of the density distribution ($Q_\lambda^{\text{matter}}$):

$$Q_\lambda^{\text{DOP}} = Q_\lambda^{\text{matter}}, \tag{2.26}$$

$$Q_\lambda^{\text{matter}} \equiv \frac{\int \rho(r) r^\lambda Y_{\lambda 0}(\theta) \, dr}{\int \rho(r) \, dr}. \tag{2.27}$$

Therefore, if the multipole moments of the neutron density are equal to those of the proton density, the multipole moments of the DOP should be equal to those of the charge density in the limit that the DOP is equivalent to such a simple folding model potential, since the multipole moments of the charge density are equal to those of the proton distribution. The many-body effects which are not included in the simple folding model described

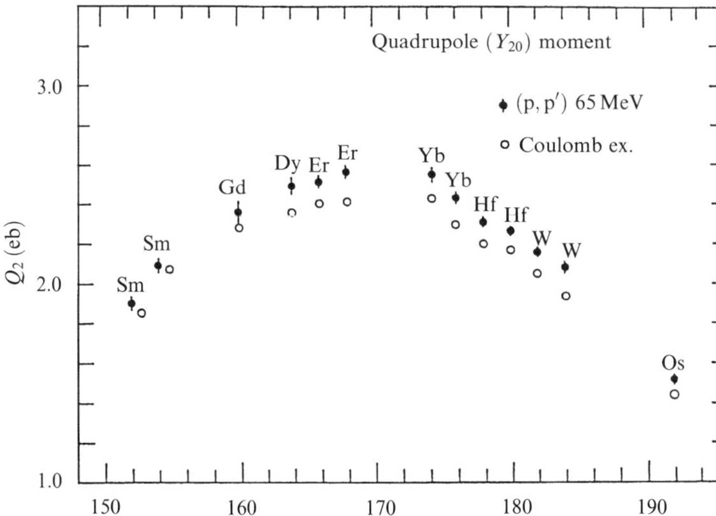

Fig. 2.16 Quadrupole moments for deformed nuclei. The closed circles represent the multipole moments of the deformed optical potential derived from the proton inelastic scattering. The open circles correspond to the charge multipole moments obtained by electron scattering and Coulomb excitation. The open squares show the multipole moments of the folded potential using the JLM density dependence (DD), and the open diamonds represent those using Green's DD. Some symbols have been shifted slightly horizontally for the sake of clarity. (Taken from Ichihara *et al.* [10].)

above will give rise to differences between the multipole moments of the DOP and the charge density.

In Figs 2.16 and 2.17, a systematic 4–6% larger value of the DOP quadrupole moment in comparison with the charge density is observed in the rare earth region. The Q_2 moments for the DOP represented in Fig. 2.15 by the black circles are larger than the Q_2 moments for the charge distribution denoted by the white circles. The squares and the diamonds in Fig. 2.17 correspond to the results of the folding model calculation based on the density-dependent effective NN interaction of the JLM type and Green's type, respectively. As shown in Fig. 2.16, for all the nuclei in this mass region, the quadrupole moments of the potential are larger than those of the charge distribution measured by the electromagnetic probes. As Brieva and Geogiev [42] suggested, however, other many-body effects, such as the non-locality of the exchange term, will also cause a difference in the multipole moments, because such effects cannot be included in the simple folding model.

For the states in the γ-vibrational band, we have introduced a γ-vibrational coupling (represented by the dashed lines in Fig. 2.18), between states in different rotation bands. An example of the fit is shown in Fig. 2.19 for the case of the ^{192}Os nucleus. The solid curves in the figure correspond to the calculation with the $Y_{22} + Y_{42}$-mode

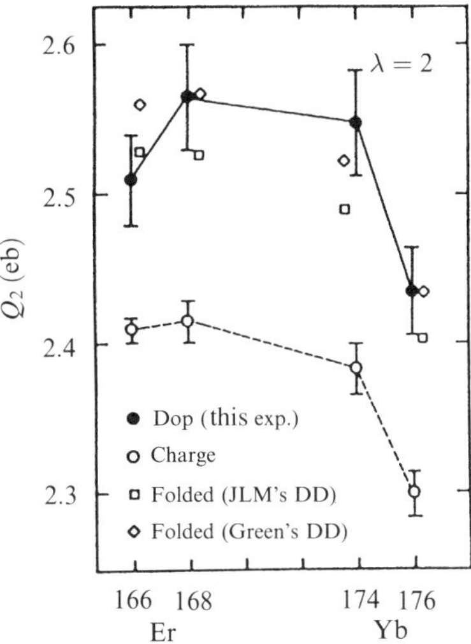

Fig. 2.17 Quadrupole moment of the deformed optical potential in the rare earth region as a function of the mass number. The closed circles represent the quadrupole (Y_{20}) moments of the deformed optical potential, while the open circles represent those obtained by electron scattering and Coulomb excitation. The closed-circle values are larger than the open-circle values for nuclei in the rare earth region. (Taken from Ichihara *et al.* [13].)

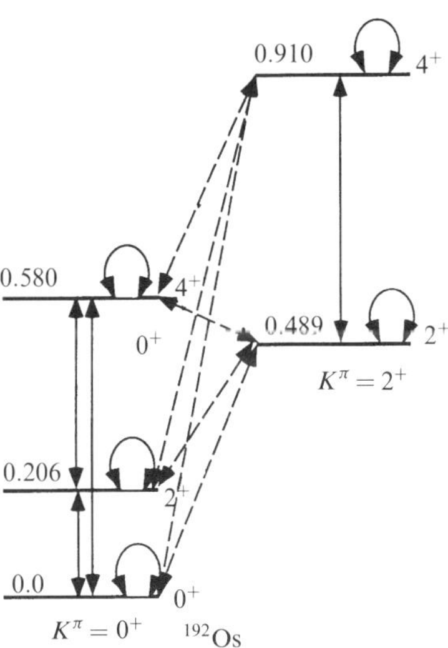

Fig. 2.18 The level scheme for ^{192}Os and the coupling scheme of the coupled-channel calculation assuming the γ-vibrational model. Solid lines with arrows indicate the rotational coupling and dashed lines with arrows indicate the vibration–rotation coupling.

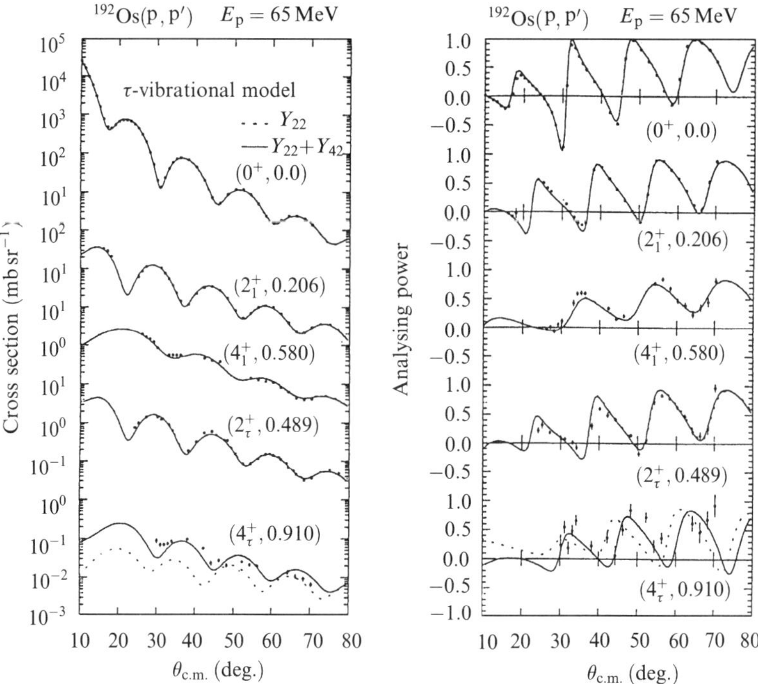

Fig. 2.19   Cross sections and analysing powers for the $0^+, 2^+, 4^+$ states of the ground rotational band, and $2^+, 4^+$ states of the γ-band of ^{192}Os by inelastic scattering of polarized protons at 65 MeV. The solid (dotted) curves represent the best-fit result of the coupled-channel calculation assuming the $Y_{22} + Y_{42}$-mode (Y_{22}-mode) γ-vibration. (Taken from Ichihara *et al.* [13].)

γ-vibration, while the dashed curves correspond to the calculation with only the Y_{22}-mode γ-vibration. Thus, we need both Y_{22}- and Y_{42}-mode γ-vibrations to explain the experimental data. In the rare earth region, the coupling between the ground rotational band and the γ-band is not negligible [10]. Quadrupole (Y_{20}) moments and hexadecapole (Y_{40}) moments were calculated from the best-fit parameters for the experimental data of the ground rotational band and the γ-vibrational band, as shown in Fig. 2.20. In addition to these, strengths of transitions to the γ-vibrational states are also shown. The γ-vibration is a collective motion associated with the surface vibration in deformed nuclei. It is treated, also in the coupled-channel formalism, by deforming the potential surface as

$$r_j(\theta, \varphi) = r_0^j \left(1 + \sum_\lambda \beta_\lambda^j Y_{\lambda 0}(\theta, \varphi) + \sum_{\lambda'\mu'} \alpha_{\lambda'\mu'} Y_{\lambda'\mu'}(\theta, \varphi) \right), \tag{2.28}$$

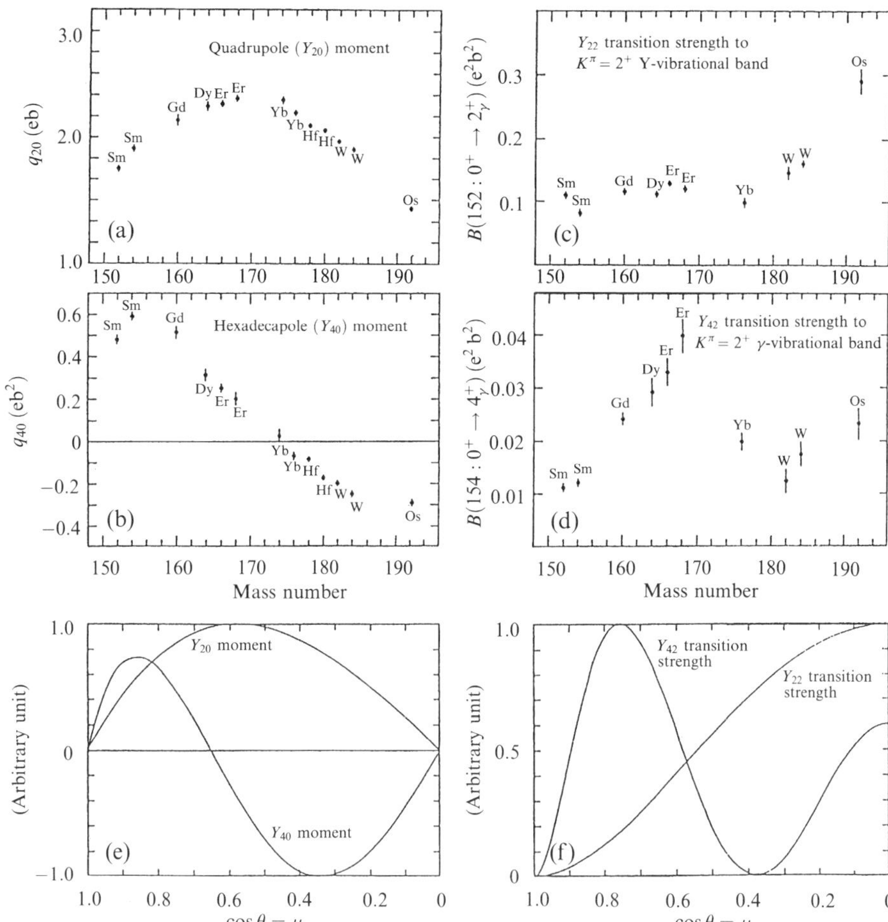

Fig. 2.20 (a) and (b): The quadrupole (Y_{20}) and hexadecapole (Y_{40}) moments of the deformed optical potential obtained from the coupled-channel analysis of inelastic scattering of polarized protons at 65 MeV. (c) and (d): The quadrupole (Y_{22}) and hexadecapole (Y_{42}) transition strength to the γ-vibrational band obtained from the coupled-channel analysis of inelastic scattering of polarized protons at 65 MeV. (e): The quadrupole (Y_{20}) and hexadecapole moments of a nucleus when the valence particles are filled from the z-axis to $\mu = \cos\theta$ according to Bertsch [31]. (f): The quadrupole (Y_{22}) and hexadecapole (Y_{42}) transition strengths to the γ-vibrational band according to the extension of Bertsch's model. (Taken from Ichihara *et al.* [13].)

where $\alpha_{\lambda\mu}$ represents the collective coordinates corresponding to the $Y_{\lambda\mu}$ γ-vibration. The transition strength is given by

$$B(IS\lambda, 0 \to \lambda) = (J_\rho/J_U)^2 (Z/A)^2 \left(\int \left(\sum_{\lambda'\mu'} \eta_{\lambda'\mu'} Y_{\lambda'\mu'} \right) R_0 \frac{\delta U}{\delta R} Y_{\lambda\mu}^* \, d\boldsymbol{r} \right)^2 , \quad (2.29)$$

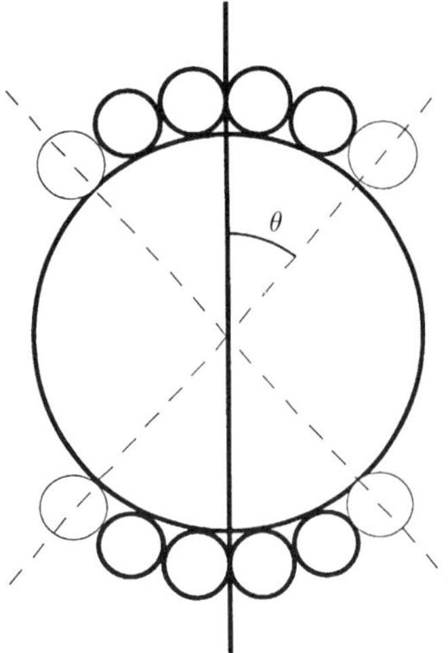

Fig. 2.21   Results from the extended polar cap model for a deformed nucleus in which the valence nucleons are filled from the z-axis to $\mu = \cos\theta$. The small circles indicate the nucleons and the broken lines correspond to the Fermi surface.

where

$$(J_\rho/J_U) = \int \rho(\boldsymbol{r})\, d\boldsymbol{r} \Big/ \int U(\boldsymbol{r})\, d\boldsymbol{r} = A \Big/ \int U(\boldsymbol{r})\, d\boldsymbol{r}, \qquad (2.30)$$

and the $\eta_{\lambda\mu}$ represent the amplitudes of the vibration of the optical potential associated with $\alpha_{\lambda\mu}$, which have been determined by the coupled-channel analysis assuming the γ-vibrational model. In Fig. 2.20, the calculated Y_{22} and Y_{42} transition strengths are plotted. We have used the real central part of the optical potential for U in eqn (2.30). This figure reflects a gross systematic trend in the transition strength as a function of target mass number. This is explained in the next section by using a simple model.

2.5.1   *Polar cap model*

To understand the overall trend of the above multipole moments and transition strengths as well as hexadecapole moments of the ground rotational band, let us introduce the polar cap model of Bertsch [31] and its extension [13,41]. At the beginning of the major shell, the first four valence particles define the z-axis in the intrinsic frame. When filling the major shell, additional particles are placed in orbits as close to this z-axis as possible. Eventually, the orbits are again filled to make a spherically symmetric distribution. Thus the Q_λ moment of a nucleus in which the valence particles are filled from the z-axis to $\mu = \cos\theta$ is proportional to the integral

$$\int_\mu^1 P_{\lambda 0}(x)\, dx, \qquad (2.31)$$

where $P_{\lambda 0}(x)$ is the λth Legendre polynomial, and $\mu\,(=\cos\theta)$ is the polar angle from the symmetry axis. Figure 2.21 shows the quadrupole (Y_{20}) and hexadecapole (Y_{40}) moments predicted by this model. This model explains the general trends of the experimental values.

We now extend the model to the case of γ-vibrations. The quadrupole (Y_{22}) and hexadecapole (Y_{42}) transition strengths to the γ-vibrational band are given by $|\langle\gamma|r^{\lambda}Y_{\lambda 2}|0\rangle|^2$, where $\lambda = 2, 4$. When the valence particles are filled from the z-axis to $\mu = \cos\theta$, the Fermi surface lies around $\mu = \cos\theta$. Figure 2.21 presents this model schematically. The γ-vibration can be interpreted as the superposition of the particle–hole states with $K^{\pi} = 2^{+}$ near the Fermi surface $(\mu\pm\delta\mu)$. Then the transition strength is proportional to

$$\left(\int_{\mu-\delta\mu}^{\mu+\delta\mu} P_{\lambda 2}(x)\,\mathrm{d}x\right)^2 \cong 2\,(2\delta\mu)^2\,(P_{\lambda 2}(\mu))^2\,, \tag{2.32}$$

where the integration over ϕ has been replaced by a constant number, because the ϕ-dependence of the $K^{\pi} = 2^{+}$ γ-vibrational state is identical to that of $Y_{\lambda 2}$. Figure 2.15 gives the prediction of the quadrupole (Y_{22}) and hexadecapole (Y_{42}) transition strengths provided by this model.

The success in explaining the Y_{42} and Y_{22} strengths by such a simple model implies that the spatial location of the particles occupying orbits near the Fermi surface changes gradually with the mass number, owing to the very stable deformation in this mass region. This model is a simple representation of shell filling in the Nilsson potential for a single-j orbit. The energy of a single particle orbit becomes larger as the oscillator quanta n_z become smaller in a single-j orbit for a prolate deformation. This is the reason the valence particles are placed in orbits as close to the z-axis as possible in the polar cap model. Although many single-j orbits exist in a major shell, the overall trend for the strength is successfully predicted by the polar cap model.

2.6 Elastic scattering at intermediate energies

When we increase the incident nucleon energy to several hundred MeV, we need to alter the above-described treatment. In this case we must take into account the relativistic kinematics for the mean-field treatment. Since the Schrödinger equation itself is non-relativistic, we must consider the equation of motion of a nucleon in the nuclear medium, which will be reduced to a non-relativistic Schrödinger equation in the limit of low incident energy. At present, the most promising approach is to employ the Dirac equation, in which a nucleon obeys this equation in a mean field of mesons. Since the Dirac equation can be reduced mathematically to the Schrödinger equation, it is in general very difficult to discuss the superiority of one theory over the other. Thus, we have to compare theories globally, including the basic equation and the approximation used.

At intermediate energies, the NN collision cross section decreases and the mean free path of the nucleon inside the nucleus becomes large. Thus, systems at intermediate energies offer a very attractive opportunity for the study of hadron behaviour and correlations deep inside the nucleus.

Experimentally, measurement of elastic and inelastic scattering from many nuclei requires a total energy resolution of approximately 200–300 keV. This is relatively difficult to realize at intermediate energies, because we need good energy resolution for both the beam and the spectrometer for the scattered particles. In this energy region the impulse approximation is expected to describe the elastic and quasi-elastic scattering of protons qualitatively, because here the velocity of the projectile is much larger than the velocity of Fermi motion of the nucleon in the nucleus. In the Relativistic Impulse Approximation (RIA), the mean field is obtained directly by folding the NN interaction in the relativistic formalism with the density distribution. Thus, the RIA seems to relate the coupling constants and the masses of the exchanged mesons directly to nuclear reactions. By using the RIA formalism, we can now investigate the medium effects of the NN interaction inside the nucleus.

In Fig. 2.22 we display data for elastic scattering of polarized protons at incident proton energies of 200, 300, and 400 MeV for a ^{58}Ni target. Measurements were performed by using the high-resolution magnetic spectrograph GRAND RAIDEN and a polarized proton beam from the RCNP ring cyclotron. In ^{58}Ni the number of protons and neutrons are almost equal and similar density distributions are expected for them. In the energy region between 200 and 400 MeV, the central part of the optical potential in the Schrödinger form changes drastically from the so-called 'wine bottle bottom' shape to a repulsive shape. Thus, this energy region offers a good opportunity to test various theories and models. Experimental data are shown in Fig. 2.22 together with the predictions of various model calculations.

Experimental data have been analysed using RIA models and a microscopic non-relativistic Schrödinger-type optical potential. Solid curves in the figure represent calculations using the code of the RIA of Murdock and Horowitz (MH) [32,43]. The density distributions (vector and scalar densities for protons and neutrons) of the target nuclei were calculated using the Relativistic Hartree (RH) approximation [36]. The NN interaction used in this model is the relativistic Love–Franey model [33], which explains the free NN scattering by using the first Born approximation to one-boson exchange potentials of ten mesons. Although the solid curve reproduces the analysing power precisely, the angular distribution of the cross section is poorly reproduced, especially at backward angles. These tendencies are the same for all three energies.

The dashed curves in the figure represent the calculation using the RIA IA2 of Tjon and Wallace [37]. The IA2 interaction is a matrix representation of the NN interaction including the negative energy sector deduced from a five-term approximation. The density distribution used here was calculated with the RH approximation using the parameters of TM2 [38], which have also been used for the calculation of unstable nuclei. The IA2 seems to predict the cross sections better than MH at all three energies, and especially at 400 MeV. However, the IA2 predictions for analysing powers are not as good as those for the MH.

The dotted curves represent the results of calculating the scattering using a non-relativistic Schrödinger optical potential. We have calculated the non-relativistic Schrödinger optical potential using the G-matrix of von Geramb, according to the procedure of Rikus and von Geramb [34]. The density distribution used was unfolded from

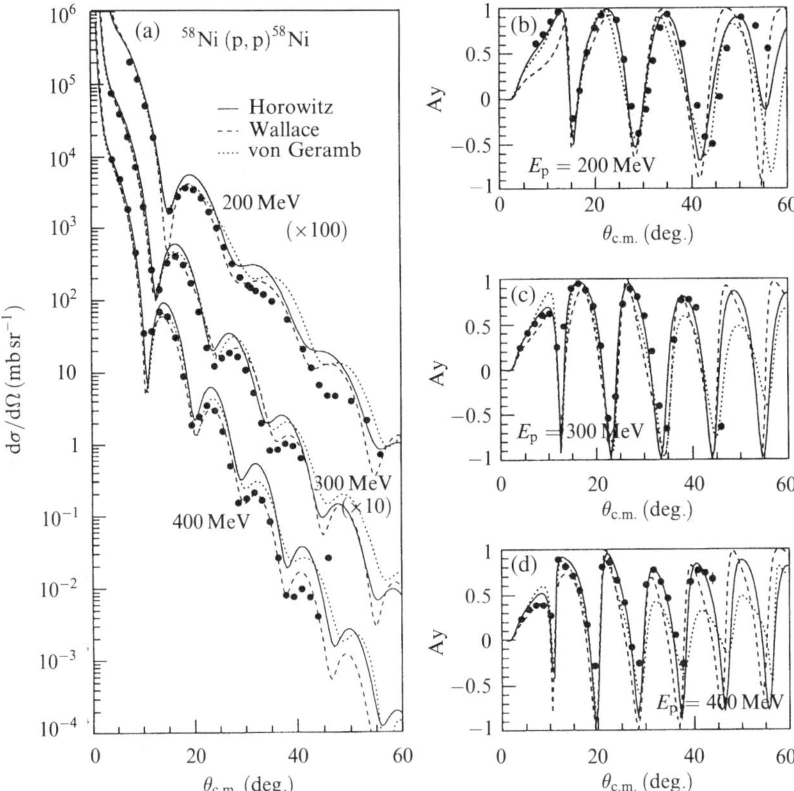

Fig. 2.22 Angular distribution of differential cross sections (a) and analysing powers [(b)–(d)] of protons from ^{58}Ni (p, p) ^{58}Ni for incident proton energies of 200, 300, and 400 MeV. The dotted curves represent results of the RIA calculation using the code of Horowitz *et al.* [32] with the parameters of the relativistic Love–Franey model [33]. The dashed curves are the results of the calculation using the relativistic impulse calculation of IA2 by Tjon and Wallace. The solid lines are the results of non-relativistic Schrödinger-type calculations using the G-matrix of von Geramb *et al.* [18,34] and the density distribution deduced from the charge distribution. (Taken from Sakaguchi *et al.* [35].)

the charge distribution [39] of the sum of the Gaussian type measured by electron scattering. At 200 MeV, the non-relativistic optical potential describes the experimental data fairly well, giving results almost equivalent to those of the RIA approaches. However, as the incident energy increases, its predictions deviate from the experimental values, especially with regard to the analysing powers at 400 MeV, which are inferior to results given by the IA2.

In summary, none of the three models can explain the experimental data fully. This is true when we consider the differential cross section data. Only the model of MH explains the analysing power precisely.

2.6.1 *Effects of medium in elastic scattering*

In order to explain the experimental data, it is neccesary to use correct density distributions for the target nuclei and modify the NN interaction inside the nucleus. For the RIA we need four types of density distributions, namely vector and scalar densities for both protons and neutrons. Since from the electron scattering we can deduce only point-proton vector densities, we need to estimate the other three types of densities with the help of a RH calculation. In Fig. 2.23(a) we display point-proton densities of the RH code of Horowitz *et al.* (dashed curve) and the RH calculation of Kaki [40] using the TM2 parameters [38] (dotted curve). The RH calculations are different inside the nucleus. We adopted the density distribution unfolded from the charge distribution of the sum of the

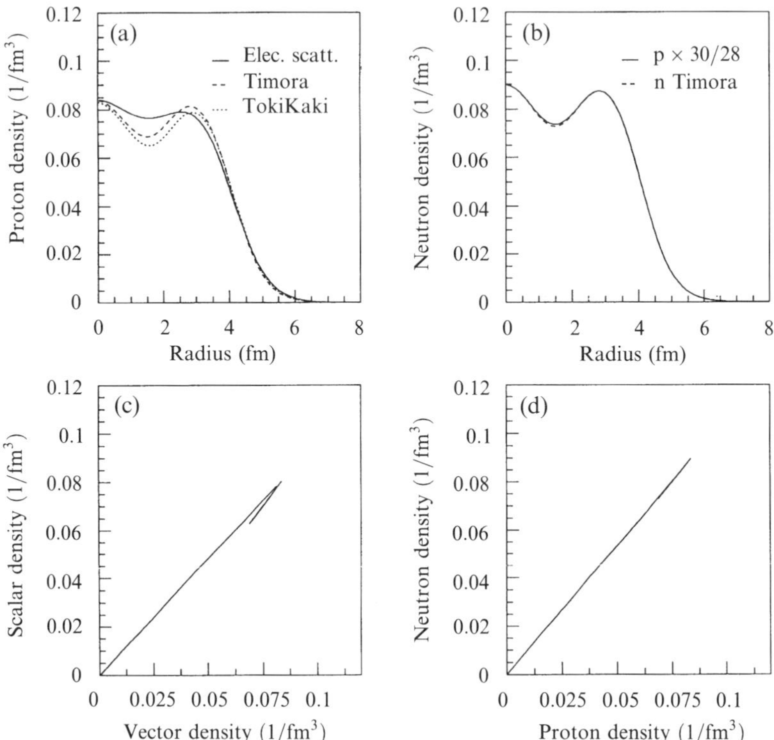

Fig. 2.23 (a) Comparison of various proton densities. The solid curve marked Elec. scatt. represents the point-proton distribution deduced from electron scattering, while the dashed curve marked Timora presents the results of the RH calculation using the code of Horowitz *et al.* [32]. The dotted curve marked TokiKaki exhibits the result of the RH calculation with the TM2 parameters [38]. (b) Comparison of shapes of the proton and neutron densities from the RH calculation. (c) Comparison of the scalar density with the vector density from the RH calculation. The ratio of the two densities seems to be almost constant. (d) Comparison of the neutron density with the proton density from the RH calculation. The ratio of the two densities seems to be almost constant. (Taken from Sakaguchi *et al.* [35].)

Gaussian type measured by electron scattering. According to the RH calculation of the density distribution for ^{58}Ni, the ratio of the scalar density to vector density is almost constant at 0.96, as is shown in Fig. 2.23(c). In Fig. 2.23(b) and (d) we compare vector densities for protons and neutrons in the RH calculation [32]. The ratio of these values is also almost constant. Thus, by using the results of the RH calculation partially, we can obtain the four types of density distribution necessary for RIA calculation from the electron scattering experiments.

In Fig. 2.24 we show the results of calculations of the differential cross sections and the analysing powers at 300 MeV. The dashed curves represent the original RIA calculation of MH. The dotted curves represent the calculation using the density distributions

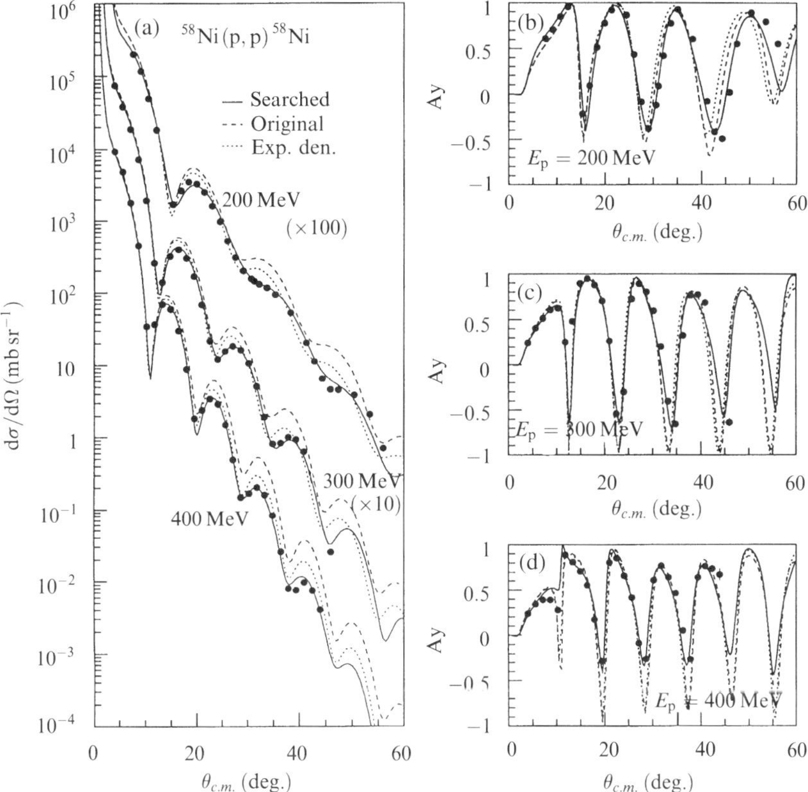

Fig. 2.24 Angular distribution of differential cross sections (a) and analysing powers [(b)–(d)] of protons from ^{58}Ni (p, p) ^{58}Ni for incident proton energies of 200, 300, and 400 MeV. The dashed curves are results of the original RIA calculation using the code of Horowitz *et al.* [32] with the parameters of the relativistic Love–Franey model. The dotted curves represent the results of a similar calculation, but using a density distribution deduced from electron scattering. The solid lines represent calculations with modified coupling constants and meson masses for the NN interaction. (Taken from Sakaguchi *et al.* [35].)

deduced from the charge distribution. The dotted curve describes the experimental data much better, while maintaining the fitting quality for analysing power unchanged from the original one. There still exist large discrepancies between the experiment and theory, and we thus need to modify the coupling constants and the masses of exchanged mesons, etc., since their character can be changed in the nuclear medium. For this reason we have replaced the coupling constants f_i^2 and the masses m_i^2 of the exchanged mesons by density-dependent ones as follows:

$$f_i^2(1 + a_i(\rho(r)/\rho_0))^{-1}, \quad m_i^2(1 + b_i(\rho(r)/\rho_0)), \qquad i = \sigma, \omega,$$

where a_i and b_i are chosen to fit the experiment. At these three incident energies, we have obtained values of approximately 0.2 for a_i and 0.01 for b_i; that is, the coupling constants are reduced by about 20% and the masses of the exchanged mesons are smaller by about 1% [35]. Thus, there seem to be some effects of the medium observed in the elastic scattering. By changing the NN interaction we can describe the scattering; nevertheless, further experimental and theoretical studies are necessary.

References

1. R. Beurtey, *Proc. 2nd inter. symp. on pol. phen. of nucl.*, Kahlsruhe, P. Huber and H. Schopper, eds, Birkhäuser, Basel, 1966, p. 33.
2. W. Haeberli, Sources of polarized ions, *Ann. Rev. Nucl. Sci.* **17** (1967) 373.
3. K. A. Brueckner, R. J. Eden, and N. C. Francis, *Phys. Rev.* **100** (1955) 891.
4. H. V. von Geramb, ed., *Microscopic optical potentials*, Lecture Notes in Physics, Vol. 89, Springer-Verlag, Berlin (1979), p. 1.
5. F. A. Brieva and J. R. Rook, *Nucl. Phys.* **A291** (1977) 219; 317; **A297** (1978) 206; **A307** (1978) 493.
6. RCNP annual report, 1978.
7. T. Noro, H. Sakaguchi *et al.*, *Nucl. Phys.* **A366** (1981) 189.
8. H. Sakaguchi, M. Nakamura *et al.*, *Phys. Rev.* **C26** (1982) 944.
9. F. Ohtani, H. Sakaguchi *et al.*, *Phys. Rev.* **C28** (1983) 120.
10. T. Ichihara, H. Sakaguchi *et al.*, *Phys. Rev.* **C29** (1984) 1228.
11. H. Ogawa, H. Sakaguchi *et al.*, *Phys. Rev.* **C33** (1986) 834.
12. Y. Takeuchi, H. Sakaguchi *et al.*, *Phys. Rev.* **C34** (1986) 493.
13. T. Ichihara, H. Sakaguchi *et al.* *Phys. Rev.* **C36** (1987) 1754.
14. Madison Convention 1967, *Proc. 4th polarization conf.*, Wisconsin.
15. S. Kato *et al.*, *Nucl. Instr. Meth.* **169** (1980) 589.
16. M. Ieiri, H. Sakaguchi *et al.*, *Nucl. Instr. Meth.* **A257** (1987) 253.
17. H. Sakaguchi *et al.*, *J. Phys. Soc. Japan* **55** (Suppl.) (1986) 61.
18. H. V. von Geramb, *AIP conf. proc.*, No. 97, p. 44.
19. N. Yamaguchi, S. Nagata, and T. Matsuda, *Prog. Theor. Phys.* **70** (1983) 459.
20. N. Yamaguchi, S. Nagata, and J. Michiyama, *Prog. Theor. Phys.* **76** (1986) 459.
21. N. Olsson *et al.*, *Nucl. Phys.* **A472** (1987) 237.
22. S.Nagata, *Butsuri* **43** (1988) 594.
23. A. Sommerfeld, *Z. Phys.* **47** (1928) 1.
24. J. P. Jeukenne, A. Lejeune, and C. Mahaux, *Phys. Rev.* **C15** (1977) 10; **C16** (1977) 80.

25. J. P. Jeukenne, A. Lejeune, and C. Mahaux, *Phys. Rev.* **C10** (1974) 1391.

26. J. P. Jeukenne, A. Lejeune, and C. Mahaux, *Phys. Rep.* **25C** (1976) 83.

27. G. Bertsch, J. Borysowics, H. McManus, and W. G. Love, *Nucl. Phys.* **A284** (1977) 399.

28. A. M. Green, *Phys. Lett.* **24B** (1967) 384.

29. F. Fabrici, S. Micheletti *et al.*, *Phys. Rev.* **C21** (1980) 830; **C21** (1980) 844.

30. G. R. Satchler, *J. Math. Phys.* **13** (1972) 1118.

31. G. F. Bertsch, *Phys. Lett.* **26B** (1967) 130.

32. C. J. Horowitz, D. P. Murdoch, and B. D. Serot, *Computational nuclear physics 1*, Chapter 7. Springer-Verlag (1991).

33. C. J. Horowitz, *Phys. Rev.* **C31** (1985) 1340.

34. L. Rikus, K. Nakano, and H. V. von Geramb, *Nucl. Phys.* **A414** (1984) 413; L. Rikus and H. V. von Geramb, *Nucl. Phys.* **A426** (1984) 496.

35. H. Sakaguchi *et al.*, *Proc. RCNP inter. mini-workshop*, Genshikaku Kenkyu, Vol. 42 (1997), p. 99; *Phys. Rev.* **C57** (4) (1998) 1749.

36. C. J. Horowitz and B. D. Serot, *Nucl. Phys.* **A368** (1981) 503.

37. J. A. Tjon and S. J. Wallace, *Phys. Rev.* **C38** (1988) 2272.

38. H. Toki, *TMU/RIKEN summer institute on unstable nuclei* RIKEN-AF-NP-173, p. 37; D. Hirata, H. Toki, T. Watabe, I. Tanihata, and B. V. Carlson, *Phys. Rev.* **C44** (1991) 1467.

39. I. Sick, *Phys. Lett.* **116B** (1982) 212.

40. K. Kaki, *Nucl. Phys.* **A531** (1991) 478.

41. T. Ichihara, H. Sakaguchi *et al.*, *Phys. Lett.* **182B** (1986) 301.

42. F. A. Brieva and B. Z. Georgiev, *Nucl. Phys.* **A308** (1978) 27.

43. D. P. Murdock and C. J. Horowitz, *Phys. Rev.* **C35** (1987) 1442.

3

Spin–isospin giant resonances by charge-exchange reactions

Mamoru Fujiwara

3.1 Introduction

A nucleus primarily consists of protons and neutrons. This microscopic world is governed by a strong force, i.e. the nuclear interaction mediated by the exchange of pions and other heavy mesons between the nucleons. If the interactions between the constituents in a nucleus are known, the behaviour of this microscopic world can, in principle, be described. But, the actual situation is not so simple since the many-body problem can be solved exactly only for two- and three-body systems with some basic physical assumptions. Thus, all theoretical analyses of nuclei need a model, the validity of which should be checked by experimental tests in order to confirm the underlying approximations or to improve them.

The atomic nucleus has characteristic features which indicate that it is bound by the strong nuclear force. Nuclear giant resonances are typical examples of this. Similar to the case of an elastic ball, a nucleus can be compressed, deformed, rotated, and can emit neutrons and protons when it is highly excited. Furthermore, when a nucleus is extremely hot, mesons appear to mediate the nuclear force.

Corresponding to the different modes of nuclear interactions, various excitations of a nucleus characterized by individual quantum numbers are possible, the electric dipole giant resonance with $J^\pi = 1^-$ being the most famous. The operator corresponding to the dipole giant resonances is $O^{(\lambda=1)} = \sum_j \tau_j r_j^1 Y_{1j}$. This resonance was observed in photo-absorption experiments [1] in 1962.

After the discovery of the dipole resonances, the giant resonances with different multipolarities (J^π), spins (S), and isospins (T) were the next exciting subjects to be studied experimentally and theoretically. The resonances are defined by the corresponding operators mediating in the excitation processes, which are expressed by $O^\lambda = \sum_j \sigma_j \tau_j r_j^\lambda Y_{\lambda j}$. According to these theoretical considerations, various nuclear giant resonances have been investigated experimentally during the last 30 years [2].

The nuclear giant resonances are sometimes characterized as broad states exhausting almost all the sum-rule strengths. A typical sample of an experimental observation of giant resonances [3,4] is shown in Fig. 3.1. Spin–isospin resonances in ^{208}Bi are strongly excited using the ^{208}Pb $(^3$He, t$)\,^{208}$Bi charge-exchange reaction at forward angles by using the ^{3}He beam at 450 MeV. The Gamow–Teller Resonance (GTR), the Isobaric Analogue States (IASs), and the Spin-flip Dipole $(\Delta L = 1)$ Resonance (SDR) in ^{208}Bi are seen in Fig. 3.1. The shapes of these resonances exhausting the major part

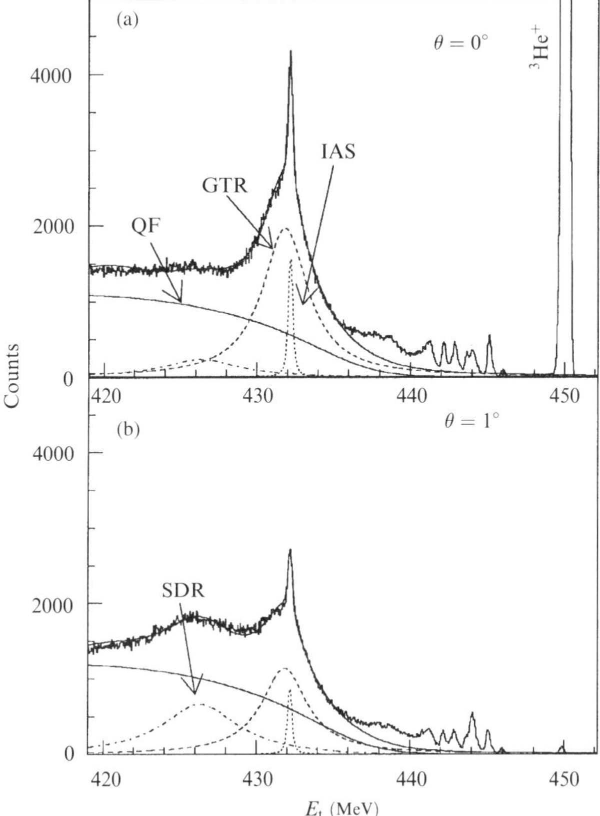

Fig. 3.1 Experimental triton energy spectra from the ^{208}Pb (^{3}He, t) reaction at $E(^3\mathrm{He})$ = 450 MeV. (a) Spectrum gated for scattering angles at $\theta = 0^\circ$. Here, the IAS, GTR, and $^3\mathrm{He}^+$ peaks are prominent. The dashed, dotted, dot-dashed, and solid lines represent the results of χ^2 fits obtained for the GTR, IAS, SDR, and non-resonant background, respectively. (b) Same as (a) but for spectrum gated for scattering angles centred at $\theta = 1^\circ$. (Taken from Akimune *et al.* [4].)

of the sum-rule values are well fitted by the Breit–Wigner-type function expressed as

$$\sigma(E) \propto \frac{1}{(E - E_\mathrm{r})^2 + (\Gamma/2)^2}. \tag{3.1}$$

Here, E_r is the centroid energy of the resonance and Γ is the resonance width.

The resonance width Γ can be written as

$$\Gamma = \Gamma^\downarrow + \Gamma^\uparrow, \tag{3.2}$$

where $\Gamma^\downarrow$ is the spreading width and $\Gamma^\uparrow$ the escape width which is connected to the microscopic particle–hole structure of the resonance. In heavy nuclei, the spreading

Table 3.1 Observed giant resonances in nuclei

Spin–isospin dependence	Multipolarities, λ	Spin and parity, J^π	Excitation energy (MeV)	Typical operator type
Isoscalar mode ($\Delta T_z = 0$)	Monopole ($\lambda = 0$)	0^+	$80A^{-1/3}$	r^2
	Quadrupole ($\lambda=2$)	2^+	$65A^{-1/3}$	$r^2 Y_{2\mu}$
	Octupole ($\lambda = 3$)	3^-	$30A^{-1/3}, 120A^{-1/3}$	$r^3 Y_{3\mu}$
Isovector mode ($\Delta T_z = 1$)	Monopole ($\lambda = 0$)	0^+	$60A^{-1/3}$	$\tau_\pm r^2$
	Dipole ($\lambda = 1$)	1^-	$31A^{-1/3} + 21A^{-1/6}$	$\tau_z r Y_{1\mu}$
	Quadrupole ($\lambda = 2$)	2^+	$130A^{-1/3}$	$\tau_z r^2 Y_{2\mu}$
Spin vibration mode ($\Delta T_z = 0$)	Monopole ($\lambda = 0$)	1^+	$40A^{-1/3}$	$\tau_z \sigma$
Gamow–Teller mode ($\Delta T_z = \pm 1$)	Monopole ($\lambda = 0$)	1^+	$E_{\mathrm{IAS}} + 7 - 30(N - Z)/A$	$\tau_\pm \sigma$
Spin-dipole mode ($\Delta T_z = \pm 1$)	Dipole ($\lambda = 1$)	$0^-, 1^-, 2^-$	$E_{\mathrm{IAS}} + 14 - 33(N - Z)/A$	$\tau_\pm \sigma r Y_{1\mu}$

process leads predominantly to statistical neutron decays because the corresponding proton decays are strongly suppressed by the Coulomb barrier.

Table 3.1 shows a summary of the nuclear giant resonances for nuclei with $A \geq 40$. Since the nuclear giant resonances are mediated by the various operators such as $f_\lambda(r)Y_{\lambda\mu}$, $f_\lambda(r)(\sigma \times Y_{\lambda\mu})_J$, $\tau f_\lambda(r)Y_{\lambda\mu}$, and $\tau f_\lambda(r)(\sigma \times Y_{\lambda\mu})_J$, one can expect various kinds of new nuclear giant resonances which are not yet observed [2,5]. For example, if the operator $r^2 \sigma \tau$ acts on the nuclear ground state, a new giant resonance called IsoVector Spin Monopole resonance (IVSM) is predicted theoretically [6]. The spin-parity of this resonance is 1^+, and its transition carries the quantum numbers $\Delta N = 2$, $\Delta S = 1$, and $\Delta L = 0$.

As mentioned above, the Giant Dipole Resonances (GDRs), which were first observed in the γ-ray absorption experiments, correspond to the collective vibration of neutrons and protons moving in the out-of-phase mode. Giant Quadrupole Resonances (GQRs) were found in the early 1970s via the electron, proton, and α-particle inelastic scattering. The Giant Monopole Resonances (GMRs) are the compression oscillations. Their excitation energy depends on the softness of nuclei against the compressive force from outside. The nuclear compressibility is of particular interest from an astrophysical viewpoint, because its excitation energy is directly related to a key quantity important in understanding the supernova explosions in the cosmos [7,8], and also in the analysis of high-energy heavy-ion collisions [9].

The Gamow–Teller Giant Resonance (GTGR) is the oscillation that exchanges neutrons and protons and also reverses the spin directions. This mode of resonances was found in the mid-1970s [10] using (p, n) reactions, and their systematic studies were carried out in the 1980s [11,12] using (p, n) reactions. The GTRs were first introduced in 1963 to explain the common retardation of the Gamow–Teller (GT) transitions in allowed β-decay compared with single-particle estimates [13,14]. Subsequently, analyses of the first forbidden β-decay were made in connection with the nuclear core polarization effect, and the SDRs were predicted at excitation energies higher than those for the GTGRs [15]. It was found much later that the existence of the GTGR was first reported in 1975 in the ^{90}Zr(p, n) reaction at the incident proton energy of 35 MeV [10]. Historically, the presence of a broad bump corresponding to the GTGR had already been seen by Bowen *et al.* [16] in 1962 in the $0°$ (p, n) spectra measured for several nuclear targets at $E_p = 143$ MeV. Unfortunately, most nuclear scientists did not understand, for a long time, the true nature of the bumps seen in the (p, n) spectra. The broad bumps corresponding to the GTGR were also observed in 1978 in the (p, n) spectra at $E_p = 800$ MeV [17]. On this occasion, the bumps were not interpreted correctly, but were identified as the quasielastic charge-exchange peaks.

In 1980, the GTGR bumps were correctly interpreted to be excited preferentially at $0°$ in (p, n) reactions at bombarding energies higher than 100 MeV [11,12,18]. More important is that the (p, n) reaction cross sections leading to IAS and GT states can be related, respectively, to the Fermi-type and GT-type β-decay matrix elements using the similarities between the $0°$ charge-exchange reactions and the charge current weak processes;

$$\left(\frac{d\sigma}{d\Omega}\right)_{\mathrm{F,GT}} (q \approx 0,\ \theta = 0°) = \left(\frac{\mu}{\pi\hbar^2}\right)^2 \left(\frac{k_f}{k_i}\right) \left[N_\tau\left(J_\tau\right)^2 B(\mathrm{F}) + N_{\sigma\tau}\left(J_{\sigma\tau}\right)^2 B(\mathrm{GT})\right],$$

$$(3.3)$$

$$\sigma = (\pi c^3 \hbar^4)^{-1} \left[G_V^2 \langle 1 \rangle^2 + G_A^2 \langle \sigma \rangle^2\right] P_e W_e F(eZ, W_e). \quad (3.4)$$

Here, σ is the neutrino capture cross section, $\langle 1 \rangle$ and $\langle \sigma \rangle$ the reduced matrix elements for the Fermi and GT transitions, respectively, P_e the emitted electron momentum, W_e the energy of the electron, and $F(eZ, W_e)$ is known in the β-decay theory as the Fermi function, which just corresponds to the distortion factors $(N_\tau, N_{\sigma\tau})$ in the case of the charge-exchange reactions. J_τ and $J_{\sigma\tau}$ in charge-exchange reactions correspond well to the coupling constants of the weak processes (G_V, G_A).

In many electro-weak processes in astrophysics, the transition strength of GT absorption cross sections must be calibrated by processes analogous to charge-exchange reactions. These calibrated transition matrix elements are used for the solar evolution processes in network chain calculations. Thus, the spin–isospin excitations are sometimes discussed in relation to astrophysics, particularly to neutrino physics.

In the GT transitions, there are two transition modes: the β^- GT transitions mediated by the $\sigma\tau^-$ operator, and the β^+ GT transitions mediated by the $\sigma\tau^+$ operator. The sum rule for the β^- and β^+ GT transitions [13,14] is

$$S_{\beta^-} - S_{\beta^+} = 3(N - Z). \quad (3.5)$$

This sum rule, sometimes called the Ikeda sum rule, is model independent. The $\sigma\tau^-$ and $\sigma\tau^+$ operators can flip the spins of quarks in nucleons. Even if the quark spin degree of freedom is taken into account, the Ikeda sum rule is still conserved [5].

3.2 Charge-exchange reactions

The presence of the GTGR was first claimed in ^{90}Zr (p, n) ^{90}Nb experiments at 35 MeV at the Michigan State University cyclotron [10]. However, the observed excitation strength was not evident to give a clear enough proof for the presence of the giant resonance. This situation completely changed after the installation of a neutron time-of-flight facility at the Indiana University Cyclotron Facility (IUCF). For bombarding energies higher than $E_p = 100$ MeV, it was found that the GTRs were strongly excited and systematically observed in almost all nuclei [11,12,19]. The strong excitations of the GT states at intermediate energies are due to the advantage that the isovector central components of the effective interactions $V_{\sigma\tau}$ and V_{τ} associated with charge-exchange reactions are relatively strong at $E_p \geq 100$ MeV and that the contributions of the tensor and spin–orbit components are generally small at forward angles [20,21]. The GTGRs were studied systematically for many target nuclei, and the SDRs were also found at excitation energies above the GTGR.

The GTGR was also studied by alternative charge-exchange reactions like (^{3}He, t) and (^{6}Li, ^{6}He) taking advantage of the high resolution and high detection efficiency. In particular, the (^{3}He, t) reaction has been used to observe the GT strengths at various bombarding energies $E \leq 100$ MeV/u [22–29]. However, it is now understood that the analysis of the (^{3}He, t) reaction at $E \leq 100$ MeV is more difficult than for the (p, n) reaction at $E \geq 100$ MeV due to strong contributions from the non-central interaction. The (^{3}He, t) reaction becomes an important alternative to investigate spin–isospin excitations for bombarding energies $E \geq 100$ MeV.

This situation has been demonstrated by the (^{3}He, t) work at Saclay [30], and recently at RCNP [31]. A comparison between the ^{58}Ni(^{3}He, t) and ^{58}Ni(p, n) spectra measured at 450 and 160 MeV, respectively, shows a close correspondence [32]. Therefore, the (^{3}He, t) reaction at 450 MeV provides an alternative probe to determine GT strengths in nuclei. Actually, it is a powerful probe compared to the one using the (p, n) reaction. This is because triton particles can be detected in a magnetic analyser with 100% detection efficiency and high energy resolution. The (^{3}He, t) reaction cross sections at $0°$ are used to extract the isospin information from the comparison with the M1 strength distribution obtained by (e, e$'$) measurements. Such a comparison [33] is shown in Fig. 3.2. One can see that analogous transition strengths are observed in the two spectra, especially in the region of $E_x \sim 8$ MeV, where the $T = 1$, 1^+ states are expected.

The ground state of ^{58}Ni has isospin $T_0 = 1$. In charge-exchange reactions, the excitation strength ratio of $2:3:1$ is expected for GT states with isospin $T_0 - 1$, T_0, and $T_0 + 1$, respectively. In the case of the M1 excitations in ^{58}Ni, the corresponding ratio is $1:1$. Thus, in order to compare the M1 and GT strengths, the strengths $B(\text{M1})$ must be corrected. After these corrections, the global shapes agree with each other. These similarities in excitation strengths of M1 and GT states indicate that the GT and M1 transitions are dominated by the same $\sigma\tau$ operator. In general, the M1 states in the (e, e$'$) reaction are excited predominantly by the spin and orbital parts $\frac{1}{2}(g_s^p - g_s^n)\vec{\sigma} \cdot \vec{\tau} + (g_l^p - g_l^n)\vec{l} \cdot \vec{\tau}$, while the GT operator is $(g_s^p - g_s^n)\vec{\sigma} \cdot \vec{\tau}$. In addition, the contributions from isobar and meson exchange currents are expected in the case of the M1 excitation.

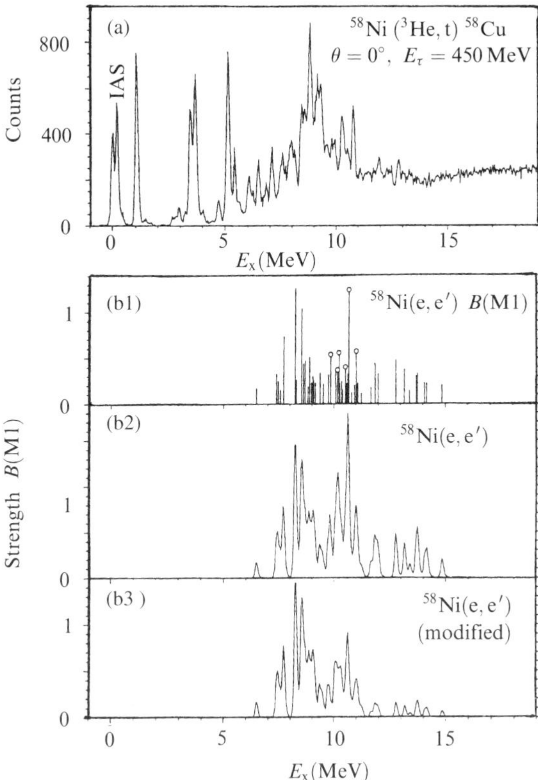

Fig. 3.2 Comparison between the ^{58}Ni(^{3}He, t) spectrum at $0°$ and the M1 strength distribution obtained in the (e, e') experiment. (a) The $0°$ ^{58}Ni(^{3}He, t) spectrum. (b1) The B(M1)↑ strength distribution in ^{58}Ni. (b2) The reconstructed B(M1) spectrum convoluted with the experimental energy resolution of the (^{3}He, t) measurement. (b3) The same as (b2), but the M1 strengs with $T = T_0 + 1$ are reduced by a factor of three. For details, see Ref. [33].

Thus, the detailed comparisons of the M1 and GT states excited via (e, e'), (p, p'), and charge-exchange reactions would be very useful to study the non-negligible role of the orbital part, meson-exchange parts, and the Δ-hole contribution [34,35], although we need a stringent test of the wave function of the states through theoretical calculations.

3.3 Gamow–Teller strengths at high excitation energy: quenching or 2p–2h configurations?

The spin-dependent interactions are based on the pion-exchange fields. The magnetic properties of the nucleus are affected by these spin-dependent interactions. Therefore, the spin excitation modes provided one with several motivations for experimental studies in the 1980s. Toki and Weise [36] predicted that a critical enhancement occurs in the magnetic excitation at the momentum transfer corresponding to the pion mass ($\Delta q \sim m_\pi c$).

The quenching of the GT excitations was also predicted; it was hoped that this phenomenon would provide good evidence for the importance of sub-nucleonic degrees of freedom. The importance of the Δ-excitation, which is the spin-flip state of one quark in a nucleon, was emphasized strongly.

One of the serious problems concerning the missing M1 states was the absence of the strong M1 states in ^{208}Pb, where simple M1 states are possible with the configurations $(\pi h_{11/2}^{-1} h_{9/2})$ and $(\nu i_{13/2}^{-1} i_{11/2})$ for proton and neutron orbitals, respectively. We can construct an isoscalar M1 state and an isovector M1 state from these two configurations. Contradicting this simple prediction, there was no evidence for the presence of the M1 state in ^{208}Pb. In the region of 7.5 MeV excitation, only one-fifth of the isovector M1 strength was identified [37]. The isoscalar M1 strength was not identified at all in ^{208}Pb as indicated in Ref. [38]. In 1982, the RCNP group [39] and the Giessen group [40] found that there is an isoscalar M1 state at 5.84 MeV excitation in ^{208}Pb. Especially, the RCNP group used the ^{208}Pb(p, p$'$) and ^{209}Bi (d, ^{3}He) ^{208}Pb reactions to identify the wave function of the 5.84 MeV 1^+ state. The (d,^{3}He) result shows that the 5.84 MeV 1^+ state has a large configuration of $(\pi h_{11/2}^{-1} h_{9/2})$, which is a good evidence for the isoscalar M1 state. The isoscalar nature of the 5.84 MeV 1^+ state was firmly established later by the microscopic DWBA analysis [41] of the (p,p$'$) data at 65 MeV and also by inelastic electron and proton scattering experiments [42,43].

Many groups at different laboratories in the world competed to search for the M1 strengths in various magic nuclei such as ^{12}C, ^{40}Ca, ^{48}Ca, ^{50}Ti, ^{52}Cr, ^{54}Fe, ^{58}Ni, ^{60}Ni, ^{90}Zr, and ^{208}Pb.

Impressive experimental evidence [44] was presented on the basis of the systematic (p, n) measurements of the GT strengths. The GT strength in the GTGR region was found to exhaust only about 70% of the sum-rule value; so where is the missing GT strength? This missing GT strength was discussed in terms of couplings with other 2p–2h states and the Δ-isobar. The isobar is a kind of the quark spin degree of freedom; can we see the quark degree of freedom in the low-energy experiments by observing the spin-flip processes? Because of these arguments, many nuclear physicists were attracted to make great efforts to look for evidence of quark degrees of freedom in nuclei.

The quenching of the GT and M1 strengths is claimed to be mainly due to the coupling with 2p–2h states at high excitation energies and partly due to the Δ–hole effect (less than 5%) [45]. However, the problem of understanding the quenching mechanism of the spin-flip strength is considered as one of the important issues. Experimental and theoretical efforts are still being devoted to clarify the role of the Δ–hole state, and to find the spin–isospin mode in the highly excited energy region of nuclei.

Figure 3.3 shows how the GT strengths continue up to the high excitation energies. Wakasa *et al.* [46] measured the double differential cross sections at scattering angles from $0°$ to $12.3°$ and the polarization transfer coefficients D_{NN} at $0°$ for the ^{90}Zr(p, n) reaction at $E_p = 295$ MeV. After applying the multipole decomposition technique to deduce the GT strengths, they conclude that the GT strength in the energy region up to 50 MeV excitation in ^{90}Nb is $93 \pm 5\%$ of the sum rule $3(N - Z) = 30$. This result suggests that the contribution of the Δ–hole admixture to the GT strength is very

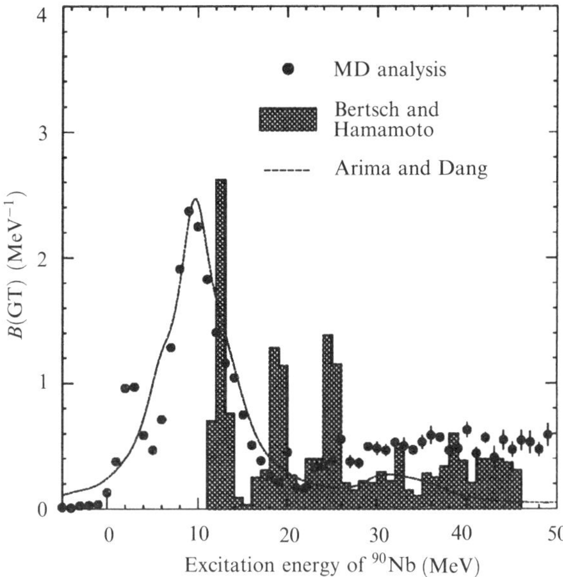

Fig. 3.3 Gamow–Teller strength distribution (filled circles) obtained from the multipole decomposition analysis. The hatched histogram shows the theoretical calculation by Bertsch and Hamamoto [47]. The solid line shows the results of the calculations by Dang *et al.* [48]. For details of the experimental data analyses, see Ref. [46].

small, and that appreciable amounts of the B(GT) strength do not go to the energy region of the Δ-isobar. A recent calculation by Dang *et al.* [48] shows that the results of the theoretical calculation without taking into account the Δ–hole effects agree well with the experimental data. It is interesting to examine whether or not the same conclusion for the GT strength distribution is also valid for the case of ^{208}Pb.

3.4   Gamow–Teller transitions to the β^+-side

Over the last ten years, the interest in developing scientific opportunities for the use of Radioactive Nuclear Beams (RNBs) [49] has increased. Relating to astrophysical problems, exciting new phenomena are observed which also address intrinsic questions in nuclear physics. Spectroscopic studies via nuclear reactions with RNBs can provide a unique way to answer questions of drip line nuclei. A great effort is under way to open this new field where one of the interesting issues is to understand the halo properties of neutron-rich nuclei.

The (n, p)-type charge-exchange reactions with high resolution are especially important in astrophysics. For example, in electron capture processes in supernova explosion, the low-lying discrete GT (GT$_+$) states play an important role. The GT$_+$ states are pushed down to low excitation energies due to the attractive force of the particle–particle residual interactions, and have relatively strong GT$_+$ strengths near the ground state. This makes a major contrast between the (p, n)- and (n, p)-type transitions since the neutron–hole and proton–particle interactions are repulsive, and thus the major part of the GT (GT$_-$) strength is pushed up to high excitation energies in the case of the (p, n)-type reactions.

The (t, ^{3}He) charge-exchange reaction has advantages for use as an important extension of the (n, p) charge-exchange reaction [50–52], primarily because of improved beam qualities compared to those of neutron beams. If high-resolution measurements of charge-exchange reactions of the (n, p)-type ($\Delta T_z = +1$) become possible at intermediate energies, significant contributions are expected towards the understanding of astrophysical [53] as well as the $\beta\beta$ decay processes. However, there exists no accelerator with a dedicated tritium beam with $E_t \geq 100$ MeV/u to make such studies possible.

Recently, such (t, ^{3}He) experiments were performed at the MSU NSCL radioactive-beam facility [54]. The primary 620 MeV ^{4}He beam from the K1200 cyclotron bombarded the metallic beryllium production target with a thickness of 9.25 g/cm^2 mounted in the target position located at the entrance of the A1200 system. The A1200 system was used in a dispersion-matched mode [55], and the triton particles were transported in a dispersive mode to the targets for (t, ^{3}He) reactions at this intermediate image position. The energy spread of tritons passing through the first half of the A1200 system to the reaction target was estimated to be $\sim$23 MeV. This energy broadening was cancelled out and the final (t, ^{3}He) spectra were obtained with a good resolution.

Figure 3.4 shows the 0° spectra for the (t, ^{3}He) reactions on ^{9}Be, ^{10}B, ^{11}B, ^{12}C, and ^{13}C. The GT states in ^{10}Be are interesting. In the ^{10}B (t, ^{3}He) ^{10}Be spectrum (shown in Fig. 3.4(b)), we recognize three strong peaks at $E_x = 3.37, 5.96$, and 9.40 MeV, and one broad peak at 12 MeV. The peaks at 3.37 and 5.96 MeV correspond to the GT transitions from the 3$^+$ ground states of ^{10}B to the 2$^+_1$ and 2$^+_2$ states in ^{10}Be. It is very natural to assume that the spin-parity of the 9.40 MeV state is 3$^+_1$, since the shell model calculation predicts the presence of a strong GT transition to a 3$^+_1$ state at around 9 MeV in the mirror nucleus ^{10}C [56].

After the examination of the proportionality relation between the 0° (t, ^{3}He) cross sections and the GT strengths B(GT) deduced from the β-decays, the B(GT) values for the GT states in ^{10,11}B are obtained as listed in Table 3.2 together with the known B(GT) values from β-decay studies.

The GT and M1 transitions in the $A = 10$ system are especially interesting in view of isobaric mirrors. In the ^{10}B (t, ^{3}He) ^{10}Be reaction at $\theta = 0°$, three states at 3.37, 5.96, and 9.4 MeV are excited (see Fig. 3.4(b)). The observed spectrum is very similar to that obtained from the (d, ^{2}He) reaction at $\theta = 0°$, which has been independently measured using the 200 MeV deuteron beam at RCNP. Thus, these three states in ^{10}Be are inferred to be spin–isospin flip GT ($\Delta S = 1, \Delta T = 1$) states. In the DWBA analyses of the 2$^+_{1,2}$ states ($E_x = 3.35$ and 5.3 MeV) in ^{10}C observed in the mirror ^{10}B (p, n)^{10}C reaction at $E_p = 186$ MeV, Wang et al. [56] show that the corresponding transitions to these two states are dominated by $\Delta J^\pi = 1^+$ transition. Thus, it is natural to conclude that the corresponding two 2$^+$ mirror states in ^{10}Be at $E_x = 3.36$ and 5.96 MeV are due to GT transitions and are the IASs of the 2$^+_1$ and 2$^+_2$ states in ^{10}C.

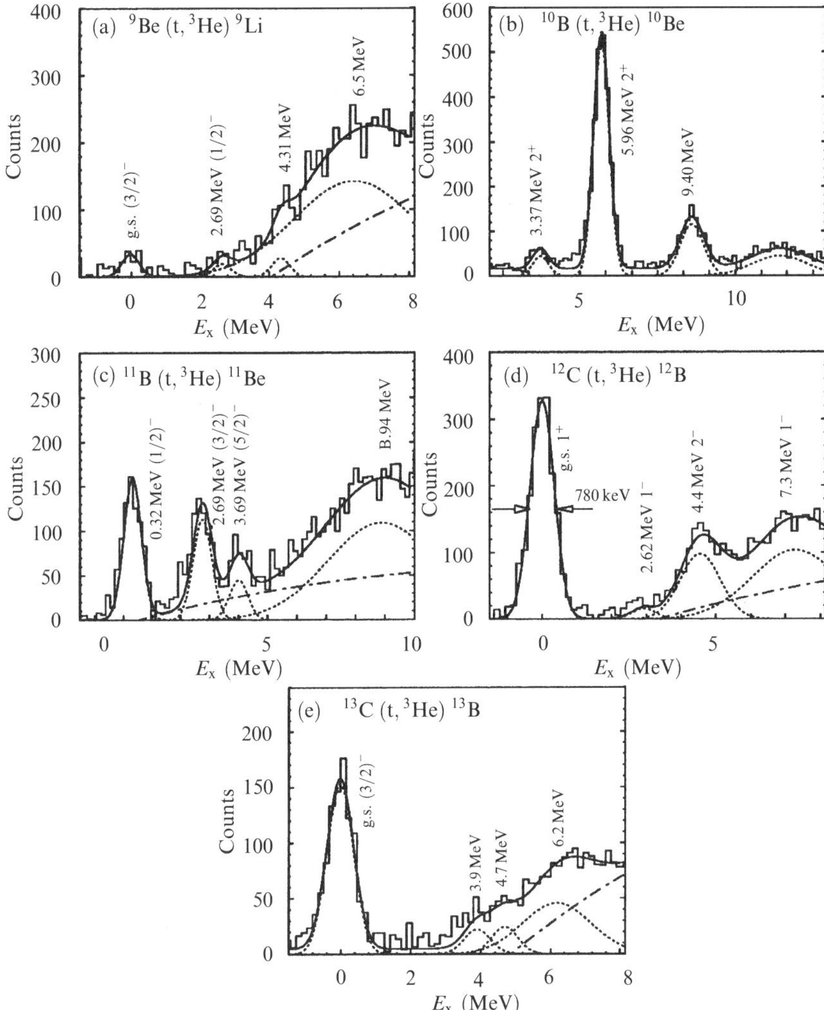

Fig. 3.4 (t, ^{3}He) spectra at $0°$ taken with a triton beam of 381 MeV. The dispersion matching method in charged-particle optics is applied to correct the energy spread of $\sim$23 MeV for the initial triton beam from the breakup (^{4}He, p $-$ t) reaction. The solid and dotted curves are the results of peak fitting. The dot-dashed curves show the non-resonant quasi-free background.

The $A = 10$ system may be one of the best cases for full isotopic spin multiplet in light nuclei. As shown in Fig. 3.5, we indicate a new isobar diagram for mirror GT and M1 states with $T = 1$ for the $A = 10$ system by taking into account the energy level compilation data [57]. The 2^+_1 states at 3.37 MeV in ^{10}Be, at 5.17 MeV in ^{10}B, and at 3.35 MeV in ^{10}C are the mirror states. The 2^+_2 states at 5.9 MeV in ^{10}Be, at

Table 3.2 Comparison of the $0°$ (t, ^{3}He) cross sections and B(GT) values

Target (J^π)	Residual (J^π, E_x) (MeV)	B(GT)	$d\sigma/d\Omega$ (mb sr^{-1})
^{9}Be($(3/2)^-$)	^{9}Li($(3/2)^-$, g.s.)	0.019^a	0.27 ± 0.07
^{10}B(3^+)	^{10}Be(2_1^+, 3.37)	0.08 ± 0.03^b	1.0 ± 0.3
^{10}B	^{10}Be(2_2^+, 5.96)	0.95 ± 0.13^b	12.0 ± 0.8
^{10}B	^{10}Be($(2^+$ or $3^+)$, 9.4)	0.31 ± 0.08^b	3.9 ± 0.7
^{11}B($(3/2)^-$)	^{11}Be($(1/2)^-$, 0.3)	0.23 ± 0.05^b	2.8 ± 0.4
^{11}B	^{11}Be($((3/2)^-)$, 2.7)	0.19 ± 0.05^b	2.3 ± 0.5
^{11}B	^{11}Be($((5/2)^-)$, 3.8)	$0.06 + 0.03^b$	$0.74 + 0.29$
^{12}C(0^+)	^{12}B(1^+, g.s.)	0.999 ± 0.005^a	11.8 ± 1.4
^{13}C($(1/2)^-$)	^{13}B($(3/2)^-$, g.s.)	0.759 ± 0.018^a	9.6 ± 0.9

^{a} Deduced from the β-decay log ft values [57]

^{b} Experimental results from the present (t, ^{3}He) measurements

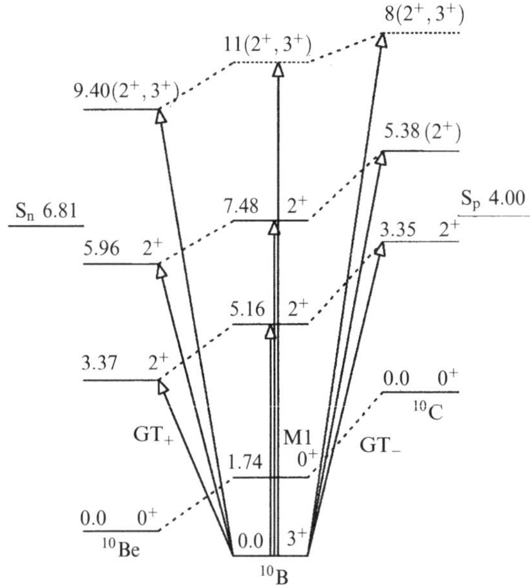

Fig. 3.5 Isobar diagram for states with the GT$_+$, M1, and GT$_-$ transitions from the ^{10}B ground state. The strong transitions in ^{10}Be, ^{10}B, and ^{10}C are indicated by the arrows. The mirror candidates of the 9.4 MeV state in ^{10}B are indicated at 11 MeV in ^{10}B, and at 8 MeV state in ^{10}C.

7.48 MeV in ^{10}B, and at 5.3 MeV in ^{10}C also comprise the mirror states. These 2_2^+ states are excited with strong GT and M1 strengths. The M1 strength from the 3^+ to 2_2^+ states in ^{10}B is measured as B(M1) $= 1.745\mu_N^2$ [58]. From this value, we can estimate the expected B(GT) values for the corresponding mirror states in ^{10}Be and ^{10}C, by neglecting possible contributions from the orbital excitation and from the isobar and meson exchange currents [35]. The value of $B(\mathrm{GT}) = B(\mathrm{M1})/2.64\mu_N^2 = 0.66$ agrees with the $B(\mathrm{GT}_-)$ value of 0.68 ± 0.02 obtained in the $^{10}\mathrm{B}(\mathrm{p,n})^{10}\mathrm{C}$ experiment [56]. On the other hand, the $B(\mathrm{GT}_+)$ value for the 5.9 MeV, mirror 2_2^+ state is 0.95 ± 0.13, which is much larger than the $B(\mathrm{GT}_-)$ one. This isospin symmetry violation would be possible if the nuclear structure differs very much between $^{10}\mathrm{C}$ and $^{10}\mathrm{Be}$ due to the presence of the Coulomb force. It is interesting to consider the question whether or not the recent elaborate theories can predict the observed difference [59,60].

3.5 Application of the (^{3}He, t) reaction

3.5.1 *Neutrino detector ^{71}Ga*

One of the most interesting subjects in modern physics is the neutrino mass problem [61]. Recently, the missing strength of solar neutrino flux has been discussed in relation to the possible neutrino oscillations [62–65]. In order to determine the neutrino flux, experiments using ^{71}Ga as a neutrino detector are in progress [66]. The reliability of the neutrino absorption cross sections using the ^{71}Ga detector was improved by measuring the ^{71}Ga (^{3}He, t) ^{71}Ge reaction at 0°.

Figure 3.6 shows the ^{71}Ga (^{3}He, t) ^{71}Ge spectrum at $\theta = 0°$. The overall shape of this spectrum is very similar to the (p, n) spectrum measured at 120 MeV [67]. The transition strength of the first excited $(3/2)^-$ state at 0.175 MeV in ^{71}Ge is very small, consistent with the (p, n) result [67] at 120 MeV. The (^{3}He, t) result is not consistent with the low-energy (p, n) result [68]. This suggests that 0° (p, n) differential cross sections at 35 MeV may not be related to the GT strengths in a simple way.

The transition strengths $B(\mathrm{GT})$ for the ground state and all excited states up to 1.36 MeV are deduced by fitting the peaks with a Gaussian shape (see Fig. 3.6). The strength of the 0.175 MeV $(3/2)^-$ state is smaller by a factor of about ten than that for the ground state. Transition strengths for the excited states in the energy region between 1.5 MeV and the neutron decay threshold energy of 7.42 MeV were obtained by binning the continuum spectrum into intervals of 250 keV width each and by summing up the counts in each bin. The $B(\mathrm{GT})$ values are summarized in Table 3.3.

The $B(\mathrm{GT})$ values obtained in the (^{3}He, t) work are normalized by using the $B(\mathrm{GT})$ value for the ground state transition which is deduced from the β-decay [69]. Using the Standard Solar Model (SSM) flux by Bahcall [70], the capture rates of the solar neutrinos are obtained. The results are listed in Tables 3.3 and 3.4. The uncertainty of the $B(\mathrm{GT})$ value for the first excited state (Table 3.4) is large, but its contribution to the total neutrino capture rate is about 1.3%, which is small. If the SSM flux is assumed, the total solar neutrino capture rate amounts to 130.3 ± 1.57 (stat.) ± 16.8 (syst.) in Solar Neutrino Units (SNU). The statistical error in the result is due to the error in the experimental $B(\mathrm{GT})$ values, and the systematic error comes from the ambiguities of the neutrino flux estimates in the SSM. The total capture rate deduced via the present experiment is close to that calculated by Bahcall [70] on the basis of the (p, n) experiment at 120 MeV.

Table 3.4 gives the solar neutrino capture rates (SNU) for the ^{71}Ga detector. In the present calculation, we used the GT strengths of the excited states lower than the

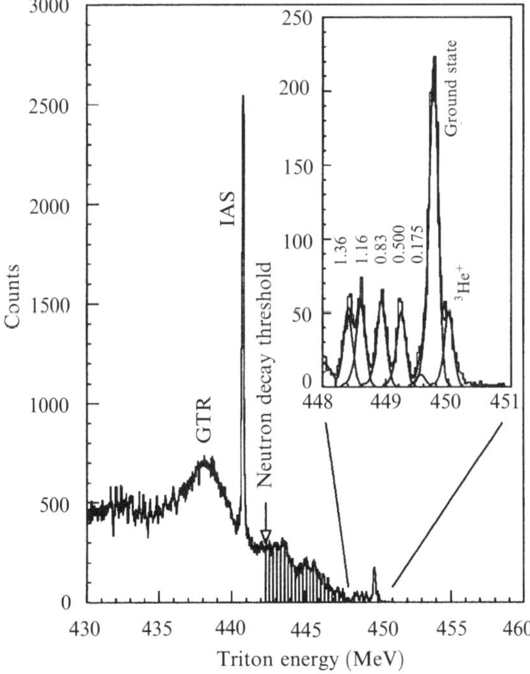

Fig. 3.6 A $0°$ ^{71}Ga (^{3}He, t) ^{71}Ge spectrum at 450 MeV. A small peak at the right side of the ground state peak is caused by ^{3}He particles at full incident energy from the atomic charge-exchange reaction of ^{3}He^{2+} $\rightarrow$ ^{3}He$^+$. (Taken from Fujiwara *et al.* [31].)

Table 3.3 The B(GT) values and solar neutrino capture rates obtained via the ^{71}Ga (^{3}He, t) ^{71}Ge reaction at 450 MeV. The B(GT) values were normalized to the B(GT) of the ground state in ^{71}Ge. The neutrino capture rate was calculated using the solar neutrino flux given by the standard solar model of Bahcall [70]

E_x (MeV)	B(GT)	Capture rate (SNU)
0.000	$(8.91 \pm 0.13) \times 10^{-2}$	114.9 ± 11.8
0.175	$(4.91 \pm 1.77) \times 10^{-3}$	1.75 ± 1.01
0.500	$(2.08 \pm 0.21) \times 10^{-2}$	3.35 ± 1.10
0.83	$(2.37 \pm 0.23) \times 10^{-2}$	$(7.51 \pm 3.37) \times 10^{-1}$
1.16	$(2.33 \pm 0.24) \times 10^{-2}$	$(8.11 \pm 1.89) \times 10^{-1}$
1.36	$(2.01 \pm 0.23) \times 10^{-2}$	$(2.78 \pm 1.31) \times 10^{-1}$
1.58–7.42	2.88 ± 0.09	9.52 ± 3.73
Total	3.07 ± 0.11	130.3 ± 18.4

Table 3.4 Neutrino capture rate for each neutrino source. All the values are obtained with the assumption of the neutrino flux given by the SSM of Bahcall. The uncertainties of the present results are caused by those from the experimental B(GT) values and from the standard solar model, respectively

Neutrino source	Present result (SNU)	Bahcall [70] (SNU)
pp	$70.8 \pm 0.2 \pm 1.4$	70.8
^{8}B	$12.6 \pm 0.4 \pm 4.7$	14.0
hep	$(5.34 \pm 0.15) \times 10^{-2}$	0.06
^{13}N	$3.74 \pm 0.07 \pm 1.87$	3.8
^{15}O	$5.81 \pm 0.14 \pm 3.37$	6.1
^{17}F	$(6.01 \pm 0.15 \pm 2.76) \times 10^{-2}$	0.06
pep	$2.94 \pm 0.09 \pm 0.15$	3.0
^{7}Be	$35.0 \pm 0.7 \pm 5.3$	34.3
Total	$130.3 \pm 1.57 \pm 16.8$	132^{+20}_{-17}

threshold energy of 7.42 MeV in ^{71}Ge. The present data agree well with the calculations given by Bahcall [70].

In order to show that the neutrino oscillation mechanism with the MSW effect [62,63] explains the deficit of solar neutrinos correctly, we should clarify the nuclear physics problem. Some problems remain to be solved for deducing the reliable SNU values concerning ^{71}Ga.

Two competing decay processes appear above the excitation energy of 7.4 MeV: neutron emission and γ-decay. The strong Fermi transition to the IAS at an excitation energy of 8.932 MeV from ^{71}Ga contributes if there is γ-decay from the levels above the threshold. The upper limit of this process is known only within 10% accuracy [71]. The γ-transitions from the GT states above 7.4 MeV to the low-lying states in ^{71}Ga might possibly contribute. Estimates of this contribution are not made, but it is important to reduce the uncertainty of the solar neutrino detection sensitivity for ^{71}Ga. The γ-decay from the GT states at high excitation energies has not yet been measured. This measurement is also important from the viewpoint of nuclear physics. Generally speaking, particle decay dominates when a particle threshold opens. In this regard, the recent observation of a strong γ transition from particle unbound states in ^{47}K is interesting. Trinder *et al.* [72] found that the γ branching ratio of the 3.239 MeV level in ^{47}K is surprisingly high compared to the proton decay width. This finding eliminated a major portion of the discrepancy between (p, n) and β-decay results concerning the GT strength values.

In the estimate of neutrino absorption by ^{71}Ga, we have usually assumed the contribution from the levels above the particle threshold, 7.4 MeV to be negligible. This assumption is unnatural since the γ-branch should smoothly decrease with increase in the excitation energy. In fact, a statistical model calculation indicates that a large γ-branch still remains in the region above the neutron threshold, 7.4 MeV. At present, the contribution of γ-transitions has not been estimated. If there are significant γ-transitions from the levels above 7.4 MeV, the SNU value for the ^{8}B neutrinos will increase. Such studies are now in progress at RCNP [73].

3.5.2 *The $\beta\beta$ decay*

Neutrinoless $\beta\beta$ decays ($0\nu\beta\beta$) have never been observed. These processes are energetically allowed, but are forbidden because of lepton number conservation. The search for $\beta\beta$ decays has been on for over 30 years [74]. Initial studies aimed at confirming the conservation law or checking the neutrino properties. However, recent interest in $\beta\beta$ decays has been enhanced in view of particle physics, nuclear physics, and astrophysics [75–77]. The observation of $0\nu\beta\beta$ decays shows directly the presence of the Majorana mass of neutrinos, thus providing clear evidence for physics towards the Grand Unified Theories (GUTs).

The $\beta\beta$ decay is a very rare process. Its measurements require a laboratory with an ultralow radiation background. The current upper limit of the electron neutrino mass is about 1 eV. The new experimental facilities at Gran Sasso (Italy), Oto (Japan), and Frejus (France) now aim at reducing this upper limit of the neutrino mass to 0.1 eV [78].

In connection to these ultralow background measurements of $\beta\beta$ decays, the effort to give the most reliable theoretical estimate of $\beta\beta$ decays continues. Two-neutrino $\beta\beta$ decays ($2\nu\beta\beta$), which are allowed in the standard model, have been observed and used for evaluating the matrix elements $M^{0\nu}$. Theoretical calculations for $\beta\beta$ decays are not easy, since they require detailed information on nuclear structures, which is quite sensitive to the interaction parameters used.

In order to check the details of the theoretical calculations, comparison with the experimental data for the GT transitions is important. ^{100}Mo and ^{116}Cd are interesting cases. The experimental data are available for ^{100}Mo and ^{116}Cd $2\nu\beta\beta$ decays [79–82].

Figure 3.7 shows the (^{3}He, t) energy spectra for ^{100}Mo and ^{116}Cd at scattering angles of $\theta = 0°$ and $1°$. The global structures of the two (^{3}He, t) spectra are essentially the same. We recognize some low-lying discrete states and IASs on the GTGRs. In addition, the presence of the SDR is observed more clearly in the (^{3}He, t) spectra at $\theta = 1°$.

These global structures in the GT excitation seem to be common for all nuclei with mass $A = 70$–120 [29,31]. A major GTR (GTGR$_1$), which consists mainly of the spin-flip $(j_>)_\mathrm{n}(j_<)_\mathrm{p}$ configurations, appears at the excitation energy above IAS. A small GT strength (GTGR$_2$), which consists of the non-spin-flip configurations of $(j_>)_\mathrm{n}(j_>)_\mathrm{p}$, appears below IAS. This behaviour could be very useful to check theoretical calculations or to improve the reliability of the theories of $\beta\beta$ decay.

The observed structures are well decomposed into the low-lying states, GTGR$_1$, GTGR$_2$, IAS, SDR, and the non-resonance continuum background as shown in Fig. 3.7 [83]. The sum of the measured GT strengths up to the excitation energy of $\sim$19 MeV is 72% for ^{100}Tc and 60% for ^{116}In, respectively, compared to the sum-rule limit of $3(N - Z)$.

In the $\beta\beta$ decay calculations for the transition from the parent nucleus to the daughter nucleus $_Z\mathrm{A}_N \rightarrow _{Z+2}\mathrm{C}_{N-2}$, the matrix element of the GT transitions $_Z\mathrm{A}_N \rightarrow _{Z+1}\mathrm{B}_{N-1}$, is required in the first step of the calculation. In the second step, the matrix element for transition from the GT states to the daughter nucleus $_{Z+1}\mathrm{B}_{N-1} \rightarrow _{Z+2}\mathrm{C}_{N-2}$ is also needed. In these calculations, a severe cancellation commonly appears between the two matrix elements [84]. In fact, Griffiths and Vogel [85] have found that a Quasi-particle

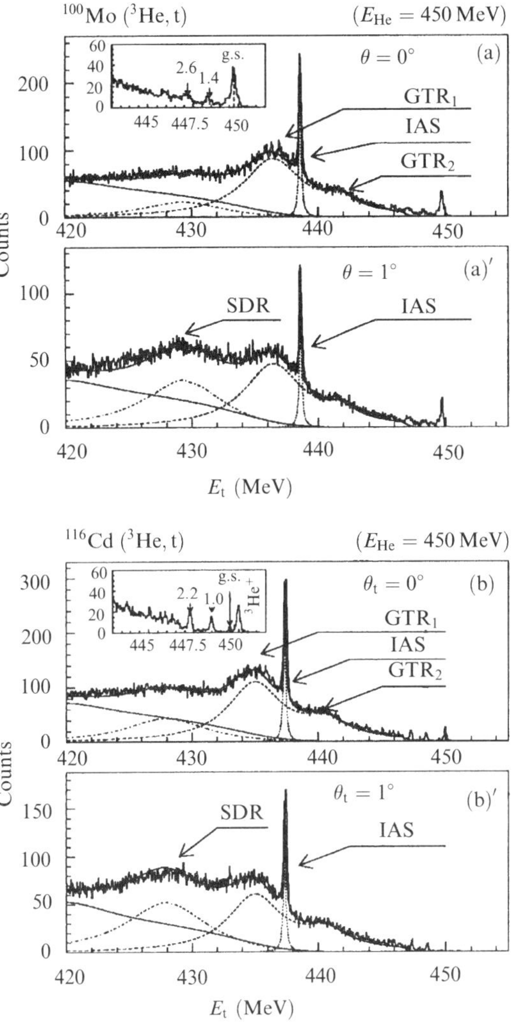

Fig. 3.7 Energy spectra of the ^{100}Mo (^{3}He, t) ^{100}Ru and ^{116}Cd (^{3}He, t) ^{116}In reactions at 450 MeV. The upper and lower panels show the 0° and 1° spectra, respectively. The dotted lines are the results of the peak fitting, and the solid line is the background due to the quasi-free process. For details, see Ref. [83].

Random-Phase Approximation (QRPA) calculation cannot reproduce the experimental values relevant to ^{100}Mo. This situation indicates that the strong GT transitions into the GTGR as the first step makes no major contribution to the $\beta\beta$ process.

The nuclear matrix element $M^{2\nu}$ for $2\nu\beta\beta$ decays is written as

$$M^{2\nu} = \sum \frac{\langle A\|\sigma\tau^-\|B(i)\rangle\langle B(i)\|\sigma\tau^-\|C\rangle}{(Q_{EC} + Q_{\beta-})/2},\tag{3.6}$$

where $\langle A\|\sigma\tau^-\|B(i)\rangle$ and $\langle B(i)\|\sigma\tau^-\|C\rangle$ are matrix elements for single weak transitions through ith intermediate state. Taking into account the severe cancellation for the transitions through the intermediate states the $M^{2\nu}$ matrix elements can be estimated using a simple argument. In fact, it was found [86] that the $M^{2\nu}$ matrix element for the $0^+ \rightarrow 0^+$ ground state transitions given by

$$M^{2\nu} \sim \frac{\langle A\|\sigma\tau^-\|B(\text{g.s.})\rangle\langle B(\text{g.s.})\|\sigma\tau^-\|C\rangle}{(Q_{\text{EC}} + Q_{\beta^-})/2} \tag{3.7}$$

reproduces well the measured experimental values for several medium-heavy nuclei. In the case of ^{100}Mo, the estimated matrix element is $0.1 + 0.01$, which agrees well with the experimental value, 0.096 ± 0.10, from the observed $2\nu\beta\beta$ half-life [79]. The estimated value for ^{116}Cd is 0.03 ± 0.003, which is also near the experimental value of 0.069 ± 0.009.

The experimental data of GT transitions for the $\beta\beta$ decay nuclei are still poor. In order to understand the mechanism of the severe cancellation in the theoretical calculations and to reach a deeper level of understanding of the $\beta\beta$ decay phenomena, More data are necessary, especially for the (n, p)-type reactions.

3.6 Decay of giant resonances

The nuclear giant resonance is explained in the microscopic shell model as a coherent superposition of many one-particle–one-hole (1p–1h) excitations, which exhaust a large part of the sum-rule limit. When the resonance is below or near the threshold of particle emission, many fragmented states are observed. The resonance has a large particle-decay width when it is well above the threshold. Understanding of the origin of the width and the fragmentation is an interesting subject since the spreading width is relevant to the coupling of 1p–1h states with 2p–2h states and many-particle–many-hole states, while the escape width relates to the coupling to the continuum. The global microscopic nature of giant resonances concerning this spreading has already been established. But, due to the lack of experimental data, the direct particle decay is not well understood. The study of the particle decay of the resonances is, therefore, a unique way to obtain information on their microscopic structure.

A pioneering work concerning the decay measurement of the giant resonances was performed by Moalem *et al.* [87]. They measured the decay of protons and α-particles from the GQR in ^{40}Ca excited by photo-absorption. In the early 1980s, decay measurements of neutrons and protons were believed to give precious information on the damping of the resonance in nuclei [88,89]. For this purpose, some interesting experiments have been performed [90–93]. One such experiment is the neutron decay measurement of the GQRs and GMRs excited via the (α, α') reactions. The Osaka group [94] clarified the microscopic structure of GQR in ^{118}Sn and ^{92}Zr by measuring angular correlations of emitted neutrons. A characteristic correlation pattern indicated that 20% of the total width ($\sim 5\,\text{MeV}$) is due to direct neutron decays.

Although these studies were very difficult and were not extended to various target nuclei, the success of such measurements sheds light on the damping mechanism of the resonances. Since the statistical decay process dominates in the case of neutron decay, a key ingredient in clarifying the damping mechanism of the resonances would be the knowledge of a reliable portion of the direct decay.

3.6.1 *Particle decay from spin–isospin resonances*

3.6.1.1 *Decay from the spin–isospin resonance in ^{208}Bi*

Particle decays from the GTGR were studied, for the first time, in Groningen [95]. The result was surprising, since the GTGR in ^{208}Bi seemed to have a vanishing spreading width. The final experimental answer to solve this point was given by the (^{3}He, t) reaction at 450 MeV using the specific advantages of this reaction at high energies. It was found that an attempt to measure proton decays of the GTGR in ^{208}Bi at the low bombarding energy failed because the reaction mechanism was too complex to excite the spin-flip GTGR in a simple way.

The (^{3}He, t) reaction at high energies is a more suitable tool to study the microscopic structure of the GTGR through particle decay measurement since, as for the (p, n) reaction at intermediate energies (≥ 100 MeV/u), it preferentially excites spin–isospin-flip modes [31,30]. Furthermore, there is the advantage that tritons can be detected at $0°$ with a magnetic spectrometer with essentially 100% efficiency and with resolution superior to that achievable in the (p, n) reaction.

Figure 3.8 shows triton energy spectrum in coincidence with decay protons emitted at backward angles. In the final triton energy spectrum in Fig. 3.8, the (^{3}He, t) spectrum was gated by the decay protons feeding to the residual neutron-hole states in ^{207}Pb. The continuum background from the quasi-free knockout process seen in the singles spectra in Fig. 3.1 disappears. This means that the coincident final triton spectra contain only the proton emission from the resonance states in ^{208}Bi. In Fig. 3.8, a sharp peak corresponding to the IAS in ^{208}Bi is observed significantly since the proton decay width $\Gamma_p^{\uparrow}$ is 136.8 ± 21.8 keV and is comparable with the total width of 232 keV. On the other hand, the broad peak corresponding to the GTGR in ^{208}Bi is seen only weakly. This is because the proton escape width for 184 ± 49 keV is of the same order of magnitude as that of IAS and thus much smaller than the total width of 3720 keV. It is worth noting that the yields of the SDR relative to that of the GTGR are bigger in the coincidence triton spectra (see Fig. 3.8(b)). The total proton decay branching ratio for the SDR has been determined to be $14.1 \pm 4.2\%$, which is much bigger than the value of $4.9 \pm 1.3\%$ for the proton escape width for the GTGR. This indicates that the penetration of protons due to the Coulomb barrier is not very much suppressed for the SDR.

The observed SDR is expected to consist of three spin components—2^-, 1^-, and 0^-. The order of the expected excitation energies is $E_x(2^-) \leq E_x(1^-) \leq E_x(0^-)$ with the strength ratio $5 : 3 : 1$. Therefore, it is not possible to discuss the proton decay width as if it were from one resonance. However, it should be noted that the shape of the SDR in the coincident triton spectrum seen in Fig. 3.8(b) is slightly different from that obtained in the singles spectrum in Fig. 3.1. The proton decay width from the 2^- SDR is observed seemingly with some enhancement. A better understanding of these observations awaits further theoretical efforts.

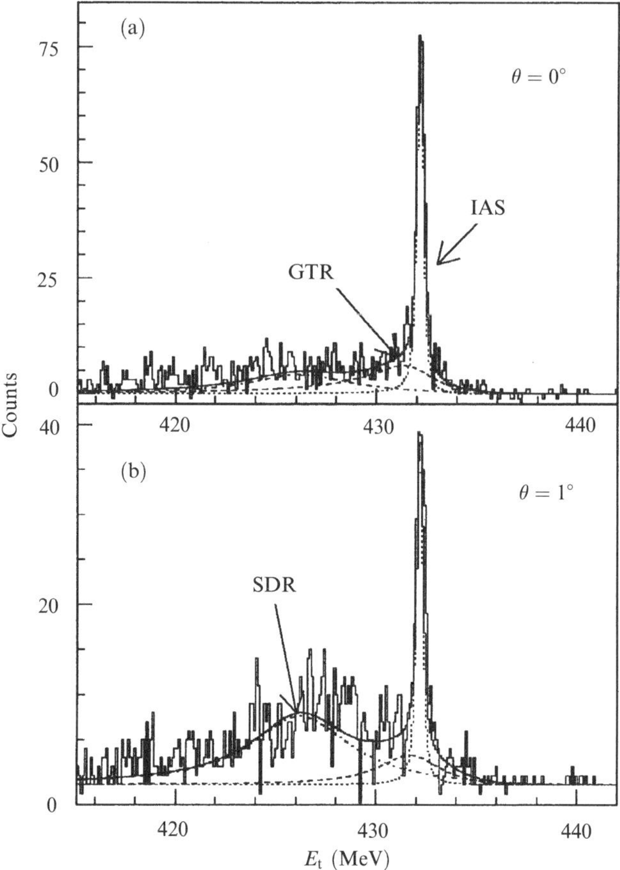

Fig. 3.8 Triton energy spectra of the ^{208}Pb(^{3}He, t) reactions gated on proton decays to neutron-hole states in ^{207}Pb. The dashed, dotted, and dot-dashed lines are the best fits obtained for the GTR, IAS, and SDR in ^{208}Pb, respectively. Spectra (a) and (b) are measured at the scattering angles at $0°$ and $1°$, respectively. For singles ^{208}Pb (^{3}He, t) ^{208}Bi spectra, see Fig. 3.1.

However, the understanding of the proton decay from the GTGR and IAS in ^{208}Bi was much improved by the theoretical inputs. The theoretical calculations by Knobles *et al.* [96] and by Coló *et al.* [97] reproduce well the experimental data for proton decays including the partial proton decay widths into the individual residual hole states in ^{207}Pb, although the theoretical controversy still continues concerning the evaluation of the nuclear response in the continuum [98,99].

Apart from the proton decay, it would be interesting to consider the γ-decay between the GTGR and the IAS. The GTGR and IAS are mediated by operators $\sigma\tau$ and τ from the same target nucleus. The GTGRs are, therefore, strongly connected to the IAS by the M1 γ-decay. Sagawa *et al.* [100] estimated the magnetic dipole transition rates from the

GTGR to the IAS. Non-energy-weighted sum-rule arguments indicate that M1 transitions between GT states and IAS are enhanced by a factor as large as $\sim$2.5 compared to their analogues in the parent nucleus, suggesting that the measurements of these M1 decays may be feasible.

A similar logic is applied to the case of E1 γ-decays from the SDR to the low-lying GT states. The SDR is excited by the operator $\sigma\tau r Y_1$, which is expected to induce strong E1 transitions into the low-lying GT states. The search for possible γ-decays, by measuring the $(^3\mathrm{He}, \mathrm{t})$ reactions in coincidence with decay γ-rays, from both GTGR and SDR, has started at RCNP [101,102].

3.6.1.2 *Decay from the spin–isospin resonance in $^{13}\mathrm{N}$*

The levels in $^{13}\mathrm{C}$ and $^{13}\mathrm{N}$ are isobaric mirrors. The concept of isospin was often discussed in the context of decay modes of excited states in these two nuclei. Furthermore, in view of the α-cluster structure, the ground state of $^{13}\mathrm{C}$ is considered to have a structure consisting of three α's and one neutron. Thus, it is very interesting to observe proton decays from the levels in $^{13}\mathrm{N}$ excited by the $(^3\mathrm{He}, \mathrm{t})$ reaction since information obtained from the data [103] gives a severe limit to check the validity of the nuclear models.

Figure 3.9 shows a two-dimensional scatter plot of decay protons obtained in coincidence with tritons gated on the $(^3\mathrm{He}, \mathrm{t})$ reaction at 450 MeV at $\theta = 0°$. A singles spectrum measured at the same time is compared with the scatter plot of decay protons (see Fig. 3.9(b)). Several sharp states in $^{13}\mathrm{N}$ are strongly excited. The overall shape of the singles $(^3\mathrm{He}, \mathrm{t})$ spectrum at $\theta = 0°$ is quite similar to that of the $^{13}\mathrm{C}\,(^3\mathrm{He}, \mathrm{t})\,^{13}\mathrm{N}$ reaction at 200 MeV [104]. There are several clear loci of proton decays from the discrete levels in $^{13}\mathrm{N}$. It is worth noting that proton decay to the low-lying states in $^{12}\mathrm{C}$ does occur even from highly excited continuum states. This indicates that there are some resonances above $E_x = 15$ MeV. In fact, there is clear indication of the presence of the broad peaks in the excitation energy region at $E_x = 17$–24 MeV, which are not mentioned in the (p, n) work by Mildenberger *et al.* [104].

On the basis of the schematic model, the possible resonances should have the structure with one neutron in the $p_{1/2}$ orbital coupled to the SDRs with $J^\pi = 0^-, 1^-,$ and 2^-. Thus, the expected resonances above the 15.6 MeV$(3/2)^-$, $T = 3/2$ state are the $2^- \otimes p_{1/2}$, $1^- \otimes p_{1/2}$, and $0^- \otimes p_{1/2}$ states. This leads to five states as the spin-flip dipole resonances in $^{13}\mathrm{N}$. These resonances correspond to the SDR in $^{12}\mathrm{N}$ coupled with one neutron, which apparently should be strongly excited by charge-exchange reactions.

The observed two-dimensional plot in Fig. 3.9(a) clearly shows that there are proton decays from the resonances above $E_x = 15$ MeV to some specific states in $^{12}\mathrm{C}$. Among them, the decay strength to the ground state is relatively weak compared with that to the 4.4 MeV 2^+ state. The proton decay also feeds to the 1^+ states at 12.71 and 15.1 MeV. This clearly indicates that the resonance above 15 MeV has a configuration consisting of one proton in the sd orbitals and one neutron in the $p_{1/2}$ orbital. It is worth noting that the Stanford group found a strong proton decay from the E1 resonance in $^{13}\mathrm{N}$ to the 15.1 MeV 1^+ state in $^{12}\mathrm{C}$ via the pion charge-exchange experiment of the $^{13}\mathrm{C}\,(\pi^+, \pi^0 + \mathrm{p})^{12}\mathrm{C}$ reaction [105].

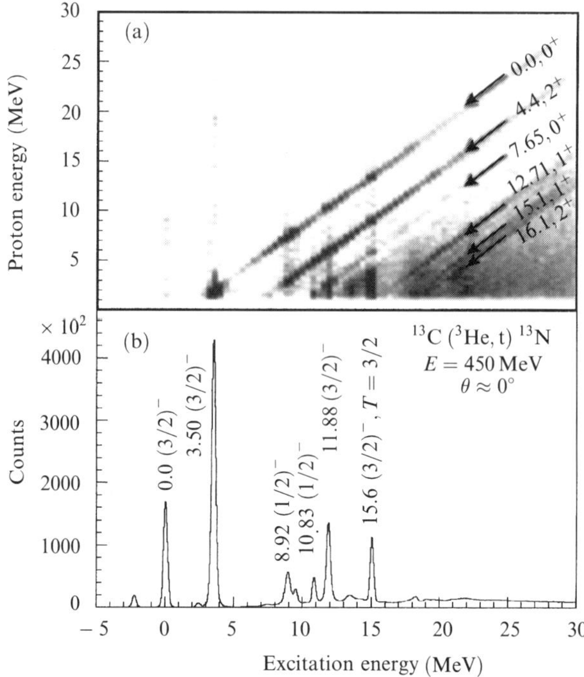

Fig. 3.9 (a) A two-dimensional scatter plot for coincidence events of proton energy versus triton energy (in units of excitation energy) gated on scattering angles centred at $\theta = 0°$. The proton decays are measured at $\theta = 160°$. The loci indicate decay of states in ^{13}N by protons to final states in ^{12}C. (b) Singles ^{13}C (^{3}He, t) ^{13}N spectrum measured at $E(^3$He$) = 450$ MeV, $\theta = 0°$.

It is interesting that the proton decay from the 10.83 MeV $(1/2)^-$ state does not occur to the ground state in ^{12}C, and this state has a strong-decay width corresponding to the decay to the second 0^+ state in ^{12}C. Since the second 0^+ state is inferred to have a three-α-cluster structure different from the ground 0^+ state of ^{12}C, the 10.83 MeV $(1/2)^-$ state is considered to have a similar α-cluster structure. We again need to await further theoretical efforts for a better understanding of these new observations.

3.7 Summary

The study of giant resonances in nuclei has a long history. In the early stage of the investigations, interest was concentrated on the discovery of the resonances. Many competitive studies were made in the past to establish the excitation energies of the GR resonances. Various types of nuclear resonances have already been established. Interesting questions are still addressed for new resonances. The isovector giant monopole resonances, isovector giant quadrupole resonances and isovector giant octupole resonances are not established. Does the double giant resonance exist with the 2p–2h nature? At present, the double dipole resonance is believed to be found by using relativistic heavy-ion collisions with the advantage of the strong Coulomb excitations. Can we find the double GT resonance? Does the spin–dipole resonance in a deformed nucleus split into two components like the giant dipole resonance? What is the mass dependence of these resonances mentioned above? There are still several questions which need further experimental efforts.

One of the common directions of investigations at present is to understand the damping mechanism of the resonances (the origin of the width) and the microscopic structures (wave functions). The study of 'microscopic structures' of the giant resonances is addressed as the next generation of developments in nuclear physics, where the efficient coincidence measurements are indispensable for detecting the proton-, neutron-, and γ-decays from the selected states. They are tools to guide us towards new discoveries. For this purpose, multi-detector systems are developed by combining high-speed computers at many laboratories. These experimental efforts will serve to improve the basic understanding of the spin–isospin resonances. We require a special development of these experimental instruments to make coincidence experiments and their analyses very easy. Some new results associated with the proton decay measurements in coincidence with the (^{3}He, t) reactions at 450 MeV are given in this article. Other results discussed here are applications of the nuclear structure information and are useful for different areas in physics. The study of the spin–isospin giant resonances is important to astrophysics and neutrino physics. A typical example is the GT strengths in nuclei, which have many applications in our attempts to understand the cosmic evolution—past and future.

Acknowledgements

The present work was supported in part by a Grant-in-Aid of Scientific Research, Ministry of Education, Science, Sports and Culture (Monbusho) under grant number 07404012. The author is grateful to Dr. Wakasa for providing the postscript file of their (p, n) results. Finally, he would like to thank many collaborators in the (^{3}He, t) work done at RCNP. He especially thanks Professors Ejiri and Toki, who encouraged him to write the present manuscript.

References

1. S. C. Fultz *et al.*, *Phys. Rev.* **127** (1962) 1273.
2. J. Speth, ed., *Electric and magnetic giant resonances in nuclei* (World Scientific, 1991).
3. H. Akimune, I. Daito, Y. Fujita, M. Fujiwara, M. B. Greenfield, M. N. Harakeh, T. Inomata, J. Jänecke, K. Katori, S. Nakayama, H. Sakai, Y. Sakemi, M. Tanaka, and M. Yosoi, *Phys. Lett. B* **323** (1994) 107.
4. H. Akimune, I. Daito, Y. Fujita, M. Fujiwara, M. B. Greenfield, M. N. Harakeh, T. Inomata, J. Jänecke, K. Katori, S. Nakayama, H. Sakai, Y. Sakemi, M. Tanaka, and M. Yosoi, *Phys. Rev. C* **52** (1995) 604.
5. F. Osterfeld, *Rev. Mod. Phys.* **64** (1992) 491.
6. N. Auerbach and A. Klein, *Phys. Rev. C* **30** (1984) 1032.
7. J. P. Blaizot, *Phys. Rep.* **64** (1980) 171.

8. S. Shlomo and D. H. Youngblood, *Phys. Rev. C* **47** (1993) 529, and references therein.

9. G. F. Bertsch and S. Das Gupta, *Phys. Rep.* **160** (1988) 189.

10. R. R. Doering, A. Galonsky, D. M. Patterson, and G. F. Bertsch, *Phys. Rev. Lett.* **35** (1975) 1691.

11. D. E. Bainum, J. Rapaport, C. D. Goodman, D. J. Horen, C. C. Foster, M. B. Greenfield, and C. A. Goulding, *Phys. Rev. Lett.* **44** (1980) 1751.

12. C. D. Goodman, C. A. Goulding, M. B. Greenfield, J. Rapaport, D. E. Bainum, C. C. Foster, W. G. Love, and F. P. Petrovich, *Phys. Rev. Lett.* **44** (1980) 1755.

13. K. Ikeda, S. Fujii, and J. I. Fujita, *Phys. Lett.* **3** (1963) 271.

14. J. I. Fujita and K. Ikeda, *Nucl. Phys.* **67** (1965) 145.

15. H. Ejiri, K. Ikeda, and J. I. Fujita, *Phys. Rev.* **176** (1968) 1277; H. Ejiri and J. I. Fujita, *Phys. Rep.* **C38** (1978) 85.

16. P. H. Bowen, C. Cox, G. Huxtable, J. P. Scanlon, J. J. Thresher, and A. Langsford, *Nucl. Phys.* **30** (1962) 475.

17. B. E. Bonner, J. E. Simmons, C. R. Newsom, P. J. Riley, G. Glass, J. C. Hiebert, Mahavir Jain, and L. C. Nortcliffe, *Phys. Rev. C* **18** (1978) 1418.

18. D. J. Horen *et al.*, *Phys. Lett. B* **95** (1980) 27.

19. C. Gaarde, *J. Phys.* **45** (1984) C4-405.

20. F. Petrovich, J. A. Carr, and H. MacManus, *Ann. Rev. Nucl. Part. Sci.* **36** (1986) 29.

21. W. G. Love, K. Nakayama, and M. A. Franey, *Phys. Rev. Lett.* **28** (1981) 1401.

22. A. Galonsky, J. P. Didelez, A. Djaloeis, and W. Oelert, *Phys. Lett. B* **74** (1978) 176.

23. D. Ovazza, A. Willis, M. Morlet, N. Marty, P. Martin, P. de Saintingnon, and M. Buenerd, *Phys. Rev. C* **18** (1978) 2438.

24. C. Gaarde, J. S. Larsen, M. N. Harakeh, S. Y. van der Werf, M. Igarashi, and A. Müller-Arnke, *Nucl. Phys. A* **334** (1980) 248; C. Gaarde, J. S. Larsen, A. G. Drentje, M. N. Harakeh, S. Y. van der Werf, and A. Müller-Arnke, *Nucl. Phys. A* **346** (1980) 497.

25. S. L. Tabor, G. Neuschaefer, J. A. Carr, F. Petrovitch, C. C. Chang, A. Guterman, M. T. Collins, D. L. Friesel, C. Glover, S. Y. van der Werf, and S. Raman, *Nucl. Phys. A* **422** (1984) 12.

26. W. A. Sterrenburg, M. N. Harakeh, S. Y. van der Werf, and A. van der Woude, *Nucl. Phys. A* **405** (1983) 109.

27. S. Y. van der Werf, S. Brandenburg, P. Gradijk, W. A. Sterrenburg, M. N. Harakeh, M. B. Greenfield, B. A. Brown, and M. Fujiwara, *Nucl. Phys. A* **496** (1989) 305.

28. J. Jänecke, F. D. Becchetti, A. M. van der Berg, G. P. A. Berg, G. Brouwer, M. B. Greenfield, M. N. Harakeh, M. A. Hofstee, A. Nadasen, D. A. Roberts, R. Sawafta, J. M. Schippers, E. J. Stephenson, D. P. Stewart, and S. Y. van der Werf, *Nucl. Phys. A* **526** (1991) 1; J. Jänecke, K. Pham, D. A. Roberts, D. Stewart, M. N. Harakeh, G. P. A. Berg, C. C. Foster, J. E. Lisantti, R. Sawafta, E. J. Stephenson, A. M. van der Berg, S. Y. van der Werf, S. E. Muraviev, and M. H. Urin, *Phys. Rev. C* **48** (1993) 2828.

29. K. Pham, J. Jänecke, D. A. Roberts, M. N. Harakeh, G. P. A. Berg, S. Chang, J. Liu, E. J. Stephenson, B. F. Davis, H. Akimune, and M. Fujiwara, *Phys. Rev. C* **51** (1995) 526.

30. I. Bergqvist, A. Brockstedt, L. Carlén, L. P. Ekström, B. Jakobsson, C. Ellegaard, C. Gaarde, J. S. Larsen, C. D. Goodman, M. Bedjidian, D. Contrado, J. Y. Grossiord, A. Guichard, R. Haroutunian, J. R. Pizzi, D. Bachelier, J. L. Boyard, T. Hennino, J. C. Jourdain, M. Roy-Stephan, M. Boivin, and P. Radvanyi, *Nucl. Phys. A* **469** (1987) 669.

31. M. Fujiwara, H. Akimune, I. Daito, H. Ejiri, M. B. Greenfield, M. N. Harakeh, T. Inomata, J. Jänecke, S. Nakayama, N. Takemura, A. Tamii, M. Tanaka, H. Toyokawa, and M. Yosoi, *Nucl. Phys. A* **599** (1996) 223c.

32. H. Akimune, I. Daito, Y. Fujita, M. Fujiwara, M. B. Greenfield, M. N. Harakeh, T. Inomata, J. Jänecke, K. Katori, S. Nakayama, H. Sakai, Y. Sakemi, M. Tanaka, and M. Yosoi, *Nucl. Phys. A* **567** (1994) 245c.

33. Y. Fujita, H. Akimune, I. Daito, M. Fujiwara, M. N. Harakeh, T. Inomata, J. Jänecke, K. Katori, C. Lüttge, S. Nakayama, P. von Neumann-Cosel, A. Richter, A. Tamii, M. Tanaka, H. Toyokawa, H. Ueno, and M. Yosoi, *Phys. Lett. B* **365** (1996) 29.

34. Y. Fujita, H. Akimune, I. Daito, M. Fujiwara, M. N. Harakeh, T. Inomata, J. Jänecke, K. Katori, C. Lüttge, S. Nakayama, P. von Neumann-Cosel, A. Richter, A. Tamii, M. Tanaka, H. Toyokawa, H. Ueno, and M. Yosoi, *Phys. Rev. C* **55** (1997) 1137.

35. A. Richter, A. Weiss, O. Häusser, and B. A. Brown, *Phys. Rev. Lett.* **65** (1990) 2519.

36. H. Toki and W. Weise, *Phys. Rev. Lett.* **42** (1979) 1034.

37. R. J. Holt *et al.*, *Phys. Rev. C* **20** (1979) 93.

38. A. Bhor and B. Mottelson, *Nuclear structure* (Benjamin, Reading, MA, 1975), Vol. II, p. 638.

39. S. I. Hayakawa *et al.*, *Phys. Rev. Lett.* **49** (1982) 1624.

40. K. Wienhard *et al.*, *Phys. Rev. Lett.* **49** (1982) 18.

41. M. Fujiwara *et al.*, *J. Phys.* **45** (1984) 453.

42. J. L. Tain *et al.*, *Phys. Rev. C* **35** (1987) 1288.

43. S. Muller *et al.*, *Phys. Rev. Lett.* **54** (1985) 293.

44. C. Gaarde, J. S. Larsen, and J. Rapaport, Spin excitations in nuclei, in *Proc. telluride conf.*, 1982 (Plenum Press, New York, 1984) p. 65.

45. A. Arima, K. Shimizu, W. Bentz, and H. Hyuga, *Advances in Nuclear Physics* **18** (1987) 1.

46. T. Wakasa, H. Sakai, H. Okamura, H. Otsu, S. Fujita, S. Ishida, N. Sakamoto, T. Uesaka, T. Satou, M. B. Greenfield, and K. Hatanaka, *Phys. Rev. C* **55** (1997) 2909.

47. G. F. Bertsch and I. Hamamoto, *Phys. Rev. C* **26** (1982) 1323.

48. N. D. Dang, A. Arima, T. Suzuki, and S. Yamaji, *Phys. Rev. Lett.* **79** (1997) 1638.

49. S. Kubonom, T. Kobayashi, and I. Tanihata, eds, *Proceedings of the fourth international conference on radioactive nuclear beams*, Omiya, Japan, 3–7 June 1996; *Nucl. Phys. A* **616** (1997) 1c-530c.

50. K. P. Jackson and A. Celler, *Proc. of spin observables of nuclear probes*, C. J. Horowitz, C. D. Goodman, and G. E. Walker, eds (Plenum, New York, 1988), p. 139.

51. T. Rönnqvist, H. Condé, N. Ollson, E. Ramström, R. Zorro, J. Blomgren, A. Håkansson, A. Ringbom, G. Tibell, O. Jonsson, L. Nilsson, P.-U. Renberg, S. Y. van der Werf, W. Unkelbach, and F. D. Brady, *Nucl. Phys. A* **563** (1993) 225.

52. N. Olsson, H. Condé, E. Ramstöm, T. Rönnqvist, R. Zorro, J. Blomgren, A. Håkansson, G. Tibel, O. Jonsson, L. Nilsson, P. -U. Renberg, A. Brockstedt, P. Ekström, M. Österlund, S. Y. van der Werf, D. J. Millener, G. Szeflinska, and Z. Szeflinski, *Nucl. Phys. A* **559** (1993) 368.

53. H. A. Bethe, G. E. Brown, J. Applegate, and J. M. Lattimer, *Nucl. Phys. A* **324** (1979) 487.

54. B. M. Sherrill, *Nucl. Phys. A* **616** (1997) 145.

55. I. Daito, H. Akimune, S. M. Austin, D. Bazin, G. P. A. Berg, J. A. Brown, D. S. Davis, Y. Fujita, H. Fujimura, M. Fujiwara, R. Hazama, T. Inomata, K. Ishibashi, J. Jänecke, S. Nakayama, K. Pham, D. A. Roberts, B. M. Sherrill, M. Steiner, A. Tamii, M. Tanaka, H. Toyokawa, and M. Yosoi, *Nucl. Inst. Methods*, **397** (1997) 465.

56. L. Wang, X. Yang, J. Rapaport, C. D. Goodman, C. C. Foster, Y. Wang, R. A. Lindgern, E. Sugarbaker, D. Marchlenski, S. de Lucia, B. Luther, L. Rybarcyk, T. N. Taddeucci, and B. K. Park, *Phys. Rev. C* **47** (1993) 2123.

57. F. Ajzenberg-Selove, *Nucl. Phys. A* **490** (1988) 1; **506** (1990) 1; **523** (1991) 1.

58. E. J. Ansaldo, J. C. Bergstrom, R. Yen, and H. S. Caplan, *Nucl. Phys. A* **322** (1979) 237.

59. T. Hosino, H. Sagawa, and A. Arima, *Nucl. Phys. A* **523** (1991) 228, and references therein.

60. Y. Kanada-En'yo and H. Horiuchi, *Phys. Rev. C* **55** (1997) 2860.

61. M. Fukugita and T. Yanagida, Physics of neutrinos in *Physics and astrophysics of neutrinos*, M. Fukugita and A. Suzuki, eds (Springer-Verlag 1994), p. 1.

62. S. P. Mikheyev and A. Yu. Smirnov, *Yad. Fiz.* **42** (1985) 1441 [*Sov. J. Nucl. Phys.* **42** (1985) 913]; *Prog. Part. Nucl. Phys.* **23** (1989) 41.

63. L. Wolfenstein, *Phys. Rev. D* **17** (1978) 2369.

64. W. C. Haxton, *Nucl. Phys. A* **570** (1994) 125, and references therein.

65. V. Castellani, S. Degl'innocenti, G. Fiorentini, M. Lissia, and B. Ricci, *Phys. Rep.* **281** (1997) 309.

66. For example, V. Berezinsky, *Comments Nucl. Part. Phys.* **21** (1994) 249.

67. D. Krofcheck *et al.*, *Phys. Rev. Lett.* **55** (1985) 1051.

68. H. Orihara *et al.*, *Phys. Rev. Lett.* **51** (1983) 1328.

69. W. Hampel *et al.*, *Phys. Rev. C* **31** (1985) 666; W. Hampel, in *Solar neutrino and neutrinos astronomy*, Proc. No 126 (API, New York, 1985), p. 162.

70. J. N. Bahcall, *Neutrino physics and astrophysics*, Cambridge University Press, 1989.

71. A. E. Champagne *et al.*, *Phys. Rev. C* **38** (1988) 900.

72. W. Trinder *et al.*, *Phys. Lett. B* **349** (1995) 267.

73. H. Ejiri and M. Fujiwara, Proposal for experiments at RCNP, 1995.

74. T. D. Lee and C. S. Wu, *Ann. Rev. Nucl. Sci.* **15** (1965) 381.

75. W. C. Haxton and G. J. Stephenson, *Prog. Part. Nucl. Phys.* **12** (1984) 409.

76. M. Doi, T. Kotani, and E. Takasugi, *Prog. Theor. Phys.* **83** (suppl.) (1985) 1.

77. H. Ejiri, *Nucl. Phys. A* **522** (1991) 305c; H. Ejiri, in *Physics and astrophysics of neutrino*, M. Fukugita and A. Suzuki, eds (Springer-Verlag, 1994), p. 500.

78. A brief article is given in *Science*, **276** (1997) 1795.

79. H. Ejiri *et al.*, *Phys. Lett. B* **258** (1991) 17.

80. S. R. Elliot *et al.*, *Phys. Rev. C* **46** (1992) 1535.

81. D. Dassie *et al.*, *Phys. Rev. D* **51** (1995) 2090.

82. H. Ejiri *et al.*, *J. Phys. Soc. Japan* **64** (1995) 339; K. Kume *et al.*, *Nucl. Phys. A* **577** (1994) 405.

83. H. Akimune, H. Ejiri, M. Fujiwara, I. Daito, T. Inomata, R. Hazama, A. Tamii, H. Toyokawa, and M. Yosoi, *Phys. Lett. B* **394** (1997) 23.

84. J. Engel, P. Vogel, and M. R. Zairnbauer, *Phys. Rev. C* **37** (1988) 731.

85. A. Griffiths and P. Vogel, *Phys. Rev. C* **46** (1992) 181.

86. H. Ejiri, *Nucl. Phys. A* **577** (1994) 399c; H. Ejiri, in *Proc. int. nucl. phys.* conf. (Beijing, 1995); H. Ejiri, in *Proc. WEIN'95* (*weak and electromagnetic interactions in nuclei*) (Osaka, 1995), H. Ejiri, Kishimoto, and Satoh, eds (World Scientific, 1995).

87. A. Moalem *et al.*, *Phys. Lett. B* **65** (1976) 167.

88. G. F. Bertsch, P. F. Bortignon, and R. A. Broglia, *Rev. Mod. Phys.* **55** (1983) 287.

89. S. Yoshida, *Prog. Theor. Phys.* 74 (suppl.) (1983) 142.

90. K. Okada *et al.*, *Phys. Rev. Lett.* **48** (1982) 1382.

91. A. Bracco *et al.*, *Phys. Rev. Lett.* **60** (1988) 2603.

92. S. Brandenburg *et al.*, *Phys. Rev. C* **39** (1989) 2448.

93. F. Zwarts, A. G. Drentje, M. N. Harakeh and A. Van der Woude, *Nucl. Phys. A* **439** (1985) 2448.

94. H. Ejiri, *J. Phys.* **45** (1984) 135, and reference therein.

95. C. Gaarde, J. S. Larsen, A. G. Drentje, M. N. Harakeh, and S. Y. van der Werf, *Phys. Rev. Lett.* **46** (1981) 902.

96. D. P. Knobles, S. A. Stotts, and T. Udagawa, *Phys. Rev. C* **52** (1995) 2257.

97. G. Coló, N. Van Giai, P. F. Bortignon, and R. A. Broglia, *Phys. Rev. C* **50** (1994) 1496.

98. P. F. Bortignon and N. Van Giai, *Phys. Rev. C* **54** (1996) 2088.

99. D. P. Knobles, S. A. Stotts, and T. Udagawa, *Phys. Rev. C* **54** (1996) 2090.

100. H. Sagawa, T. Suzuki, and N. Van Giai, *Phys. Rev. Lett.* **75** (1995) 3629.

101. M. N. Harakeh and M. Fujiwara *et al.*, Research proposal to RCNP for a study of the γ-decay from the Gamow–Teller resonance in ^{90}Nb (1993/1994).

102. M. Fujiwara, A. Kraznahorkay *et al.*, Research proposal to RCNP for a study of the γ-decay of the spin dipole resonance in ^{208}Pb (1996).

103. Preliminary data are shown by M. Fujiwara *et al.*, in *Proceedings of international workshop on physics with recoil separators and detector arrays*, R. K. Bhowmik

and A. K. Sinha, eds (New Delhi, 30 January–2 February, 1995), Allied Publishing Ltd.

104. J. L. Mildenberger, W. P. Alford, A. Celler, O. Häusser, K. P. Jackson, B. Larson, B. Pointon, A. Trudel, M. C. Vetterli, and S. Yen, *Phys. Rev. C* **43** (1991) 1777.

105. S. S. Hanna, *Nucl. Phys. A* **577** (1994) 173c.

4

Giant resonances and highly excited states

Y. Fujita

4.1 Introduction

4.1.1 *What is a giant resonance?*

A Giant Resonance (GR) is often observed as a broad structure in an excitation spectrum above the excitation energy $E_x = 10\,\mathrm{MeV}$. It is a fundamental mode of nuclear excitation interpreted to be a high-frequency surface vibration of a nucleus if we assume a phenomenological liquid drop model of a nucleus. There, a naive picture of a GR is the one-phonon excitation on the surface of a viscous liquid drop consisting of nuclear matter. It should be noted that the excitation energies of $E_x = 10$–$20\,\mathrm{MeV}$ correspond to oscillation frequencies of $10^{21}\,\mathrm{Hz}$, which are the fastest vibrations of a many-body quantum system. Since a number of nucleons participate in the vibration in phase, the GR has a collective nature. The degree of collectivity is denoted by the fraction of the total vibrational amplitude allowed for each mode, called the sum-rule limit value, is observed in a state.

Since a nucleus is a quantum system, basically each GR, like other excitation modes of a quantum system, is distinguished by a few quantum numbers. Representative ones are angular momentum, spin, and isospin as summarized in Fig. 4.1. Different angular momenta of the specific mode are associated with different shapes of vibration if we imagine that a nucleus is a liquid drop. The modes of excitation depending on the transferred angular momentum L are: $L = 0$, monopole modes; $L = 1$, dipole modes; $L = 2$, quadrupole modes; and so on. In this simple picture, modes associated with isospin excitation, i.e. an excitation in which protons and neutrons oscillate out of phase, are called IsoVector (IV) modes, while those without isospin excitation, i.e. protons and neutrons oscillating in phase, are called IsoScalar (IS) modes. Similarly, modes associated with spin excitation, i.e. spin-up and spin-down nucleons oscillating out of phase, are called magnetic modes, while those not associated with spin excitation are called electric modes—by analogy with electromagnetic transitions. In the multipole decomposition of the electromagnetic transitions, modes carrying angular momentum λ are called Mλ and Eλ transitions, respectively, for the magnetic and electric transitions. Thus, the corresponding GRs excited with angular momentum λ are also called Mλ and Eλ modes.

We should, however, be careful to note that this kind of visual scheme is neither always correct nor clear for a quantum system like a nucleus, especially when the mode is associated with spin or isospin excitations. In such cases, we have to come back to a

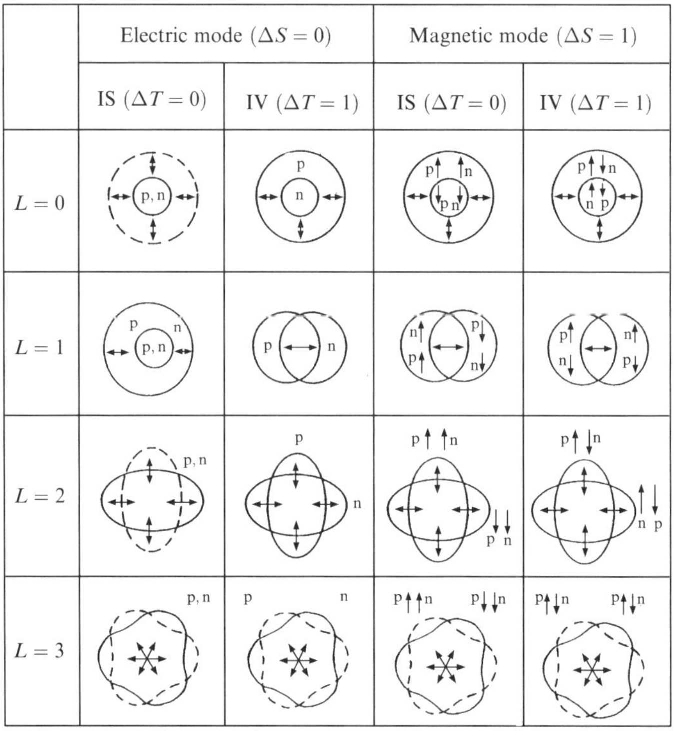

	Electric mode ($\Delta S = 0$)		Magnetic mode ($\Delta S = 1$)	
	IS ($\Delta T = 0$)	IV ($\Delta T = 1$)	IS ($\Delta T = 0$)	IV ($\Delta T = 1$)
$L = 0$	p, n	p n	p↑ ↑n p n	p↑ ↓n n p
$L = 1$	p n p, n	p n	n↑ p↓ p↑ n↓	p↑ n↑ n↓ p↓
$L = 2$	p, n	p n	p↑ ↑n p↓ n↓	p↑ ↓n n p
$L = 3$	p, n	p n	p↑↑n p↓↓n	p↑↓n p↑↓n

Fig. 4.1 Schematic illustration of various GR modes. Each mode is identified roughly by the transfer of angular momentum L, spin excitation ΔS and isospin excitation ΔT. For simplicity, the illustrations show the excitations of GR modes in the ground state of a spherical nucleus.

microscopic viewpoint assuming that a GR is described as a coherent superposition of one-particle–one-hole (1p–1h) excitations resulting from an action of a specific interaction. Thus, the mode of GR should be distinguished by the type of operator exciting the specific GR. For example, Eλ operators excite electric modes with angular momentum λ. These modes are further divided into IS and IV modes, where the former (the latter) is associated with (without) isospin operator. For example, for IS electric excitations, it is usual to assume an operator [1] with a form

$$O(Lm) = (Z/A) \sum_i r_i^L Y_m^L(\Omega), \quad L \geq 2, \tag{4.1}$$

where the summation runs over all the nucleons i, and L is the angular momentum (multipolarity) of the transition. Additionally, the τ operator, which can change isospin T, is associated with IV-type excitations, and the σ operator, which can flip spin S, with magnetic-type excitations. The form of the operators for each type of oscillation shown in Fig. 4.1 is given in Fig. 4.2. It is always important to think of an operator (or more

Microscopic classification of giant resonances

	$\Delta S = 0$ $\Delta T = 0$	$\Delta S = 1$ $\Delta T = 0$	$\Delta S = 0$ $\Delta T = 1$	$\Delta S = 1$ $\Delta T = 1$
$L = 0$		$\sum \tau_i$ IAS		$\sum \vec{\sigma}_i \tau_i$ GTR
Second order	$\sum r_i^2$ ISGMR	$\sum r_i^2 \tau_i$ IVGMR	$\sum r_i^2 \vec{\sigma}_i$ ISSMR	$\sum r_i^2 \vec{\sigma}_i \tau_i$ IVSMR
$L = 1$		$\sum r_i Y_m^1 \tau_i$ IVGDR	$\sum r_i Y_m^1 \vec{\sigma}_i$ ISSDR	$\sum r_i Y_m^1 \vec{\sigma}_i \tau_i$ IVSDR
Second order	$\sum r_i^3 Y_m^1$ ISGDR			
$L = 2$	$\sum r_i^2 Y_m^2$ ISGQR	$\sum r_i^2 Y_m^2 \tau_i$ IVGQR	$\sum r_i^2 Y_m^2 \vec{\sigma}_i$ ISSQR	$\sum r_i^2 Y_m^2 \vec{\sigma}_i \tau_i$ IVSQR
$L = 3$	$\sum r_i^3 Y_m^3$ ISGOR	$\sum r_i^3 Y_m^3 \tau_i$ IVGOR	$\sum r_i^3 Y_m^3 \vec{\sigma}_i$ ISSOR	$\sum r_i^3 Y_m^3 \vec{\sigma}_i \tau_i$ IVSOR

Fig. 4.2 Operators responsible for exciting representative modes of GRs and their abbreviated names. These are: IAS, isobaric analogue state; GTR, Gamow–Teller resonance; ISGMR, isoscalar giant monopole resonance; IVGMR, isovector spin monopole resonance; ISSMR, isoscalar spin monopole resonance, IVSMR, isovector spin monopole resonance ISGDR, isoscalar giant dipole resonance; ISGQR, isoscalar giant quadrupole resonance; ISGOR, isoscalar giant octupole resonance. The list is not complete, and only some of them have been found experimentally. (Courtesy of J. Jänecke.)

practically an active interaction) corresponding to a transition, in addition to having a rough idea of the oscillation shape given in Fig. 4.1, in order to understand the nature of a specific collective excitation.

Assuming a harmonic-oscillator shell model, one sees that the operator $O(Lm)$ can only induce 1p–1h excitations with $\Delta N < L$. Since the parity changes as $(-)^{\Delta N}$, its conservation requires that an odd-L transition is allowed only for odd ΔN and vice versa. For example, E2 (electric quadrupole) transition is allowed for $0\hbar\omega$ and $2\hbar\omega$ jumps and E3 for $1\hbar\omega$ and $3\hbar\omega$ jumps. This is shown in Fig. 4.3. A more detailed microscopic description of GR is given in a review paper [2].

4.1.2 *Sum rule and general features of a giant resonance*

In the naive picture of a nucleus as consisting of a limited number of neutrons and protons filling specific orbits in the nuclear potential, the number of nucleons (degrees of freedom) allowed to participate in each excitation mode is limited, depending on the quantum numbers of the excitation mode. Thus, the transition amplitude of each mode of a nuclear vibration has a maximum value. Correspondingly, the excitation strength of

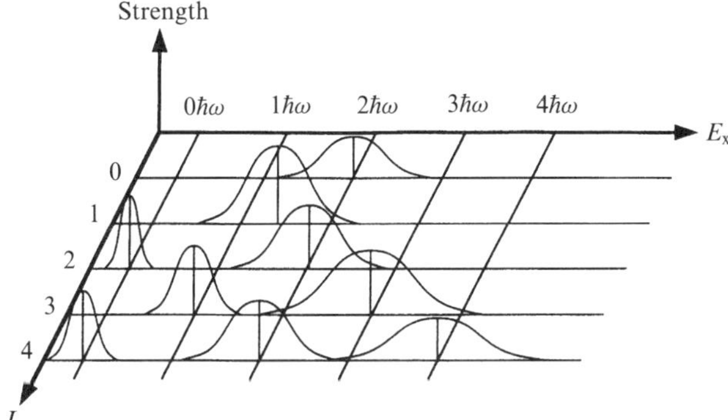

Fig. 4.3 Schematic strength distributions of nuclear excitations for $L = 0\text{--}4$ modes as a function of excitation energy.

the mode observed by using a specific nuclear reaction should have a certain limit. The maximum strength is called the sum-rule limit. It is, therefore, convenient to express the observed strength of a GR by the percentage of the sum-rule limit that is exhausted by a state or by a bump-like structure of GR. In order to calculate the sum-rule limit, it is usually necessary to assume some nuclear model, and thus the sum-rule limit may depend on the nuclear model being used. In order to avoid this, an Energy-Weighted Sum Rule (EWSR) is often used for the electric-type transitions. It is known that the total EWSR value S, i.e. each transition strength multiplied by the excitation energy and summed over for all final states, can be calculated in a model-independent way by using a double commutation relationship between the total Hamiltonian and the relevant operator. For the magnetic transitions, the sum rule is more complicated and more dependent on each type of transition. More details on sum rule can be found in some of the review papers [3,4].

Keeping the image of GRs as representing the vibrations of nuclear matter, the main features of a vibrational mode to be called GR, especially of electric type, are summarized [3] as follows: (a) The excitation of the mode is a general feature observed in a wide variety of nuclei (with the possible exception of light nuclei). (b) The observed strength exhausts some large fraction of the strength expected by the sum rule for a specific vibrational mode of a quantum nuclear system. (c) The strength concentrates in a rather small excitation-energy interval, i.e. the width Γ is smaller than the centroid energy E_c. (d) E_c and Γ are more or less smooth functions of the mass number A. Particularly, E_c is nearly proportional to $A^{-1/3}$, reflecting the fact that the resonance frequency of an elastic body decreases as the radius of the body increases.

All of these features are usually true for GRs in a larger-A nucleus and for smaller-L transitions ($L \leq 3$). As will be reviewed in Sections 4.2 and 4.5, IS giant monopole resonance ($L = 0$), IV giant dipole resonance ($L = 1$), and IS giant quadrupole resonance ($L = 2$) have these features. In $L = 3$ resonances, the strength is divided into $1\hbar\omega$ and $3\hbar\omega$ components, but each of them still seems to retain the features (see Sections 4.2, 4.4, and 4.5). For $L \geq 4$ transitions, the expected transition strength fragments over a large range of excitation energy. For example, $L = 4$ strength fragments over $0\hbar\omega$, $2\hbar\omega$, and $4\hbar\omega$ transitions. As described in Section 4.4, $0\hbar\omega$ mode shows a clear resonance shape in the heavy nucleus ^{208}Pb, where the existence of $2\hbar\omega$ mode is also known. However, whether or not the $2\hbar\omega$ mode exists in lighter nuclei is not so clear; probably it does not. It seems that the degree of freedom of a nuclear system, i.e. the nucleon number, is too small to form a complicated shape of a large-L vibration at the surface of a lighter nucleus. For a lighter nucleus, nuclear shell effects and the influence of the nuclear surface become more important and the specific features of each individual nucleus become more apparent.

Isovector-type magnetic resonances show peculiar features, which is unlike the IS-type resonances. The isospin structures of Gamow–Teller and M1 resonances are discussed in Section 4.7.

4.1.3 *The width of a giant resonance*

As we have seen, a GR is usually observed as a broad bump sitting on a background. A rather large energy-width Γ of about a few to several MeV is related to the relaxation time (decay time) Δt of an oscillation through the uncertainty relationship $\Delta t \times \Gamma = \hbar$. If a vibration is harmonic, it lasts forever, but this is not the usual case. Like the vibration of a solid body, a vibration of a nucleus also damps due to the anharmonic effects. Microscopically, a GR is described as a collective superposition of 1p–1h excitations built on a base state (in a usual experiment on the ground state), i.e. 1p–1h excitations occur in-phase (coherently). Since each 1p–1h excitation has its own oscillation frequency (excitation energy) different from the others, it is easy to expect that 'in-phase' oscillations do not last long.

There are two kinds of mechanisms that destroy the coherence of the 1p–1h excitations. The first is particle decay. Usually, a GR is situated at a high excitation energy above the proton or neutron particle threshold energy. If particle decay occurs, the 1p–1h nature of the GR is lost and thus the GR decays in a short period. Correspondingly, the decay width denoted by $\Gamma^{\uparrow}$ occurs. The second mechanism is the decay of the 1p–1h excitation by spreading its energy to a more complicated 2p–2h excitation by nuclear collisions among nucleons, and then finally to an equilibrium many-particle–many-hole excitation. The width caused by such a decay process is called the spreading width and it is denoted by $\Gamma^{\downarrow}$. Due to the strong nuclear force acting among the nucleons, the nuclear vibration shrinks after several oscillations. The number of oscillations is roughly estimated to be E_{c}/Γ.

After emitting particles, a nucleus becomes rather cold and the excitation energy is below particle separation energy. Then the γ-radiation (caused by the electromagnetic interaction) is the dominant deexcitation process of a nucleus. (After the emission of γ-rays, a β-decay can follow.) Since the γ-radiation process is very slow compared to the particle decay, the decay width caused by it is usually very small, of the order of 1 eV. Then a resonance situated below or around the particle separation energy should be observed as clustering discrete states. As an example of such a resonance, features of low-energy octupole resonance are described in Section 4.4.

The observation of particle decay from a GR is discussed in Section 4.6. In some cases, particle emission is largely hindered by some selection rule, i.e. by the mismatch of quantum numbers before and after the decay, even if the decay is allowed energywise. Observation of sharp peaks at excitation energies much higher than the particle separation energy is discussed in Section 4.7 in relation to isospin selection rules. A description of the width and decay of GRs can be found, for example, in a review paper by Knöpfle [5].

4.2 Experimental findings of giant resonances

From an experimental side, the most important point in establishing a new mode of GR is how to observe a specific mode selectively from many possible nuclear excitation modes overlapping each other in their excitation energies. Thus, for the study of each mode, a selective way of excitation and a selective way of observation have been considered. Here, the experimental findings of representative GRs are reviewed briefly mainly for the electric (non-spin exciting) modes.

4.2.1 *The isovector giant dipole resonance*

The first GR observed in nuclei was the Giant Dipole Resonance (GDR). It is an $L = 1$, IV-type electric (spin-non-flip) ($\Delta T = 1$ and $\Delta S = 0$) GR with spin and parity $J^\pi = 1^-$, if the excitation is on $J^\pi = 0^+$ ground state in an even-Z and even-N nucleus. It was established already in the mid-1960s [6]. The photo-absorption cross section was measured as a function of incident photon energy. It was found that the cross section has a huge bump structure with centroid of the excitation energy (E_c) at 10–20 MeV, dependent on the mass number.

The high-energy photon necessary to excite the GDR was usually created by the bremsstrahlung process using electrons from electron accelerators already much developed in the 1950s. When the wavelength of a photon is large compared to the nuclear radius, the main interaction of the photon with the nucleus is of electric dipole type, because a massless photon can bring in only very small angular momentum. Additionally, since the electric field associated with a photon interacts only with the protons of the nucleus, the photon is an ideal medium for the excitations of the IV-type, electric $L = 1$ (E1) mode of the GR. Since the main body of the GDR is situated above the neutron threshold, a photo-excited GDR decays mainly by neutron emission. Thus, in an actual measurement, the absorption cross section is usually deduced from a measurement of the yields of neutrons decaying from the GDR (for details of particle decay from a GR, see Section 4.6). The GDRs with spin-parity $J^\pi = 1^-$ were observed in nuclei for almost full range of A as a bump-like structure (see Fig. 4.4). Fine structures due to shell effects are observed for the bumps in very light nuclei, but in heavier nuclei the bumps show shapes which fit well to Breit–Wigner- (Lorentzian-) type resonance distributions,

$$y(E) = y_0 \frac{(\Gamma/2)^2}{(E - E_c)^2 + (\Gamma/2)^2}, \tag{4.2}$$

where Γ shows the full-width-at-half-maximum of the resonance shape. The E_c values follow the $80A^{-1/3}$ systematics rather well ($E_c = 15$–25 MeV) particularly in heavy-mass nuclei, and the widths Γ are 3–5 MeV. It should be noted that the GDR is an

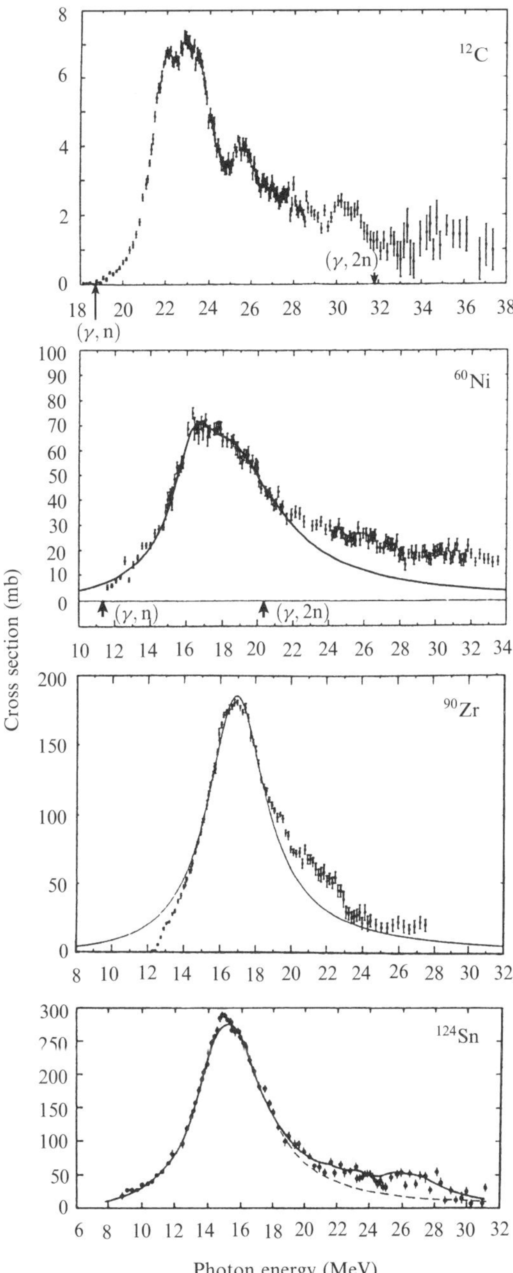

Fig. 4.4 The total photo-neutron cross sections for the nuclei ^{12}C, ^{60}Ni, ^{90}Zr, and ^{124}Sn. Structureless resonance shapes are observed for larger-A nuclei. For details of the curve-fits, see Ref. [6]. (From Ref. [6].)

excitation mode with $1\hbar\omega$ jump. Since the energy difference $1\hbar\omega \approx 41 A^{-1/3}$ MeV, we see that the observed excitation energy is much higher than that expected from a naive shell model.

The IS-type electric dipole excitation corresponds to a centre-of-mass (translational) motion of a nucleus. Thus, the mode is spurious in the sense of a first-order effect. If the second-order effect considering the density oscillation is taken into account, the ISGDR can exist as the 'squeezing mode' [7]. Recently, the existence of ISGDR was suggested at $E_x = 22.5$ MeV in ^{208}Pb by using a small-angle inelastic α-scattering including the measurement at $0°$ [8].

4.2.2 The isoscalar giant quadrupole resonance

In order to excite a large-L mode, such as the $L = 2$ quadrupole mode, not only the energy to excite the mode but also the angular momentum to orient the nucleons should be given to a nucleus. Further study on the GR was stimulated from results of inelastic-type reactions. In the early 1970s, electron inelastic-scattering experiments suggested the existence of higher-L modes [9–11] in addition to the strongly excited GDR. But soon it was recognized that the inelastic experiments using light 'hadron' beams, like proton (p), deuteron (d), or α-beams, were more suited for the excitation of usual surface-vibration modes of nucleons in which protons and neutrons are not discriminated and thus move together (IS mode) [12]. Electrons interact with a nucleus through the Coulomb force (and thus interact only with protons directly), while hadrons interact mainly through the strong force. The charge symmetry nature of the strong force is suited for the excitation of the IS mode without being disturbed by the IV-type GDR so much.

For the study of IS electric ($\Delta T = 0$ and $\Delta S = 0$) excitations, light-ion beams with an energy of 20–70 MeV/nucleon (MeV/u) are often used. At these energies, it is known that the reaction is favourable for exciting these modes, because (1) nuclear interaction is suited for the IS and spin-non-flip excitations, and (2) the reaction mechanism is mainly one-step dominant so that the excitation is mainly of 1p–1h nature. Among light-ion beams, the IS and spinless particle α ($T = 0$ and $S = 0$) is often favoured. A typical energy spectrum observed in an (α, α') reaction, showing a few GRs at the same time [13], is given in Fig. 4.5. Each bump structure with different transition L shows different angular distribution of cross section. The analysis of the angular distribution is very important for the identification of the L nature of each GR as well as for the estimation of the fraction of the EWSR [14].

Among the $L = 2$ modes of GRs, the IS electric quadrupole (E2) mode is called simply the Giant Quadrupole Resonance (GQR). It has spin and parity $J^\pi = 2^+$, if the excitation is on the $J^\pi = 0^+$ ground state. A systematic study of the GQR was performed using the IS particle α as an incident beam [15]. For $A \geq 40$ nuclei, the GQR is observed as a broad bump at $E_x \approx 15$ MeV as shown in Fig. 4.6, while in $A < 40$ nuclei its existence was not so clear, suggesting that the collectivity of the mode is low. The strength exhausts 60–100% of the IS E2 EWSR in heavier nuclei, while in lighter nuclei, the EWSR fraction is less than 50%, as shown in Fig. 4.7. The observed width Γ of the GQR is large (≈ 5 MeV) in light nuclei, while it is small (≈ 3 MeV) in heavy nuclei. It was found that the width is roughly proportional to $A^{-2/3}$. A significantly larger

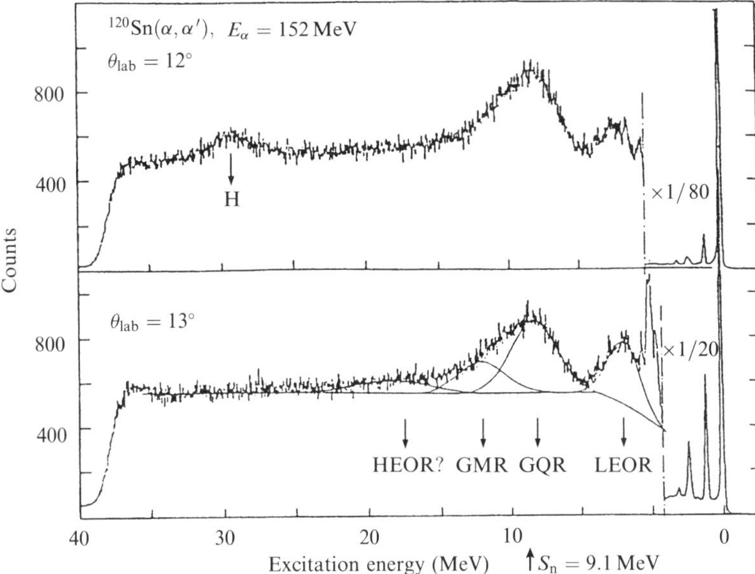

Fig. 4.5 Isoscalar electric GRs excited in the ^{120}Sn by (α, α') reaction at $E_\alpha = 152$ MeV. A possible decomposition into a continuum and different multipoles is indicated in the lower spectrum. (From Ref. [13].)

width is seen in the rare-earth region of deformed nuclei. The E_c value roughly follows the $65A^{-1/3}$ MeV systematics. The GQR is an excitation mode with $2\hbar\omega$ jump. If we remember that the energy difference $1\hbar\omega$ is approximately $41A^{-1/3}$ MeV, the E_c value is much lower than that expected from a naive shell model. We have seen that the strength of the IVGDR is raised higher in energy, while that of the ISGQR is pushed lower. The upward and downward redistributions of strengths in IV and IS modes, respectively, are due to repulsiveness of the IV interaction and the attractiveness of the IS interaction.

The IV E2 mode is expected at $E_c \sim 130A^{-1/3}$ MeV. Experimentally, however, the existence of the IV E2 (and also higher-L modes) is not so well established as that of the IS modes.

4.2.3 *The isoscalar octupole resonance*

Through a similar analysis of the angular distribution of the (α, α') reaction performed for the GQR, it was found that about 20% of the E3 EWSR is localized in a peak at $E_x \approx 7$ MeV with a width $\Gamma = 1$–3 MeV [16]. As shown in Fig. 4.8, the broad peak was commonly observed in heavier nuclei from ^{66}Zn to ^{197}Au at $E_c = 30A^{-1/3}$ MeV, but not in ^{40}Ca and lighter nuclei. It is strange that in the heaviest nucleus ^{208}Pb, no broad peak was observed at this excitation. The strength has been identified as the low-energy ($1\hbar\omega$ part) E3 excitation ($J^\pi = 3^-$, if the excitation is on $J^\pi = 0^+$ ground state) and is called the Low-Energy Octupole Resonance (LEOR).

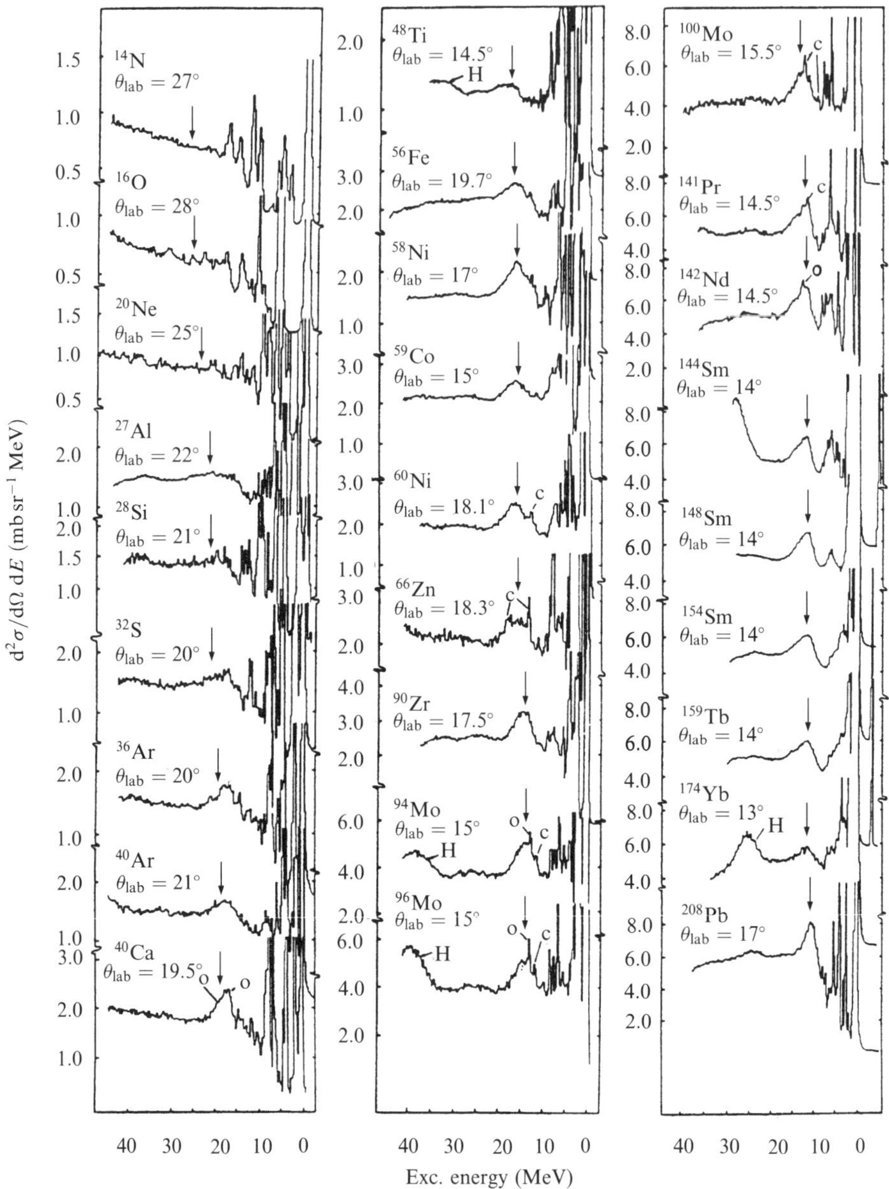

Fig. 4.6　Inelastic α-spectra for various nuclei at incident energies $E_\alpha = 96$ or $115\,\mathrm{MeV}$. A broad GQR bump is observed clearly in each nucleus heavier than ^{36}Ar. The arrow marks the centre of the GQR bump E_c, which is expected from the energy systematics. (From Ref. [15].)

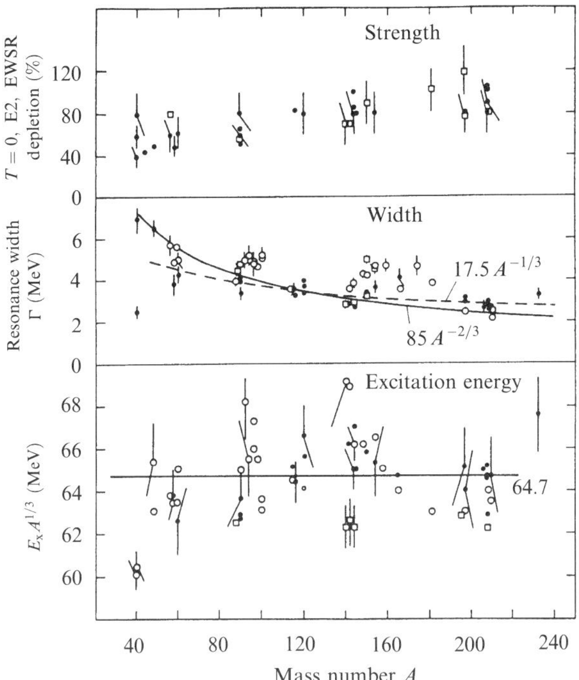

Fig. 4.7 Systematics of the experimentally observed values for the GQR in nuclei with $A \geq 40$. These are excitation energy E_x (= centre of resonance structure E_c), width Γ, and the summed fraction S of the EWSR. (From Ref. [3].)

The energy region around LEOR ('LEOR region') was studied further using high-resolution (p, p′) and (α, α') reactions. Later development of the study on the LEOR, including the reason for the missing strength in ^{208}Pb, is described in Section 4.4.

The main part (about 75%) of the $L = 3$ excitation is expected at the excitation energy of $3\hbar\omega$, i.e. around the 25–30 MeV energy region. At this high excitation energy, the width becomes greater and the peak height becomes lower, and thus the detection becomes much more difficult than for the previously mentioned GRs. Experimental efforts are still needed for the analysis of this mode [17]. More details on this mode are also given in Section 4.5.

4.2.4 *The isoscalar monopole resonance*

The E_c value of the IS E0 mode, the $J^\pi = 0^+$ Giant Monopole Resonance (GMR), roughly follows the $80A^{-1/3}$ MeV systematics in heavy, $A > 150$, nuclei ($E_c = 13.7$ MeV in ^{208}Pb). The value is somewhat larger than that of the GQR, but the bump structures of these two resonances overlap each other. The angular distribution of an $L = 0$ mode is expected to show its maximum at $0°$ and to decrease drastically toward backward angles. As seen from Fig. 4.5, which shows the spectra at $\theta_{lab} = 12°$ and $13°$,

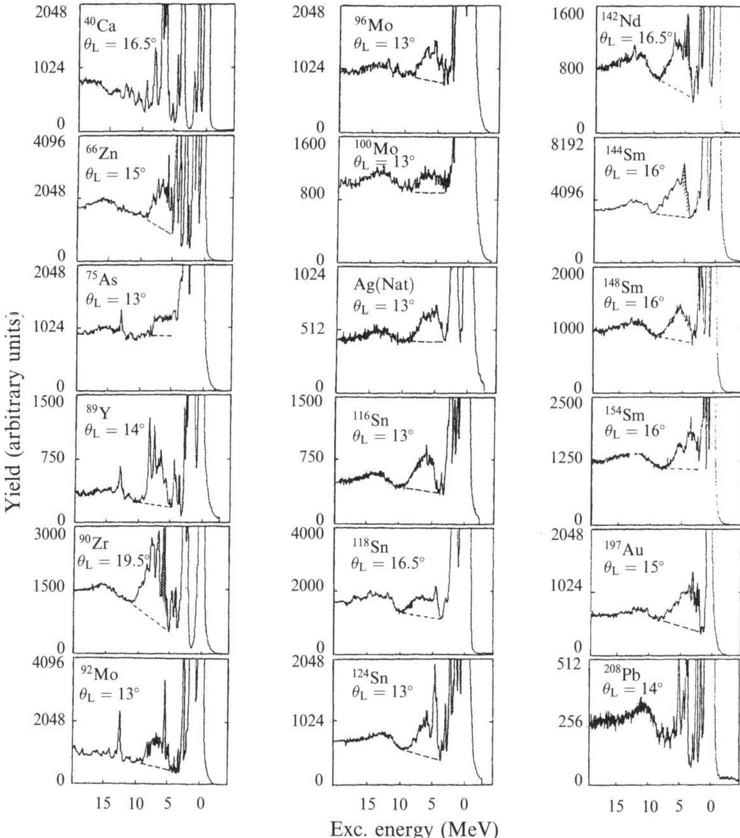

Fig. 4.8   Inelastic α-spectra at incident energies $E_\alpha = 96$ or 115 MeV, showing the LEOR region for various nuclei at the θ_{lab} showing a maximum for $L = 3$ transition. A LEOR peak, when present, is above the dashed background line. (From Ref. [16].)

it is difficult to identify the GMR bump without measuring the spectra at very forward angles.

Measurements of small-angle (including $0°$) inelastic scattering, however, are not at all easy, because both inelastically scattered particles and the incident beam travel in the same direction and the former should be distinguished from the latter—a million times more intense beam. With an improvement of the experimental techniques to measure an excitation spectrum at small angles in an inelastic reaction, i.e. mainly owing to the efforts of reducing the background associated with a small-angle inelastic scattering measurement, the GMR was established by using the (α, α') reaction [18,19] as well as by using ^{3}He inelastic scattering [20]. The two-component fits of GR region in the ^{116}Sn(α, α') spectra at small angles are shown in Fig. 4.9.

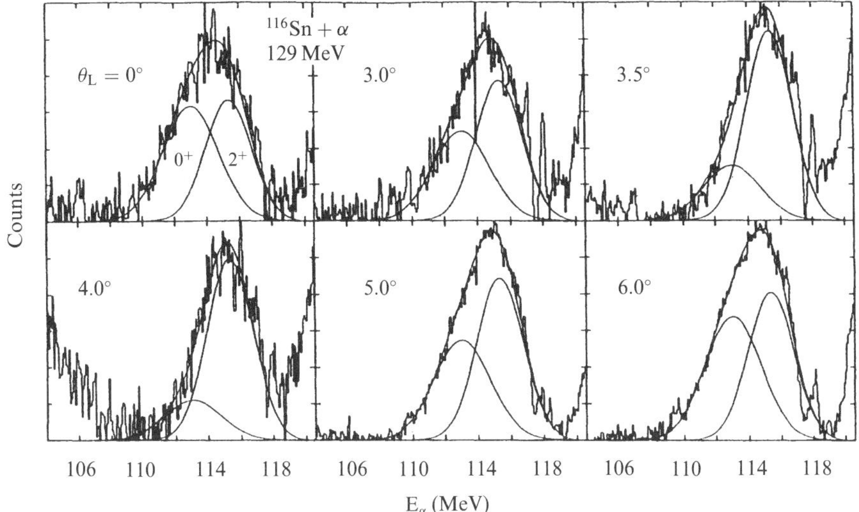

Fig. 4.9 The GR region of the ^{116}Sn(α, α') spectra at small angles including at $0°$. The continuum background is subtracted. The bump structure is decomposed into two components assuming the existence of the GMR (0^+) and the GQR (2^+). (From Ref. [19].)

Since the GMR is a 'breathing' mode of the nucleus, the oscillation frequency, or excitation energy, is related to the nuclear compressibility at the surface of a nucleus, which remains an important problem [21].

4.3 Characteristic incident energies for giant resonance studies

Hereafter, the works performed at RCNP are reviewed. Sections 4.4 and 4.5 present the results of studies on electric modes. The AVF cyclotron at RCNP provides various light ions with an energy of 20–80 MeV/u. It is known that in this energy region of the incident particle, the nuclear force is favourable for the excitation of IS electric modes rather than the spin and/or isospin modes.

Section 4.4 describes the fine structure of GR-like collective modes situated in the region of excitation energy below or around the particle separation energy. There the broadening effect of particle decay for a state is not so large, i.e. the decay width $\Gamma^\uparrow$ is small, and thus each state forming the GR-like structure is still sharp. In these modes, the strength distribution is decided by the spreading width $\Gamma^\downarrow$.

On the other hand, Section 4.5 is devoted to the description of GRs situated at much higher excitation energy. Since the decay width $\Gamma^\uparrow$ becomes large for these modes, they are usually observed as a broad bump structure. Study of decay modes also becomes important in addition to the study of their excitation. Section 4.6 discusses the decays of GRs and highly excited states.

At beam energies exceeding 100 MeV/u, on the other hand, the part of the effective interaction exciting the spin and isospin modes ($V_{\sigma\tau}$ interaction) becomes more prominent. At RCNP, these beams are provided by a separated sector cyclotron (so-called ring cyclotron) boosting the energy of the beam from the AVF cyclotron by about a factor of 4–6.

The $L = 0$ Gamow–Teller and M1 giant resonances ($J^{\pi} = 1^{+}$) are the well-known spin–isospin excitation modes. In these modes, the isospin quantum number T is closely connected with their structures as well as with their decay. In Section 4.7, isospin structures revealed through the study of fine structure are discussed for these $L = 0$ spin–isospin excitation modes.

4.4 Fine structure of electric giant resonances

4.4.1 *Observations of resonance structures consisting of discrete levels*

The magnetic spectrometer RAIDEN [22,23] is a momentum analyser for particles ejected out after a nuclear reaction. It is designed for the energy region of the AVF cyclotron. Owing to its large dispersion and high resolution, it is used for the fine-structure study of nuclear levels. The spectrometer consists of several magnetic components having different functions, in total realizing a good focus and a high resolving power at the focal plane. The schematic drawing of the magnet configuration is shown in Fig. 4.10.

As the excitation energy increases, the nuclear level density increases. Owing to the high resolving power of the spectrometer, the region of excitation energy that could be analysed was extended from the so-called low-lying region to the higher excitation region around the particle separation energy, i.e. the region of some of the GRs, like the LEOR. In the particle unbound region, the intrinsic level-width of each state usually becomes greater and even high resolution is unable to resolve the states, except for some cases where the particle decay is hindered by some conservation law.

4.4.2 *The low-energy octupole resonance*

From the above overview, it is clear that the interesting energy region for the fine-structure study is the region below or around the particle separation energy, i.e. the region of LEOR. As already mentioned, LEOR is the $1\hbar\omega$ mode of the E3 excitation and lies in the excitation energy region 5–10 MeV. It was observed as a bump structure on top of the so-called background (see Fig. 4.8), like the case of other GRs when it was originally found in an (α, α') experiment with a resolution of more than 100 keV [16]. From the consideration of the decay mechanism of states, however, we can expect that the strength of LEOR should be found in discrete levels up to around the excitation energy of 10 MeV, where the particle decay is forbidden or hindered. A systematic study was performed using high-resolution (p, p′) reaction with an energy resolution of 10–20 keV to investigate the fine structure of states in the LEOR region.

The LEOR in ^{90}Zr had been observed as a bump at $E_x = 7$ MeV in the (α, α') experiment [16]. The simple $L = 3$ assignment of the bump structure of LEOR in ^{90}Zr, however, was questioned on the basis of a (p, p′) study by the Grenoble group from the analysis

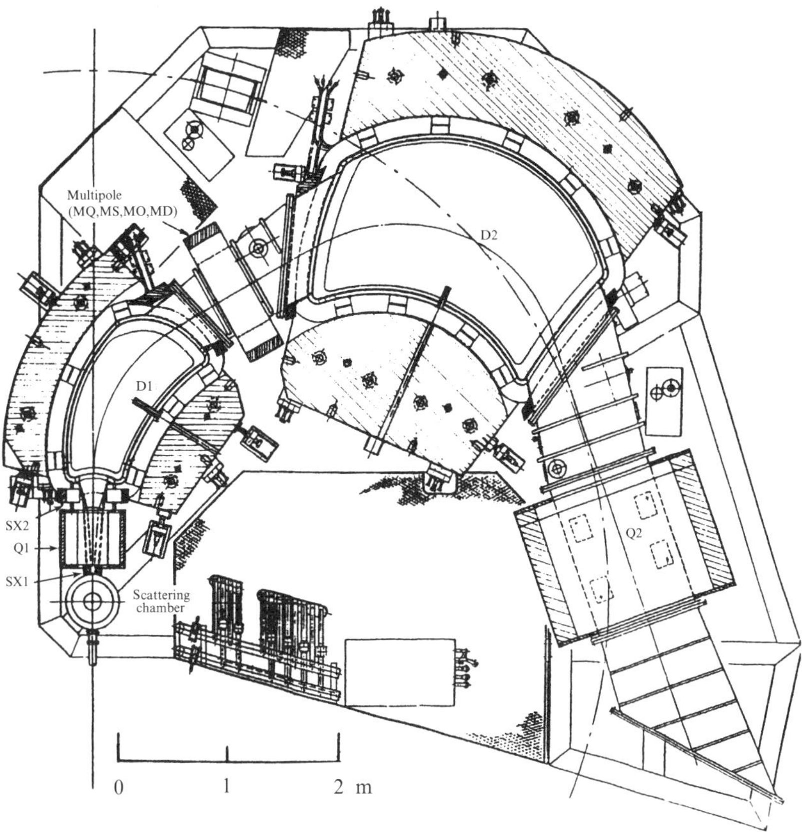

Fig. 4.10 Geometry of magnetic spectrometer RAIDEN. It consists of two dipole magnets D1 and D2, and quadrupole magnets Q1 and Q2 which diverge the beam in the horizontal direction. The multipole magnet MP is placed at the middle focusing point in the vertical y-direction and is used for efficient correction of kinematic broadening effects. (From Ref. [22].)

of the differential cross section and the analysing power of the bump [24]. The answer to this question about the nature of the 'LEOR bump' in ^{90}Zr was given by a high-resolution (p, p′) study [25], and was later confirmed in the study on ^{48}Ca [26]. In these studies: (a) a highly structured spectrum was observed up to more than $E_x = 10$ MeV (an example of a structured spectrum is shown in Fig. 4.11 for ^{48}Ca); (b) the width of each level was not much greater than that of a low-lying state; (c) the amount of 'background' was much lower than that observed in a low-resolution experiment observing the LEOR as a bump (for example, compare the $E_x = 5$–12 MeV region in Figs 4.5 and 4.11); (d) angular distribution analysis of each level revealed that the number of $L = 3$ states is the largest, but many $L = 4$ states and some $L = 5$ states coexist in the LEOR region (examples of angular distribution are shown in Fig. 4.12); (e) the envelope of the EWSR

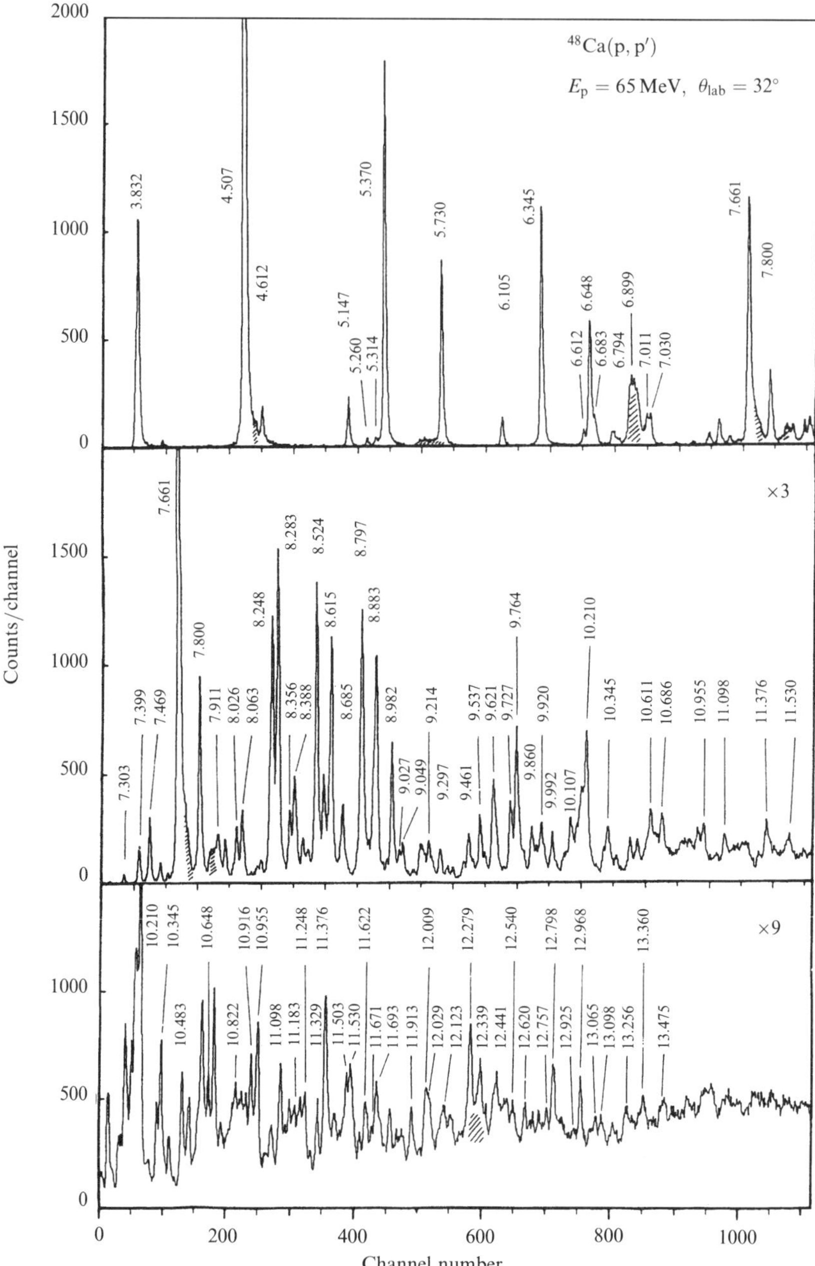

Fig. 4.11   The 15 keV resolution spectra of inelastically scattered protons on a ^{48}Ca target from the first excited 2^+ state to the $E_x = 14$ MeV region. With the high resolution, discrete peaks are observed for the whole energy range examined here. The peaks are labelled with their excitation energies in MeV. (From Ref. [31].)

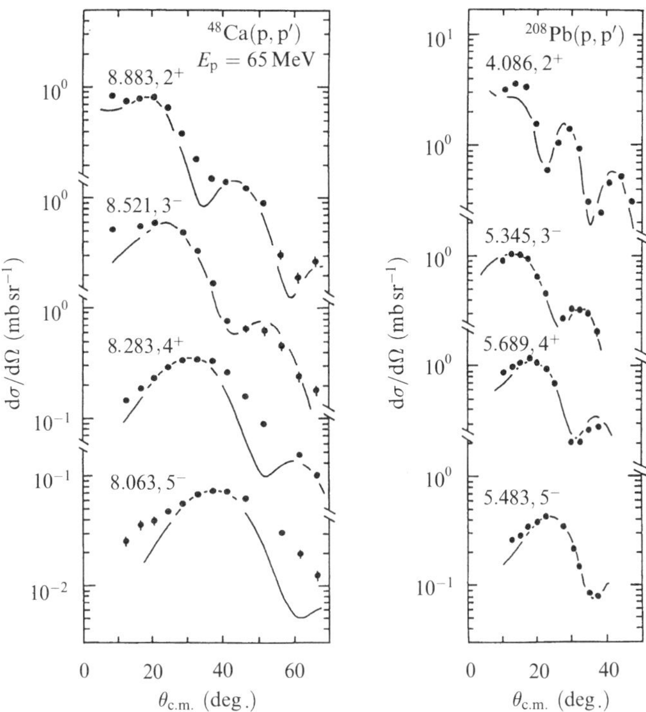

Fig. 4.12 The angular distributions of the differential cross sections for various L states in the LEOR region obtained in ^{48}Ca and ^{208}Pb(p, p$'$) reactions, and the fit with the DWBA calculations (solid curves). (From Ref. [26].)

strength distribution of the $L = 3$ states showed a bump-like structure (see Fig. 4.13). The findings (b) and (c) are consistent with the fact that the particle decay is not allowed in the LEOR region of ^{90}Zr or of ^{48}Ca, because the neutron separation energy S_n is as high as 12.0 and 9.9 MeV, respectively, in the nuclei. Thus, no decay width is expected for each level, and additionally a real physical background caused by the quasi-free scattering should not be observed below an excitation energy S_n.

The systematic observation of LEOR as clustering $L = 3$ states was extended to the doubly magic nucleus ^{208}Pb [26], to neutron number $N = 28$ singly closed shell nuclei ^{50}Ti, ^{52}Cr, and ^{54}Fe [27], and to ^{58}Ni [28]. A clear feature of the octupole strength in all these nuclei was the following: (a) there exists a (or a few) low-lying (so-called collective) 3$^-$ state(s) at $E_x = 2.5$–4.5 MeV exhausting 5–20% of the EWSR value; (b) fragmented $L = 3$ states (LEOR) are observed at around $E_x = 6$–9 MeV, the centre of which follows the systematics $E_x = 32A^{-1/3}$ well; (c) they exhaust, in total, 10–15% of the EWSR value; (d) the envelope of the fragmented state shows a bump structure of a usual Breit–Wigner-type distribution with a width $\Gamma = 0.7$–2.0 MeV.

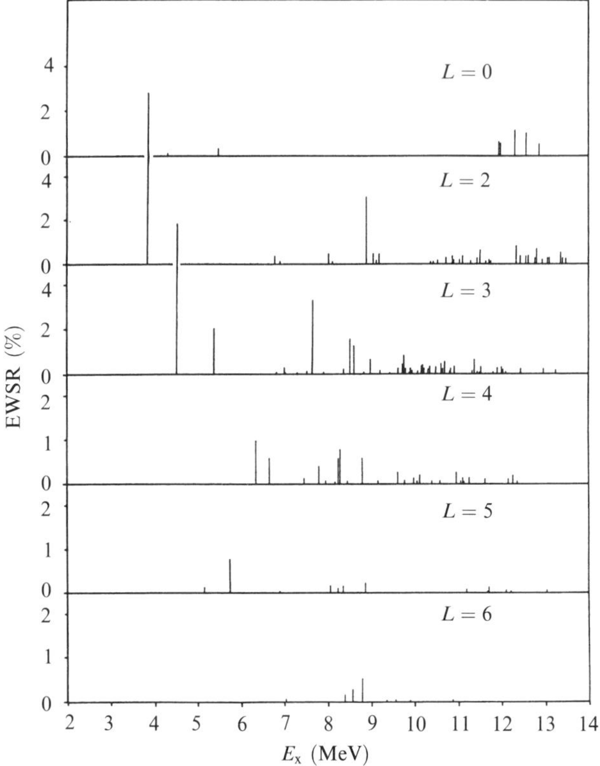

Fig. 4.13 The distribution of the EWSR percentages observed in the ^{48}Ca(p, p′) reaction for various L transfers. (From Ref. [31].)

A clever way to distinguish a bump-like resonance structure consisting of discrete states is to make a cumulative-sum plot of strengths. Then the information on the distribution and the strength scattered in the strength distribution is integrated. In this plot, a resonance structure is expected to have a shape like a stretched S. Such a plot is shown in Fig. 4.14 for ^{48}Ca as a typical example. The stretched-S shape is clearly observed for the cumulative sum of the $L = 3$ strength. The cumulative-sum curve is fitted with the integrated Breit–Wigner function

$$Y(E) = Y_0 \left[\frac{1}{2} + \frac{1}{\pi} \tan^{-1/2} \left(\frac{E - E_c}{\Gamma} \right) \right] \tag{4.3}$$

in Fig. 4.14 to obtain E_c as well as Γ. We found $E_c = 9.7\,\text{MeV}$ and $\Gamma = 2.3\,\text{MeV}$.

As we have seen, high-resolution experiments made it possible to observe a resonance structure, not as a bump structure but as a clustering of states. Through high-resolution studies on the high-level-density LEOR region near $E_x = S_n$, interesting aspects in

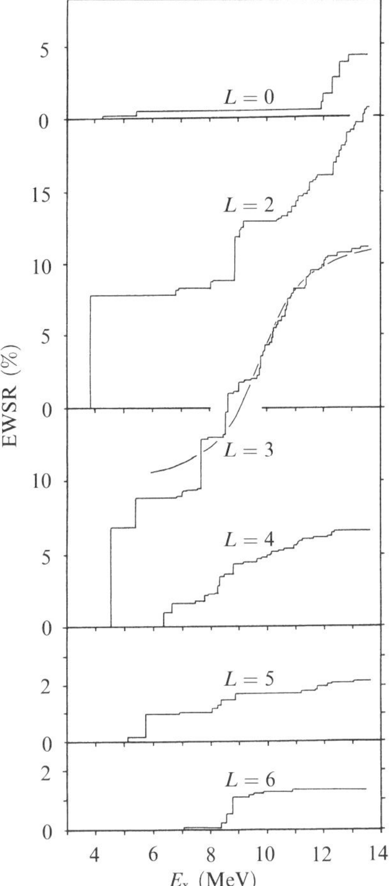

Fig. 4.14 The cumulative sum of EWSR percentages for $L = 0, 2, 3, 4, 5, 6$ strengths observed in the ^{48}Ca(p, p$'$) reaction. (From Ref. [31].)

this region became clear: (a) the hidden LEOR structure in ^{208}Pb; (b) the nature of the underlying so-called background associated with usual low-resolution experiments.

In the original systematic study using the (α, α') reaction [16], the LEOR was observed in a wide range of nuclei from fp-shell nuclei to gold, but, strangely enough, not in ^{208}Pb. In the high-resolution study on ^{208}Pb using the (p, p$'$) reaction [26], however, many $L = 3$ states were observed in the LEOR region in the range $E_x = 4.5$–$7.5\,$MeV. The total strength was about 15% of the EWSR, a typical value for a LEOR. The study triggered a discussion as to why the LEOR was observed in the (p, p$'$) reaction but not in the (α, α') reaction in ^{208}Pb. Theoretically, the difference in the isospin nature of the two probes was discussed [29], because the (α, α') can excite only IS strength, while the (p, p$'$) can excite both IS and IV strengths. Experimentally, however, it was suspected that the key point was the difference in the resolution of the two experiments. An (α, α') experiment with a higher resolution of 25 keV was performed [30]. As a result of angular distribution analysis, it became clear that the major $L = 3$ states found in the (p, p') experiment were also observed in the (α, α') experiment (see Fig. 4.15). Thus, the existence of the LEOR in ^{208}Pb was finally confirmed.

As mentioned above, real continuous background caused by the quasi-free scattering process is expected above the excitation energy of some kind of particle separation energy (usually neutron separation energy S_n). In a method that analyses a GR as a bump, the amount of background subtracted was usually much more than that expected from the real continuous background (below S_n, no real background is expected!). Then there arose the question as to the nature of the subtracted background in such an analysis. The answer was that the $L = 3$ strength was not the only one in the LEOR region. A typical example of such a spectrum is ^{48}Ca for which a rather low level density is expected [31]. The lowest particle separation energy is that of a neutron, which is found at $S_n = 9.9$ MeV. It is clear from the spectrum shown in Fig. 4.11 that practically no background is found below $E_x = S_n$, and, as expected, only above this excitation energy the continuous background underlying the sharp peaks increases. The strength distribution given in terms of the EWSR is shown in Fig. 4.13. Although $L = 3$ is the major strength in the LEOR region at $E_x = 6.5$–12 MeV, it is clear that many other

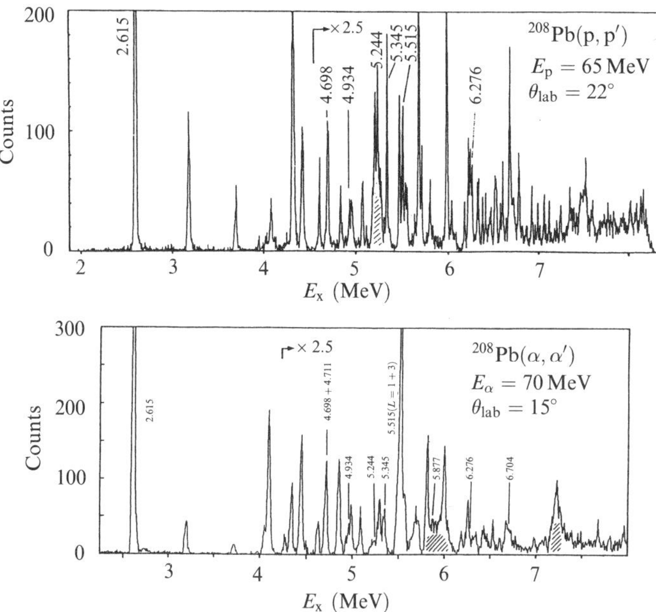

Fig. 4.15 A comparison of high-resolution spectra obtained in ^{208}Pb(p, p') and ^{208}Pb(α, α') reactions. A good correspondence of $L = 3$ states between the two spectra is observed up to $E_x = 7$ MeV. (From Ref. [30].)

minor strengths coexist in the same region. It is plausible to think that a mixture of states, with different multipolarities having angular distributions out of phase with each other, was observed as a structureless background in a low-resolution experiment which observed the LEOR as a bump.

The systematic study by using a (p, p') reaction at $E_\mathrm{p} = 65\,\mathrm{MeV}$ was extended not only to ^{50}Ti but also to deformed Ti isotopes ^{48}Ti and ^{46}Ti [32,33]. In accordance with the increase of the deformation of the ground state as the neutron number N decreases, the lowering of the excitation energy of the LEOR from the systematics was observed (Fig. 4.16). Additionally, splitting of the low-lying (first) 3^- state(s) made the distinction between the low-lying state(s) part and the LEOR part rather ambiguous. The less collective nature of a GR in deformed nuclei can be attributed to the divergence of the energies of particle and hole states. Then the particle–hole energy of each configuration constituting a GR becomes more diverged and the GR is expected to have less harmonic nature.

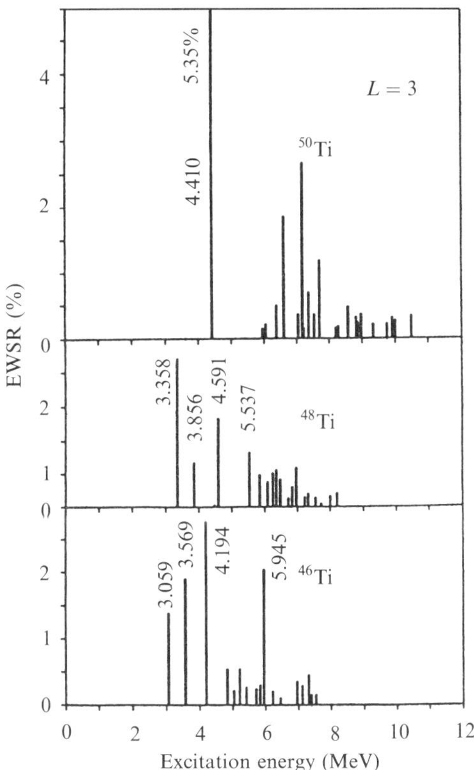

Fig. 4.16 The distribution of the EWSR fractions for the octupole ($L = 3$) strengths observed in the ^{50}Ti, ^{48}Ti, and ^{46}Ti(p, p') reactions. (From Ref. [33].)

4.4.3 $L = 4$ and other multipolarities

As already mentioned, many discrete states with various multipolarities coexist in the so-called LEOR region. In ^{208}Pb, a resonance-like concentration of the $L = 4$ strength was found to completely overlap with the $L = 3$ (LEOR) distribution (see Fig. 4.17) [34]. Furthermore, a notable feature is that the resonance-like concentration of $L = 4$ strength is commonly observed for all spherical nuclei (Fig. 4.18). The fraction of the EWSR in the resonance-like structure is 5–10%. This value appears to be very small, but it should be noted that the non-energy-weighted intensity is rather large because of the relatively low excitation energy of this resonance-like structure. Furthermore, since direct experimental information on the location and distribution of the $L = 4$ strength, except for the knowledge of a few low-lying 4^+ states, is so far rather scarce [3], this was an interesting finding. The $L = 4$ hexadecapole vibration is expected at excitations $0\hbar\omega$, $2\hbar\omega$, and $4\hbar\omega$. Since some hexadecapole strength has been reported in the GQR region ($2\hbar\omega$ excitation) of ^{208}Pb as an intruder strength distorting the $L = 2$ angular distribution of the GQR [3], it is natural to think that the strengths observed in the lower excitation are those of $0\hbar\omega$.

It is clear from the upper part of Fig. 4.17(a) that the $L = 4$ strength is divided into two groups, i.e. the well-known low-lying collective state and a bump consisting

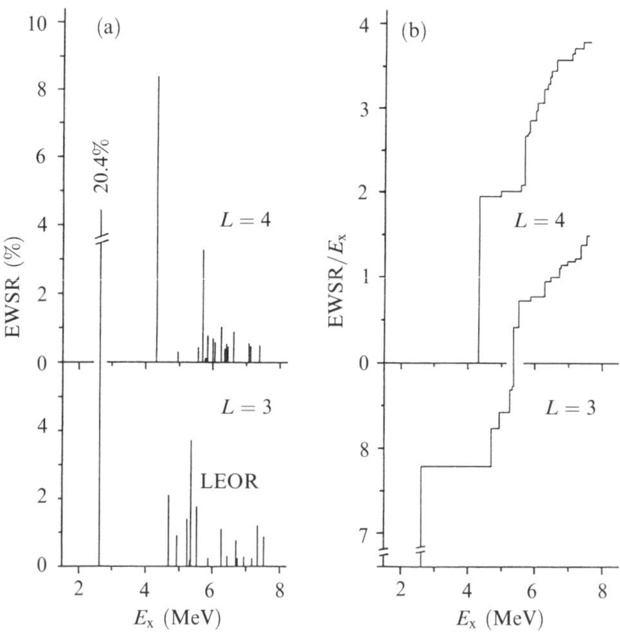

Fig. 4.17 (a): The distributions of the EWSR percentages for the $L = 3$ and 4 strengths in ^{208}Pb. (b): Cumulative sum of the strengths. The ordinate gives the percentage of EWSR divided by E_x (MeV), showing the strength proportional to single-particle strength. (From Ref. [34].)

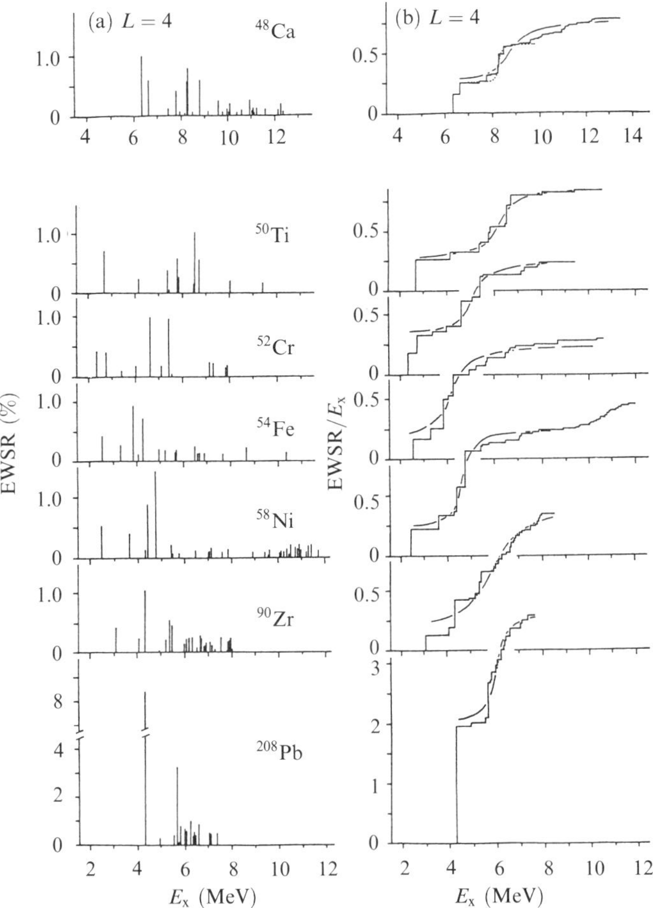

Fig. 4.18 (a): The distributions of the EWSR percentages for the $L = 4$ strengths in various nuclei with mass numbers ranging from $A = 48$ to 208. (b): Cumulative sum of the strengths. The ordinate gives the percentage of EWSR divided by E_x (MeV), showing the strength proportional to single-particle strength. Solid curves show the fit of the cumulative-sum strengths to the integrated Breit–Wigner function. (From Ref. [34].)

of many states clustering around $E_x = 6$–8 MeV. This structure observed for the $0\hbar\omega$ $L = 4$ strength is quite similar to that of the $1\hbar\omega$ $L = 3$ strength, i.e. the strength is divided typically into a low-lying collective state and a bump-like structure called LEOR. The feature is also clear in the cumulative sum shown in Fig. 4.17(b). Corresponding to the clustering states at around 6 MeV in the upper part of Fig. 4.17(a), we see the stretched-S shape characteristic of a resonance structure on top of the first rise showing the low-lying collective state. Because of the similarity of the resonance structure with that of the LEOR, the $L = 4$, $0\hbar\omega$ resonance-like structure can be called the Low-Energy Hexadecapole Resonance (LEHR) [34].

Again, by applying the integrated Breit–Wigner function to the cumulative sum of the $L = 4$ strength, the widths Γ and the central values of the excitation energy E_c were obtained for the LEHR. This feature of the fit is shown in Fig. 4.18 by solid lines. The results of the fit are summarized as follows: (a) a small width Γ is obtained for the doubly magic nuclei ^{48}Ca and ^{208}Pb; (b) the E_c values vary by a factor of about two.

Since the excitation energy of a GR is usually assumed to be proportional to $A^{-1/3}$, the $E_c A^{-1/3}$ values will be rather constant irrespective of the mass number A. Values of $E_c A^{-1/3}$ are plotted for the LEOR and the LEHR as a function of mass number A in Fig. 4.19. Although the values are slightly high in doubly magic nuclei, rather constant behaviour is seen for the LEOR. Therefore, $E_c = 30A^{-1/3}$ is suggested for the excitation-energy systematics for the LEOR. For the LEHR, on the other hand, the fluctuation is rather large, reflecting the low collectivity of the $L = 4$ strength in the $0\hbar\omega$ region, except for ^{208}Pb, in which the EWSR fraction of as much as 20% is observed in the $0\hbar\omega$ region. In the deformed nuclei $^{74-82}$Se, low-lying 4^+ states are simply fragmented, and

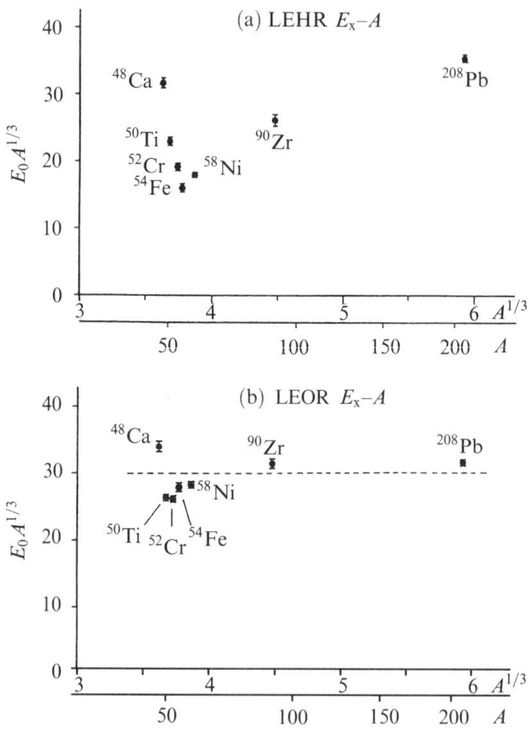

Fig. 4.19 Values of $E_0 A^{1/3}$ for LEHR and LEOR as a function of mass number A. The dashed line in the LEOR plot shows the mean value $E_0 A^{1/3} = 30$. (From Ref. [34].)

neither a concentration of the $L = 4$ strength to the first 4^+ state, nor to a resonance-like structure described here is reported [35].

In ^{48}Ca, fragmented $L = 2$ and $L = 0$ states were observed above $E_x = 10.5$ and 12 MeV, respectively, as shown in Fig. 4.13. It is natural to think that these levels are the lower tail of the GQR and the GMR, whose centres are predicted at $E_x = 17$ and 22 MeV, respectively. The corresponding cumulative sums (see Fig. 4.14) increase continuously toward the end of the energy region investigated here. Similar fragmented $L = 2$ states are observed in ^{208}Pb [26].

4.5 Giant resonances in highly excited region

4.5.1 *Giant resonances observed as a bump*

In the highly excited region above the particle separation energy, the decay width becomes large and giant resonances should be observed as the usual bump structures. For the excitation of IS and electric modes of GR, it was believed to be best to use a real IS (i.e. $A = Z$) and electric (i.e. spin $S = 0$) particle, viz. the α-particle. The energies used were roughly in the range of 20–60 MeV/u, because nuclear interaction is most favourable for the excitation of those modes at these energies. Actually, as described in Section 4.2, by late 1970s, using the α-particle as a projectile, the GQR and the GMR were established for a wide range of nuclei heavier than mass $A > 50$.

Toward the 1980s, however, it was noticed that the ^{3}He beam with a similar energy/nucleon was a better choice. The reasons are as follows: (a) IS, electric modes are excited by the ^{3}He particle as well; (b) the oscillation pattern of the angular distribution is wider in the $(^3$He$, ^3$He$')$ reaction than in the (α, α') reaction at the same incident energy/nucleon, and thus it is easier to distinguish the L-transfer peculiar to each GR; (c) the amount of background caused by the breakup of ejectile after pickup reactions, which was observed sometimes as a spurious GR-like bump in heavy-ion inelastic scatterings, is much smaller. Using these properties, the $3\hbar\omega$ mode of the $L = 3$, the main part of the $L = 3$ excitation, called a High-Energy Octupole Resonance (HEOR), was clearly distinguished, as described in detail in Section 4.5.2.

By combining the $(^7$Li$, ^7$Be$)$ reaction with the measurement of γ-decay from the first excited state of ^{7}Be, it is possible to distinguish the spin-flip and spin-non-flip modes excited in charge-exchange-type reactions. The IV-type electric modes were studied by using this unique tool, as described in Section 4.5.4.

4.5.2 *The high-energy giant octupole resonance*

In a schematic nuclear model based on the harmonic-oscillator shell model, the IS electric giant octupole resonance is predicted to be split into two components at excitation energies of $E_x = 1\hbar\omega$ and $3\hbar\omega$. The $L = 3$ strength is shared roughly in the ratio $1 : 3$ by these resonances, respectively [36]. As already described in detail, the $1\hbar\omega$ component was observed in collective low-lying state(s) and in clustering $L = 3$ states at $E_c = 30A^{-1/3}$ MeV. They consumed about 25% of the EWSR in total. It was interesting to look for the $3\hbar\omega$ component carrying the remaining main part of the IS E3 strength. There were a few indications from the inelastic reactions using the electron, proton, or heavy ions, but no clear results could be extracted, mainly due to the large amount of underlying background caused by the excitation of various modes and by the contributions of various reaction mechanisms like quasi-free scattering or pickup–breakup phenomena.

Clearer results were obtained by using the $(^3\mathrm{He}, {}^3\mathrm{He}')$ reaction with incident energy of 37–47 MeV/u. Since at least a few MeV width was expected for the bump, the energy resolution of around 100 keV was enough. Thus, the scattered $^3\mathrm{He}$ was detected by a few sets of silicon detector telescopes consisting of ΔE and E detectors for particle identification and energy determination. The silicon detectors have an advantage of covering a wider range of excitation energy.

As shown in Fig. 4.20, in the experiments for target nuclei heavier than $A = 90$, up to $A = 208$, and at scattering angles of around $20°$, a few bump-like structures were commonly observed in addition to the discrete low-lying states [37]. From the analysis of the angular distribution of the scattered particles, the bumps at around 7 and 15 MeV were identified as the LEOR and the GQR, respectively. The third smaller bump showing the $L = 3$ angular distribution consistent with that of the LEOR was identified to be the HEOR. Assuming the IS E3 transition, it was found that this bump carries about 50–90% of the EWSR. It should be noted that the observed cross section was nearly the same as that of the LEOR, but the fraction of the EWSR is nearly three times greater due to the

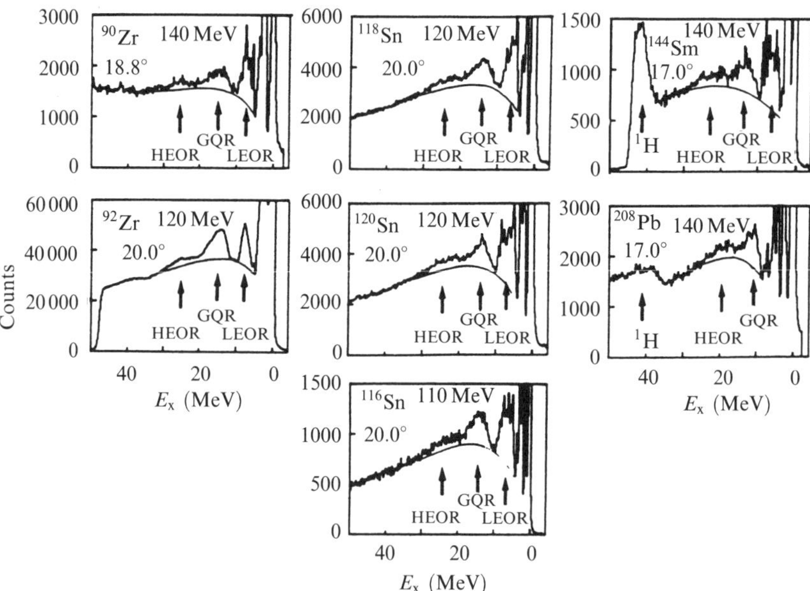

Fig. 4.20 The energy spectra of inelastically scattered $^3\mathrm{He}$ particles from medium-heavy and heavy target nuclei. The positions of the LEOR, the GQR, and the newly observed HEOR resonances are indicated. The solid curves in the figures are the continuum shapes assumed in deriving the strengths of the resonances. (From Ref. [37].)

proportionately higher excitation energy of the HEOR. Although the EWSR fractions obtained are uncertain by about 20% due to the ambiguous background subtraction, the total amount is almost 100%, if the numbers are added to those observed for the LEOR (about 20–30% of the EWSR). Thus, it is clear that almost complete E3 strength has been found by the observation of the HEOR in addition to the LEOR. The centroid energies of the resonance HEOR follow $E_x = 118A^{-1/3}$ MeV systematics as a function of A. The energy systematics and the exhausted EWSR fractions are consistent with those predicted from the schematic model [36].

4.5.3 A few topics in giant resonances in high excitation region

The ^{3}He inelastic scattering experiment on ^{208}Pb gave information on the $2\hbar\omega$ component of the ($L = 4$) HexaDecapole Resonance (HDR) [38]. A considerable fraction of the HDR strength is expected at an excitation energy of about $2\hbar\omega$, so the $L = 4$ strength should be found in the GQR region. Actually, a few experimental results using the (α, α') reaction [39,40] suggested the admixture of $L = 4$ strength with the GQR bump structure, but the similarity of the $L = 2$ and $L = 4$ angular distributions hindered a definite conclusion. In the ^{3}He inelastic scattering experiment, owing to the almost out-of-phase nature of $L = 2$ and $L = 4$ angular distributions, it was possible to deduce the existence of $L = 4$ strength with about 25% of the EWSR overlapping with the GQR bump centred at $E_c = 10.9$ MeV. The admixture of the $L = 4$ strength with the $L = 0$ GMR centred at $E_c = 13.7$ MeV, however, was small, suggesting that the $L = 4$ strength is rather concentrated in ^{208}Pb. Adding the 20% found in the discrete states at the $0\hbar\omega$ region [34], about half of the IS E4 EWSR fraction is found in ^{208}Pb.

Another interesting finding was the threshold in the excitation function of the GMR [41]. Small-angle (α, α') and (^{3}He, ^{3}He$'$) experiments were performed at various incident energies (in units of MeV/u), and an excitation function of the GMR was measured. As shown in Fig. 4.21, a rather clear threshold was observed for the excitation curve of the GMR. Although this phenomenon is still not well explained, it is interesting to note that the threshold energy of 20 MeV/u roughly corresponds to the velocity of sound in nuclear matter ($v/c \sim 0.2$).

4.5.4 Observation of isovector electric modes

The (^{7}Li, ^{7}Be) reaction combined with the measurement of γ-decay from ^{7}Be gives us a unique tool to distinguish the IV electric and IV magnetic modes. For the projectile ^{7}Li, $J^\pi = (3/2)^-$, while the nucleus ^{7}Be has two particle stable states. They are the ground state ^{7}Be$_0$ ($J^\pi = (3/2)^-$) and the first excited state ^{7}Be$_1$ ($E_x = 0.43$ MeV, $J^\pi = (1/2)^-$), and the excited state decays by γ-transition. The (^{7}Li, ^{7}Be$_0$) reaction, therefore, can excite both spin-flip and spin-non-flip states in a target nucleus, while the (^{7}Li, ^{7}Be$_1$) reaction can excite only the spin-flip states. Those states associated with spin-flip are selected by making a coincidence with the transition γ-ray.

Figure 4.22 shows a series of spectra for the ^{12}C(^{7}Li, ^{7}Be) reaction, taken at the RAIDEN spectrometer using 26 MeV/u ^{7}Li beam [42–44]. In the singles spectrum shown in Fig. 4.22(a), both contributions through the (^{7}Li, ^{7}Be$_0$) and (^{7}Li, ^{7}Be$_1$) reactions are contained. The peaks appearing through the (^{7}Li, ^{7}Be$_1$) reaction, which can be identified

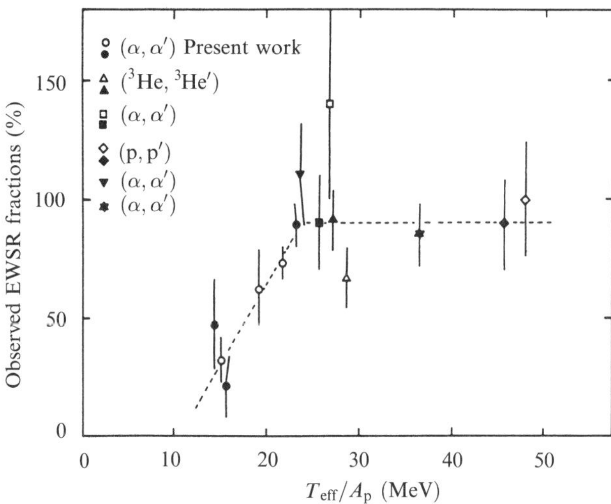

Fig. 4.21 The observed EWSR fractions of the GMR in ^{144}Sm and ^{208}Pb as a function of bombarding energy per nucleon. The dashed line (serving only as a guide to the eye) has a kink, suggesting the existence of threshold for exciting the GMR. (From Ref. [41].)

by the coincidence spectrum shown in Fig. 4.22(c), are shadowed. Their peak positions are shifted by the 0.43 MeV energy lost by the γ-emission. The contribution from the $(^{7}\mathrm{Li}, {}^{7}\mathrm{Be}_0)$ reaction can be found by subtracting the Fig. 4.22(c) spectrum from the Fig. 4.22(a) spectrum. The spin-non-flip strength is extracted by further subtracting the Fig. 4.22(c) spectrum (after adjusting the excitation energy by 0.43 MeV) from the Fig. 4.22(b) spectrum. The spin-non-flip strengths thus extracted are shown in the upper part of Fig. 4.23 and compared with the spin-flip strength. As can be seen, the spin-flip states (unnatural parity states) such as Gamow–Teller $J^{\pi} = 1^{+}$ state or the $L = 1$ spin-flip excitations called SDRs, and the natural parity states are clearly separated from each other. The broad bump at around 7 MeV in the spin-non-flip spectrum is identified as the isobaric analogue of the $J^{\pi} = 1^{-}$ GDR, the shape of which, determined by the (γ, n) reaction, is indicated by the dotted line. It is interesting to note that above $E_{\mathrm{x}} = 11$ MeV the spin-non-flip strength shows a surplus above the dotted line, suggesting the existence of new GRs with electric IV nature. They might be the candidates for the IVGQR or the IVGMR expected theoretically at $2\hbar\omega$ excitation. Further experiments (probably at different incident energies) are needed to identify the nature of these bumps.

4.6 Formation and decay of giant resonances

4.6.1 *Particle decay from highly excited states*

Compared with other quantum systems, nuclei are unique in the variety of their decay modes. Corresponding to the three different forces 'active' in the nuclear system, three types of decays are possible: (1) the strong force induces particle decay and also fission;

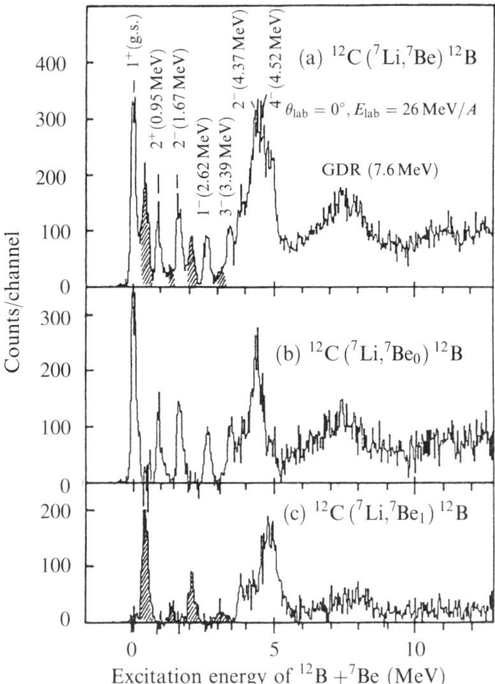

Fig. 4.22 (a): Singles spectrum for the ^{12}C(^{7}Li, ^{7}Be) reaction at $E_{in} = 26$ MeV/u and at $0°$. Two kinds of target excitation spectra going through both ^{7}Be$_0$ and ^{7}Be$_1$ channels of the projectile overlap each other. (b): Expected contribution from the (^{7}Li, ^{7}Be$_0$) channel. The spectrum is obtained by subtracting the contribution of the (^{7}Li, ^{7}Be$_1$) channel shown in (c) from the singles spectrum shown in (a). (c): Contribution from the (^{7}Li, ^{7}Be$_1$) channel known by taking the coincidences between the projectile ^{7}Be and the 0.43 MeV γ-ray emitted in the ^{7}Be $\rightarrow$ ^{7}Be$_0$ transitions. The shift of peak positions by 0.43 MeV from corresponding peaks in (b) by the γ-ray emission is clearly seen. (From Ref. [42].)

(2) the electromagnetic force induces γ-decay; and (3) the weak force induces β-decay. Because of the large difference in the strengths of these forces, particle decays are favoured whenever allowed and the probability of the γ-decay is usually about three orders of magnitude smaller than that of the particle decay. Therefore, from broad bump-like GRs at highly excited region, a nucleus deexcites mainly through particle decay processes and later through γ-decay. The direct γ-decay probability is small, and E1 is dominant if it is allowed.

Among the particle decays from GRs, proton and neutron decays are the ones most commonly observed. Basically, these decays become possible above the proton and neutron separation energies, respectively. Practically, however, particles have to penetrate the following potentials by means of tunneling, which hinders the decay to some extent.

106   Y. FUJITA

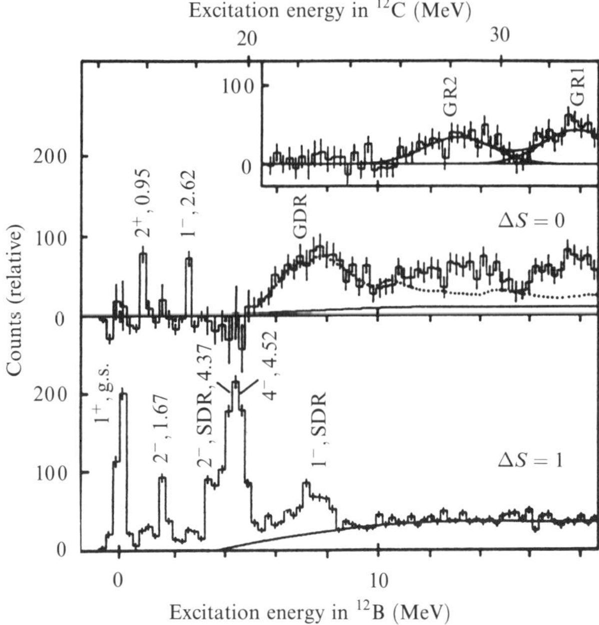

Fig. 4.23   The spectra showing the spin-non-flip ($\Delta S = 0$) and the spin-flip ($\Delta S = 1$) strengths excited in the ^{12}C(^{7}Li, ^{7}Be) reaction. These spectra were produced by the recombination of the spectra from the (^{7}Li, ^{7}Be$_0$) and (^{7}Li, ^{7}Be$_1$) channels shown in Fig. 4.3(b) and (c). (From Ref. [44].)

The first kind of potential which both protons and neutrons feel is the centrifugal potential

$$E_{\mathrm{cent}} = \frac{\hbar^2 L(L+1)}{2mR^2},\tag{4.4}$$

where L is the angular momentum of the decaying particle and R is the radius of the mother nucleus. For the nuclei in the Ni region and $L = 2$ particles, the magnitude of E_{cent} is roughly 5 MeV. The other potential which only protons experience is the Coulomb potential. If the atomic number of mother nucleus is Z_{m} and that of the decaying particle is Z_{p} ($= 1$ for proton), the potential is

$$E_{\mathrm{Coul}} = \frac{Z_{\mathrm{m}} Z_{\mathrm{p}} e^2}{R},\tag{4.5}$$

where e is the unit of electric charge. For proton decay of nuclei in the Ni region, the magnitude is roughly 10 MeV. Large Coulomb potentials, especially for large-Z nuclei, make it difficult for low-energy protons to come out. On the other hand, an $L = 0$ neutron can decay if the kinetic energy of the Fermi motion slightly exceeds the nuclear potential. This makes a large difference between the features of proton and neutron decays.

4.6.2 *Excitation and decay of giant resonances*

In inelastic scattering and charge-exchange reactions induced by relatively light projectiles, like protons, ^{3}He, or α, with an energy not too low ($>30\,\text{MeV/u}$), it is expected that the nuclear reaction proceeds mainly in one step and that the energy is given to a relatively simple excitation, e.g. a 1p–1h excitation. As already explained, GRs are excitations of this kind. It should be noted that both inelastic scatterings and charge-exchange reactions are similar to each other in the sense that they both give rise to 1p–1h-type excitations, although the nucleons contributing to the former are of the same kind (either protons or neutrons), while those contributing to the latter are different (protons and neutrons).

From a naive view, the reaction processes can be classified into non-resonant-process and resonant-process by which states with definite spin and parity are excited. In the high-excitation region above the particle separation energy, these processes show themselves as the underlying continuum part and the resonant peaks (i.e. GRs), respectively, in the inelastic or charge-exchange 'singles' spectra.

As shown in the upper part of the Fig. 4.24 [45], the continuum part is caused by the direct interaction of an incident particle with one nucleon in the target nucleus giving rise to emission of the nucleon mainly in the forward (transferred momentum) direction. The reaction process is called a Knock-Out (KO) process or a Quasi-Free Scattering (QFS), because only one nucleon participates in the process and others behave as spectators. The broad energy distribution is caused mainly by the Fermi motion of the nucleon moving inside the nuclear potential. In cases where the incident particle interacts with a few

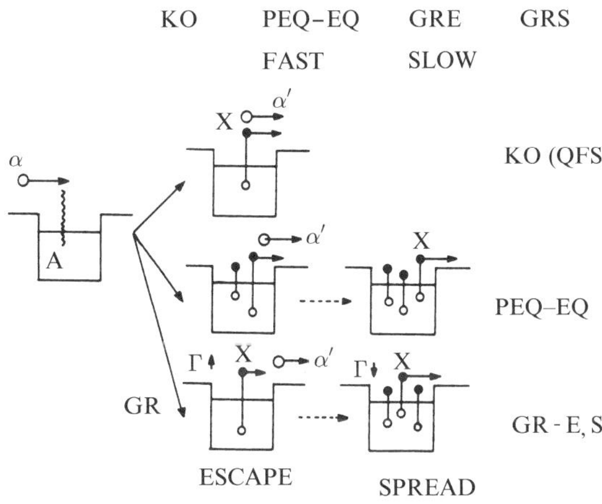

Fig. 4.24 Schematic view of the initial states (doorways) highly excited in an (α, α') reaction. The doorways are roughly categorized into the knock-out (KO) process, the pre-equilibrium (PEQ) stage, and the (giant) resonance state (GR) with definite spin and parity. The initial 1p–1h nature of the GR may decay either by emitting a particle, giving escape width (ESCAPE) to the GR, or by damping into more complicated multi-particle multi-hole states, giving spreading width (SPREAD) to the GR, and finally into equilibrium state (EQ). (From Ref. [45].)

nucleons in the target nucleus, Pre-EQuilibrium (PEQ) excitation can be formed (middle part of Fig. 4.24). One nucleon may be emitted immediately in the PEQ stage and/or later in the EQuilibrium (EQ) stage.

The various GRs are the resonant peaks caused by resonant processes. Macroscopically a GR is a damping oscillation at the nuclear surface. The oscillation usually persists for some time in a resonant manner, and then shrinks. Microscopically, a GR is the coherent superposition of 1p–1h states with definite spin and parity (bottom part of Fig. 4.24). A particle can escape directly from the resonant state, i.e. from a GR. The decay is called a direct escape (GR-E) which adds the decay width (escape width) $\Gamma^\uparrow$ to a GR. Since nucleons execute Fermi motions inside the nucleus, a high-energy nucleon can interact with some other nucleon and make the second excitation. Then the simple excitation decays (or 'damps') into a slightly more complicated configuration, like 2p–2h, which adds the spreading width $\Gamma^\downarrow$ to the GR (GR-S) as described in detail in the review article [46]. The coupling to the more complicated states can proceed until the EQ stage is reached. It should be noted that particle decay is possible at every stage. The width of the GR is inversely proportional to the life of the GR as deduced from the uncertainty principle.

Naturally, the particle coming out directly from the initial 1p–1h stage has a large energy. This directly decaying '1p' nucleon is special, because it carries rich information on the microscopic structure of the GR. If the energy of a particle decaying directly from a GR (E_{pa} in Fig. 4.25) and the excitation energy of the GR (E_x^{GR} in Fig. 4.25) is known, the binding energy of the '1h' state forming the partner of the 1p–1h excitation is calculated. Since properties like excitation energies (E_{ho} in Fig. 4.25) or J^π values of low-lying hole states (j_{ho} in Fig. 4.25) are usually well studied, the configuration of the

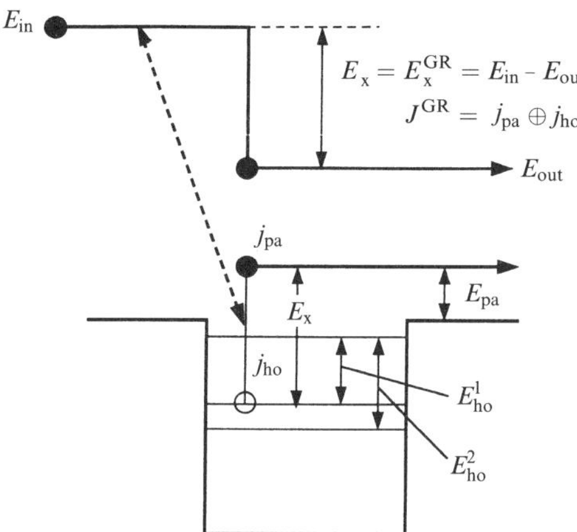

Fig. 4.25 Energy relationships before and after the excitation of a GR consisting of 1p–1h configuration, and its particle decay. By knowing the E_x of the GR and the energy of an emitted particle, E_{pa}, directly coming out of the particle state, the excitation energy of the corresponding hole state, E_{ho}, is estimated.

hole states contributing to form the GR can be determined. If the nature of the hole state is known, then the information can be used (1) in restricting the possible configurations of the partner particle states, if spin and parity values of the GR are known; and (2) in restricting the possible spin and parity values of a new GR, if the spin of the decaying particle (j_{pa} in Fig. 4.25) is studied, for example, by measuring the angular correlations.

With the increase of element number Z, the Coulomb potential becomes larger and the proton decay is hindered, which makes it difficult to measure the proton decay from a heavier element. If the EQ state is reached by spreading the initial energy into many-particle–many-hole states, the emitted nucleons, if any, are mostly slow. The emission of a nucleon from the EQ stage is described by a statistical evaporation process. At this stage, the emission of the slow nucleon is practically limited to neutrons with smaller angular momentum, because of the centrifugal potential and also the Coulomb barrier for protons. Particle decay takes the excitation energy and the initial angular momentum away and proceeds until the excitation energy is low so that no particle decay is possible. Then the decay proceeds further by emission of γ-rays and then possibly by β-decay.

4.6.3 *Neutron decay of GQR induced by α-particles*

It is expected that the measurement of 'fast' particles carrying information on the initial stage of a nuclear reaction is useful for the study and separation of the resonant part (GR) and the non-resonant underlying continuum (the so-called background). On the other hand, slow particles, emitted mainly in the EQ stage, are expected to have a statistical nature. Decay studies of GRs are favourably performed for medium and heavy nuclei, because GRs are well defined and have a resonance shape. Also, in these nuclei most of the GRs are situated well above the neutron threshold energy. Then GRs deexcite preferentially by emitting neutrons, because protons have to penetrate a large Coulomb barrier. On the other hand, measurement of neutrons is not so easy; for their detection, solid or liquid scintillation counters, containing a large percentage of hydrogen, are often used. For the measurement of neutron energy Time-Of-Flight (TOF) methods are used frequently [47–49].

Correlation studies of decaying neutrons following α-particle-induced reactions were performed on medium-heavy nuclei ^{92}Zi and ^{119}Sn at a beam energy of 109 MeV by using TOF method and also Position-sensitive Solid state Detectors (PSDs) for measuring the angular distribution and energy of the inelastically scattered α-particles [50,51]. The arrangement of the counter systems is shown in Fig. 4.26. Figure 4.27 shows the singles spectrum of inelastically scattered α-particles measured at $\theta_{lab} = 15.8°$ for ^{119}Sn and two different two-dimensional coincidence spectra with neutron detectors placed at $\theta_{lab} = -70°$ (here the minus sign specifies a neutron counter placed across the beam axis from the α-detector) and $\theta_{lab} = 110°$, respectively. Since PSDs are placed at the second maximum of the angular distribution for the $L = 2$ excitation, the GQR bump is clearly observed in the singles spectrum shown in Fig. 4.27(a). In the contour map of coincidence spectra shown in Fig. 4.27(b) and (c), two different kinds of elevations are observed. One is at low neutron-energy region increasing toward the low-energy side, below about $E_n = 2$ MeV. The other is the broad ridge-like structure of fast neutrons having an energy corresponding to the constant excitation energy of the residual nucleus

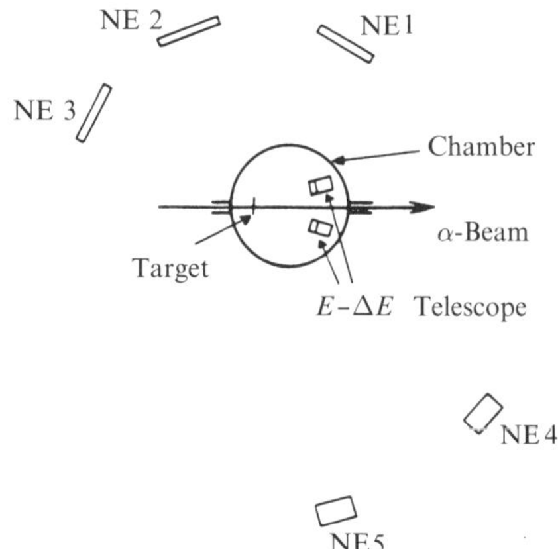

Fig. 4.26 The arrangement of the particle and neutron counters used for the correlation study between inelastically scattered α-particles and emitted neutrons. The α-particles are detected by the E–ΔE telescope detectors, while the neutrons are detected by neutron scintillation counters named NE1-5. The energies of neutrons are deduced by measuring the flight time from the target. (From Ref. [45].)

^{118}Sn. It is expected that the former consists of neutrons emitted in the PEQ and the EQ stages, because the neutrons emitted in these stages have low energy and their number decreases exponentially as the energy increases. On the other hand, the ridge is caused by direct neutrons, i.e. neutrons emitted in KO (QFS) and direct escape from the GQR (GR-E).

It is noted here that the ridge in Fig. 4.27(b) ranges over a wide region of target excitation, while the ridge in Fig. 4.27(c) ranges only in the region of the GQR. The difference can be explained by (1) the sign of detector angle, and (2) the largeness of angle itself. The KO process emits recoil neutrons only in forward directions and also in directions across the beam axis from the inelastically scattered α-particles. On the other hand, the neutrons from GR-E process are emitted rather symmetrically in forward and backward directions. A neutron counter placed at $\theta_{\text{lab}} = -70°$ detects neutrons from both GR-E and KO processes, while neutrons emitted in the KO process are not detected by the counter placed at $\theta_{\text{lab}} = 110°$. From the spectra taken at $\theta = 110°$, it became clear that there exist fast neutrons directly escaping from GQR and that the residual nucleus is left with definite excitation (a definite hole state).

Various processes contributing to the neutron emission spectra are well distinguished by comparing the neutron spectra taken at different angles and at different excitation energies of the target nucleus in the ^{92}Zr$(\alpha, \alpha'n)$ measurements. The neutron spectrum taken at the recoil direction ($\theta_{\text{lab}} = -60°$) in coincidence with the incoming α-particles having energies corresponding to the excitation region about 3–4 MeV above the GQR region has the EQ, PEQ, and KO components (Fig. 4.28(a)), while that taken at the same direction as the α-counter ($\theta_{\text{lab}} = 60°$) does not have the KO component (Fig. 4.28(b)). At backward directions, the contributions from non-resonant processes become small. The

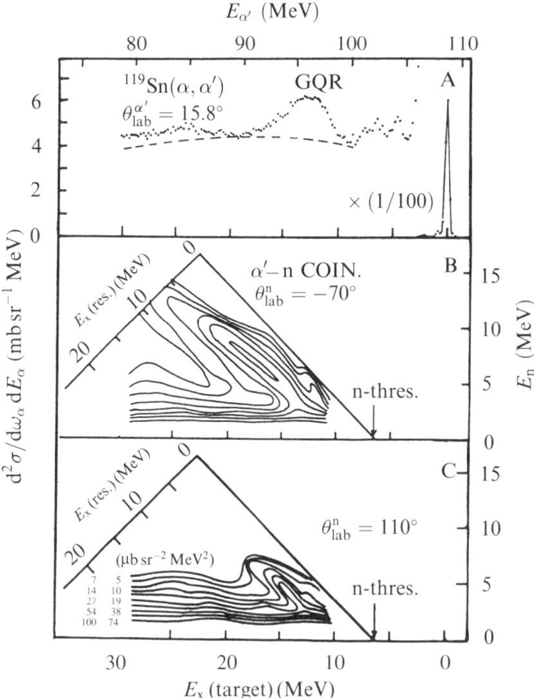

Fig. 4.27 Singles and correlation contour maps obtained for the ^{119}Sn(α, α'n) reaction. The inelastically scattered α-particles were detected at $\theta_{\text{lab}} = 15.8°$, the GQR peak angle. (a): Singles energy spectrum. E_x(target) shows the excitation energy of the target nucleus. (b): Contour map of neutron yield measured by a neutron counter placed at $\theta_{\text{lab}} = -70°$ in coincidence with α-particles. E_n is the neutron energy, and E_x(res.) shows the excitation energy in the residual nucleus (^{118}Sn) produced by the neutron emission. (c): Contour map of neutron yield measured by a neutron counter placed at $\theta_{\text{lab}} = 110°$ in coincidence with α-particles. (From Ref. [50].)

neutron spectrum taken at the anti-recoil direction ($\theta_{\text{lab}} = 110°$) again in coincidence with the α-particles corresponding to the excitation region about 3–4 MeV above the GQR region (Fig. 4.28(d)) shows mainly the EQ contribution, while that in coincidence with the α-particles corresponding to the GQR region (Fig. 4.28(c)) shows both EQ and GR-E contributions.

In this ^{92}Zr(α, α'n) measurement, the GR-E contribution had two components marked by 1 and 2 in Fig. 4.28(c); from the energy systematics, they were attributed to the neutron escape, leaving the residual nucleus ^{91}Zr in the hole states $2d_{5/2}$ and $1g_{9/2}$, respectively. In this review, it was often mentioned that a GR consists of the coherent superposition of 1p–1h states. This neutron decay measurement indicates directly the existence of 1p–1h components in the GQR. By further analysing the data, it was concluded that the escape probability of fast neutrons from the GQR was of the order of 20–25%.

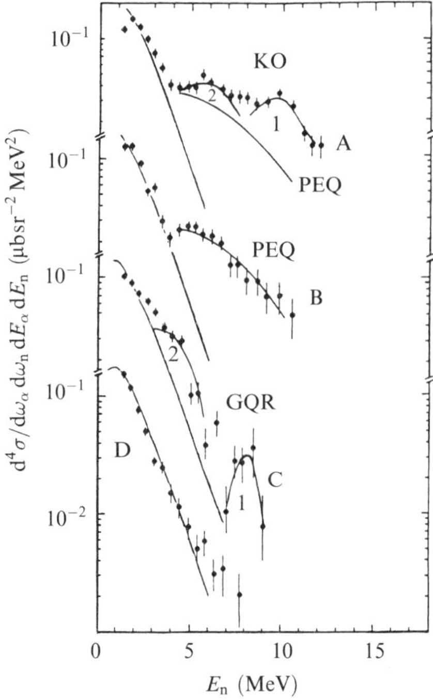

Fig. 4.28 The neutron energy measured in coincidence with α-particles emitted at the peak angle of the GQR ($\theta_{\text{lab}} = 15.9°$) in the $^{90}\text{Zr}(\alpha, \alpha'\text{n})$ reaction. (a): Neutron spectrum at $\theta_{\text{lab}} = -60°$ (the recoil neutrons direction) in coincidence with α-particles with energies corresponding to the excitation of 3–4 MeV above the GQR. (b): Neutron spectrum measured at $\theta_{\text{lab}} = +60°$ in coincidence with α-particles with energies corresponding to the excitation of 3–4 MeV above the GQR. (c): Neutron spectrum measured at $\theta_{\text{lab}} = 110°$ (the anti-recoil direction) in coincidence with α-particles with energies corresponding to the approximate excitation of the GQR. (d): Neutron spectrum measured at $\theta_{\text{lab}} = 150°$ (backward angle) in coincidence with α-particles with energies corresponding to the excitation of 3–4 MeV above the GQR. (From Ref. [45].)

Finally, it should be mentioned that a resonance structure of 1p–1h nature can be observed selectively by making a coincidence measurement with fast neutrons emitted in the backward (anti-recoil) direction. In such a spectrum the contributions from KO, PEQ, and EQ are minimized. Neutron, and also proton, coincidence measurements, depending on the situation, can therefore be an effective tool in extracting the resonance structure superposed on a continuum caused by the PEQ and/or EQ.

4.7 The isospin structure of magnetic giant resonances

4.7.1 *The isospin structure of* M1 *and Gamow–Teller resonances*

The isospin T is a unique quantum number in nuclear systems. The notion of isospin is based on the formalism that regards the neutron and proton as two different states of a single particle, the nucleon. For nuclei consisting of Z protons and N neutrons, the

z-component of isospin is defined as $T_z = (1/2)(N - Z)$ and the ground state usually has the isospin $T = T_0 = T_z$ due to the fact that in stable nuclei like-particle pairing is favoured over the neutron–proton IV pairing. On the other hand, excited states like GRs, can have higher values of T than the ground state. In this section, the identification of the quantum number isospin T is discussed for the case of Gamow–Teller (henceforth GT) mode and M1 mode of nuclear excitation, as an example. It is shown that in some cases it leads to a better understanding of the nuclear excitations through the recognition that a nuclear structure is characterized in terms of mass A and isospin T, instead of the conventional view that a nuclear structure is inherent in a specific nucleus defined by Z and N.

The GT mode is characterized as a transferred-angular-momentum $L = 0$ spin–isospin ($\Delta S = 1$, $\Delta T = 1$) excitation observed in charge-exchange reactions. The name 'Gamow–Teller' originally comes from the β-decay associated with spin flip. In various charge-exchange reactions using hadronic probes as excitation media, the GT mode has been most clearly observed at $0°$ because of its $L = 0$ nature, and at bombarding energies exceeding $100\,\mathrm{MeV/u}$, where the nuclear interaction favours the spin–isospin excitations. At such bombarding energies, a rather simple reaction mechanism is also expected. The main part of the GT mode, the GT giant resonance predicted by Ikeda *et al.* [52], is at $E_x = 5$–$10\,\mathrm{MeV}$. It should be noted that such a high-excitation region cannot be reached in a usual β-decay because of the limited Q-value (Q-window) associated with the decay process. The GT giant resonances have systematically been observed for most nuclei heavier than the fp-shell nuclei in pioneering (p, n) reactions.

The $L = 0$ spin excitations ($\Delta S = 1$) observed by inelastic scatterings, on the other hand, are called the M1 modes by analogy with the M1 γ-transitions. In the excitations of the M1 mode, both isospin changes, $\Delta T = 0$ and $\Delta T = 1$, are allowed. In a simple Shell Model (SM) picture, M1 states consist of both proton and neutron 1p–1h configurations within a major shell. The GT states excited by (p, n)-type (or (n, p)-type) charge-exchange reactions consist of proton-particle–neutron-hole (or neutron-particle– proton-hole) with the same shell configurations as those of the M1 states. Due to the same shell configurations contributing to the excitation, these two modes are the isobaric analogue states of each other for the parts of these modes characterized by the same isospin.

We examine the isospin structure of $L = 0$ spin–isospin excitations constructed in nuclei with ground state isospin $T_0 = 0$ and $T_0 \geq 1$. General cases of isospin structure of nuclei are treated by Bohr and Mottelson in Ref. [36], and general information on spin excitations can be found in a review paper by Osterfeld [53]. Details on GT resonances are given, for example, in an article by Rapaport and Sugarbaker [54].

4.7.2 *The isospin structure in $T_0 = 0$ nucleus*

4.7.2.1 *Isospin structures of M1 and Gamow–Teller resonances*

We study the GT and M1 excitations in an even–even $N = Z$ self-conjugate target nucleus having spin and parity $J^\pi = 0^+$ as an example of a $T_0 = 0$ nucleus. Then both GT and M1 states have $J^\pi = 1^+$.

The M1 states excited in an inelastic scattering can take both isospin values, $T = 0$ and $T = 1$. They become prominent at forward angles including $0°$ because of their $L = 0$ nature. Since the ground state isospin is $T_0 = 0$, the M1 states with $T = 0$ and $T = 1$ are excited by the IS and the IV interactions, respectively, and thus they can, respectively, be called IS states and IV states (see Fig. 4.29). In hadron reactions at intermediate energies ($100\,\mathrm{MeV/u}$ or more), it is known that the central part of the effective nuclear interaction $V_{\sigma\tau}$, causing the spin–isospin excitation, is prominent at small momentum transfer [53–55]. Therefore, it is usually a good approximation to assume that the excitation of the IV M1 states is dominated by the $V_{\sigma\tau}$ interaction. It should be noted that the $V_{\sigma\tau}$ interaction cannot contribute to the excitation of the IS states, because of the isospin selection rule. It is known that they are excited by the V_σ interaction and also by the tensor-type interaction through an exchange term [56]. Therefore, we should note here the fact that, due to the different types of interactions contributing to the excitations, IS and IV M1 excitations in $T_0 = 0$ nuclei are categorized by completely different modes independent of each other, even though they are conventionally called the same M1 modes. Remember that the name 'M1 mode' comes from the electromagnetic interaction causing both IS and IV transitions.

Charge-exchange reactions, on the other hand, are caused solely by IV interactions (in this case, mainly $V_{\sigma\tau}$) leading to final nuclei with ground states of isospin $T_\mathrm{f} = 1$. Therefore only $T = 1$ excitations are allowed for the $J^\pi = 1^+$ states of the GT modes.

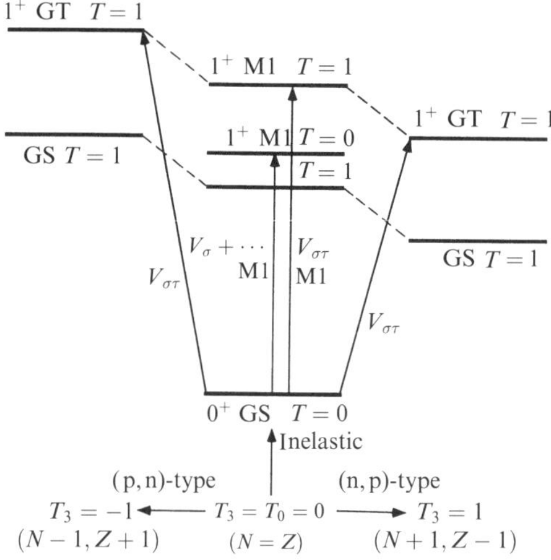

Fig. 4.29   Isospin of $J^\pi = 1^+$ states excited from the ground state of an even–even nucleus with $T_0 = 0$ ($N = Z$). The interactions mainly responsible for each excitation are shown along the arrows indicating the transitions. The isobaric analogue relationships among states are shown by broken lines. No quantitative significance is to be attached to the relative positions of the levels. (From Ref. [57].)

The main interactions exciting these 1^+ states are shown in Fig. 4.29 along the arrows indicating the transitions.

As the shell-model configurations of GT modes excited by (p, n)-type and (n, p)-type reactions are same as those of the M1 mode, the $T = 1$ GT mode is an isobaric analogue mode of the IV ($T = 1$) M1 mode. The analogous relationships among states are indicated in Fig. 4.29 by dashed lines. Moreover, in hadron charge-exchange and inelastic scattering reactions the GT and $T = 1$ M1 states are excited mainly by the $L = 0$ spin- and isospin-flip part of the effective nuclear interaction $V_{\sigma\tau}$. It is therefore expected that not only the Coulomb-energy-corrected excitation energies are similar, but also that the excitation cross sections observed in these reactions are closely related to each other. Using these expectations, it should be possible to identify the isospin of an M1 state by studying the existence or the non-existence of the corresponding analogous GT state. For further details on this subject, see Ref. [57].

4.7.2.2 *Isospin structure of* M1 *states in* ^{28}Si

Based on the fundamental knowledge and expectations described above, the isospin structure of M1 and GT states in the $N = Z$ target nucleus ^{28}Si was studied by comparing, level-by-level, the results from two characteristic reactions, namely, hadron charge-exchange, and hadron inelastic scattering reactions. In order to make a fruitful comparison, as already mentioned, hadron results performed at intermediate energies should be used. In the past, such a comparison, however, was limited by the relatively poor energy resolution obtainable in a charge-exchange reaction at intermediate energies. Recently, it has become possible to measure (^{3}He, t) reaction using a 150 MeV/u ^{3}He beam with a resolution of 150 keV or less [57]. Since high-energy tritons have a large magnetic rigidity, a QQDD-type spectrometer, GRAND RAIDEN, with a bending radius of 3 m was used for the momentum analysis [58]. In order to improve the energy resolution, the so-called dispersion-matching technique was used (see, e.g. Ref. [59]). The good resolution obtained allowed us to make a comparison between a (p, n)-type charge-exchange reaction and an inelastic (p, p$'$) reaction having a typical resolution of 40–80 keV.

As already mentioned, in the comparison of M1 states with GT states observed, respectively, in (p, p$'$) and (^{3}He, t) reactions, only the $T = 1$ M1 states should be seen in correspondence with the GT states, and the ratios of cross sections should be about the same for all pairs of correspondingly observed states. The comparison was made between the $0°$ ^{28}Si (^{3}He, t) ^{28}P spectrum and those from the ^{28}Si(p, p$'$) reaction at $E_\mathrm{p} = 200$ MeV [60]. The strength distribution for such modified $B(\sigma)$ values is shown in Fig. 4.30(b1). Figure 4.30(b) is shifted by an excitation energy of 9.28 MeV relative to Fig. 4.30(a), since the isobaric analogue state of a well-established 1^+ state at 10.59 MeV state in ^{28}Si is observed at 1.31 MeV in ^{28}P. In order to make the (p, p$'$) results directly comparable with the (^{3}He, t) spectrum, the $B(\sigma)$ distribution shown in Fig. 4.30(b1) was convoluted with the peak shape of the (^{3}He, t) spectrum. Then the difference between the reconstructed (p, p$'$) spectrum shown in Fig. 4.30(b2) and the (^{3}He, t) spectrum shown in Fig. 4.30(a) should be attributed to the $T = 0$ states excited in the (p, p$'$) reaction. A careful comparison showed that the best agreement with the (^{3}He, t) spectrum was

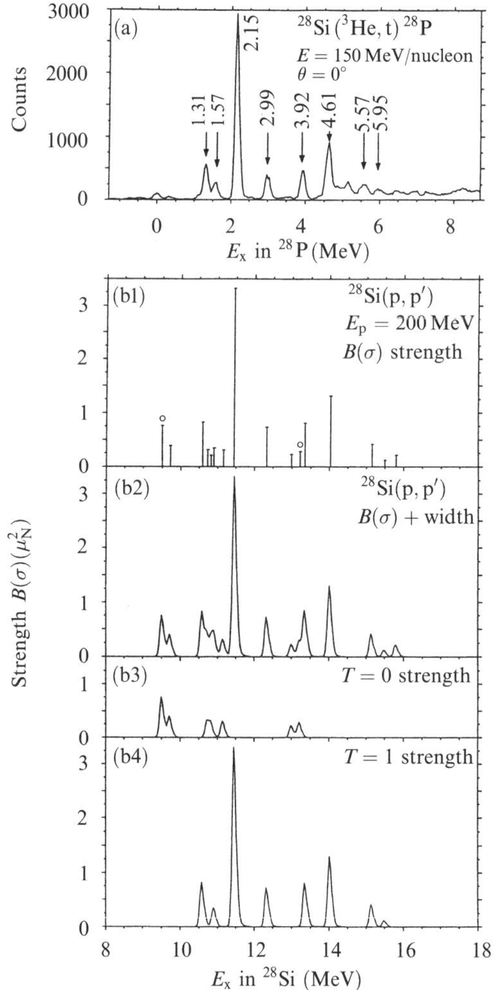

Fig. 4.30 Comparison between the $0°$ ^{28}Si(^{3}He, t) spectrum and the strength distribution $B(\sigma)$ obtained in the ^{28}Si(p, p$'$) experiment [60]. (a): The $0°$ ^{28}Si(^{3}He, t) spectrum. (b1): $B(\sigma)$ distribution observed in the ^{28}Si(p, p$'$) experiment (see text for details). (b2): Reconstructed (p, p$'$) spectrum after convoluting with the experimental energy resolution of the (^{3}He, t) experiment. (b3): Reconstructed (p, p$'$) spectrum for the $T = 0$ states after the isospin assignment by making a comparison between the spectra shown in (a) and (b2). (b4): Same as (b3), but for the $T = 1$ states. Figure (b) is shifted relative to (a) with an excitation energy of 9.3 MeV, the Coulomb displacement energy. (From Ref. [57].)

achieved by assuming the $T = 0$ and $T = 1$ strength distributions shown in Fig. 4.30(b3) and (b4), respectively. As a result, each M1 state was identified by its isospin T.

The quenching of the strengths of the magnetic excitations is a long-standing subject of theoretical and experimental interest. Theoretically, the quenching is attributed to

(a) the ground state correlation, and to (b) the contribution of Δ–h excitation, where the mechanism (a) is common to the quenching of both IV and IS strengths. On the other hand, the Δ–h excitation mechanism contributes only to the IV excitations, and not to the IS excitations. In light $N = Z$ nuclei, isospin is usually a good quantum number, and in inelastic reactions $T = 0$ and $T = 1$ M1 states in $N = Z$ nuclei, respectively, are excited by IS- and IV-type interactions. Only after an unambiguous identification of $T = 0$ and $T = 1$ strengths, is it possible to discuss the quenching mechanism.

4.7.3 *The isospin structure in $T_0 \geq 1$ nuclei*

4.7.3.1 *The isospin structure of* M1 *and Gamow–Teller resonances*

For simplicity, again we study the $L = 0$ spin–isospin excitations in an even–even target nucleus but for one with $N > Z$ having ground state isospin $T_0 \geq 1$ and spin and parity $J^\pi = 0^+$. For this case, again the GT states are characterized as $J^\pi = 1^+$, $L = 0$ spin–isospin excitations. In a (p, n)-type reaction, the GT strength in the daughter nucleus is distributed over states with final isospin values $T = T_0 - 1$, T_0, and $T_0 + 1$. The three T components split by the symmetry energy (see Fig. 4.31).

In first-order approximation, the transition strength of each T component is proportional to the square of the Clebsch–Gordan (CG) coefficient of isospin coupling, denoted by C in Fig. 4.31. For a heavy nucleus with large $T = T_0$, the CG coefficient for the transition to the $T = T_0 - 1$ component is much larger than the others. Most of the GT strength, therefore, is exhausted by the $T = T_0 - 1$ component forming a bump-like structure commonly denoted the GT Resonance (GTR). Since the pioneering (p, n)

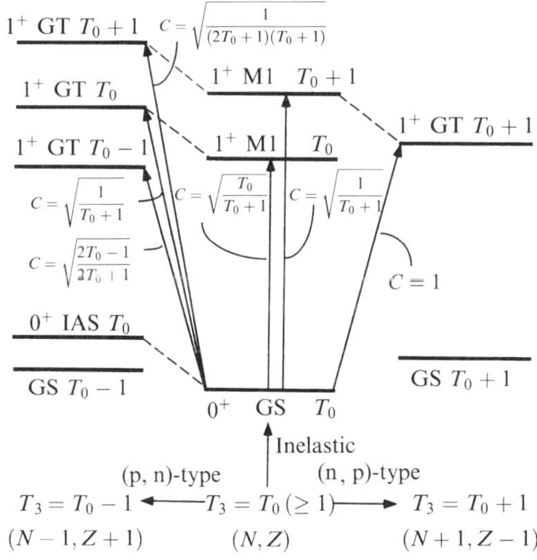

Fig. 4.31 The analogous relationship of M1 and GT states constructed on even–even nucleus with groundstate isospin $T = T_0 = (N - Z)/2 \geq 1$. The CG coefficients for isospin transitions are denoted by C.

experiments on ^{90}Zr and ^{208}Pb targets at $E_p = 200\,\text{MeV}$ [61,62] and (^{3}He, t) experiments at intermediate energies [63], the GT strengths in light-heavy and heavy nuclei have been perceived as a single broad peak (GTR) with often a few associated low-lying discrete states, as reviewed in Refs [53,54,64].

The situation is different for a target nucleus with small T_0. For example, ^{58}Ni has a ground state isospin $T_0 = 1$, and the square of the CG coefficient of isospin coupling leads to a transition strength ratio of $2:3:1$ for the $T = T_0 - 1$, $T = T_0$, and $T = T_0 + 1$ components, respectively. The $J^\pi = 1^+$ GT strength excited in ^{58}Cu is, therefore, expected to spread more or less evenly over the three isospin components.

Inelastic scattering reactions, like (p, p$'$) or (e, e$'$), can populate the $L = 0$ spin excitation states called the M1 states. Taking into consideration that particle–hole configurations of these M1 states are the same as those of GT states, if neutrons and protons are not distinguished (the Wigner supermultiplet scheme), the M1 states are the analogue states of either the T_0 or the $T_0 + 1$ GT states. The transition strength ratio, however, is different from that in the (p, n)-type reaction; for the $T_0 = 1$ target nucleus ^{58}Ni, it is $1:1$ for the T_0 and $T_0 + 1$ components, again from the squares of the CG coefficients. Furthermore, the (n, p)-type charge-exchange reactions populate only the GT states that are analogues to the $T_0 + 1$ part of the GT states populated in the (p, n)-type reactions.

Again, a good-resolution (^{3}He, t) experiment can be utilized to obtain information on the fine structure of the GT states. Then the observed structure should be compared with the structure of M1 states observed in inelastic scattering as well as with that of the GT states observed in (n, p)-type reactions on the same target nucleus.

4.7.3.2   *The isospin structure of GT strength observed in the* ^{58}Ni (^{3}He, t) ^{58}Cu *reaction*

As already mentioned, ^{58}Ni is a nucleus with $T_0 = 1$. Figure 4.32(a) shows the $0°$ spectrum. With an energy resolution of $140\,\text{keV}$, we see the fine structure of the main GT strength at around $E_x = 9\,\text{MeV}$, which was observed more or less as a single broad bump in pioneering (p, n)-type reactions. In addition, several sharp peaks are observed up to $E_x = 13\,\text{MeV}$. The gross feature of the spectrum, however, is quite similar to that observed in the ^{58}Ni(p, n) reaction at $E_p = 160\,\text{MeV}$ [65], suggesting again the usefulness of the (^{3}He, t) reaction as a means for studying GT excitations [66–69].

Among the $T_0 - 1$, T_0, and $T_0 + 1$ components of the GT strength, the $T_0 - 1$ part is identified through the non-existence of the corresponding M1 strength in the inelastic scattering experiments, and the T_0 and $T_0 + 1$ parts through its existence. From an (e, e$'$) experiment [70], the B(M1) strength was mapped with a resolution of $30\,\text{keV}$ in the excitation energy range $E_x = 5.9$–$15.0\,\text{MeV}$, and is shown in Fig. 4.32(b1). The analogue states of the six $T_0 + 1$ GT states studied with the ^{58}Ni (t, ^{3}He) ^{58}Co reaction at $E_t = 25\,\text{MeV}$ [71] are marked with the small circles on top of the vertical lines in Fig. 4.32(b1). They are the main strengths in the second hump. Further information on the $T_0 + 1$ component was obtained by a recent ^{58}Ni (n, p) ^{58}Co reaction at $E_n = 198\,\text{MeV}$ [72]. In order to account for the $T_0 + 1$ GT strength in the (n, p) reaction, it is natural to assume that the 1^+ states observed at $E_x > 11.5\,\text{MeV}$ in the (e, e$'$) spectra are of $T_0 + 1$ nature.

To make the (e, e′) results more comparable with the (^{3}He, t) spectrum, the B(M1) distribution (Fig. 4.32(b1)) was convoluted with the peak shape in the (^{3}He, t) spectrum (see Fig. 4.32(b2)). For a direct comparison, the spectrum was further modified so that the strengths of all $T_0 + 1$ candidates are reduced by a factor of three to compensate for the difference of CG coefficients in exciting the $T_0 + 1$ states in charge-exchange

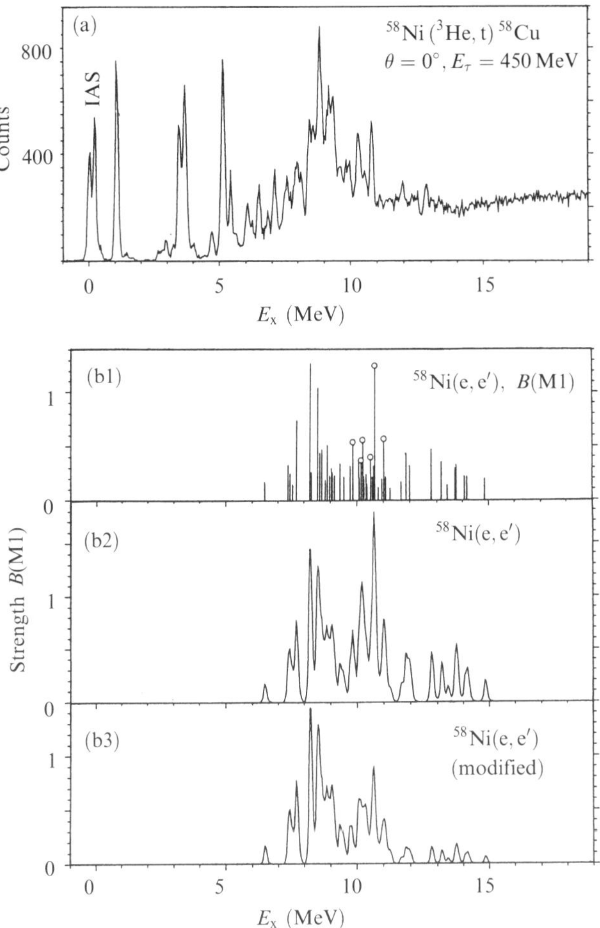

Fig. 4.32 Comparison between the $0°$ ^{58}Ni(^{3}He, t) spectrum and the M1 strength distribution obtained in the ^{58}Ni(e, e′) experiment [70]. (a): The $0°$ ^{58}Ni(^{3}He, t) spectrum. (b1): The B(M1) distribution from the ^{58}Ni(e, e′) experiment. (b2): The reconstructed (e, e′) spectrum after convoluting with the experimental energy resolution of the (^{3}He, t) experiment. (b3): The same as (b2), but the $T_0 + 1$ strength being reduced artificially by a factor of three. Figure (b) is shifted with an excitation energy of 0.2 MeV relative to (a), since the Isobaric Analogue State (IAS) of ^{58}Ni, the 0^+ ground state, is observed at $E_x = 0.20$ MeV in ^{58}Cu. (From Ref. [73].)

and inelastic reactions (Fig. 4.32(b3)). This spectrum now looks very similar to the $(^3\text{He}, \text{t})$ spectrum. It is noted that to each peak structure in the modified (e, e') spectrum, a corresponding peak is observed in the $(^3\text{He}, \text{t})$ spectrum up to nearly $E_x = 14\,\text{MeV}$.

The isospin of each level was assigned by examining whether or not the corresponding part could be found in the modified (e, e') spectrum. Below $E_x = 10\,\text{MeV}$, the isospin $T = T_0 - 1$ was assigned (with a few exceptions) to all the levels not observed in the (e, e') spectrum. To the levels above $10.0\,\text{MeV}$ and already reported in the (n, p)-type reactions, an isospin $T_0 + 1$ was assigned. The $B(\text{GT})$ strength distributions for the three different isospin components are shown in Fig. 4.33, and more detailed discussion on the isospin decomposition is given in Ref. [73].

The observation of 'sharp' GT states up to nearly $E_x = 13\,\text{MeV}$ might be very much against the commonsense expectation that the GT resonances show a broad bump-like

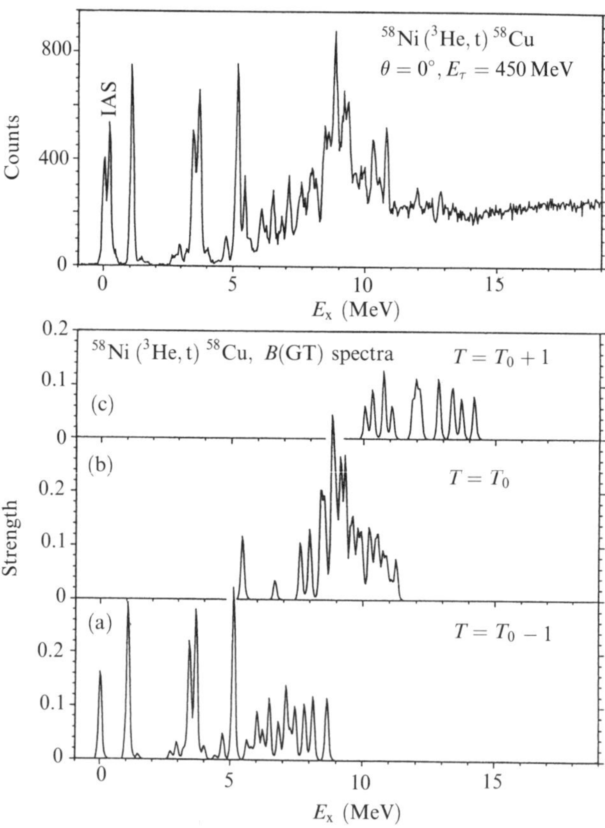

Fig. 4.33 The $B(\text{GT})$ strength distributions for the three isospin components convoluted with the experimental response function. $B(\text{GT})$ spectrum for the $T = T_0 - 1$ (a), $T = T_0$ (b), and $T = T_0 + 1$ (c) components. (From Ref. [73].)

structure. It can be qualitatively understood from the particle separation energies, and the shell and isospin structure of ^{58}Cu. The proton separation energy S_p is at $E_x = 2.9$ MeV, while that of the neutron is at 12.4 MeV (see Fig. 4.34). It is obvious that the neutron decay width does not contribute to the total width of the GT resonance in ^{58}Cu, unlike the GT resonances in heavier nuclei. Energetically, proton decay is possible, but it is largely hindered by the Coulomb barrier and the centrifugal potential, especially for a high-L particle configuration, like f$_{5/2}$, the main particle configuration of the GT strength in ^{58}Cu. More important is the fact that the sharp states observed at above 11 MeV are mainly of $T_0 + 1$ nature. Proton decay from $T_0 + 1$ ($T = 2$) states is allowed only to $T_0 + 1/2$ ($T = 3/2$) and $T_0 + 3/2$ ($T = 5/2$) states. Since the ground state isospin of the nucleus ^{57}Ni after the proton decay is $T = 1/2$, proton decays to the low-lying states are prohibited by the isospin selection rule; the decay is possible to $T = 3/2$ excited states above $E_x = 5$ MeV. Then the proton energy available for the decay becomes small and the decay rate (and thus the decay width) becomes very small. If the particle decay width is somewhat smaller than the experimental resolution of 140 keV, we should not see the

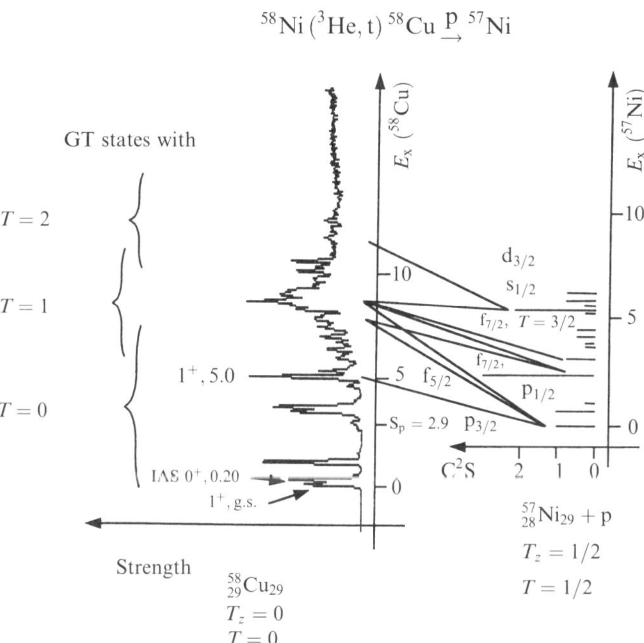

Fig. 4.34 Energy relationship showing the possible proton decay of three different T states of ^{58}Cu excited in ^{58}Ni(^{3}He, t) reaction. In the energy region of the main GT resonance, proton decay is energetically allowed, but not the neutron decay. Due to the T selection rule, $T = T_0+1$ ($T = 2$) states can decay only to $T = 3/2$ excited states in ^{57}Ni, while $T = T_0$ ($T = 1$) and $T = T_0-1$ ($T = 0$) states can decay to low-lying $T = 1/2$ states.

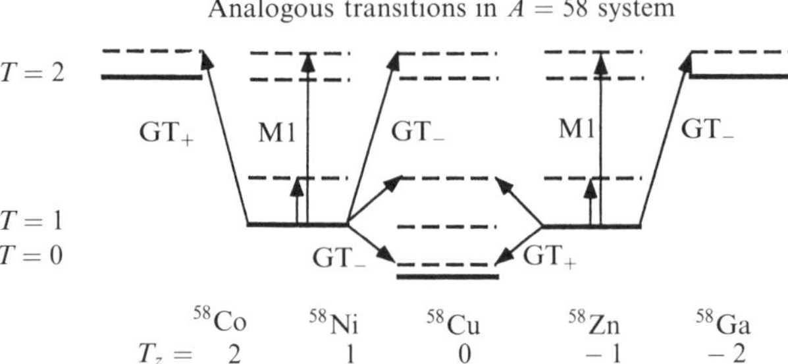

Fig. 4.35   Analogous relationship of M1 and GT transitions for the $A = 58$ system. It is seen that by assuming the Wigner supermultiplet scheme, the $T = T_0 + 1$ transition strength extracted through the isospin decomposition, for example, is the analogous transition strength expected in an exotic ^{58}Zn $\rightarrow$ ^{58}Ga charge-exchange reaction.

broadening of levels. There the fragmentation of strength is caused by the spreading width and each GT state should be observed as a discrete level.

Finally, it is of interest to consider the meaning of the $T_0 - 1$, T_0, and $T_0 + 1$ B(GT) transition strengths based on the Wigner supermultiplet scheme (see Fig. 4.35). First, the $T_0 - 1$ transition strength obtained in the β^--decay-like ^{58}Ni $\rightarrow$ ^{58}Cu transitions are the analogue transitions of the β^+-decay of ^{58}Zn, i.e. ^{58}Zn $\rightarrow$ ^{58}Cu transition to low-lying $T_0 - 1$ states. Similarly, the T_0 transition strengths are analogous to the ^{58}Zn $\rightarrow$ ^{58}Cu transition strengths to T_0 states in ^{58}Cu, and also to those of the M1 transitions from ground states of ^{58}Zn to the T_0 M1 excited states. Extending this analogy, we even come to the idea that the $T_0 + 1$ B(GT) strengths extracted in the present ^{58}Ni $\rightarrow$ ^{58}Cu reaction are not only similar (after correcting for geometric factors) to the B(GT) strengths to be observed in the ^{58}Ni $\rightarrow$ ^{58}Co reaction but also to those expected in the ^{58}Zn $\rightarrow$ ^{58}Ga charge-exchange reaction (or ^{58}Ga $\rightarrow$ ^{58}Zn β^+-decay). These exotic transitions starting from unstable nuclei are of great importance, not only in nuclear physics but also for astrophysics and nucleosynthesis. It is interesting to note that by recognizing the nuclear structure as characterized in terms of A and T, exotic transition strengths starting from far-stability nuclei can be estimated at least approximately.

4.8   Giant resonances: excitation with variety and harmony—a brief summary

A nucleus is often referred to as a micro-quantum laboratory. It consists of limited number of nucleons, i.e. it has limited degrees of freedom, which is responsible for nuclear excitations to possess great variety. This can be known through a study of the A dependence of nuclear excitations. In a larger-A nucleus, a GR which has the nature of a vibration at the surface of nuclei usually becomes more prominent and its strength concentrates in one bump-like structure. As we have seen, representative GRs, like GQR or GMR, begin to show prominent resonance shapes in nuclei heavier than $A \approx 50$. The collectivity of these GRs is high in heavy nuclei, and often the full strength expected by the EWSR is found. In lighter nuclei, on the other hand, the shell effects are larger and the strength more or less spreads among individual discrete states. The picture of collective vibration does not fit very well.

In a very rough way, each excitation mode in a nucleus is distinguished by quantum numbers L, S, and also by T, which is a unique quantum number in a nuclear system consisting of protons and neutrons. The GRs are vibrational excitation modes also distinguished by these quantum numbers. The collectivity of each mode is largely dependent not only on the mass number A of a nucleus but also on the angular-momentum transfer L; higher-L modes are usually less collective, a tendency observed in deformed nuclei as well.

The variety of collectivity in different modes and nuclei can be understood naively from the microscopic viewpoint that a GR is a coherent superposition of 1p–1h excitations. To form a collective motion, the component 1p–1h excitations should be superposed in phase. An in-phase motion is easier if the excitation energy of each 1p–1h excitation is similar, i.e. if 1p–1h excitations are more harmonic. This is better realized in heavier nuclei than in deformed or light nuclei. In higher-L modes, all 1p–1h excitations with $\Delta N \leq L$ are allowed.

The decay of a collective excitation in a quantum system is a commonly observed phenomenon. Decay by emitting a particle, on the other hand, is rather unique in nuclei. Since particle decays are caused by the strong interaction, the lifetime of a state, and thus the width of the state, is influenced largely depending on whether or not a decay channel is open. Accordingly, the appearance of a GR is also very much influenced. A resonance structure, like LEOR, situated below or around the particle separation energy, is observed as a cluster of discrete states. It is unique, because the width of such a resonance structure (the width of the envelope of the clustering states) is caused only by the spreading width $\Gamma^\downarrow$. In higher-excited GRs, on the other hand, individual states are not observed because of higher level density and decay widths $\Gamma^\uparrow$ larger than the level spacings. They are observed as a usual bump-like resonance shape. Particle decay can also be used for the study of microscopic structure of GRs. The microscopic 1p–1h nature of GRs has been examined through the detection of fast neutrons directly decaying from the GQR. The configuration of the hole state left in the nucleus is then deduced.

Structures of GRs are usually discussed for each nucleus specified by the proton number Z and the neutron number N. The isospin T, the z-component of which is defined by using Z and N, is a unique quantum number in a nuclear system. The importance of isospin in classifying nuclear structures is well known as the Wigner supermultiplet scheme. Through the study of isospin structures of $L = 0$, $\Delta S = 1$ excitations, we have seen that structures of IV GRs, GT and M1, can be understood better by classifying their structures with T and A.

In this review, only nucleonic degrees of freedom are treated for the formation of GRs. It should be stressed, however, that in some cases mesonic and sub-nucleonic degrees of freedom should also be considered. For example, the contributions of excited states of nucleon, such as Δ, are often discussed as being responsible for the reduction of the strength of magnetic excitations.

Finally, it is mentioned that we have studied only those excitation modes of GRs which are relatively easy to detect. Many unknown modes and their characteristics are still waiting to be discovered. New selectivities should be combined for the study of a new mode and new characteristics of GRs.

References

1. A. M. Bernstein, *Advances in Nuclear Physics* **3** (1969) 325.
2. J. Speth and J. Wambach, in *Electric and magnetic giant resonances in nuclei*, J. Speth, ed., World Scientific (1991), p. 1; and references therein.
3. A. van der Woude, in *Electric and magnetic giant resonances in nuclei*, J. Speth, ed., World Scientific (1991), p. 99; and references therein.
4. E. Lipparini and S. Stringari, *Phys. Rep.* **175** (1989) 103.
5. K. T. Knöpfle, in *Electric and magnetic giant resonances in nuclei*, J. Speth, ed., World Scientific (1991), p. 233; and references therein.
6. B. L. Berman and S. C. Fultz, *Rev. Mod. Phys.* **47** (1975) 713.
7. M. N. Harakeh and A. E. L. Dieperink, *Phys. Rev.* **C23** (1981) 2329.
8. B. F. Davis *et al.*, *Nucl. Phys.* **A599** (1996) 277c.
9. R. Pittham and T. H. Walcher, *Phys. Lett.* **36B** (1971) 563.
10. S. Fukuda and Y. Torizuka, *Phys. Rev. Lett.* **29** (1972) 1109.
11. M. B. Lewis and F. E. Bertrand, *Nucl. Phys.* **A196** (1972) 337.
12. A. Moalem, W. Benenson, and G. M. Crawley, *Phys. Rev. Lett.* **31** (1973) 482.
13. F. E. Bertrand, *Nucl. Phys.* **A354** (1981) 129c.
14. G. R. Satchler, *Nucl. Phys.* **A195** (1972) 1.
15. D. H. Youngblood *et al.*, *Phys. Rev.* **C13** (1976) 994.
16. J. M. Moss *et al.*, *Phys. Rev.* **C18** (1978) 741.
17. H. L. Clark, D. H. Youngblood, and Y.-W. Lui, *Phys. Rev.* **C54** (1996) 72.
18. D. H. Youngblood *et al.*, *Phys. Rev. Lett.* **39** (1977) 1188.
19. C. M. Rozsa *et al.*, *Phys. Rev.* **C21** (1980) 1252.
20. M. Buenerd *et al.*, *Phys. Lett.* **84B** (1979) 305.
21. D. H. Youngblood, H. L. Clark, and Y.-W. Lui, *Phys. Rev. Lett.* **76** (1996) 1429.
22. H. Ikegami *et al.*, *Nucl. Instrum. and Meth.* **175** (1980) 335.
23. Y. Fujita *et al.*, *Nucl. Instrum. and Meth.* **225** (1984) 298.
24. P. Martin *et al.*, *Nucl. Phys.* **A315** (1979) 291.
25. Y. Fujita *et al.*, *Phys. Lett.* **98B** (1981) 175.
26. Y. Fujita *et al.*, *Phys. Rev.* **C32** (1985) 425.
27. M. Fujiwara *et al.*, *Phys. Rev.* **C32** (1985) 830.
28. M. Fujiwara *et al.*, *Phys. Rev.* **C37** (1988) 2885.
29. W. Unkelbach *et al.*, *Phys. Rev.* **C40** (1989) 1534.
30. Y. Fujita *et al.*, *Phys. Rev.* **C45** (1992) 993.
31. Y. Fujita *et al.*, *Phys. Rev.* **C37** (1988) 45.
32. M. Fujiwara *et al.*, *Phys. Rev.* **C35** (1987) 1257.
33. A. Higashi *et al.*, *Phys. Rev.* **C39** (1989) 1286.
34. Y. Fujita *et al.*, *Phys. Rev.* **C40** (1989) 1595.
35. K. Ogino *et al.*, *Phys. Lett.* **130B** (1983) 147.
36. A. Bohr and B. R. Mottelson, *Nuclear structure*, Vol. 2, Benjamin, 1975.
37. T. Yamagata *et al.*, *Phys. Rev.* **C23** (1981) 937.

38. T. Yamagata *et al.*, *Phys. Lett.* **123B** (1983) 169.
39. M. N. Harakeh *et al.*, *Nucl. Phys.* **A327** (1979) 373.
40. H. P. Morsch *et al.*, *Phys. Rev.* **C22** (1980) 489.
41. T. Yamagata *et al.*, *Nucl. Phys.* **A381** (1982) 277.
42. S. Nakayama *et al.*, *Nucl. Instrum. and Meth.* **A302** (1991) 472.
43. S. Nakayama *et al.*, *Phys. Rev. Lett.* **67** (1991) 1082.
44. S. Nakayama *et al.*, *Phys. Rev.* **C46** (1992) 1667.
45. H. Ejiri, *J. Phys.* **C4** (1984) 135; H. Ejiri, *Nucl. Phys.* **A482** (1988) 471c.
46. G. F. Bertsch, P. F. Bortignon, and R. A. Broglia, *Rev. Mod. Phys.* **55** (1983) 287.
47. H. Ejiri *et al.*, *Nucl. Phys.* **A305** (1978) 167.
48. H. Sakai *et al.*, *Phys. Rev.* **C20** (1979) 464.
49. H. Sakai *et al.*, *Phys. Rev.* **C23** (1980) 1002.
50. K. Okada *et al.*, *Phys. Rev. Lett.* **48** (1982) 1382.
51. H. Ohsumi *et al.*, *Phys. Rev.* **C32** (1985) 1789.
52. K. Ikeda, S. Fujii, and J. I. Fujita, *Phys. Lett.* **3** (1963) 271.
53. F. Osterfeld, *Rev. Mod. Phys.* **64** (1992) 491; and references therein.
54. J. Rapaport and E. Sugarbaker, *Annu. Rev. Nucl. Part. Sci.* **44** (1994) 109.
55. W. G. Love and M. A. Franey, *Phys. Rev.* **C24** (1981) 1073.
56. Y. Sakemi *et al.*, *Phys. Rev.* **C51** (1995) 3162; and references therein.
57. Y. Fujita *et al.*, *Phys. Rev.* **C55** (1997) 1137.
58. M. Fujiwara *et al.*, RCNP report, 1991, p. 173.
59. Y. Fujita *et al.*, *Nucl. Instrum. and Meth.* **B126** (1997) 274; and references therein.
60. G. M. Crawley *et al.*, *Phys. Rev.* **C39** (1989) 311.
61. D. E. Bainum *et al.*, *Phys. Rev. Lett.* **44** (1980) 1751.
62. D. J. Horen *et al.*, *Phys. Lett.* **95B** (1980) 27.
63. C. Ellegaard *et al.*, *Phys. Rev. Lett.* **50** (1983) 1745.
64. C. Gaarde, *Nucl. Phys.* **A396** (1983) 127c.
65. J. Rapaport *et al.*, *Nucl. Phys.* **A410** (1983) 371.
66. I. Bergqvist *et al.*, *Nucl. Phys.* **A469** (1987) 648.
67. A. Brockstedt *et al.*, *Nucl. Phys.* **A530** (1991) 571.
68. H. Akimune *et al.*, *Nucl. Phys.* **A569** (1994) 245c.
69. M. Fujiwara *et al.*, *Nucl. Phys.* **A599** (1996) 223c.
70. W. Mettner *et al.*, *Nucl. Phys.* **A473** (1987) 160.
71. F. Ajzenberg-Selove *et al.*, *Phys. Rev.* **C30** (1984) 1850.
72. S. El-Kateb *et al.*, *Phys. Rev.* **C49** (1994) 3128.
73. Y. Fujita *et al.*, *Phys. Lett.* **365B** (1996) 29.

5

Breakup processes and cluster structures of light composite projectiles

Y. Sakuragi

5.1 Introduction

In nuclear reactions induced by light composite projectiles, many of which are loosely bound systems, projectile breakup takes place strongly as a real or a virtual process and is known to play an important role in various nuclear reactions. In this chapter, we review the present status of our understanding of reaction mechanisms relevant to the nuclear breakup processes and their role in nuclear reactions induced by light composite projectiles.

5.1.1 *What is breakup?*

The projectile breakup processes discussed in this chapter may be written symbolically as

$$a + A \rightarrow (b + c + \cdots) + A,$$

where A denotes a target nucleus while a is a composite projectile nucleus composed of subunit particles b, c, ... which are either nucleons or their clusters. When the target nucleus A is left in its ground state, the process is called an *elastic breakup*, while if A is left in one of the excited states (A^*), it is called an *inelastic breakup*. Experimentally, the elastic and inelastic breakup processes can be resolved only when all the energies and angles of all emitted fragment particles are measured completely (called an *exclusive measurement*). Otherwise, the data include the elastic and inelastic events as well as other possible reaction processes such as transfer reactions and incomplete fusion, which is called an *inclusive measurement*. All the breakup processes discussed in this chapter are those corresponding to the exclusive measurement with a definite final state.

If the ground state of the projectile nucleus has a well-developed cluster structure, the projectile breakup into the relevant clusters will be one of the most favourable reaction processes in nucleus–nucleus collisions induced by the projectile nucleus. Typical examples are the deuteron (d), ^{6}Li, ^{7}Li, and ^{9}Be projectiles. These nuclei are very loosely bound systems of the $p + n$, $\alpha + d$, $\alpha + t$, and $\alpha + \alpha + n$ clusters with threshold energies of 2.22, 1.47, 2.48, and 1.57 MeV, respectively, and the breakup into the relevant clusters is known to be a dominant reaction process. Thus, the breakup process is closely related to the cluster structure of the projectile nucleus and the study of the nuclear breakup provides valuable information on the nuclear cluster structure. Conversely, a precise knowledge of the nuclear cluster structure is essential to study the reaction mechanism relevant to the breakup process.

The breakup processes of these light projectile nuclei also play important role in other reaction processes induced by the projectile nuclei on a target nucleus, such as the elastic scattering, inelastic scattering, and transfer reactions [1]. A review of the breakup effects on various reaction processes is one of the main purposes of the present chapter.

5.1.2 *Types of breakup processes*

In a breakup reaction, $a + A \rightarrow (b + c + \cdots) + A$, the fragment particles are emitted with none or strong final-state correlation, depending on the phase space of the emitted particles. If the interaction between the fragment particles is strong enough, they can form a resonance at a suitable kinematical condition.

From a different viewpoint, the breakup process can be interpreted as an excitation of the projectile nucleus into continuum-energy excited states above the breakup threshold. For the lightest composite nuclei such as deuteron and ^{3}He, the continuum has less structure and to a good approximation it can be regarded as a uniform *non-resonant* continuum. For a nucleus with mass number $A \geq 6$, the spectrum of the continuum (breakup) states has a complex structure consisting of some resonance states which are surrounded by background (non-resonant) continuum states, although the border line between resonant and non-resonant background continuum becomes unclear at high excitation energies where the resonance is broader.

The process in which the fragment particles are *finally* emitted through the non-resonant continuum is called a *non-resonant breakup* process, while the process in which the fragment particles are *finally* emitted through one of the resonance states of the projectile nucleus is called a *resonant breakup* process. In the latter process, the fragmented particles are emitted with a strong correlation to each other having particular quantum numbers characteristic of the resonance state. The resonant breakup process is also referred to as the *sequential breakup* process, in which the projectile is excited to a resonance state which decays subsequently into fragment particles outside the strong interaction field of the target nucleus. Similarly, the non-resonant breakup process is also referred to as a *direct breakup* process, in which the projectile breaks up *directly* into the fragment particles without passing through any resonance state of the projectile nucleus. In other words, the projectile nucleus is assumed to be excited to a continuum state *directly* by one-step transition from the ground state to the final continuum state.

However, one should be careful in using these terminologies to discuss the breakup process and its reaction mechanism. Experimentally, the resonant/non-resonant breakup processes (or direct/sequential ones), are defined only by the *exit-channel information*, namely, the excitation energy of the projectile nucleus in the final state, which by itself tells us little about the reaction paths between the entrance and exit channels. Namely, the resonant breakup only means that, at the final stage of the reaction process, the fragments are emitted with a relative energy corresponding to a particular resonance energy of the compound system but it does not necessarily mean that the projectile was excited to the resonance state and remained there until it dissociated into fragment clusters in the final stage of reaction. The latter process is what is meant by 'sequential breakup'. Similarly, the non-resonant breakup only indicates that the fragment particles are emitted from the non-resonant continuum state at the final stage of reaction, but this does not specify the process implied by direct breakup, i.e. the one-step transition to this final state without passing through any intermediate states.

In the projectile(a)–target(A) three-body system, a(b+c)+A (or many-body system, a(b+c+$\cdots$)+A, in general), the energy of the subsystem (b+c) need not be conserved separately as long as the total energy of the three-body system is conserved; hence, transitions among various excited states of the projectile nucleus can take place freely at the intermediate stage of the reaction. For example, the fragments emitted through a resonance state may have passed through various non-resonant states in the intermediate stage of the reaction (see Fig. 5.1), and vice versa. Therefore, the terminologies of the direct/sequential breakup may not necessarily be appropriate to describe what is really going on in the breakup process with the three-body dynamics, because they specify particular reaction paths from the ground state to individual final states via one-step transitions.

The importance of multistep transitions among various breakup states including those between the resonant and non-resonant states was shown in practice in the case of ^{6}Li $\rightarrow$ α + d breakup reactions [2–4]. It was shown that the reaction paths from the ground state to low-lying resonance states via non-resonant continuum states were significant

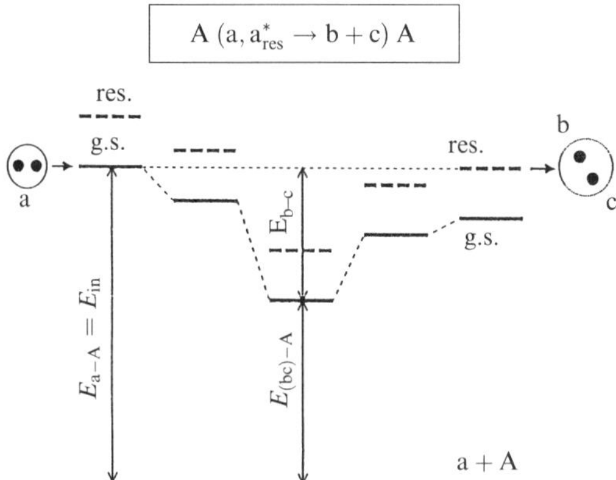

Fig. 5.1 A schematic reaction diagram for a *resonant breakup* reaction. Suppose that the projectile nucleus (a) is in the ground state (g.s.) in the entrance channel (left-hand side), which breaks up into fragments b and c through a resonance state (res.) in the final stage of the reaction (right-hand side). However, the projectile can be excited to any continuum state either above or below the resonance state in the intermediate stage of the reaction, as long as the total energy of the system is conserved; $E_{\text{in}} = E_{\text{a–A}} = E_{\text{(bc)–A}} + E_{\text{b–c}}$. Here, $E_{\text{a–A}} = E_{\text{in}}$ denotes the incident energy, which is the c.m. energy of the projectile(a)–target(A) system in the entrance channel, while $E_{\text{b–c}}$ and $E_{\text{(bc)–A}}$ are the energy of the subsystem b + c and that of the relative motion between the c.m. of the b + c and the target nucleus A in the intermediate stage where a is excited to a non-resonant continuum state.

in explaining the observed resonant breakup reaction. Similarly, the multistep processes through resonance states were found to be important to explain the non-resonant breakup reactions. The multistep transitions among non-resonant breakup states are also known to play a key role, especially in the deuteron breakup reactions [5] in which no p–n resonance states exist.

The importance of such *multistep processes among breakup states* is one of the key concepts to understand the reaction mechanism associated with the breakup process.

5.1.3 *Projectile breakup effects on elastic scattering*

As already mentioned, the breakup reaction takes place strongly in nucleus–nucleus collisions with a projectile nucleus having a well-developed cluster structure in its ground state which is a weakly bound state just below the breakup threshold. A developed cluster structure in the ground state implies a similarity of nuclear structure between the ground state and the breakup continuum which enables a strong transition between them to occur easily. This, in turn, implies that the elastic scattering of such fragile projectiles is likely to be affected by the strong coupling to the breakup process (Fig. 5.2(a)).

In fact, this was widely confirmed for many light projectiles such as deuteron [6,7], ^{6}Li and ^{7}Li [2,3,8], ^{9}Be [9], and some β-unstable nuclei like ^{7}Be [10], ^{6}He, and ^{11}Li

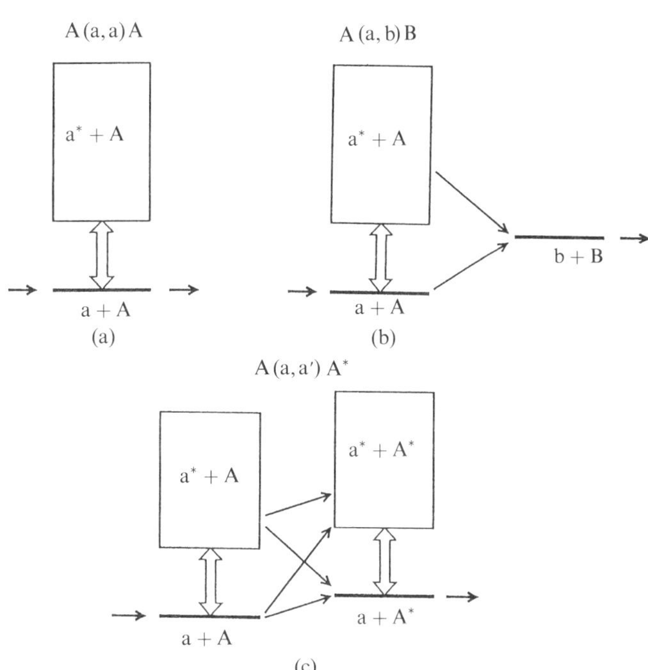

Fig. 5.2 A schematic diagram of the projectile breakup effects (a) on elastic scattering, (b) on transfer reaction, and (c) on inelastic scattering.

[11–13], the latter two being known as the *neutron halo nuclei*. In many cases, the breakup effect on the elastic scattering is known to be a repulsive effect. The *dynamical polarization potential* induced by the coupling to the breakup process has a substantial real part of repulsive nature and its strength amounts to 40–50% of the unperturbed real central potential predicted by various folding models at low energies [3,8,9].

For a projectile nucleus having a non-zero spin in its ground state, such as deuteron $(S = 1)$, ^{6}Li $(S = 1)$, and ^{7}Li $(S = 3/2)$, the coupling to the breakup process also induces *dynamical spin-dependent interactions*, such as the spin–orbit and tensor interactions. This is particularly important to understand the spin-dependent interactions for light-heavy-ions like ^{6}Li and ^{7}Li, for which a large fraction of the spin–orbit and tensor interactions are induced by the dynamical coupling to the breakup states (especially the resonance states), as shown later on. Especially for low-energy ^{6}Li and ^{7}Li scattering, almost 100% of the spin–orbit interaction has its origin in dynamical coupling [14,15]. For deuteron scattering, however, the dynamical spin-dependent interactions are not so important, although part of the tensor interaction has its origin in the coupling to the breakup states, as will be discussed below in more detail. The main difference between the deuteron and the ^{6}Li, ^{7}Li ions, is closely related to the difference of nuclear structure in the breakup continuum, viz. the existence of the low-lying resonance states in the latter and lack of enough structure in the breakup continuum in the former.

Thus, the study of the projectile breakup effects on elastic scattering is important not only to clarify the reaction mechanism relevant to the breakup processes but also to understand the nucleus–nucleus interactions for composite projectiles from a microscopic viewpoint.

5.1.4   *Breakup effects on other reaction processes*

In addition to elastic scattering, projectile breakup can also affect other reaction processes such as the inelastic scattering and transfer reactions through various multistep processes (see Fig. 5.2(b) and (c)). In order to study these effects, one needs a reaction theory which treats possible multistep transitions among various reaction processes correctly and describes the breakup reactions as well as other reaction processes consistently in a single framework. A theoretical formalism called the Continuum-Discretized Coupled-Channel (CDCC) method [1,3,6,16] is known to be probably the best for this purpose.

5.1.4.1   *The CDCC method*

The CDCC method was first introduced in the study of deuteron breakup process and its effects on elastic and inelastic scattering as well as on the (d, p) reaction [1,16]. It was later applied also to the study of projectile breakup processes of light-heavy ions such as ^{6}Li, ^{7}Li, ^{9}Be, and ^{12}C [3]. In fact, most of the results for the breakup effects on the elastic scattering mentioned above were obtained by the CDCC studies.

In the CDCC method, the breakup continuum states are described in terms of a finite number of discrete states suitably constructed from the original continuum states and the coupling among the discretized continuum states are treated exactly in a coupled-channel framework. Thus, the breakup reaction and the elastic scattering (where the projectile is in the ground state) are described simultaneously in a single CDCC framework. Furthermore, the CDCC method can easily be generalized so as to include the target excitation channels as well as the projectile–target mutual excitation channels. Therefore, the inelastic scattering, A (a, a′) A*, and inelastic breakup reactions, A (a, a*) A*, are also studied together with their elastic counterparts (Fig. 5.2(c)).

The CDCC method is known to be currently one of the most reliable as well as practical theoretical tools for a comprehensive study of the breakup process and its role in nuclear reactions. The outline of the CDCC method together with microscopic models for the nucleus–nucleus interaction will be given in the next section in some detail.

5.1.4.2 *Breakup effects on the transfer reactions*

The importance of deuteron breakup effects on the (d, p) and (p, d) reactions was already recognized in the early days of nuclear spectroscopy using these reactions and its study has a long history [16]. There have been a number of pioneering works which take into account the deuteron breakup effects, such as the sudden approximation [17] and various adiabatic theories [18] treating the three-body dynamics approximately, in addition to the standard Distorted-Wave Born Approximation (DWBA) theories [19]. However, it is only recently, after practical numerical works using CDCC became available, that the various roles of the deuteron breakup process in (d, p) and (p, d) reactions have been clarified precisely and systematically. A historical review and some applications of the CDCC method to the deuteron breakup will be found in recent review articles [1,16].

The projectile breakup affects more or less all other reaction processes since the elastic scattering, which is the entrance channel for all other reaction processes, is strongly distorted by the coupling to the breakup process. However, if the breakup effect only distorts the elastic channels, it could be included in the distorting potential of the elastic channel approximately by using an optical potential which reproduces the elastic scattering correctly. In fact, many of the (d, p) or (p, d) reactions were known to be described fairly well by standard DWBA calculations using a phenomenological optical potential as the distorting potential in the deuteron channel.

However, it is known that such DWBA analyses are often confronted with serious difficulties in reproducing some data on the (d, p) or (p, d) reactions, in which the direct transfer of a neutron to/from the deuteron ground state is kinematically mismatched due to a large momentum or angular-momentum transfer. In such cases, the neutron transfer through the deuteron breakup states, (d, d*)(d*, p) (or its inverse process for the (p, d) reaction), competes with, or even dominates over, the direct transfer to/from the deuteron ground state. In this case, the breakup effect is not only a distortion effect but plays an important dynamical role [16] as well.

Since the elastic channel (d + A) and the breakup channels (d* + A) are strongly coupled to each other, one needs to rely on the Coupled-Channel Born Approximation (CCBA) type calculations (Fig. 5.2(b)) for this type of transfer reactions. Namely, in a CDCC calculation the coupling among the deuteron channels is treated exactly and neutron transfers from the ground and breakup states of deuteron are treated perturbatively. Such CCBA analyses of the (d, p) and (p, d) reactions based on the CDCC method have been made successfully [16].

Similar dynamical effects of breakup processes are also important in α-transfer reactions such as (^{6}Li, d), (^{7}Li, t), and their inverse processes, although no practical numerical work has been carried out so far except for one analysis [20] of the (d, ^{6}Li) reaction. A systematic analysis of α-transfer reactions in a finite-range CCBA framework will be quite important in order to understand the absolute values of the spectroscopic factors for α-cluster states in various nuclei and their systematics over a wide range of nuclei.

5.1.4.3 *Breakup effects on inelastic scattering*

In the case of inelastic scattering leaving the target nucleus in one of its excited states, the projectile breakup process also plays an important role, beyond the distortion effect of the elastic scattering case. One of the additional effects is the distortion effect on the exit channel. Namely, the wave function in the exit channel (a + A*) will be distorted by the *inelastic breakup* channels (a* + A*) built on the exit channel as strongly as the entrance (elastic) channel is distorted by the elastic breakup channels (a* + A). Such a distortion effect on the exit channel could also be taken into account, although approximately, by the use of an optical potential similar to that of the entrance channel.

However, the breakup process also plays an additional dynamical role due to the multistep processes through the elastic and inelastic breakup channels (Fig. 5.2(c)). Therefore, one needs to do a large-scale CDCC calculation which takes account of all the coupling among the four kinds of channels, a + A, a* + A, a + A*, and a* + A*. A typical example of this type of analysis will be shown in Section 5.3 in the case of ^{6}Li inelastic scattering by ^{12}C to its low-lying excited states. If the inelastic transition of the target nucleus, A $\rightarrow$ A*, is weak enough to be treated perturbatively, the above large-scale Coupled-Channel (CC) calculation may be replaced by a CCBA calculation of the type (a + A, a* + A)$_{CC}$ $\rightarrow$ (a + A*, a* + A*)$_{CC}$.

5.1.4.4 *Excitation of cluster states in heavier nuclei*

In heavier nuclei such as ^{12}C, ^{16}O, ^{20}Ne, ^{24}Mg, etc., the ground state no longer has a cluster-like structure and the breakup reactions may not take place as strongly as in lighter nuclei. This is mainly because the breakup takes place only through a rather weak transition between states with very different characteristics, namely the transition from the tightly bound shell-model-like ground state to the final breakup states, the typical cluster states. So, the breakup effect on the elastic scattering of such heavier nuclei is not so strong as in the case of light fragile projectiles.

However, the breakup process of these heavier projectiles plays an important role in the *projectile excitation to the cluster states*. In this case, the breakup effects play the role of a strong distortion on the final states due to the similarity of nuclear structure between the final cluster states and the breakup states. In fact, in such heavier nuclei, it is well known that there exist a number of cluster-like states around the threshold for a particular configuration of clusters having a well-developed associated cluster structure, which is known as a *threshold rule* [21]. Some of the cluster states are bound states below the threshold and some are unbound (breakup) states above the threshold. The typical example is the ^{12}C nucleus, in which the 0_1^+ ground state and the first excited 2_1^+ and 3_1^- states are shell-model-like states having a compact structure, while the 0_2^+ state

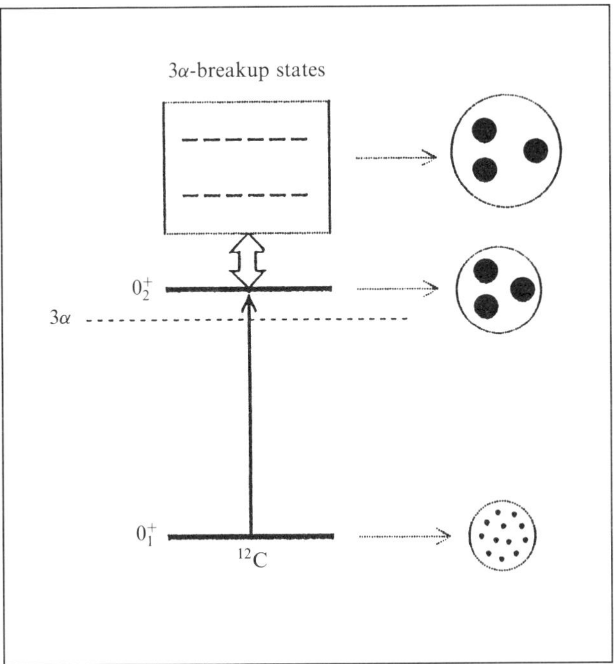

Fig. 5.3 A schematic reaction diagram of the ^{12}C $\rightarrow$ 3α breakup effect on the ^{12}C excitation to the 0_2^+ state at 7.64 MeV having a typical 3α-cluster structure.

at 7.64 MeV, which is a 3α resonance state just above the 3α threshold at 7.4 MeV, and some other states around it have a well-developed 3α-cluster structure [22]. Therefore, once one such cluster state is excited by the interaction with a target nucleus, the exit channel wave function will be affected strongly by the strong coupling to the breakup states surrounding the final cluster state (Fig. 5.3).

In this case, the breakup process comes into play as a distortion effect on the *exit channel* of the reaction. The role of the ^{12}C $\rightarrow$ 3α breakup on the inelastic scattering leading to the above-mentioned ^{12}C* (0_2^+) state will be shown below as a typical example of this kind of reaction.

5.2 Microscopic coupled-channel formalism

To study the nuclear breakup reaction and its role in various reaction processes, it is very important to rely on a realistic nuclear structure model with which we can properly describe the transitions among relevant nuclear states and to use a reliable as well as practical nuclear reaction theory which can take into account the breakup reaction dynamics as correctly as possible. From this point of view, the microscopic CC theory [2,3] recently developed for this purpose is one such ideal tool. This theory is based on the microscopic cluster model for light nuclei and the method of CDCC [1,2,6] for describing the breakup reaction dynamics. In this section, we summarize briefly the theoretical framework. Further details can be found elsewhere [3].

5.2.1 *Microscopic cluster model*

The microscopic cluster model is one of the most successful models in nuclear physics, especially for the study of nuclear structure and reactions of light nuclear systems [21]. Because it takes into account the Pauli principle among nucleons, the cluster model describes well not only the genuine cluster states of the loosely bound cluster structure but also the shell-model-like states of the spatially compact structure. The reliability of the microscopic cluster model wave functions and the corresponding transition densities were tested very carefully and extensively with available experimental data, such as the electron-scattering form factors, the $B(E\lambda)$ values and (for the unbound states) the scattering phase shifts of the relevant clusters [3,21].

5.2.2 *Reaction theory for the breakup process—the CDCC method*

The CDCC method is now established as one of the most reliable as well as practical nuclear reaction theories for describing nuclear breakup processes [1,3]. Because of the nature of the CC theory, this method can be used not only to analyse the real breakup reactions but also to study, within the same theoretical framework, the role of the breakup process at the intermediate stages of nuclear reactions leading to final states other than the breakup.

To make the essential points of the theoretical framework clear, let us assume a three-body system composed of a target nucleus (A) and a couple of nuclear clusters (b and c) forming the projectile nucleus. For simplicity of notation, we also neglect the intrinsic spins of the clusters and the target nucleus. Let the projectile nucleus have only one bound state, which is thus the ground state, all other states being unbound continuum states above the breakup threshold. Then the wave function for the three-body system may be written as

$$\Psi(\mathbf{r}, \mathbf{R}) = \hat{\phi}_0(\mathbf{r})\hat{\chi}_0(\mathbf{R}) + \sum_\ell \int \phi_\ell(k, \mathbf{r})\chi(k, \mathbf{R})\, \mathrm{d}k, \qquad (5.1)$$

where $\mathbf{r}$ and $\mathbf{R}$ denote the relative position vectors between the clusters, and that between the target nucleus and the c.m. of the clusters, respectively. Here, $\phi_\ell(k, \mathbf{r})$ is the scattering-state wave function with a relative momentum k and angular momentum ℓ of the two-cluster system. We first truncate the model space for the relative motion of the clusters by a maximum momentum $k_{\max}$ and a maximum angular momentum $\ell_{\max}$ and then we divide the k-integration range $[0, k_{\max}]$ into N bins $\{[k_{i-1}, k_i], \ i = 1\text{–}N\}$. This gives

$$\Psi(\mathbf{r}, \mathbf{R}) \simeq \hat{\phi}_0(\mathbf{r})\hat{\chi}_0(\mathbf{R}) + \sum_\ell^{\ell_{\max}} \int_0^{k_{\max}} \phi_\ell(k, \mathbf{r})\chi(k, \mathbf{R})\, \mathrm{d}k$$

$$= \hat{\phi}_0(\mathbf{r})\hat{\chi}_0(\mathbf{R}) + \sum_\ell^{\ell_{\max}} \sum_{i=1}^{N} \int_{k_{i-1}}^{k_i} \phi_\ell(k, \mathbf{r})\chi(k, \mathbf{R})\, \mathrm{d}k. \qquad (5.2)$$

If the momentum bin width $\Delta k = k_i - k_{i-1}$ is small enough, it may be reasonable to assume a weak k dependence of the wave function $\chi(k, \boldsymbol{R})$ for the projectile–target relative motion within the bin and to take this out of the k integration by replacing k by an average value for the bin, $\bar{k}_i$:

$$\Psi(\boldsymbol{r}, \boldsymbol{R}) \simeq \hat{\phi}_0(\boldsymbol{r})\hat{\chi}_0(\boldsymbol{R}) + \sum_{\ell}^{\ell_{\max}} \sum_{i=1}^{N} \left[\frac{1}{N_{\ell i}} \int_{k_{i-1}}^{k_i} \phi_\ell(k, \boldsymbol{r})\,\mathrm{d}k \right] N_{\ell i} \chi(\bar{k}_i, \boldsymbol{R})$$

$$\equiv \hat{\phi}_0(\boldsymbol{r})\hat{\chi}_0(\boldsymbol{R}) + \sum_{\ell}^{\ell_{\max}} \sum_{i=1}^{N} \hat{\phi}_{\ell i}(\boldsymbol{r})\hat{\chi}_{\ell i}(\boldsymbol{R}) \equiv \sum_{\alpha}^{N_{\max}} \hat{\phi}_\alpha(\boldsymbol{r})\hat{\chi}_\alpha(\boldsymbol{R}). \tag{5.3}$$

Here, $\hat{\phi}_{\ell i}(\boldsymbol{r})$ and $\hat{\chi}_{\ell i}(\boldsymbol{R})$ are defined as

$$\hat{\phi}_{\ell i}(\boldsymbol{r}) \equiv \frac{1}{N_{\ell i}} \int_{k_{i-1}}^{k_i} \phi_\ell(k, \boldsymbol{r})\,\mathrm{d}k, \tag{5.4}$$

$$\hat{\chi}_{\ell i}(\boldsymbol{R}) \equiv N_{\ell i} \chi(\bar{k}_i, \boldsymbol{R}), \tag{5.5}$$

where $N_{\ell i}$ is a normalization factor [2,3] such that $\hat{\phi}_{\ell i}$ is normalized as $\langle \hat{\phi}_{\ell i} | \hat{\phi}_{\ell' i'} \rangle = \delta_{\ell \ell'}\delta_{ii'}$. In eqn (5.3), the subscript α is introduced to denote combined quantum numbers $\{\ell, i\}$, including the ground state (elastic) channel and $N_{\max}$ is the total number of coupled states.

In this approximation, we now have a discrete set of internal wave functions for the projectile 'breakup states' $\hat{\phi}_\alpha(\boldsymbol{r})$, each of which is a kind of wave packet defined as in eqn (5.4), and $\hat{\chi}_\alpha(\boldsymbol{R})$ represents the associated wave function for the projectile–target relative motion. If a particular bin contains a resonance, it is more realistic to leave a weight function $f_{\ell i}(k)$ in the k integration when we take $\chi(k, \boldsymbol{R})$ out of the integration [2,3]:

$$\hat{\phi}_{\ell i}(\boldsymbol{r}) \equiv \frac{1}{N_{\ell i}} \int_{k_{i-1}}^{k_i} f_{\ell i}(k)\phi_\ell(k, \boldsymbol{r})\,\mathrm{d}k. \tag{5.6}$$

The k dependence of $f_{\ell i}(k)$ should contain information on the resonance, i.e. the resonance energy ε_R and the width Γ, and is usually taken to be a Lorentzian [2,3] $f_{\ell i}(k) = |(\mathrm{i}/2)\Gamma/(\varepsilon - \varepsilon_R + (\mathrm{i}/2)\Gamma)|$, where $\varepsilon = \hbar^2 k^2/(2\mu)$. More generally, $f_{\ell i}(k)$ is taken as $f_{\ell i}(k) = \sin \delta_\ell(k)$, where $\delta_\ell(k)$ is the b–c scattering phase shift.

By the above procedure of truncation and discretization of the breakup continuum, the three-body problem has been reduced to a normal CC problem with a finite number of discrete excited states and the resultant coupled equations (the CDCC equation) are of the form of standard CC equations,

$$(T_\alpha + V_{\alpha\alpha} - E_\alpha)\hat{\chi}_\alpha(\boldsymbol{R}) = -\sum_{\beta \neq \alpha}^{N_{\max}} V_{\alpha\beta}(\boldsymbol{R})\hat{\chi}_\beta(\boldsymbol{R}) \quad (\alpha = 1\text{–}N_{\max}). \tag{5.7}$$

The coupled equations can be solved numerically in a standard way to obtain the transition matrix elements for the elastic scattering as well as for the excitation of the *discretized breakup states.* One should, however, note that, due to the unbound nature of the breakup states, the coupling potentials between the breakup states $V_{\alpha\beta}(R)$ as well as the diagonal ones $V_{\alpha\alpha}(R)$ have very long range form factors compared to those of normal collective excitations (such as phonons and rotations). Thus, the numerical integration of the coupled equations up to a large matching radius is necessary in practical calculations. The long-range nature of the form factors strongly correlates with the discretization of the continuum states, especially with the width of the bins, Δk: the range of the form factor becomes longer as the Δk value becomes smaller. Therefore, one should be careful about the convergence of the calculated physical quantities, such as the cross sections for the elastic scattering and breakup reactions, with respect to the refinement of the discretization by decreasing Δk as well as the enlargement of the model space by increasing $k_{\max}$ and $\ell_{\max}$. In other words, the model space and its discretization should be set up so as to obtain a required quality of convergence in the calculated physical quantities. For examples of such convergence tests for the deuteron and ^{6}Li breakup see Refs [6] and [2,23], respectively.

When the unbound (breakup) states of a projectile nucleus have multi-cluster configurations, like the ^{12}C (0_2^+) state composed of 3α clusters, the above method, called the *method of momentum bins*, cannot be applied. For these systems, an alternative method, called the *method of pseudo-states* [24], is often used, in which the multi-cluster unbound continuum is discretized by a variational method. The continuum states are approximated by a discrete set of pseudo-states obtained by the diagonalization of the internal Hamiltonian matrix for the multi-cluster system in the L^2-integrable basis function space. In this method, the long-range nature of the 'discretized' unbound states may be restored by including the long-range components in a set of basis functions used in the variational calculation.

5.2.3 *Projectile–target folding model interactions*

In the above CDCC equations, $V_{\alpha\alpha}(R)$ is the diagonal potential in the αth channel, while $V_{\alpha\beta}(R)$ ($\alpha \neq \beta$) is the coupling potential between the αth and βth channels. These potentials are defined as $V_{\alpha\beta} = \langle \hat{\phi}_\alpha | V | \hat{\phi}_\beta \rangle$ using the discretized breakup-state wave functions.

In the above explanation of the CDCC procedure, we have assumed a three-body model, a(b + c) + A, for simplicity and, in this case, the projectile–target interaction potentials $V_{\alpha\beta}(R)$ are given by the so-called Cluster Folding (CF) model [25]:

$$V_{\alpha\beta}^{(CF)}(R) = \langle \hat{\phi}_\alpha(r) | V_{bA}(r_1) + V_{cA}(r_2) | \hat{\phi}_\beta(r) \rangle_r, \tag{5.8}$$

where V_{bA} (V_{cA}) is the optical potential for the b + A (c + A) system evaluated at an incident energy per nucleon equal to that of the projectile nucleus a(b+c), while $\hat{\phi}_\alpha(r)$ is the point-cluster-model wave function for the projectile nucleus in the αth channel. Such a CF model interaction can be applied to projectile nuclei like deuteron, ^{6}Li, and ^{7}Li, for which the two-cluster model is valid for all the low-lying states of interest, including the ground state. Once a suitable set of the optical potentials for the b + A and c + A systems are chosen, there is no free-parameter left in the theory.

The CDCC procedure described above can also be applied to the case that uses microscopic cluster model wave functions of the projectile nucleus [2,3]. In this case, the basic nuclear interactions should also be microscopic. For this purpose, an effective nucleon–nucleon (NN) interaction v_{NN} evaluated in the nuclear medium (G-matrix) or its parameterized form such as the M3Y interaction [26] or its density-dependent version, DDM3Y [27], are used. The projectile–target potentials are calculated by the Double-Folding (DF) model [28]:

$$V_{\alpha\beta}^{(DF)}(\boldsymbol{R}) = \left\langle \hat{\phi}_{\alpha}^{(P)} \Phi_0^{(T)} \middle| \sum_{i\in P, j\in T} v_{NN}(\boldsymbol{r}_{ij}) \middle| \hat{\phi}_{\beta}^{(P)} \Phi_0^{(T)} \right\rangle$$

$$= \int \rho_{\alpha\beta}^{(P)}(\boldsymbol{r}_1)\rho_{00}^{(T)}(\boldsymbol{r}_2)v_{NN}(\boldsymbol{r}_1,\boldsymbol{r}_2,\boldsymbol{R})\,\mathrm{d}\boldsymbol{r}_1\,\mathrm{d}\boldsymbol{r}_2. \tag{5.9}$$

Here, $\rho_{00}^{(T)}$ is the ground state nucleon density for the target nucleus, while $\rho_{\alpha\beta}^{(P)}$ is the diagonal ($\alpha = \beta$) or transition ($\alpha \neq \beta$) density of the projectile nucleus calculated from the microscopic wave functions [3,22]:

$$\rho_{\alpha\beta}^{(P)}(\boldsymbol{r}) = \left\langle \hat{\phi}_{\alpha}^{(P)} \middle| \sum_{i=1}^{A_P} \delta(\boldsymbol{r} - \boldsymbol{r}_i) \middle| \hat{\phi}_{\beta}^{(P)} \right\rangle. \tag{5.10}$$

In the DF model, all the contributions of nucleon exchange between the colliding nuclei are neglected except the Single-Nucleon Knock-on Exchange (SNKE) term, which can be taken into account by the exchange term of the NN interaction. This exchange term is usually treated in the zero-range approximation [28], although an effect of a finite-range treatment has also been discussed recently [29]. In the DF model version, the projectile nucleus need not be a two-cluster-like nucleus as the DF model interaction can be applied to any combination of colliding nuclear systems and to any kind of nuclear excitations provided the microscopic transition densities are available. This is particularly useful for studying the excitation / breakup processes associated with a multi-cluster configuration, such as $^{12}C \to 3\alpha$ and $^{9}Be \to \alpha + \alpha + n$, for example, in which no bound state exists in any binary subsystem and hence the method of pseudo-states is used for discretizing the breakup continuum.

When the effective NN interaction adopted is a real potential such as M3Y or DDM3Y, a suitable imaginary part has to be added to the DF model potentials in order to take into account the absorption of flux into other reaction channels not explicitly included in the model space of the CDCC equations. In general, the form factor of the imaginary potentials should be different from that of the folded real potentials. It is, however, often the case that the imaginary part having the same radial form factor as that of the folding real potential is used for simplicity: $V_{\alpha\beta}(\boldsymbol{R}) = (N_R + iN_I)V_{\alpha\beta}^{(DF)}(\boldsymbol{R})$, where N_I is a real constant and normally fixed so that the elastic scattering is reproduced, while N_R is kept fixed at unity unless otherwise mentioned. The above choice of the imaginary part does not cause a serious error for most heavy-ion reactions, since the strong absorptive characteristic of heavy ions prevents detailed shape of the potentials from affecting the calculated cross sections strongly [3], except for a few cases where nuclear rainbow scattering or molecular resonances are observed due to a weak absorption.

In order to include the Coulomb breakup process, the effective NN nuclear force is simply replaced by the Coulomb force between nucleons and the (diagonal/transition) nucleon densities are replaced by the corresponding charge densities. In a practical calculation, one has to integrate the CDCC equations with long-range Coulomb coupling potentials up to a large projectile–target separation and include a large number of partial waves [4] until a required quality of convergence is obtained in the calculated observables.

5.3   Elastic scattering and breakup reactions of composite projectiles

The CDCC method has been applied widely to the study of nuclear reactions induced by many light composite projectiles such as the deuteron, ^{3}He, ^{6}Li, ^{7}Li, ^{9}Be, ^{12}C, ^{16}O, some β-unstable nuclei like ^{7}Be, ^{8}B, ^{14}O, and the halo nuclei ^{6}He and ^{11}Li. In this short review, it is impossible to discuss all the cases and, hence, we focus on some typical examples.

5.3.1   *Deuteron breakup and elastic scattering*

The deuteron is the most typical and simplest composite projectile having a small binding energy (2.2 MeV) and its breakup process has been studied for many decades. However, it is only recently that we have begun to understand clearly the reaction mechanism of the deuteron breakup and its role in elastic scattering and transfer reactions in terms of extensive numerical works based on the CDCC method [5,7,30–32].

5.3.1.1   *Effects of deuteron breakup to spin-triplet ($S = 1$) states*

The deuteron–target interactions are calculated by the CF model, eqn (5.8), based on a p + n + target three-body model. The proton–target and neutron–target optical potentials evaluated at one-half of the deuteron incident energy are used as input. The nucleon optical potentials are composed of the central and spin–orbit components, which produce the corresponding components of the folding potential, respectively. In addition, the central part of the nucleon optical potentials folded with the small but finite D-state component of the deuteron ground state wave function produces a tensor component in the folding potential. The dashed curves in Fig. 5.4(a) are the results of the single-channel calculation of the d + ^{58}Ni elastic scattering at $E_{\mathrm{d}} = 56$ MeV using the folding model potential [7,30]. The calculated cross sections overshoot the experimental data at intermediate and large angles, while the gross features of the observed vector and tensor analysing powers are reproduced well by the folding model calculation. This may suggest that the folding model accounts reasonably well for the spin-dependent part of the deuteron–target interaction.

The deuteron breakup induces an additional correction term to the folding model potential. Since the deuteron ground state is a spin-triplet ($S = 1$) state, the breakup takes place predominantly into the spin-triplet breakup states by the spin-non-flip component of the coupling potential. This induces a large correction mainly to the deuteron–target central potential, the effect of which is observed primarily in the cross sections. In fact, CDCC calculations with the spin-triplet breakup states nicely reproduce the observed elastic-scattering cross sections for deuterons. An example is shown by the solid curves in Fig. 5.4(a) [7,30].

In addition to the central component, the deuteron breakup also gives corrections to spin–orbit and tensor components of the deuteron–target folding potential, called the dynamical spin–orbit and tensor interactions. However, the CDCC calculation for the present system shows that the deuteron breakup to the spin-triplet breakup states produces only minor effects on the spin observables, as shown by the solid curves in Fig. 5.4(a). This is a general tendency, especially for higher-energy scattering [7,31], which has been confirmed by a systematic CDCC analysis of the elastic scattering of polarized deuterons. The small breakup effect on the spin observables implies that the dynamical spin–orbit and tensor interactions almost vanish for some reason. In fact, this was proved by an approximate selection rule [31] for the tensor rank (K) of the dynamical polarization potential induced by the deuteron breakup. This selection rule was derived under a reasonable assumption of a weak spin and excitation-energy dependence of the intermediate breakup states, and it predicts that only the scalar ($K = 0$) component,

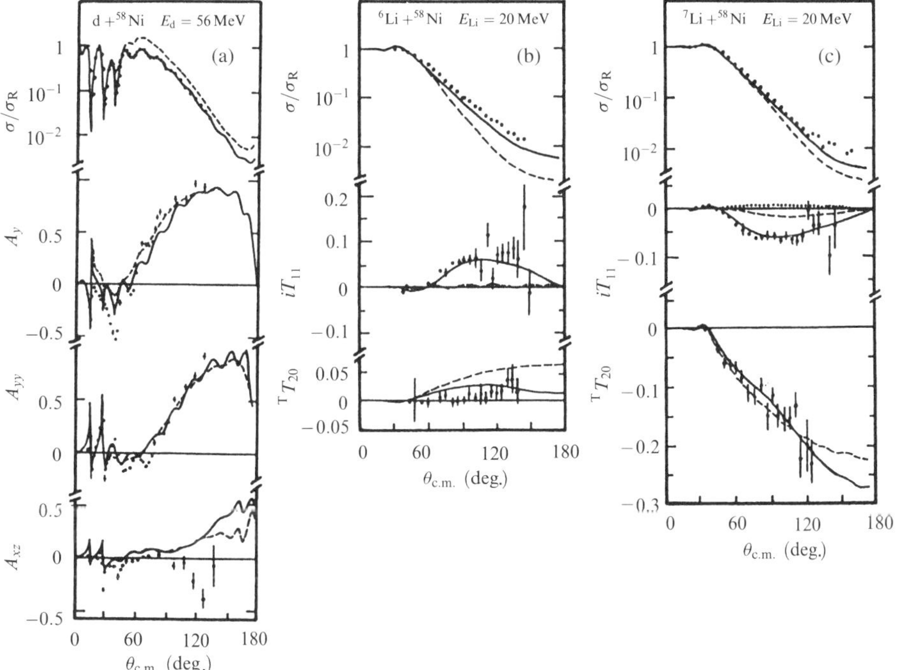

Fig. 5.4 Cross sections and vector and tensor analysing powers for the elastic scattering of (a) deuteron at $E_{lab} = 56\,\text{MeV}$, (b) ^{6}Li at $E_{lab} = 20\,\text{MeV}$, and (c) ^{7}Li at $E_{lab} = 20\,\text{Mev}$ by a ^{58}Ni target. The dashed curves are the single-channel calculations with the CF model potentials, while the solid curves are the CDCC calculations. (See text for details.) The curves are taken from Ref. [7] for the deuteron and from Refs [14,15] for ^{6}Li and ^{7}Li.

namely the central one, of the dynamical polarization potential survives and all other ($K \neq 0$) components vanish approximately [31].

5.3.1.2　*Effects of deuteron breakup to spin-singlet ($S = 0$) states*

In the folding model, the sum of the proton–target and neutron–target spin–orbit interactions has a component which induces the spin-flip transition between the spin-triplet ($S = 1$) state and the spin-singlet ($S = 0$) state of the proton–neutron system. Due to the spin-flip interaction, the incident deuteron can also be excited to the spin-singlet continuum states. A recent CDCC calculation [32] has shown that the inclusion of the $S = 0$ breakup channels greatly improves the fit of the tensor analysing powers A_{yy} at backward angles. It shows an effect very similar to one that the so-called T_{L}-type tensor interaction, introduced by hand, has on A_{yy}. In fact, it was proved in the second-order perturbation theory that the deuteron breakup to the $S = 0$ states induces the T_{L}-type tensor interaction [33].

5.3.1.3　*CDCC calculations for deuteron breakup reactions*

The deuteron breakup reactions (d $\rightarrow$ p + n) on various target nuclei have also been studied by the CDCC method [5] and it was shown to be very successful in reproducing the experimental data for the breakup cross sections. In the analyses, it was shown that a proper account of the coupling among the breakup states is extremely important. This implies that a perturbative treatment of the deuteron breakup from the ground state to the individual final (breakup) states causes a serious error and that the multistep transitions between the ground state and the breakup states as well as those among the breakup states play an important role. The validity of various types of DWBA and adiabatic theories for deuteron breakup reactions has been examined by the CDCC analyses [5].

A similar CDCC analysis was made also for the ^{3}He $\rightarrow$ d + p breakup reactions [5].

5.3.2　*Elastic scattering and breakup of light-heavy ions*

Among various heavy ions, the CDCC was first applied successfully to the ^{6}Li $\rightarrow \alpha + d$ breakup process [2] and later to other heavy ions [3]. One of the major differences between the light-ion breakup and the heavy-ion one is the existence of narrow resonance states in the continuum for the latter. The resonance states play an important role and produce unique features of the breakup effect, particularly on the spin dependence of the nucleus–nucleus interactions for non-zero-spin projectiles. For example, the coupling to the resonant breakup states induces a very large effect on the vector analysing power in the elastic scattering of ^{6}Li, which is in contrast to the deuteron breakup case. This is a result of breakdown of the above-mentioned selection rule for the tensor rank of the dynamical polarization potential due to the existence of the resonances, because the selection rule was based on the assumption of a uniform breakup continuum.

The breakup effect on the central interactions, which reflects mainly upon the cross sections is, in general, very large for the weakly bound heavy-ion projectiles as in the case of deuteron, because the central component of the dynamical polarization potential is free from the above-mentioned selection rule [30]. Again, the breakup to the resonance states plays an important role, although the non-resonant breakup process also has non-negligible effects, especially at higher energies.

5.3.2.1   *Elastic scattering of* ^{6}Li *and* ^{7}Li *at low energies*

The breakup effect on the elastic scattering of these projectiles is known to be very large in general [3]. Figure 5.5(a) shows an example of the ^{6}Li breakup effect on the elastic scattering by ^{28}Si at $E_{\text{lab}} = 99\,\text{MeV}$. The solid curve shows the result of the CDCC calculation [2] using the DF model interaction, where the ^{6}Li $\rightarrow \alpha + $ d breakup channels in the S-wave ($\ell = 0$) , P-wave ($\ell = 1$), and D-wave ($\ell = 2$) are taken into account, while the dotted curve shows the result of the single-channel calculation where the coupling to the breakup states is switched off in the same CDCC calculation. The large difference between the two curves shows a large breakup effect. Note that there exist three resonance states in the D-wave, 3^+, 2^+, and 1^+ at $E_{\text{x}} = 2.18$, 4.31, and $5.65\,\text{MeV}$, which make about 70–80% contribution to the whole breakup effect.

The dashed curve in Fig. 5.5(a) shows the single-channel calculation, same as the dotted one but using a modified real potential, the strength of which has been reduced by 45% from the original folding potential, and the resultant cross sections are very

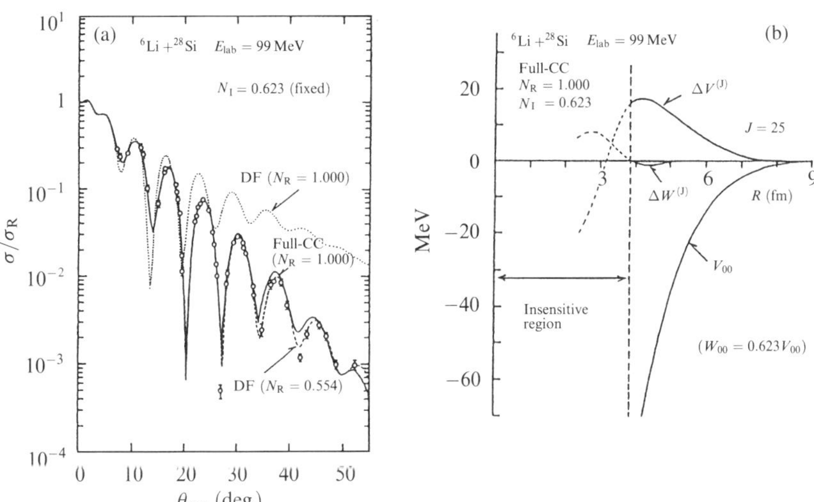

Fig. 5.5   Elastic scattering of ^{6}Li by ^{28}Si at $E_{\text{lab}} = 99\,\text{MeV}$. (a) The angular distribution of the elastic cross section. The solid curve shows the result of the CDCC calculation without renormalization to the real DF model interactions ($N_{\text{R}} = 1.0$), while the dotted and dashed curves are the single-channel DF model calculation with $N_{\text{R}} = 1.0$ and 0.554, respectively. The value of N_{I} for the imaginary part is fixed at 0.623 in all the three calculations. (b) The real part (ΔV) and the imaginary part (ΔW) of the dynamical polarization potential obtained by the CDCC calculation for a grazing angular momentum $J = 25$. V_{00} represents the real part of the elastic-channel potential obtained by the DF model with the M3Y interaction. (Taken from Sakuragi *et al.* [2] for (a) and Kamimura *et al.* [34] for (b).)

close to those of the full CDCC calculation. This result suggests that the coupling to the breakup channels induces a large repulsive correction to the real potential predicted by the DF model. This was confirmed explicity with the dynamical polarization potential, $\Delta V + i\Delta W$, calculated directly from the wave functions obtained as the solutions of the CDCC equations, as shown in Fig. 5.5(b). The real part ΔV is seen to be repulsive and its strength amounts to 40–50% of the folding potential around the nuclear surface region to which the scattering is most sensitive.

From systematic CDCC analyses [3,8] of the ^{6}Li and ^{7}Li elastic scattering, the importance of the breakup effect and its repulsive nature were found to be universal over a wide range of incident energies (up to $E/A \simeq 30\,\mathrm{MeV}$) and target nuclei. This finding explains clearly that the breakup effect is the origin of the well-known *anomalous renormalization factor* for the DF model potentials, $N_\mathrm{R} = 0.5$–0.6, found phenomenologically in the (single-channel) DF model analyses of the elastic scattering of these projectiles [28].

A similar effect of the projectile breakup is also found in the case of the ^{9}Be projectile [9].

As already mentioned, the effects of the projectile excitation / breakup on the spin observables are also significant in the elastic scattering of ^{6}Li and ^{7}Li at low incident energies [14,15]. In the elastic scattering of ^{6}Li and ^{7}Li ions, for example by a ^{58}Ni target at $E_\mathrm{lab} \simeq 20\,\mathrm{MeV}$, the spin–orbit interactions predicted by the CF model are known to produce only a tiny fraction of the vector analysing powers (iT_{11}) observed experimentally for both projectiles, which are shown by the dashed curves in Fig. 5.4(b) and (c). A large discrepancy between the folding model predictions and experimental data is resolved completely by taking into account the effects of the projectile excitation / breakup in the CDCC-like calculations, as shown by the solid curves [14,15]. For the ^{6}Li scattering, the projectile breakup to the three resonance states mentioned above plays a dominant role in this case, while for the ^{7}Li scattering the projectile excitation to the $(1/2)^-$ bound state at $0.478\,\mathrm{MeV}$ and the breakup to the α–t F-wave ($\ell = 3$) ($(7/2)^-$ and $(5/2)^-$ resonance states at $E_\mathrm{x} = 4.63$ and $6.68\,\mathrm{MeV}$) play an important role. It should be noted that a considerable improvement of the fit to the tensor analysing power ($^\mathrm{T}T_{20}$) is also observed.

In addition to the resonance states, the effect of the breakup into non-resonant continuum states was also studied in detail by the CDCC method and a significant effect comparable to that of the resonance states was identified in both projectiles above the Coulomb-barrier energies [23,35].

5.3.2.2 *Energy dependence of the dynamical effects*

At low energies, the main origin of the spin–orbit interaction for the ^{6}Li and ^{7}Li projectiles was shown to be the *dynamical* spin–orbit interaction induced by the projectile excitation to breakup states and the *static* spin–orbit interaction given by the folding models had negligible contribution. However, the dominance of the dynamical effects is no longer valid at intermediate energies, say over 100 MeV per nucleon, where the folding spin–orbit interaction dominates over the dynamical one [36]. Figure 5.6(a) shows an example of such a case; the elastic scattering of ^{6}Li at $E(^6\mathrm{Li}) = 600\,\mathrm{MeV}$ from ^{58}Ni. Here, the

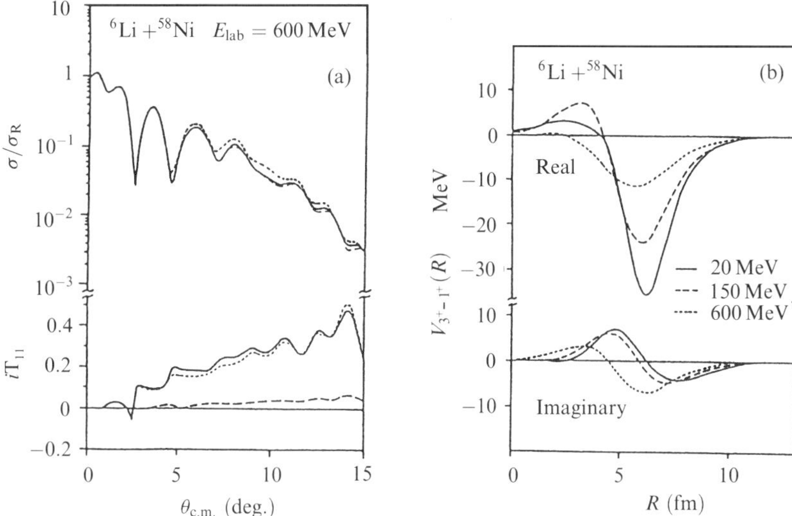

Fig. 5.6   (a) Cross section and vector analysing power of the elastic scattering of ^{6}Li by ^{58}Ni at $E_{lab} = 600$ MeV. The solid and dashed curves are the CDCC calculations [36] with and without the folding spin–orbit potential, while the dotted curves are the single-channel CF model calculation. (b) Energy dependence of the coupling potential between the 1^+ ground state and the 3^+ resonance state in ^{6}Li in the CF model. (Taken from Sakuragi [36].)

roles of the folding and the dynamical spin–orbit interactions have been reversed to those of the low-energy scattering seen in Fig. 5.4. It is also noticed that the breakup effect on the cross section becomes smaller compared to the low-energy case. These results suggest that the breakup effect itself becomes less important as a whole in this energy region.

The main origin of the drastic change of the breakup effect is related to the fact that the nuclear force itself becomes weaker as the incident energy increases until it reaches its minimal strength around 200 MeV/u. In the CDCC calculation of Fig. 5.6(a), this fact reflects indirectly upon the energy dependence of the cluster–target interactions which are the inputs of the cluster folding potentials (eqn (5)). The deuteron–target and α-target optical potentials in this energy region (100 MeV/u) are known to be much weaker than those for the low-energy scattering, which makes the CF model potentials weak. It should be noted that this energy dependence of the input optical potentials reflects not only upon the diagonal potential ($V_{00}^{(CF)}$) in the elastic channel but also upon the coupling potentials ($V_{\alpha\beta}^{(CF)}$) which induce the projectile breakup. Thus, the coupling potentials to the breakup channels themselves become weak at intermediate energies, as shown in Fig. 5.6(b), which shows the energy dependence of the coupling potential between the ground state and one of the most important breakup states in ^{6}Li, the 3^+ resonance state [36].

A similar diminishing breakup effect on the elastic scattering has also been found in the CDCC analysis based on the microscopic DF model interactions using the DDM3Y NN force, which reflects the energy dependence of the NN interaction more directly. The DF model study of the energy dependence of the breakup effect shows that the effect is still important at least in the energy range of $E/A \simeq 50\,\mathrm{MeV}$, where it induces about 15% repulsive correction to the folding potentials [37,38].

5.3.2.3 *Coulomb breakup of* ^{6}Li *and Coulomb–nuclear interference*

A loosely bound projectile like ^{6}Li incident on a heavy target nucleus may be broken up not merely by the nuclear field but also by the Coulomb field of the target nucleus. If it is a *pure Coulomb breakup* process and can be treated by the *first-order perturbation theory*, it is known that the Coulomb breakup cross section can be related directly to the cross section of the inverse process, viz. the radiative capture reaction such as $\alpha + d \rightarrow$ ^{7}Li $+ \gamma$. Based on this idea, it was proposed to use Coulomb breakup reactions to extract astrophysical information relevant to the radiative capture reactions at extremely low energies, which is called the *Coulomb dissociation method* [39]. This method has been applied to various heavy ions and is believed to be a useful one [40–42].

However, while applying this scenario, one should be very careful about the above-mentioned conditions in which the Coulomb dissociation method can be valid. In fact, in the case of ^{6}Li breakup by ^{208}Pb at 156 MeV, for which the method was applied [40] 'successfully', it was shown by the CDCC calculation [4] that the above two conditions were not satisfied at all. It was found instead that the nuclear breakup contribution was far from negligible even at the most forward angles and, in addition, the multistep transitions among the breakup channels played a significant role, indicating the breakdown of the first-order treatment of the breakup process. As for the nuclear breakup contribution, it was estimated [40] to be smaller by one order of magnitude than the Coulomb breakup one, while the CDCC predicted the same order of contribution as for both, leading to a strong interference between the two amplitudes, consistent with the experimental data (Fig. 5.7(a)).

The main difference between the two analyses comes from the nuclear model for the coupling form factors [4]. The former calculation assumed a collective surface-vibration model normally used for bound excited states, giving a coupling form factor of rather short range, while the CDCC used the DF model based on realistic cluster model wave functions which properly describe the unbound nature of the excited (breakup) states and, hence, gave a coupling form factor having a realistic long tail at the nuclear surface where the nuclear breakup took place most strongly, as shown in Fig 5.7(b). This example demonstrates clearly the importance of using a realistic nuclear-structure model which properly describes the spatial nature of the ground state as well as the final breakup states. In this sense, the cluster model may be an ideal nuclear-structure model as far as the ^{6}Li $\rightarrow \alpha + $ d breakup reaction is concerned.

The nuclear breakup contribution is also important in the ^{7}Li $\rightarrow \alpha + $ t breakup by heavy nuclei [43], despite a strong E1 transition due to the Coulomb interaction allowed for ^{7}Li. Again, this is due to a developed cluster structure in the ^{7}Li nucleus. The nuclear breakup is also known to be important in the ^{16}O $\rightarrow \alpha + {}^{12}$C breakup reaction [44].

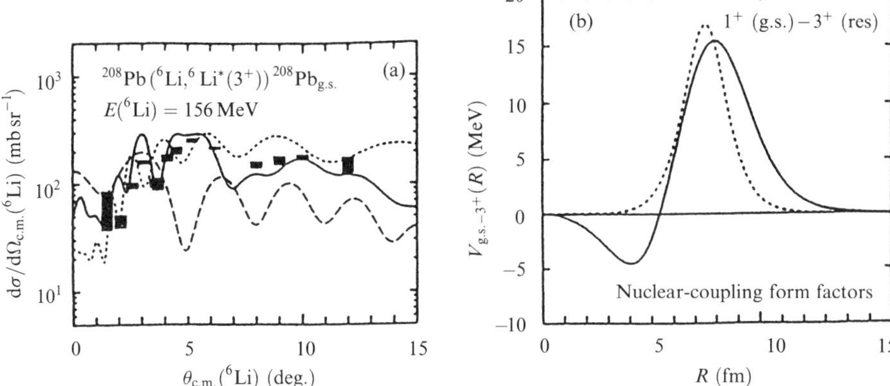

Fig. 5.7   (a) Cross section for the $^6\mathrm{Li} \to \alpha + \mathrm{d}$ breakup by $^{208}\mathrm{Pb}$ at $E_{\mathrm{lab}} = 156\,\mathrm{MeV}$ via the 3^+ resonance state in $^6\mathrm{Li}$ as a function of α–d c.m. angle [40]. The solid curve is the CDCC calculation [4] with the nuclear and Coulomb breakup contributions included, while the dashed (dotted) curve is the CDCC calculation with the nuclear (Coulomb) breakup contribution alone. (b) Comparison of nuclear coupling potentials for the $^6\mathrm{Li}(1^+_{\mathrm{g.s.}} \to 3^+_{\mathrm{res}})$ transition. The solid curve is given by the DF model using microscopic cluster model wave functions [4], while the dotted curve is given by the surface-vibrational model. (Taken from Hirabayashi and Sakuragi [4].)

This is mainly because the E1 transition is strongly suppressed and, in fact, forbidden in the first-order Coulomb excitation, due to the symmetry of the system.

On the other hand, the nuclear breakup contributions are found to be small or almost negligible for the $^{14}\mathrm{O} \to {}^{13}\mathrm{N} + \mathrm{p}$ and $^8\mathrm{B} \to {}^7\mathrm{Be} + \mathrm{p}$ breakup reactions [41,42].

5.3.2.4 $^{12}\mathrm{C} \to 3\alpha$ *breakup effect on the excitation of the 3α-cluster state*

Finally, let us discuss the breakup effect on the cluster-state excitation in a heavier nucleus like $^{12}\mathrm{C}$ or $^{16}\mathrm{O}$. As already discussed, the ground state of these composite nuclei is no more cluster-like and the breakup effect is not very important compared to that in the lighter projectile nuclei such as $^6\mathrm{Li}$ and $^7\mathrm{Li}$. However, the cluster-like excited states near the breakup threshold for the associated clusters will be affected strongly by the coupling to surrounding breakup states (see Fig. 5.3).

A typical example is given in Fig. 5.8, which shows the results of the CDCC analysis [45] of the inelastic excitation of $^{12}\mathrm{C}$ by $^6\mathrm{Li}$ to the lowest three excited states of $^{12}\mathrm{C}$, the 2^+_1, 0^+_2, and 3^-_1 states at $E_{\mathrm{x}} = 4.44$, 7.65, and 9.64 MeV, respectively. The 0^+_2 state has a well-developed 3α-cluster structure, while the other two states as well as the ground state have no such cluster nature. In the figure, the dotted curves are the result of the four-channel calculation which includes the coupling among the ground state and the three excited states of $^{12}\mathrm{C}$ but neglects the coupling to the $^6\mathrm{Li} \to \alpha + \mathrm{d}$ breakup and the $^{12}\mathrm{C} \to 3\alpha$ breakup channels, while the dashed curves are the result of CDCC calculation which further includes the $^6\mathrm{Li}$ breakup channels but not the 3α-breakup channels. The mutual excitations of the $^6\mathrm{Li}$ and $^{12}\mathrm{C}$ nuclei are also taken into account. The drastic

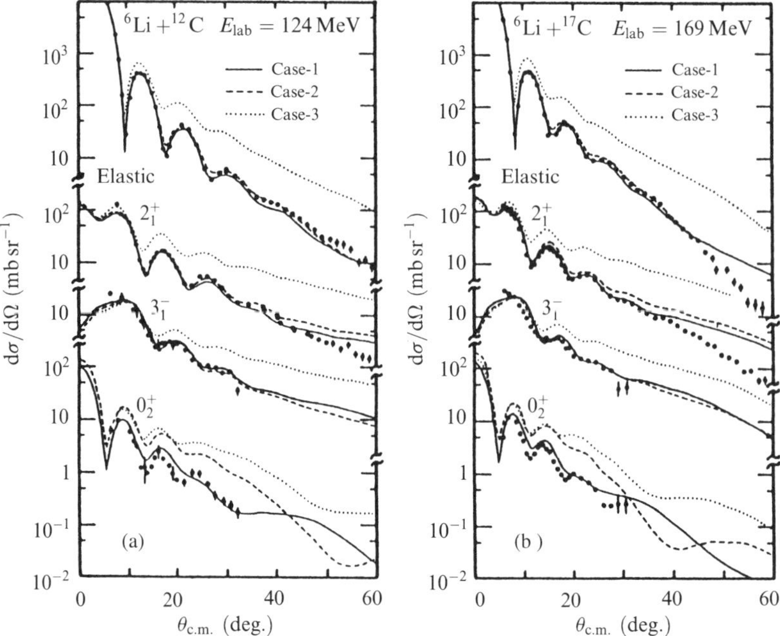

Fig. 5.8 Cross sections for the elastic and inelastic scattering of ^{6}Li by ^{12}C at $E_{\text{lab}} = 124$ and 169 MeV. The solid and dashed curves are the CDCC calculations [45] with the ^{6}Li breakup channels; the former further includes the coupling to the 3α-breakup channels but the latter does not. The dotted curves based on the CC calculation with only the four states of ^{12}C included. (Taken from Sakuragi *et al.* [45].)

change from the dotted curves to the dashed one shows a large effect of the ^{6}Li breakup not only on the elastic scattering but also on the inelastic excitation of the target nucleus (see Fig. 5.2(b)), leading to excellent fits to the experimental data, except for the 0_2^+ state where the CDCC calculations with the ^{6}Li breakup still deviate appreciably from the data.

The solid curves are the result of CDCC calculation [45] which takes into account the coupling to the 3α breakup states in addition to the channels included in the calculation of the dashed curves. The improvement of the fit to the 0_2^+-state data is impressive enough. It is also interesting to note that the inclusion of the 3α breakup states does not affect the fits to the elastic scattering and the inelastic scattering to other non-cluster excited states, reflecting the large difference of their nuclear structure from that of the 0_2^+ state.

A similar important role of the ^{12}C $\rightarrow 3\alpha$ breakup states in the cluster-state excitation has also been found for the ^{12}C + ^{12}C system, both in high-energy scattering [46] as well as in low-energy one leading to the formation of molecular resonances having ^{12}C + 3α and $3\alpha + 3\alpha$ configurations [47].

5.4 Summary and conclusions

The important role of nuclear breakup process in nuclear reactions induced by light composite projectiles has been studied systematically in terms of the CDCC method based on the cluster model wave functions for light nuclei and some of the typical examples are presented in this review article, in addition to an outline of the theoretical framework. The reaction mechanism of the breakup process and its effect on nucleus–nucleus interactions are shown to be understood clearly by referring to the relevant nuclear cluster structures of the ground and excited states of the nuclei. In other words, it is essential to use a reliable nuclear structure model to describe the nuclear excitation and breakup processes in light composite nuclei.

It is also important to note, although not discussed in detail here, that the multistep transitions among various excited states, especially those among the breakup states, are very important to be taken into account because of a large overlap of wave functions which extend beyond the nuclear surface region where nuclear breakup takes place most strongly. In this sense, the CDCC formalism based on the microscopic cluster model is an ideal tool for studying nuclear breakup processes in nuclear reactions. Although some important reaction mechanisms associated with the breakup processes have been unveiled so far and also their effects on the nucleus–nucleus interactions are becoming clear, a number of questions remain unresolved concerning these processes.

5.5 Acknowledgements

The present review is based on a series of studies with a number of collaborators, especially M. Kamimura, M. Yahiro, Y. Iseri, H. Kameyama, M. Kawai, M. Tanifuji, Y. Hirabayashi, S. Okabe, C. Samanta, Y. Abe, K. Katori, S. Kubono, and many others. The author thanks them all for stimulating discussions and fruitful collaborations.

References

1. M. Kamimura, M. Yahiro, Y. Iseri, Y. Sakuragi, H. Kameyama, and M. Kawai, *Prog. Theor. Phys. (Suppl.)* **89** (1986) 1 and references therein.
2. Y. Sakuragi, M. Yahiro, and M. Kamimura, *Prog. Theor. Phys.* **70** (1983) 1047.
3. Y. Sakuragi, M. Yahiro, and M. Kamimura, *Prog. Theor. Phys. (Suppl.)* **89** (1986) 136 and references therein.
4. Y. Hirabayashi and Y. Sakuragi, *Phys. Rev. Lett.* **69** (1992) 1892.
5. Y. Iseri, M. Yahiro, and M. Kamimura, *Prog. Theor. Phys. (Suppl.)* **89** (1986) 84.
6. M. Yahiro, M. Nakano, Y. Iseri, and M. Kamimura, *Prog. Theor. Phys.* **67** (1982) 1467.
7. M. Yahiro, Y. Iseri, H. Kameyama, M. Kamimura, and M. Kawai, *Prog. Theor. Phys. (Suppl.)* **89** (1986) 32.
8. Y. Sakuragi, *Phys. Rev. C* **35** (1987) 2161.
9. Y. Hirabayashi, S. Okabe, and Y. Sakuragi, *Phys. Lett. B* **221** (1989) 227.
10. T. Yamagata, K. Yuasa, N. Inabe, M. Nakamura, M. Tanaka, S. Nakayama, K. Katori, M. Inoue, S. Kubono, T. Itahashi, H. Ogata, and Y. Sakuragi, *Phys. Rev. C* **39** (1989) 873.

11. K. Yabana, Y. Ogawa, and Y. Suzuki, *Phys. Rev. C* **45** (1992) 2909; *Nucl. Phys. A* **539** (1992) 295.

12. I. J. Thompson, J. S. Al-Khalili, J. A. Tostevin, and J. M. Bang, *Phys. Rev. C* **47** (1993) R1364; J. S. Al-Khalili and J. A. Tostevin, *Phys. Rev. C* **49** (1994) 386; J. S. Al-Khalili, I. J. Thompson, and J. A. Tostevin, *Nucl. Phys. A* **581** (1995) 331.

13. Y. Sakuragi, S. Funada, and Y. Hirabayashi, *Nucl. Phys. A* **588** (1995) 65c; S. Funada, *Prog. Theor. Phys.* **93** (1995) 373.

14. H. Nishioka, J. A. Tostevin, R. C. Johnson, and K.-I. Kubo, *Nucl. Phys. A* **415** (1984) 230.

15. H. Ohnishi, M. Tanifuji, M. Kamimura, Y. Sakuragi, and M. Yahiro, *Nucl. Phys. A* **415** (1984) 271.

16. N. Austern, Y. Iseri, M. Kamimura, M. Kawai, G. Rawitscher, and M. Yahiro, *Phys. Rep.* **154** (1987) 125.

17. M. Tanifuji, *Nucl. Phys.* **58** (1964) 81; S. T. Butler, *Nature* **207** (1965) 1346.

18. R. C. Johnson and P. J. R. Soper, *Phys. Rev. C* **1** (1970) 976; H. Amakawa, S. Yamaji, A. Mori, and K. Yazaki, *Phys. Lett. B* **82** (1979) 13; H. Amakawa and N. Austern, *Austr. J. Phys.* **36** (1983) 633.

19. N. Austern, *Direct nuclear reaction theories* (Wiley, New York, 1970).

20. J. E. Bowsher, T. B. Clegg, H. J. Karwowski, E. J. Ludwig, W. J. Thompson, and J. A. Tostevin, *Phys. Rev. C* **45** (1992) 2824.

21. K. Ikeda, H. Horiuchi, and S. Saito, *Prog. Theor. Phys. (Suppl.)* **68** (1980) 1.

22. M. Kamimura, *Nucl. Phys. A* **351** (1981) 456.

23. Y. Hirabayashi, *Phys. Rev. C* **44** (1991) 1581.

24. M. Kawai, *Prog. Theor. Phys. (Suppl.)* **89** (1986) 11.

25. F. G. Perey and G. R. Satchler, *Nucl. Phys. A* **97** (1967) 515; H. Amakawa and K.-I. Kubo, *Nucl. Phys. A* **266** (1976) 521.

26. G. Bertsch, J. Borysowicz, H. McManus, and W. G. Love, *Nucl. Phys. A* **284** (1977) 399.

27. M. El-Azab Farid and G. R. Satchler, *Nucl. Phys. A* **438** (1985) 525.

28. G. R. Satchler and W. G. Love, *Phys. Rep.* **55** (1979) 183.

29. D. T. Khoa and W. von Oertzen, *Phys. Lett. B* **304** (1993) 8; D. T. Khoa, G. R. Satchler, and W. von Oertzen, *Phys. Rev. C* **51** (1995) 269; D. T. Khoa *et al.*, *Phys. Rev. Lett.* **74** (1995) 34.

30. Y. Iseri, H. Kameyama, M. Kamimura, M. Yahiro, and M. Tanifuji, *Nucl. Phys. A* **490** (1988) 383.

31. Y. Iseri, M. Tanifuji, H. Kameyama, M. Kamimura, and M. Yahiro, *Nucl. Phys. A* **533** (1991) 574.

32. Y. Iseri and M. Tanifuji, *Phys. Lett. B* **354** (1995) 183.

33. M. Tanifuji and Y. Iseri, *Prog. Theor. Phys.* **87** (1992) 247.

34. M. Kamimura, M. Yahiro, Y. Iseri, Y. Sakuragi, M. Nakano, and Y. Fukushima, *Proc. 1983 RCNP int. symp. on light ion reaction mechanism* (Osaka, May 1983), p. 558.

35. Y. Sakuragi, M. Yahiro, M. Kamimura, and M. Tanifuji, *Nucl. Phys. A* **480** (1988) 361.

36. Y. Sakuragi, *Phys. Lett. B* **220** (1989) 22.
37. Y. Sakuragi, M. Ito, Y. Hirabayashi, and C. Samanta, *Prog. Theor. Phys.* **98** (1997) 521.
38. C. Samanta, Y. Sakuragi, M. Ito, and M. Fujiwara, *J. Phys. G: Nucl. Part. Phys.* **23** (1997) 1697.
39. G. Baur, C. A. Bertulani, and H. Rebel, *Nucl. Phys. A* **458** (1986) 188.
40. J. Kiener *et al.*, *Z. Phys. A* **399** (1991) 489; *Phys. Rev. C* **44** (1991) 2195.
41. T. Motobayashi *et al.*, *Phys. Lett. B* **264** (1991) 259.
42. T. Motobayashi *et al.*, *Phys. Rev. Lett.* **73** (1994) 2680.
43. Y. Sakuragi and Y. Hirabayashi, *Origin and evolution of the elements*, eds. T. Kajino and S. Kubono (World Scientific, 1993), p. 220.
44. D. O'Kelly *et al.*, *Phys. Lett. B* **393** (1997) 301.
45. Y. Sakuragi, M. Kamimura, and K. Katori, *Phys. Lett. B* **205** (1988) 204.
46. Y. Sakuragi and Y. Hirabayashi, *J. Phys. Soc. Jpn.* **58** (Suppl.) (1988) 560.
47. Y. Hirabayashi, Y. Sakuragi, and Y. Abe, *Phys. Rev. Lett.* **74** (1995) 4141.

6

Nuclear clusters

S. Ohkubo, T. Yamaya, and P. E. Hodgson

6.1 INTRODUCTION

6.1.1 *The α-particle model—a long history*

Since the observation of radioactivity by Becquerel in 1896 [1], and subsequently of α-particles from heavy nuclei by Rutherford in 1903, the α-particle has continued to be of central importance in nuclear physics. The nucleus itself was discovered by Rutherford [2] using α-particles. The α-particle model is the oldest nuclear model, since Gamov showed [3,4] that α-decay in heavy nuclei can be explained by assuming that in the decay process an α-particle leaves the nucleus by tunnelling through a Coulomb barrier. The classical α-particle model, in which the compositeness of the particle, and therefore the Pauli principle, is not taken into account, was useful in understanding the structure of light nuclei [5]. However, after the success of the shell model [6] and the collective model [7], which revealed single-particle and collective (rotational and vibrational) aspects of nuclei, the α-particle model was almost forgotten for some years.

Nevertheless, one of the characteristic features of nuclear structure is that it shows several aspects that are very different from each other: nuclear clustering is one of the important aspects of nuclear structure. Unlike atoms and molecules, the nucleus is a self-bound many-body system. As the nucleus is heated, nuclear clusters (α, t, ^{3}He, ^{12}C, ^{16}O, ...) as well as its constituent neutrons and protons can be emitted above the threshold energies. In most nuclei the threshold energy of α-particle emission is very low; in medium and heavy nuclei the ground state is unstable against α-decay. This suggests that the α-particle, which is an extremely stable magic nucleus having high symmetry and large binding energy, can be considered as an important constituent subunit in the nucleus. In principle, many kinds of clusters are possible, but the probability of formation depends on the stability of the clusters, and of all possible clusters the α-particle is the most stable. This is the basic idea of the α-particle model. Unlike the shell model and the collective model, one of the advantages of the α-particle model is that it can treat the bound and scattering states of the nucleus in a unified way. The clustering is more likely to take place in the nuclear surface where the density is lower, and therefore light nuclei containing $2N$ neutrons and $2N$ protons are expected to have a cluster structure [8].

6.1.2 *The revival of the α-particle model*

The α-particle model was revived to explain the structure of ^{8}Be by assuming a phenomenological α–α potential [9]. The idea of α-clustering in light nuclei has been formulated in its present form by Wildermuth and Kanellopoulos [10] based on the equivalence of the harmonic-oscillator shell model and the cluster model representation [11]. Such a structure for ^{8}Be has been revealed by a microscopic model which takes into account the Pauli principle [12]. Brink formulated the quantum α-particle (cluster) model of light nuclei in the generator coordinate formalism [13], which fully takes into account the Pauli principle; the shell model was used in the no-clustering limit. In 1968, Ikeda extended the idea of the cluster model by introducing the *threshold rule* and formulated the Ikeda diagram, which shows what kind of cluster structure appears as the excitation energy increases [14]. The 1970s, saw a renaissance of the α-cluster model when many nuclear physicists studied nuclear cluster and heavy-ion reactions and successfully revealed the clustering aspects of light nuclei [15–20]. After the model enjoyed its prosperity in light nuclei in the 1970s, it encountered difficulties when extended to the fp-shell region. It was thought, rather pessimistically, that the model may not be valid in heavier nuclei where the strong spin–orbit force breaks the α-correlations. This difficulty has been overcome very recently and again it has become clear that the α-cluster model is of vital importance in the fp-shell nuclei as well. Also, the long-standing problem of the α-width of α-decay in heavy nuclei is now finally being solved in the context of the α-cluster model. This review emphasizes the importance and fruitfulness of the α-cluster viewpoint [21], not only in light nuclei but also in heavier nuclei, based on our research.

It is of course always possible to treat these problems microscopically by analysing each of the four constituent nucleons of the α-particle individually. Although this is a more fundamental approach, as it relates the phenomena to the Nucleon–Nucleon (NN) interactions, it is far more complicated mathematically, and quite often the data are better fitted by the much simpler α-cluster model. Thus, for example, Satchler *et al.* [22] obtained excellent fits to the neutron–α and proton–α scattering cross sections and analysing powers, whereas the very complicated microscopic analysis of Sugie *et al.* [23] did not fit the data very well. A similar comparison can be made for the analyses of α–α scattering by Buck *et al.* [24] and Tohsaki [25]. These comparisons show the usefulness of the α-cluster model.

The formalism of nuclear cluster models, such as the Generator Coordinate Method (GCM), the Resonating Group Method (RGM), the Orthogonality-Condition Model (OCM), and the Local-Potential Cluster Model (LPCM), is described in an appendix at the end of this chapter.

6.1.3 *Experimental studies of the α-cluster structure*

Experimental studies of the clustering of nucleons, particularly α-clustering, have led to the discovery of rotational cluster bands in nuclei. Evidence for α-cluster structures in nuclei comes from observations of energy levels of the rotational bands and their reduced widths or spectroscopic factors for α-transfer reactions. Measurements of α-particle capture, γ-ray, and α-particle scattering by a target nucleus have proved to be successful tools for studying the energy levels of an α + core structure. These measurements allow the observation of resonances corresponding to the α-cluster states. The energy dependence of the cross sections of elastically scattered α-particles is a significant datum for the identification of resonance states by phase shift analysis and this makes clear the character of the level structure, namely the spin-parities and reduced widths of the α-cluster states [26]. Measurements of γ-ray angular distributions, $\gamma-\gamma$ coincidence, and attenuated Doppler shifts for the (α, γ) reaction are also useful for the observation of α-cluster states in nuclei. Simpson *et al.* [27] suggested that the energy levels of the ground state band in ^{44}Ti, and the intraband E2 transitions are very well described by an asymmetric-rotor model with asymmetry parameter $\gamma \sim 20$. Such a quasi-rotational behaviour is attributed to an α-cluster outside a closed core. For the investigations of high-spin cluster states, heavy-ion induced γ-ray reactions, such as the ^{28}Si $(^{19}$F, p2n$\gamma)$ ^{44}Ti reaction, are efficient and appropriate. The E2 transitions observed in such reactions often show members of a rotational band starting from the bandhead state to the high-spin state, which, however, are not necessarily the α-cluster rotational bands.

In addition to these reactions, the clustering of nucleons has been studied by a variety of transfer reactions, including α-transfer reactions such as $(^{6}$Li, d), $(^{7}$Li, t), and $(^{16}$O, ^{12}C) or their inverse reactions for α-stripping or α-pickup, respectively. In particular, the $(^{6}$Li, d) and (d, ^{6}Li) reactions are important for experimental studies of α-clustering. The advantages of these reactions are as follows:

(1) a large overlap of the wave functions of deuteron and an α-particle in ^{6}Li,
(2) a well-known effective interaction between deuteron and the α-particle in ^{6}Li,
(3) diffractive j-dependent angular distributions arising from a single allowed L-transfer,
(4) favourable L-transfer to $J^{\pi} = 0^{+}$ as well as to high-spin states, i.e. a wide Q-window,
(5) no excited states in ^{6}Li,
(6) the applicability of the Distorted Wave Born Approximation (DWBA) analysis for the direct α-transfer reaction.

Features (3) and (4) are particularly important, as many α-cluster levels with a wide range of angular momenta comprise a rotational band. A DWBA calculation for the α-particle transfer in the $(^{6}$Li, d) and (d, ^{6}Li) reactions shows that the reactions occur in a region of the nuclear surface where the amplitudes of the overlap integral in the reaction show a sharp maximum related to only a few partial waves. Since the DWBA analysis is a good approximation for the $(^{6}$Li, d) reaction, the optical potential parameters obtained from the analysis of the elastic scattering data are expected to be appropriate for the DWBA calculations. Consequently, the $(^{6}$Li, d) reaction is one of the most suitable ones for a DWBA analysis.

Ever since its discovery, α-decay has been used as a probe for studying nuclear structure since its energy, mass, and decay width relations are very simple. Furthermore, measurements of α-decay lifetimes and α-branching ratios of states are very useful for investigating the core–α-cluster potential. In particular, these measurements are indispensable for studies of a significant α-cluster component in low-lying states of heavy nuclei in the region of mass number $A \sim 200$. Some heavy nuclei in the actinide region are known to decay by both α- and exotic-cluster emissions.

6.2 The α-cluster structure of light nuclei

6.2.1 ^{8}Be *and* ^{12}C: 2α *and* 3α *structures*

^{8}Be is an unstable nucleus. The microscopic RGM, which takes into account the Pauli principle, reproduces the observed phase shifts of α–α scattering [12], and the resonance property of the ground state rotational band is described well by various α-cluster models [24,25,28–31]. The local potentials with a hard core [28,32] which reproduce the α–α scattering were shown to be derived equivalently from a deep local supersymmetric-theory potential [33]. The simplest model which takes into account the Pauli principle in the sense of the OCM [29] is a cluster model with a deep local potential between the α-particles [24]. This local potential is almost exactly the same as the double-folding model potential [34]. The α–α interaction can be used in the study of nuclear reactions involving α-particles and cluster structures of other nuclei [35]. Non-$4N$ nuclei such as ^{6}Li, ^{7}Li, ^{7}Be, ^{9}Be, ^{10}B, and ^{11}B in this mass region are also described well by the cluster model.

The structure of ^{12}C is described by various α-cluster models [36–38]. As shown in Fig. 6.1, the energy levels below $E_{\mathrm{x}} = 15\,\mathrm{MeV}$ are reproduced well by the 3α-cluster model calculations [36].

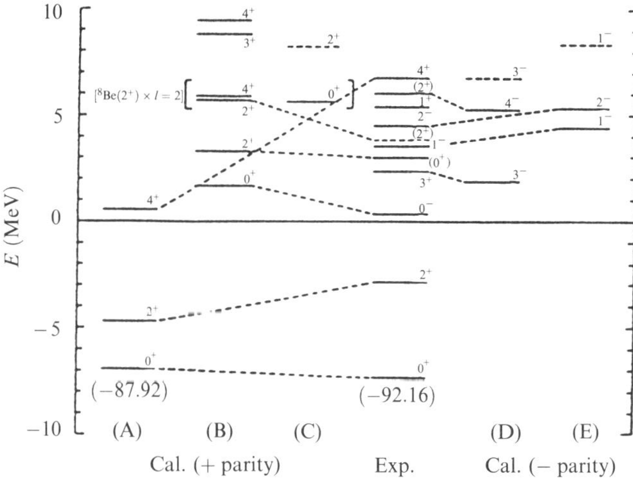

Fig. 6.1 The energy spectra of ^{12}C calculated in the 3α GCM cluster model are classified into several bands and are compared with experiment. (A) $K = 0^+$ ground state band, (B) the first family of the excited positive-parity states including the 0_2^+ and 2_2^+, and (C) the second excited positive-parity band. Uncertain higher members are shown by dashed lines. (D) $K = 3^-$ band and (E) $K = 1^-$ band. (Taken from Uegaki *et al.* [36].)

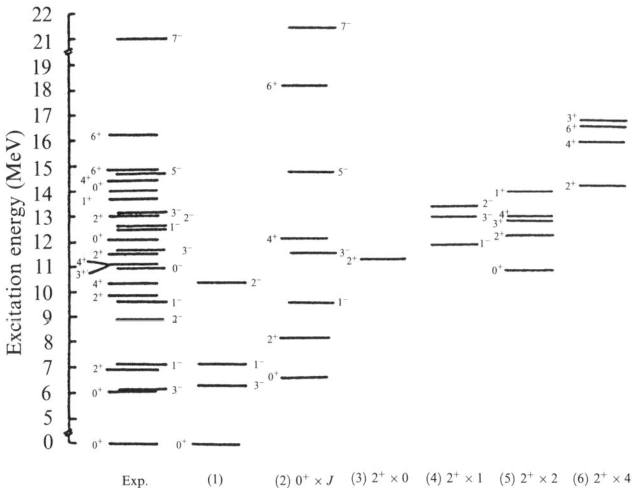

Fig. 6.2 The energy levels of ^{16}O calculated in the $\alpha + {}^{12}$C model [39] are classified according to the coupling scheme and are compared with experiment. (1) Shell model states, (2) $\alpha + {}^{12}$C(g.s.) coupling, (3)–(6) states of the $\alpha + {}^{12}$C(2^+) coupling with its relative angular momentum 0, 1, 2, and 4, respectively. (Taken from Ikeda *et al.* [17].)

6.2.2 *The structure of ^{16}O: double-closedness and α-clustering*

The structure of ^{16}O has been studied successfully using the $\alpha + {}^{12}$C OCM, which takes the excitation of the core into account [39]. The $K = 0^+$ band starting from the mysterious 0^+ state at 6.06 MeV and the $K = 0^-$ band whose bandhead state is 1^- at 9.59 MeV are described as parity doublet bands which have a common $\alpha + {}^{12}$C configuration. The α-clustering is more developed for the $K = 0^-$ band than for the $K = 0^+$ band. Other states, including the shell-model-like states, are also described by the model (Fig. 6.2). Many states are classified in the α-cluster model as a weak coupling of the relative motion and the excited states of ^{12}C. It is noted that the mysterious 0^+ state and the ground state are simultaneously reproduced by the α-cluster model. This means that the double magic nature of the ground state is closely related to the appearance of the α-cluster 0^+ state as the first excited state. The OCM calculation assuming realistic oscillator width parameters for α-particle and ^{12}C further improves the agreement with experiment: its precise wave functions are applied to the astrophysical problem of ^{4}He-burning of ^{12}C at the very low energy region [40]. The energy levels and transition probabilities of ^{16}O are also described well by the $\alpha + {}^{12}$C cluster model with a local potential, except that the ground state is beyond its scope [41].

6.2.3 *The α-cluster structure of ^{20}Ne: unification of bound and scattering states*

The ground state band and the $K = 0^-_1$ band of ^{20}Ne are described as parity doublets which have an $\alpha + {}^{16}$O cluster configuration. The $K = 0^-$ band has a well-developed

α-cluster structure, while the ground state band has a transient character between shell-like and cluster configurations. Although the ground state band alone is also described by the shell model, the large α-widths of the $K = 0^-$ band are difficult to explain in the model. The $K = 0_2^+$ band at 6.72 MeV has a shell-like character and the $K = 0_3^+$ band at 7.19 MeV has predominantly a core $+ 2\alpha$ configuration. These two bands and the $K = 2^-$ band are described well in the framework of the α-cluster model in which the dissociation of the ^{16}O core is taken into account (Fig. 6.3) [42]. The most characteristic mode peculiar to the cluster structure in ^{20}Ne is the appearance of the $K = 0_4^+$ band (0^+: $\sim$8.7 MeV, 2^+: $\sim$8.8 MeV, 4^+: 10.80 MeV), which is located near the Coulomb barrier and has a very large α-width. This band is the so-called higher nodal band, in which the relative motion between the α-particle and ^{16}O is excited. As for the high-spin members of the band, by investigating the angular distributions of α-particle scattering from ^{16}O between 20 and 25 MeV using the OCM and the folding model [43], with a potential derived from the effective two-body force, it has been shown that the Backward Angle Anomaly (BAA) or Anomalous Large Angle Scattering (ALAS) in α-particle scattering from ^{16}O (the anomaly consists in back-angle cross sections that are one to

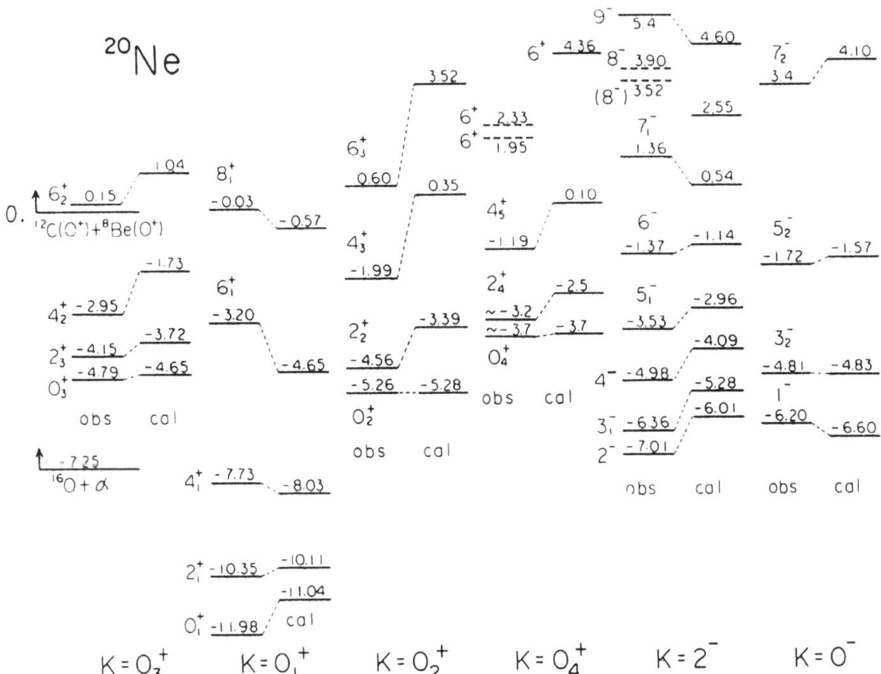

Fig. 6.3 The observed rotational bands in ^{20}Ne are compared with the α-cluster model calculation in which the α-excitation of the core is included. Energies are measured in MeV from the threshold into the subunit clusters ^{12}C and ^{8}Be. (Taken from Fujiwara [42].)

     S. OHKUBO *ET AL.*

two orders of magnitude greater than those found in scattering from neighbouring nuclei) is caused by the broad resonances of the 6^+ and 8^+ members of the higher nodal band. This gives a unified explanation of the α-cluster structure at bound and scattering states. As shown in Fig. 6.4, the same interaction potential between the α-particle and ^{16}O can describe the α-particle scattering up to much higher energies [44]. A phenomenological global potential with Woods–Saxon squared form factor, which is very similar to the above folding potential, can also reproduce the $\alpha + {}^{16}$O scattering in a wide range of incident energies up to 166 MeV [45]. It is important that the local potentials can describe in a unified way the characteristic features of the $\alpha + {}^{16}$O system from very high energies to the bound state: rainbow scattering, ALAS, higher nodal band, parity doublet $K = 0^-$

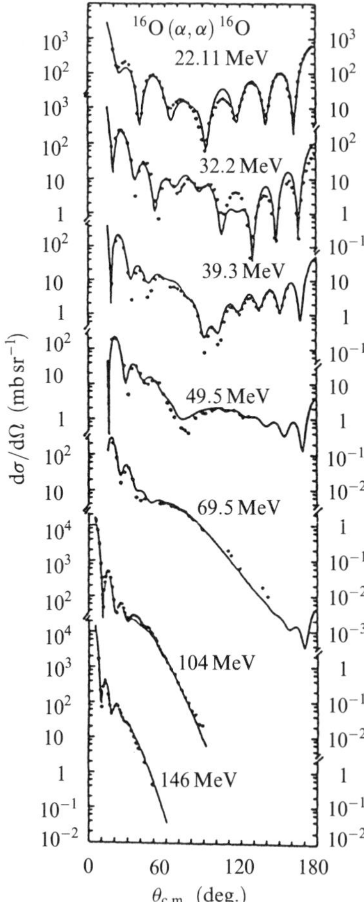

Fig. 6.4 Comparison of folding model cross sections of ^{16}O (α, α) ^{16}O scattering at various energies with experiment. (Taken from Ohkubo *et al.* [44].)

band, and the shell-like ground band. As to the high-spin member 10^+ of the higher nodal band, by studying the fusion excitation function of the $\alpha + {}^{16}\text{O}$ system with the global optical potential [46], it is shown to be located at about $E_x = 29\,\text{MeV}$. Buck *et al.* [47] showed that the parity doublet bands are reproduced well by an α-cluster model with a simple local potential with hyperbolic cosine form factor; the potential is too short-ranged to describe α-particle scattering from ${}^{16}\text{O}$ [45].

6.3   Extension of the α-cluster model to the ${}^{44}\text{Ti}$ region

6.3.1   *The coexistence model and the quartet model*

The low-lying intruder states in the fp-shell region, such as the 0^+ (3.35 MeV) state in ${}^{40}\text{Ca}$ and the 0^+ (1.84 MeV) state in ${}^{42}\text{Ca}$, as well as in the sd-shell region are difficult to describe by the conventional shell model. Gerace and Green [48] reproduced the low-lying states by using the coexistence model, in which the spherical shell model states are strongly admixed with the deformed states, including the 4p–4h excitation. This suggested, from the shell model point of view, the importance of the four-body correlations in the intruder states. The four-body correlations or α-correlations in the fp-shell region were studied theoretically in the quartet model [49,50]. To observe the quartet states, the $({}^{16}\text{O}, {}^{12}\text{C})$ reaction was studied extensively [51]. However, it seems that no conclusive result on the definite existence of α-cluster states in the fp-shell region was obtained from the $({}^{16}\text{O}, {}^{12}\text{C})$ reactions.

6.3.2   *The α-cluster model and its dilemma*

The nucleus ${}^{44}\text{Ti}$ is the analogue of ${}^{20}\text{Ne}$ and many theoretical and experimental studies have been devoted to it. In the α-transfer reaction ${}^{40}\text{Ca}\,({}^{6}\text{Li, d})\,{}^{44}\text{Ti}$ studied by Strohbush *et al.* [52], the state at 8.54 MeV (tentatively assigned as 0^+) is strongly enhanced. Also, the state at 11.73 MeV (1^-), which was identified by Frekers *et al.* [53] by inspection of their ${}^{40}\text{Ca}(\alpha, \alpha)$ excitation function, was proposed as an α-cluster state. These two excited states were later interpreted as the bandhead states of the parity doublet bands which have the $\alpha + {}^{40}\text{Ca}$ cluster structure.

The microscopic $\alpha + {}^{40}\text{Ca}$ cluster model in the RGM, which was very successful in the ${}^{20}\text{Ne}$ region, was applied to study the α-cluster aspects of ${}^{44}\text{Ti}$ [54,55]. Although the energy levels of the ground state band and the $B(E2)$ values were reproduced by the model, it was difficult to reproduce the excited α-cluster bandhead 0^+ (8.54 MeV) and 1^- (11.73 MeV) states. On the other hand, by using a phenomenological cluster model of Buck *et al.* [47] with a hyperbolic cosine local potential, it was shown by Pal and Lovas [56] that the energy levels and the large $B(E2)$ values of the ground state band of ${}^{44}\text{Ti}$ can be reproduced well without the effective charges. Also the energy levels and the large $B(E2)$ values were reproduced well in the hybrid (Nilsson model + cluster model) model calculations [57] by coupling the ground state band with the $K = 0^+$ α-cluster band state at 8.54 MeV. It was difficult to reproduce the ground state band and the excited α-cluster bands *simultaneously* in the cluster model. Based on the quantitative microscopic study of the energy dependence of the volume integrals of the equivalent local potential between the α-particle and ${}^{40}\text{Ca}$, it was argued [55] that the situation for

α-clustering in ^{44}Ti is different from that in the ^{20}Ne region and that the good agreement, of the α-cluster model, with the experiment for the ground state band is superficial, since the α-cluster of the ground state band is broken by the strong spin–orbit force. This seems to suggest that the α-cluster model established in light nuclei is difficult to apply in the heavier-mass region above the fp-shell.

6.3.3 *The unification of the bound and scattering states of* ^{44}Ti

In contrast to the controversial situation in the α-cluster study of ^{44}Ti, the interaction potential between the α-particle and ^{40}Ca was studied thoroughly, both theoretically and experimentally, to understand the ALAS in α-particle scattering from ^{40}Ca. An optical potential with a Woods–Saxon squared form factor for the real part was found by Delbar *et al.* [58] to describe the α-particle scattering from ^{40}Ca over a wide range of incident energies, $E_\alpha = 24$–166 MeV. By using a semiclassical method in which scattering waves are decomposed into barrier waves reflected at the Coulomb potential barrier at the surface region and the internal waves penetrating the barrier, it was shown that the ALAS is caused by the internal waves under weak absorption [59]. In the energy range $E_\alpha = 10$–27 MeV, the measured excitation function of the fusion cross sections for the $\alpha + {}^{40}$Ca system shows an oscillatory structure. As was the case in ^{20}Ne, it was very important to understand the bound and scattering states in a unified way.

As shown in Fig. 6.5(a), Michel *et al.* [60] found that the oscillations of the fusion excitation function can be reproduced by using a simplified form of the optical potential of Delbar *et al.* [58]. The calculation also gives a good overall fit to the excitation function for elastic scattering at several angles from 12 to 18 MeV. A decomposition of the fusion cross section data into the contributions of individual partial waves (Fig. 6.5(b)) shows that it is the even partial waves of the $N = 14$ ($N = 2n + L$; n being the number of the nodes of the wave function) band that are responsible for the structure; the odd partial waves give a fusion cross section that shows no structure. It is noted that the

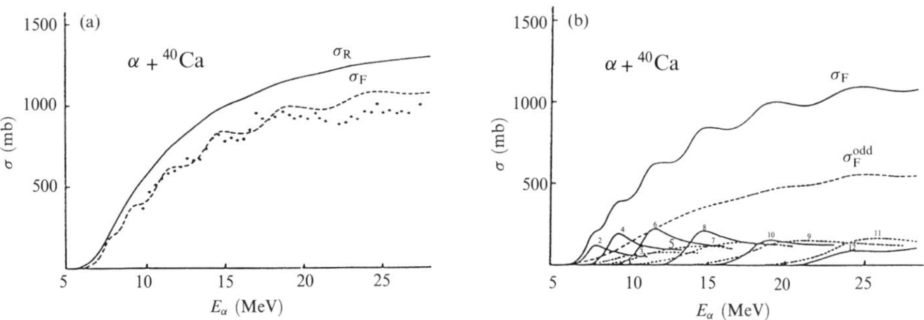

Fig. 6.5 (a) The experimental excitation function of the fusion cross sections for α-particle on ^{40}Ca compared with the calculated reaction cross section (solid line) and the fusion cross section (dashed line). (b) The calculated excitation function of the fusion cross section for α-particles on ^{40}Ca showing the contributions of odd and even partial waves to the observed structure. (Taken from Michel *et al.* [60].)

energy where the fusion oscillations appear is just above the quasi-bound energy region where the controversial situation of the α-cluster structure study was confronted. Michel *et al.* [61] applied the potential to the α-cluster structure study of ^{44}Ti and calculated the energy levels by using the real part of the optical potential. As shown in Fig. 6.6, quite remarkably, the bandhead of the $N = 12$ band, which is the first Pauli-allowed band, systematically falls below the $\alpha + {}^{40}$Ca threshold, within a few MeV from that of the ground state band of ^{44}Ti; moreover, the band terminates at $J^{\pi} = 12^{+}$. The calculated energy levels, B(E2) values, and rms radii are given in Table 6.1, together with the potential depth U_0 finely tuned around the original value $U_0 = 180$ MeV for each state and the experimental B(E2) values. The calculated transition probabilities are in very satisfactory agreement with experiment, especially if one takes into account that no effective charge has been introduced in the calculation [large effective charges ($\delta e \sim 0.5e$) are needed in (fp)^4 shell model calculations [57] in order to reproduce the B(E2) experimental values]. On the other hand, the rms intercluster distance is seen to decrease from $R = 4.50$ fm for the ground state to $R = 3.83$ fm for the $J^{\pi} = 12^{+}$ state (anti-stretching effect). These values should be compared with the sum of the ^{4}He and ^{40}Ca rms radii, which amounts to 5.16 fm. The overlap between the clusters, which is moderate for the low-spin members of the band, is seen to become severe at high spin;

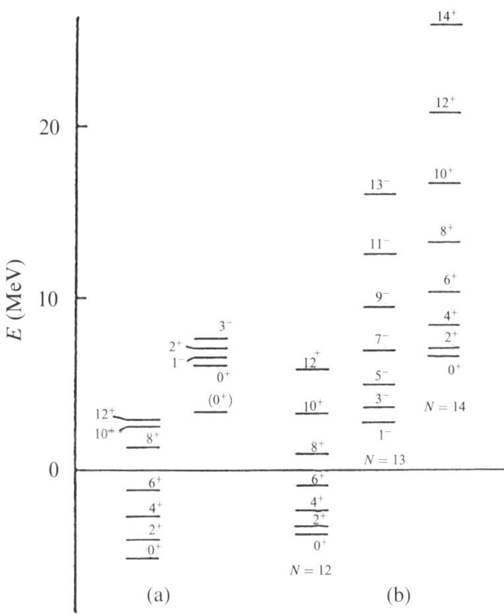

Fig. 6.6 (a) Experimental ground state band and α-particle cluster state candidates in ^{44}Ti; (b) ^{44}Ti $N = 12$, 13, and 14 states supported by the local potential (the potential depth is fixed at the average value $U_0 = 180$ MeV). Energies are given with respect to the $\alpha + {}^{40}$Ca threshold. (Taken from Michel *et al.* [61].)

Table 6.1 Theoretical and experimental B(E2) values for the $J \to J-2$ transitions (in $e^2\,\mathrm{fm}^4$), and intercluster rms radii for the ^{44}Ti $N = 12$, 13 states. The local potential depth given for the $N = 12$ states is one that reproduces the experimental energies with respect to the α-threshold; for the $N = 13$ states U_0 is fixed at the average value of 180 MeV (Taken from Michel *et al.* [61])

		$N = 12$				$N = 13$	
J^π	U_0	B(E2)		$\langle R^2 \rangle^{1/2}$	J^π	B(E2) Theor.	$\langle R^2 \rangle^{1/2}$
	(MeV)	Theor.	Expt.	(fm)			(fm)
0^+	184.1			4.50	1^-		5.35
2^+	182.5	107.3	120 ± 30	4.51	3^-	264.7	5.28
4^+	181.2	146.4	280 ± 60	4.47	5^-	279.1	5.15
6^+	180.7	140.2	160 ± 20	4.38	7^-	243.6	4.95
8^+	179.0	118.1	>14	4.27	9^-	184.5	4.71
10^+	181.8	74.9	140 ± 30	4.06	11^-	118.1	4.43
12^+	186.8	33.6	40 ± 8	3.83	13^-	54.9	4.14

thus, the $N = 12$ band as expected, does not have a strong cluster character, although, as in the ^{20}Ne case, a proper consideration of the α-clustering seems to be important for a correct reproduction of the intraband transition probabilities.

The calculation predicts two additional quasi-rotational bands of opposite parities corresponding to $N = 13$ and 14 [61]. These states are shown in Fig. 6.6 together with those of the ground state band. The $N = 14$ excited positive-parity band is composed of broad states and its head lies about 6.5 MeV above the threshold. An examination of their wave functions shows that the separation between the clusters is considerably larger than that in the ground state band; it is concluded that this band is in fact the analogue of the higher-nodal $K = 0_4^+$ band of ^{20}Ne, and it has a strong α-character. The broad oscillations observed in the experimental $\alpha + {}^{40}$Ca fusion excitation function between $E_\alpha = 10$ and 27 MeV (Fig. 6.5) provide a direct manifestation of states of this band with spins ranging from 6 to 12. This is one of the only examples, along with the well-established case of ^{20}Ne, where an excited vibrational mode of the α-cluster degree of freedom has been observed experimentally. The origin of the fusion oscillations can be reinterpreted in the semiclassical picture [62] that it is due to interference between the waves reflected by the internal and external barriers.

6.3.4   *The prediction of the parity doublet $K = 0^-$ band in ^{44}Ti*

The calculations invariably locate the $N = 13$ negative-parity band halfway between the $N = 12$ and $N = 14$ positive-parity bands, with its bandhead slightly above the $\alpha + {}^{40}$Ca threshold [61]. This band (which is the analogue of the parity doublet $K = 0^-$ band of ^{20}Ne) also shows much stronger α-cluster properties than does the ground state band. The states of this band, which are predicted to lie considerably below their respective barriers, are much narrower than those of the $N = 14$ band and, thus, appear as genuine molecular states, in that their lifetime is much longer than their rotational period; in contrast, these characteristic times are of comparable magnitude for the $N = 14$ states, while the states of the $N = 12$ band appear to have a transitional character between the shell and the cluster phases. Consequently, Michel *et al.* [61] concluded from the unified study of bound and scattering states that the $N = 13$ α-cluster band should be actively searched for since their experimental detection would definitely confirm their picture and pave the way for an extension of the α-cluster spectroscopy in the fp-shell. They also noted that the spin of the 8.54 MeV state might not be 0^+ since the spin assignment was only tentative. It was also suggested [61] that the interpretation of the state at 11.73 MeV as a simple $\alpha + {}^{40}$Ca cluster state is seriously hampered by the incompatibility between its small width ($\Gamma = 40$ keV) and its position with respect to the $L = 1$ barrier, because local potential calculations systematically predict widths of about 1 MeV when this state is located at the correct experimental energy.

Horiuchi [55] has argued that the α-cluster model should not be applied to the ground state band because it always puts the 1^- state at a lower energy (around 5 MeV) than the 0^+ state (at 8.54 MeV). Instead, he suggested that the model should be applied to the 0^+ (8.54 MeV) and 1^- (11.7 MeV) states, but not to the ground state band. This has been investigated by Ohkubo [63] using a folding model potential chosen so that the first Pauli-allowed $N = 12$, 0^+ state corresponds to the 0^+ state at 8.54 MeV. He then calculated the fusion excitation function, which agreed quite well with the data, and also the elastic scattering cross sections for a range of energies from 18 to 100 MeV. These agreed with the data at lower energies but failed badly at higher energies, indicating that this potential is unacceptable.

Another possibility is that, as in the proposal of Horiuchi [55], the analysis should omit the ground state band and should be applied to the excited mixed-parity cluster states (11.2 MeV (0^+), 11.7 MeV (1^-), 12.17 MeV (2^+), and 12.76 MeV (3^-)) [64–68]. This possibility has also been investigated by Ohkubo [63] using a folded potential fitted to the 0^+ state at 11.2 MeV. He found results very similar to those just mentioned, namely that it fails to fit the higher-energy scattering data. The essential physical defect in both these potentials is that the rainbow scattering begins at too low an energy. This could perhaps be obtained by allowing the potential to become deeper as the energy increases, but this is physically unacceptable. It is not possible to improve the fit by adjusting the imaginary potential.

It has been shown by Ohkubo [63] that only by using a folding potential fitted to the ground state 0^+ the fusion excitation function, elastic scattering cross sections over a wide range of energies involving ALAS, and the rainbow scattering can all be reproduced well (Fig. 6.7). Merchant *et al.* [69] also confirmed the prediction of Refs [61,63] by using a finite-range potential. Their calculated results are shown in Fig. 6.8. The experimental high-spin members above the excitation energy 20 MeV in Fig. 6.8, which were derived by fitting the differential cross sections in the backward direction, show the linearity as a function of $L(L + 1)$. This suggests a rotational band, which may be supported by the extrapolation to the 1^- state at 11.7 MeV; however, it seems difficult [63,70] to assign a simple $\alpha + {}^{40}$Ca cluster structure to this state because of its small width (40 keV), as mentioned before. The observed structure is attributable to the combined effects of the broad resonant high-spin states [71]. The same pitfalls have already been discussed in the case of $\alpha + {}^{16}$O scattering [43,45]. The importance of a unified description of

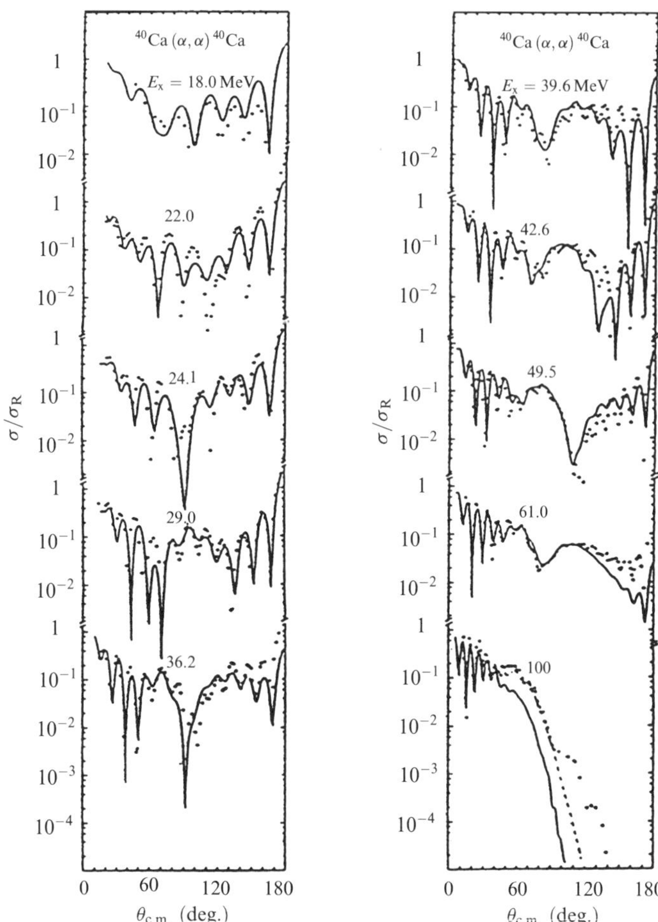

Fig. 6.7 Comparison of the theoretical cross sections (ratio with respect to the Rutherford cross section σ_R) calculated using the folding potential (solid lines) with the experimental angular distributions (black dots) at various incident α-particle energies. For $E_\alpha = 100$ MeV, the calculated results with two different imaginary potentials are shown by solid and dashed lines. (Taken from Ohkubo [63].)

bound and scattering states was emphasized by Hodgson [72], who noted the similarity between the α + nucleus system and the nucleon + nucleus system.

6.3.5 *Observation of the parity doublet band in ^{44}Ti*

It was important to confirm or deny the above predictions especially those concerning the negative-parity band. Although the α-transfer experiments on ^{40}Ca using the (^{6}Li, d) reaction were carried out in the 1970s, no indications had then been reported on the existence of the negative-parity α-cluster band. An experiment to search for the states

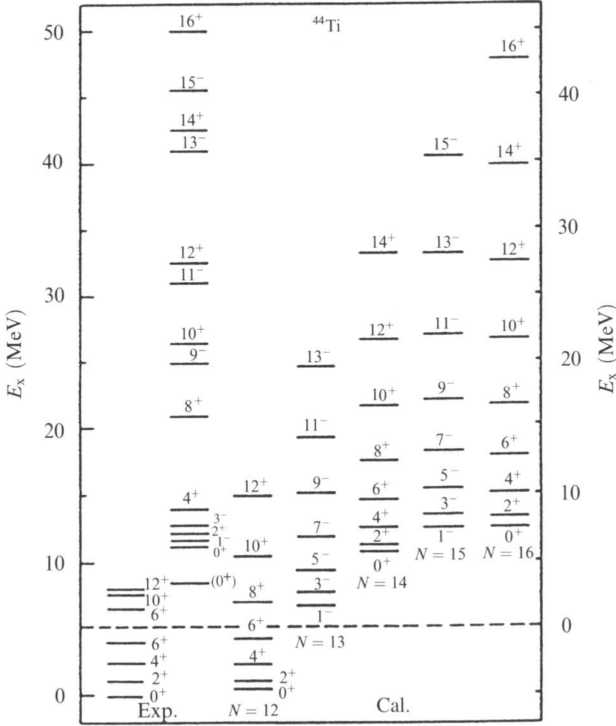

Fig. 6.8 A comparison of the calculated energies, of the bands with $N = 12$–16 of the α-cluster states in ^{44}Ti generated by a finite-range folded potential, with the experimental data. (Taken from Merchant *et al.* [69].)

of the $N = 13$, $K = 0^-$ band using the ^{40}Ca (^{6}Li, d) ^{44}Ti reaction was performed by Yamaya *et al.* [73,74] at RCNP. The experiment started in 1988. A 50 MeV ^{6}Li^{2+} beam was provided by the isochronous cyclotron at RCNP. Deuterons emitted from the (^{6}Li, d) reaction were analysed with the magnetic spectrometer RAIDEN [75].

The energy spectra of deuterons in the above reaction are shown in Fig. 6.9. Firstly, it was found that the spin of the controversial 8.54 MeV state is not 0^+ and that the 11.7 MeV 1^- state is not clearly populated in the α-transfer reaction [73]. This indicates that both states are unsuitable candidates for the lowest states of the parity doublet band of ^{44}Ti in the α-cluster model. Figure 6.10 displays the angular distributions of deuterons among the members of the positive-parity $K = 0^+$ and negative-parity bands. These angular distributions exhibit shapes which are characteristic of the transferred orbital angular momentum. The curves present the results of α-transfer DWBA calculations, normalized to the experimental data, to deduce the α-spectroscopic factors S_α.

As shown in Fig. 6.10, the DWBA curves reproduce the experimental angular distributions at $E_L = 50$ MeV as well as the results of the analyses of the data at $E_L = 37$ MeV. For the members of the positive-parity band ($N = 12$), the results of the DWBA

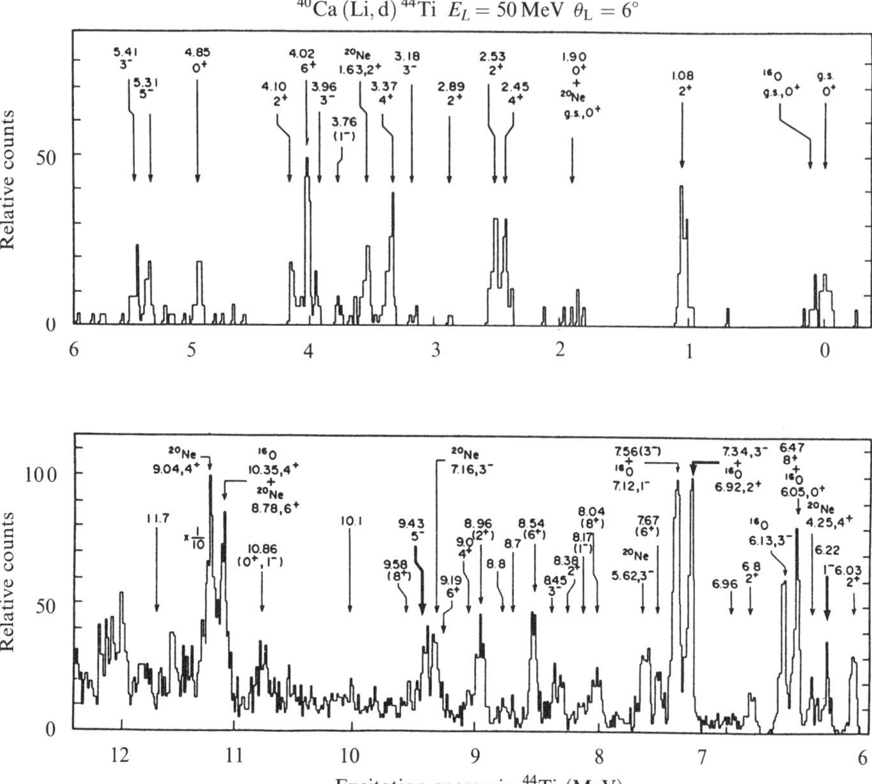

Fig. 6.9 Deuteron spectra of the excited states below and above the α-threshold in the ^{40}Ca (^{6}Li, d) ^{44}Ti reaction at $E(^6$Li$) = 50\,\mathrm{MeV}$. The peaks assigned as the negative-parity bands are indicated by bold arrows. (Taken from Yamaya *et al.* [74].)

calculations for the states of $J^\pi = 0^+, 2^+, 4^+, 6^+$, and 8^+ are in good agreement with data and the deduced spectroscopic factors S_α are about 0.1–0.25. The 7.67 and 8.04 MeV states were assigned to $J^\pi = (10^+)$ and (12^+) as the members of the positive-parity band in the heavy-ion induced γ-decay reaction [76]. But, in the α-transfer reaction at both incident energies $E_L = 37$ and $50\,\mathrm{MeV}$, the discrepancies between the experimental data and the DWBA calculations in the angular distributions with $L = 10$ and 12 are undoubtedly, for these states. These angular distributions for the 7.67 and 8.04 MeV states are fitted by the DWBA curves of $L = 6$ ($N = 12$) and $L = 3$ ($N = 13$), respectively. These states excited by the α-transfer reaction may be different from the states observed by the heavy-ion-induced reaction.

Secondly, for the negative-parity band, as shown in Fig. 6.10(b), three members, 1^-, 3^-, and 5^-, of the $N = 13$, $K = 0^-$ band were observed at 6.22, 7.34, and 9.43 MeV, respectively, in the experiment through the analysis of the angular distributions with the

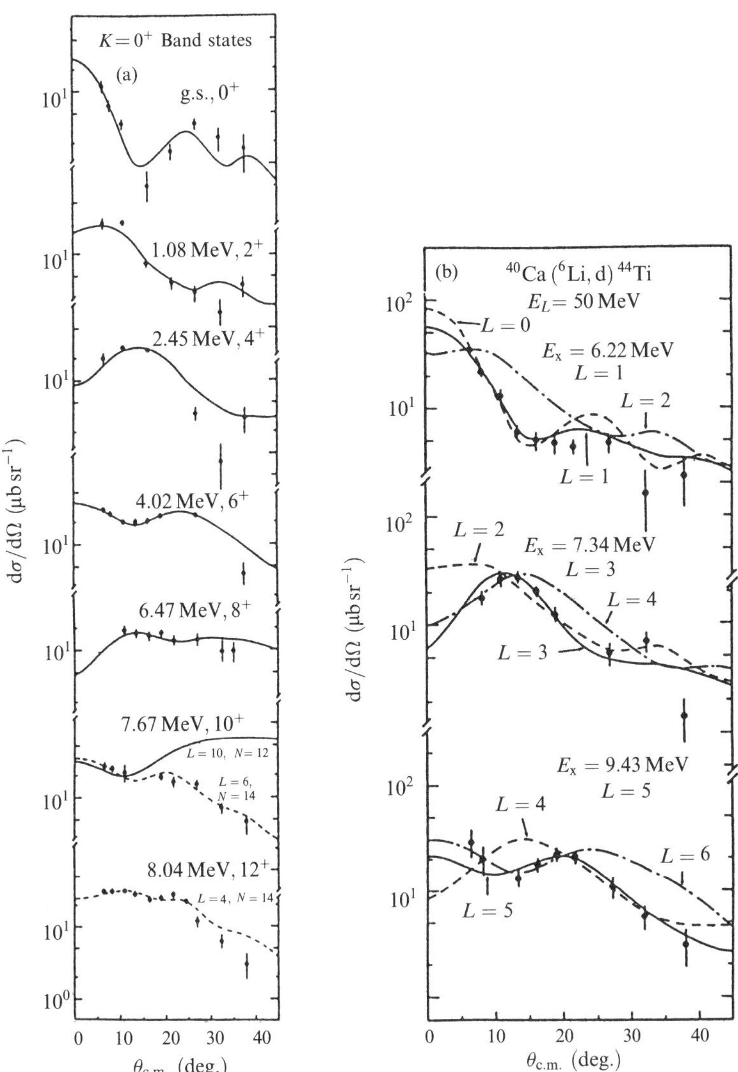

Fig. 6.10 Angular distributions of the ^{40}Ca (^{6}Li, d) ^{44}Ti reaction leading to (a) the $N = 12$ ground state band and (b) the $N = 13$, 6.22, 7.34, and 9.43 MeV states at $E_L(^{6}\text{Li}) = 50$ MeV. The solid curves indicate the best fit to the data by the DWBA calculations. The dashed curves in (a) are the results calculated assuming $N = 14$. In (b) the calculated results with neighbouring L values are shown, for comparison, by dashed and dot-dashed lines. (Taken from Yamaya *et al.* [74].)

finite-range DWBA [74]. The spin assignment of 1^- for the 6.22 MeV state was decisively confirmed by analysing the additional angular distribution at 28 MeV by Fulbright *et al.* [77]. The observed ground state band and the parity doublet $K = 0^-$ band states are compared with the theoretical calculations [61,63,69,78] in Fig. 6.11. The experimental results and analyses by Yamaya *et al.* were reconfirmed by independent analyses using a breakup fusion approach [79] and by independent ^{40}Ca (^{6}Li, d) ^{44}Ti experiments with $E_L = 60.1$ MeV [80]. As shown in Table 6.2, α-spectroscopic factors deduced from the DWBA calculations [81] are consistent with those predicted by the α-cluster theory. Although the S_α for the $K = 0^-$ band is smaller than that of the $K = 0^+$ band, this is due to the fragmentation of the α-strength over the nearby states. The S_α summed over the states for the negative-parity states are 0.25, 0.37, and 0.30 for the 1^-, 3^-, and 5^- states [82]. This is consistent with the theoretical predictions and the α-clustering of the $K = 0^-$ band is more developed than the $K = 0^+$ band.

To know why the α-cluster structure persists in ^{44}Ti in spite of the strong spin–orbit force, competition between α-clustering and spin–orbit force was investigated by Yamada [83] using the core $+ 3N + N$ cluster model. He showed that although the α-correlations are strongly broken for the high-spin members such as 8^+, 10^+, and 12^+ by the strong spin–orbit force, they are preserved for the low-spin states. This also

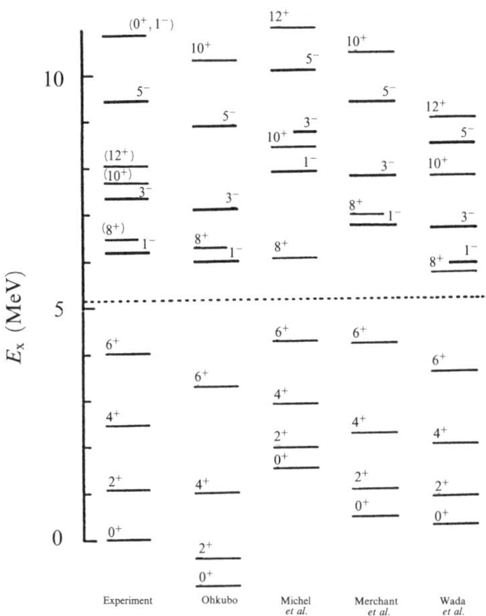

Fig. 6.11 Comparison of the observed energy levels of the parity doublet bands in ^{44}Ti with the theoretical predictions by the α-cluster models: semi-microscopic folding potential (Ohkubo [63]), phenomenological Woods–Saxon squared potential (Michel *et al.* [61]), folding potential (Merchant *et al.* [69]) and microscopic model (Wada *et al.* [78]). The negative-parity states observed by Yamaya *et al.* [74] are indicated by thick lines. (Taken from Yamaya *et al.* [74].)

Table 6.2   The α-spectroscopic factors for the members of the $N = 12$, $K = 0^+$ and $N = 13$, $K = 0^-$ bands in ^{44}Ti deduced from the DWBA analyses using several theoretical α-wave functions are compared with the theoretical predictions by Michel *et al.* [61]. The wave functions used are obtained by using a Woods–Saxon potential with a large diffuseness parameter (WS-II), Woods–Saxon squared potential (MRO [61]), microscopic RGM wave functions obtained with the same size parameters for α and ^{40}Ca (RGM-I) and different realistic size parameters (RGM-II). The S_α values in the brackets are the results obtained from the data at $E = 28$ MeV (Taken from Yamaya *et al.* [81])

States (MeV)	J^π	WS-II potential	MRO potential	RGM-I	RGM-II	Theory (MRO)
$K = 0^+$ band						
g.s.	0^+	0.20	0.24(0.25)	0.40		0.20
1.08	2^+	0.17	0.12	0.43		0.20
2.45	4^+	0.18	0.10	0.42		0.18
4.02	6^+	0.18	0.11			0.16
6.47	8^+	0.23	0.20			0.13
$K = 0^-$ band						
6.22	1^-	0.14	0.16(0.10)	0.14	0.12	0.42
7.34	3^-	0.11	0.11	0.13	0.09	0.41
9.43	5^-	0.12	0.09	0.15	0.08	0.38

explains why the ground state band deviates from the pure rotational spectrum as the spin increases. Yamada and Ohkubo [84] pointed out that other intruder states, especially the low-lying $K = 0^+$ band starting from 1.90 MeV (which may be an 8p–4h state in the shell model) and the $K = 2^+$ band starting from 2.53 MeV state, can be understood in the schematic α-cluster model in which the excitation of the core ^{40}Ca to the $K = 0^+$ α-cluster band whose bandhead state is the mysterious 0^+ state (3.35 MeV) is taken into account. The calculated energy levels are shown in Fig. 6.12. The calculated energy spectra and B(E2) values are in good agreement with the experimental data. All the low-lying rotational bands are described in the α-cluster model. The validity of the α-cluster viewpoint in ^{44}Ti has thus been confirmed theoretically and experimentally.

6.3.6   *The prediction of the α-cluster $K = 0^-$ band in ^{40}Ca and its observation at RCNP*

The ^{40}Ca nucleus is analogous to ^{16}O, so experimental and theoretical studies from the α-cluster viewpoint were made in the 1970s. The energy levels and the large intraband B(E2) values of the $K = 0^+$ band, which is enhanced in the α-transfer reactions, were reproduced in the α-cluster model. However, since the parity doublet negative-parity band with an α-cluster structure was not found in a microscopic calculation [17,85] or experimentally, the persistence of the α-cluster viewpoint in ^{40}Ca was questioned. On the other hand, from the viewpoint of a unified description of bound and scattering states, the real parts of the optical potentials which reproduce the ALAS in $\alpha + ^{36}$Ar scattering were applied to study the α-cluster structure of ^{40}Ca [86,87]. The α-cluster model with such

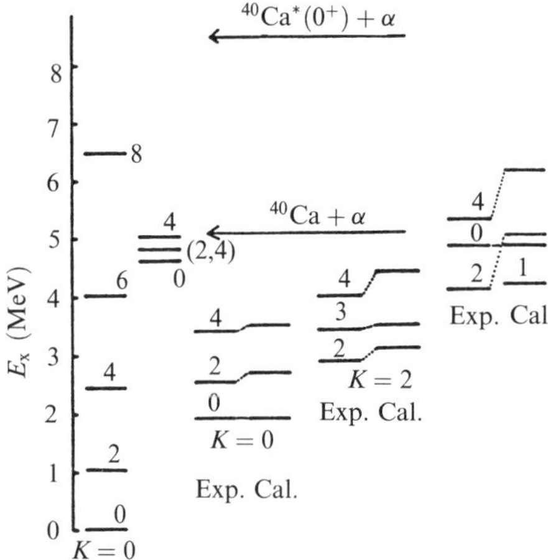

Fig. 6.12 Comparison of the experimental positive-parity energy spectra of ^{44}Ti below $E_x = 5.3$ MeV with the results calculated (right) in a core-excited $\alpha + {}^{40}\text{Ca}^*$ model. (Taken from Yamada *et al.* [84].)

a potential [86,87] could give enhanced $B(E2)$ values in the $K = 0^+$ band. As shown in Fig. 6.13, as in the case of ^{44}Ti, the model inevitably predicted the $K = 0^-$ band with the α-cluster structure which is the parity doublet partner of the $K = 0^+$ band.

To observe this $K = 0^-$ band in ^{40}Ca, a measurement of the α-transfer reaction ^{36}Ar $(^6\text{Li, d})\,^{40}$Ca was performed by Yamaya *et al.* [88,89] using a 50 MeV $^6\text{Li}^{2+}$ beam provided by the cyclotron at RCNP. The members of the $K = 0^-$ band were observed in the experiment (Fig. 6.14). What is interesting is the fragmentation of the α-strengths of the $K = 0^-$ band as shown in Fig. 6.15. The centroid for the negative-parity states corresponds well to the theoretical predictions by Ohkubo and Umehara [86] and Reidemeister *et al.* [87]. Such a fragmentation was not known in the parity doublet band in the sd-shell region.

Microscopic α-cluster model calculations were performed using the OCM, in which excitation of the ^{36}Ar core and the Pauli principle are taken into account [90]. As shown in Fig. 6.16, the model reproduces the $K = 0^+$ and $K = 0^-$ cluster bands successfully. Not only the α-cluster bands but also the shell-model-like states, including the ground state, are reproduced well. The fragmentation of the α-strengths, which is due to the coexistence of cluster states and shell-model-like states in ^{40}Ca, is also reproduced (Fig. 6.17). The $B(E2)$ values for electric transition between the intracluster bands and between the shell-model states are well reproduced (Fig. 6.18).

6.3.7 *The higher-nodal positive-parity band in ^{40}Ca and ^{44}Ti*

In spite of its α-cluster structure, the higher-nodal band in ^{20}Ne has not been observed in the α-transfer reactions because of the very large width of the excited states near or above the Coulomb barrier. In the reactions ^{36}Ar $(^6\text{Li, d})\,^{40}$Ca at $E_L = 50$ MeV, 17

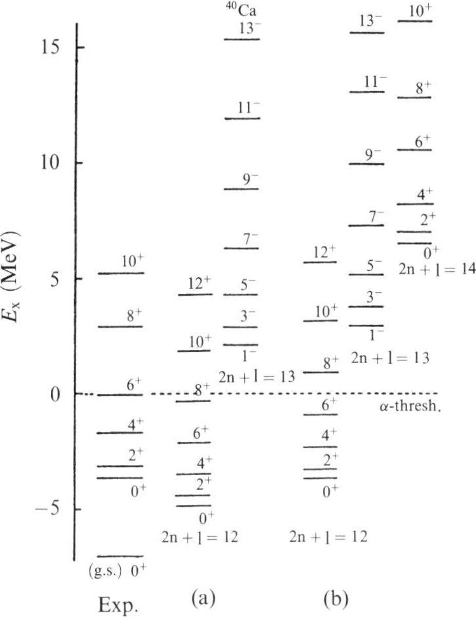

Fig. 6.13 Comparison of energy levels calculated with a Woods–Saxon squared potential (a) $V = 185\,\text{MeV}$ (strength used to analyse $\alpha + {}^{36}\text{Ar}$ scattering), (b) $V = 181.9\,\text{MeV}$ (strength adjusted to fit the 0^+ (3.35 MeV) state), with the experiment. (Taken from Ohkubo *et al.* [86].)

levels, indicated by the underlined spin members in Fig. 6.14, are strongly excited in the excitation energy region higher than 11 MeV in ^{40}Ca [89]. These levels are assigned as the 0^+, 2^+, 4^+, and 6^+ states of a $K = 0^+$ band with $N = 14$. The observed energy levels and those calculated as the higher-nodal band [86] are compared in Fig. 6.19. The large spectroscopic factors, S_α's, for these levels are deduced from the DWBA analyses for the experimental data. This fact shows the much stronger cluster character of the higher-nodal band in ^{40}Ca.

The 10.86 MeV state in ^{44}Ti is strongly excited in the ^{40}Ca $(^6\text{Li}, \text{d})\,^{44}$Ti reaction at $E_L = 32\,\text{MeV}$ [52] , 37 MeV [82], and 50 MeV [73,74]. This state is identified as $J = 0^+$ with $N = 14$ from the $E_L = 37$ and 50 MeV data and can be regarded as a candidate for the bandhead state of the higher-nodal $K = 0^+$ band in ^{44}Ti. This S_α-value is also about 1.0.

In the sd-shell nuclei, the well-known higher-nodal band ($N = 10$) is built on the 0_4^+ state near 8.7 MeV in ^{20}Ne. The width of this state is about 1 MeV, but the widths of the states of the higher-nodal $N = 14$ band in ^{40}Ca and ^{44}Ti are less than 100 keV. These small energy widths are due to the high Coulomb barrier in the core $+ \alpha$ system. The higher-nodal bands were found for the first time in the α-transfer reactions by Yamaya *et al.* [82,89].

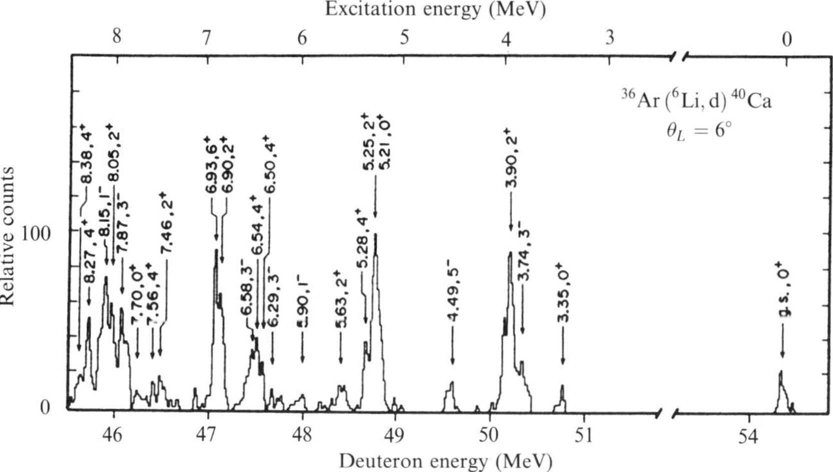

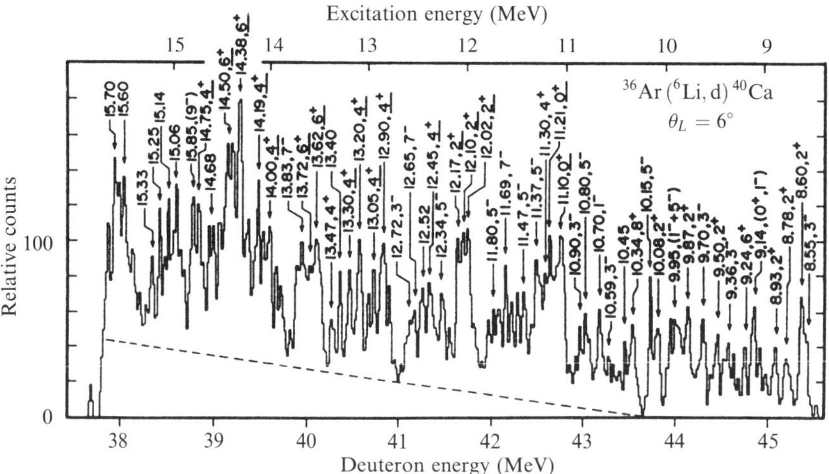

Fig. 6.14   Deuteron energy spectra at $\theta_L = 6°$ from the ^{36}Ar (^{6}Li, d) ^{40}Ca reaction. The dashed line indicates the maximum background from the ^{6}Li breakup process. Assigned spin-parities are indicated for each peak. The members of the $N = 14$ band are indicated by underlining the spin of each state. (Taken from Yamaya *et al.* [88].)

6.3.8   *The α-cluster structure of* ^{42}Ca

By using the microscopic $\alpha + {}^{38}$Ar cluster model with the OCM, the structure of ^{42}Ca was also investigated [91]. It was shown that not only the intruder $K = 0^+$ band starting from 1.84 MeV but also the ground state band are well explained in the $\alpha + {}^{38}$Ar cluster model (Fig. 6.20). Furthermore, not only the energy levels but also the electric transitions

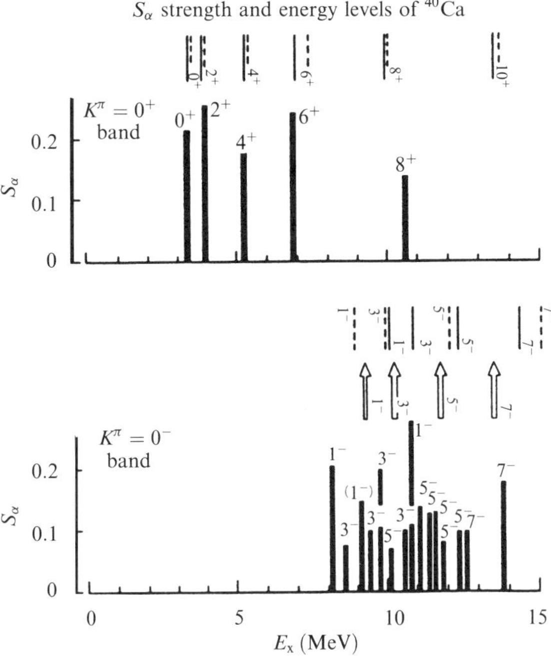

Fig. 6.15 The α-spectroscopic factors for the members of the $K = 0^+$ and $K = 0^-$ bands in ^{40}Ca. The theoretically predicted energy levels calculated with a Woods–Saxon squared potential [86] (solid line) and with the spline potential [87] (dashed line) are compared with the energy positions of the observed levels. The centroid for the negative-parity states fragmented from the $K = 0^-$ band is shown by arrows. (Taken from Yamaya *et al.* [88].)

(Fig. 6.21) and charge form factors are reproduced well by the model. The coexistence of α-cluster and the shell-like structure is essential in understanding the various properties of this nucleus in a unified way. It was also shown by studies of ^{43}Sc [92] that coexistence of the α-cluster and the shell-like structure is also important in non-4N nuclei. Weak coupling features based on the α-cluster structure persist in the ^{40}Ca–^{44}Ti region. It has been pointed out from the weak coupling point of view that the α-cluster structure can persist in ^{38}Ar [93]. This suggests that α-clustering is also important in the latter half of the sd-shell region across the $N = Z = 20$ double shell closure. There are indications that 8p–nh states exist not only in ^{44}Ti but also in ^{42}Ca and ^{40}Ca. These states in the low excitation energies in ^{40}Ca–^{48}Cr may be understood in the core $+ \alpha + \alpha$ cluster model as was the case in ^{44}Ti.

It is now established that α-cluster structure persists in the fp-shell region in spite of the strong spin–orbit force. Further systematic theoretical studies and measurements of reactions such as ^{38}Ar (^{6}Li, d) ^{42}Ca and ^{36}Ar (^{12}C, α) ^{44}Ti are expected. The diagram of the molecular viewpoint proposed by Ikeda for light nuclei is now extended to much heavier nuclei including those in the fp-shell region as shown in Fig. 6.22 [94]. In

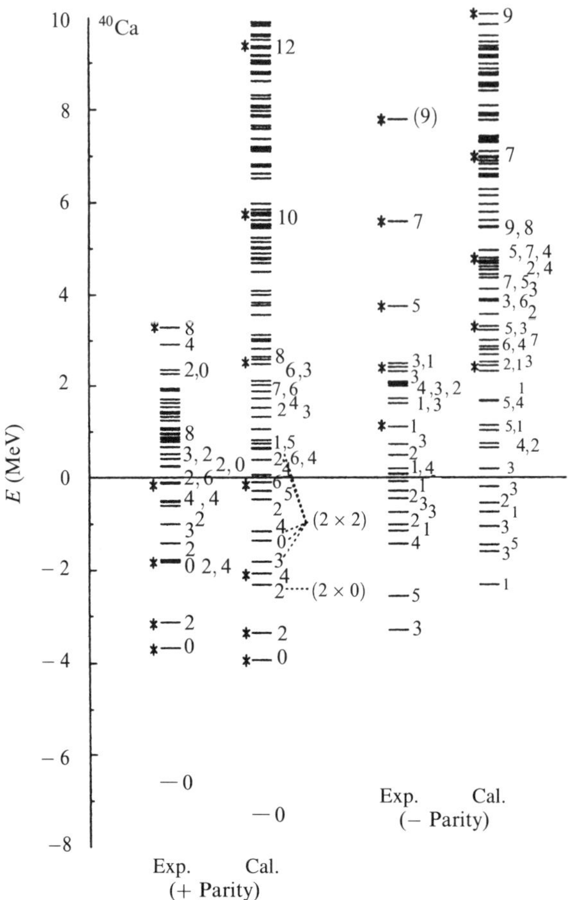

Fig. 6.16 The experimental energy spectra in ^{40}Ca compared with the calculated results in the OCM. The energy scale is measured from the α-threshold. The states marked by an asterisk are the members of $\alpha + {}^{36}$Ar cluster bands. The multiplets of weak coupling ^{36}Ar$(2^+) \times L$ bands are indicated by dotted lines. (Taken from Sakuda and Ohkubo [90].)

fact, at very high excitation energies of 60–70 MeV in ^{48}Cr and ^{56}Ni, the existence of the molecular structure, respectively, of ^{24}Mg $+$ ^{24}Mg and ^{28}Si $+$ ^{28}Si, which has hyperdeformation, has been shown [95–97]. The α-chain structure, such as 10α in ^{40}Ca and 15α in ^{60}Zn, as well as other exotic α-states [98] are also of great interest.

6.4 The α-clustering in ^{94}Mo region

6.4.1 *The α–^{90}Zr interaction*

The success of the α-cluster model in the lighter fp-shell nuclei encourages us to study further the α-clustering aspects in much heavier nuclei, where not only the spin–orbit

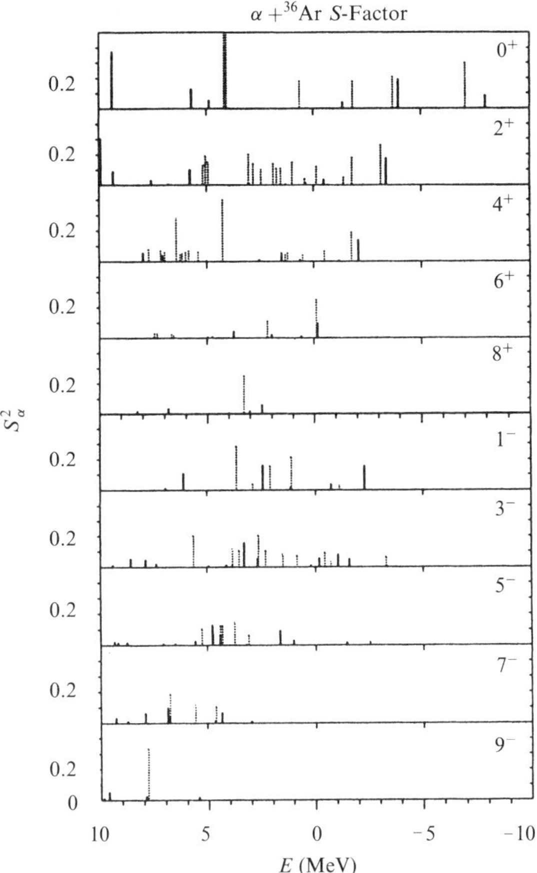

Fig. 6.17 The α-spectroscopic factors (solid lines) calculated in the OCM compared with the experimental data of ^{40}Ca [88,89]. (Taken from Sakuda and Ohkubo [90].)

force becomes much stronger but also the outermost protons and neutrons occupy different major shell orbits. ^{94}Mo is a typical nucleus which has two protons and two neutrons outside the double closed shell; however, α-clustering aspects in this region are scarcely understood. It has been found by a systematic analysis of $\alpha + {}^{90}$Zr scattering over a wide energy range that the shape of the real part of optical potentials in the low-energy region should be different from that in the high-energy region. This makes it difficult to extend an optical potential determined uniquely by a phenomenological analysis at higher energies to the very low energy region while preserving its shape. It is noted that the potentials obtained for $\alpha + {}^{16}$O scattering by Michel *et al.* [45] and for $\alpha + {}^{40}$Ca scattering by Delbar *et al.* [58] are very similar to the folding model potentials [43,63,99].

In fact, $\alpha + {}^{90}$Zr scattering and the structure of ^{94}Mo can be reproduced well in a unified way with a double-folding model [100] using a two-body interaction of DDM3Y,

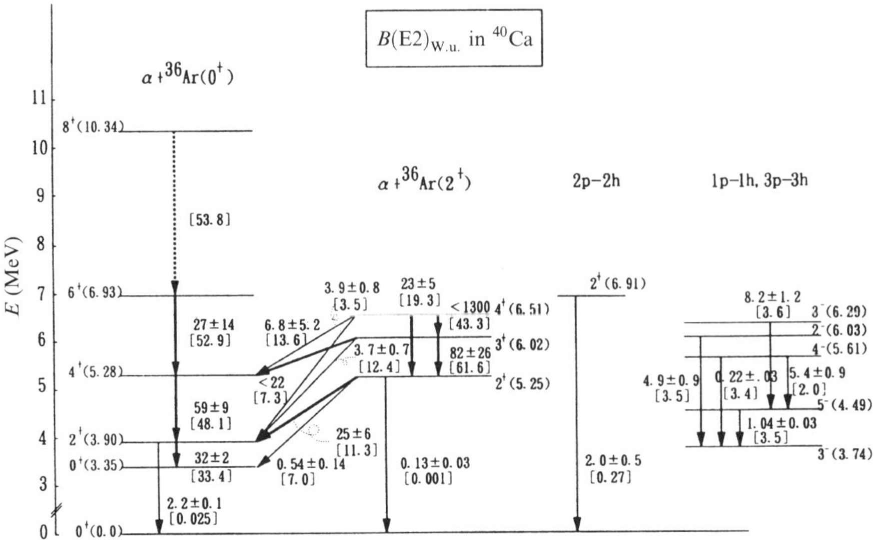

Fig. 6.18 Calculated and experimental E2 transitions of ^{40}Ca (transitions are in the Weisskopf unit). The numbers are the experimental rates and those in parentheses are the calculated rates in the OCM. Classification of the spectra into several bands has been done according to their main components. (Taken from Sakuda and Ohkubo [90].)

which has density and energy dependence. The folding model is given by

$$U(\boldsymbol{R}) = \lambda \int d\boldsymbol{r}_1 \int d\boldsymbol{r}_2 \, \rho_{\text{T}}(r_1)\rho_\alpha(r_2) t(E_\alpha, \rho_{\text{T}}, \rho_\alpha, s = \boldsymbol{R} + \boldsymbol{r}_2 - \boldsymbol{r}_1), \qquad (6.1)$$

where ρ_α and ρ_{T} are the matter density distributions of α-particle and target nuclei, respectively, and t is the effective NN interaction which depends on the local densities and bombarding energy E_α. The departure of the normalization factor λ from unity indicates a necessity to take into account other effects not included in the bare folding model. By introducing a phenomenological imaginary potential, $\alpha + {}^{90}$Zr scattering is reproduced well by the model [100].

6.4.2   *The structure of ^{94}Mo*

Using the potential obtained from the analysis of $\alpha + {}^{90}$Zr scattering, the calculated bound states with $N = 16$, which is the first Pauli-allowed band, corresponds to the experimental ground band of ^{94}Mo. Usually, the calculated ground band shows a rotational spectrum. It has been shown that by introducing a small L-dependence in λ, $\lambda = \lambda_0 - cL$ ($c/\lambda_0 = 0.0035$), the experimental spectrum is reproduced well (Fig. 6.23). This L-dependence is consistent with the value needed in the ^{44}Ti case. Such an L-dependence

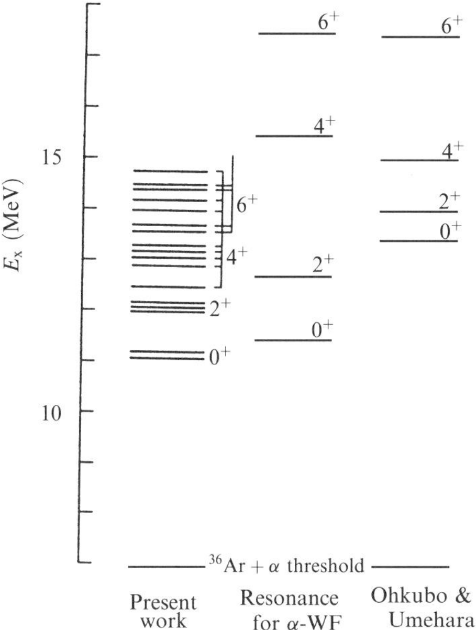

Fig. 6.19   The observed energy levels of the $K = 0^+$ band with $N = 14$ in ^{40}Ca compared with the results of the theoretical calculation by Ohkubo and Umehara [86]. The middle figure shows the resonance energies for the relative wave functions between an α-particle and ^{36}Ar core with the Woods–Saxon squared potential. (Taken from Yamaya *et al.* [89].)

is quite reasonable. In fact, microscopic studies of interaction between composite particles have shown that the equivalent local potential should have an L-dependence (although small) due to the Pauli principle and that the strength of the potential should decrease as L increases.

The calculated rms radii and B(E2) values of the ground band are shown in Table 6.3. The calculated intercluster rms radius of the ground state amounts to 89% of the sum of the experimental radii for α-particle and ^{90}Zr, which is as large as that in ^{44}Ti [61]. This suggests that the ground state has a compact α-cluster structure. The calculated B(E2) values are very much enhanced and by introducing a small effective charge, $\delta e = 0.2e$, a good agreement with the experiment is obtained. These values for high-spin states with enhanced shell-model-like character may be overestimated [102]. In the shell model [103], large effective charges are needed to reproduce the experimental B(E2) values.

The calculations locate the α-cluster states with the $N = 17$, $K = 0^-$ band, which is a parity doublet partner of the ground band, at about 6 MeV, and the higher-nodal $N = 18$, $K = 0^+$ band just below the Coulomb barrier. It is very interesting to observe the parity doublet bands and the higher-nodal band in the α-transfer experiment. An α-cluster model study of the ground state band of ^{94}Mo with a phenomenological potential [104] gives a result consistent with the folding model calculation.

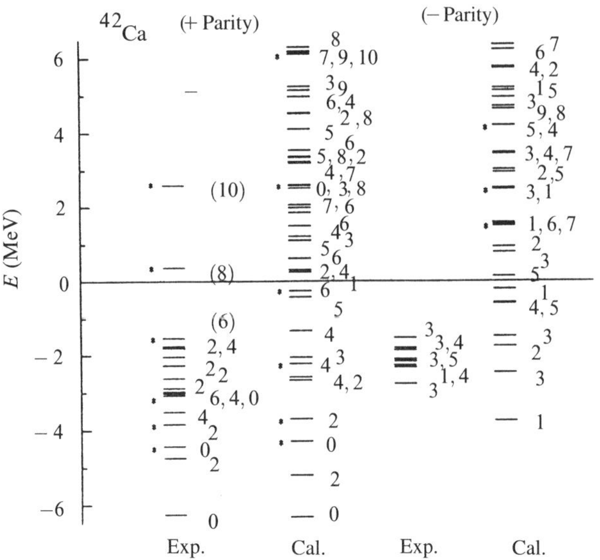

Fig. 6.20 Energy spectra of ^{42}Ca calculated in the OCM compared with the experimental data. The energy scale is measured from the α-threshold. The states marked by an asterisk are the members of $\alpha + {}^{38}$Ar(0^+) cluster bands. (Taken from Sakuda and Ohkubo [91].)

Table 6.3 Theoretical and experimental B(E2) values for the $J \to J - 2$ transitions (in Weisskopf units) and intercluster rms radii for the ground band of ^{94}Mo. The normalization factor λ given is one that reproduces the experimental energies with respect to the α-threshold (Taken from Ohkubo [100])

J^π	E_x (MeV)	λ	$\sqrt{\langle R^2 \rangle}$ (fm)	B(E2) (W.u.) Exp.	B(E2) (W.u.) Cal.
0^+	0.0	1.161	5.28		
2^+	0.871	1.150	5.30	17.2	17.4
4^+	1.574	1.143	5.26	26.6	24.1
6^+	2.423	1.136	5.20		24.3
8^+	2.955	1.135	5.09		21.6

6.5   The α-clustering in ^{212}Po region

6.5.1   *The α-width of ^{212}Po and α-clustering*

^{212}Po is a typical nucleus in the very heavy mass region. Most of the studies done up to now have focused on how the observed large magnitude of the α-width of the ground state

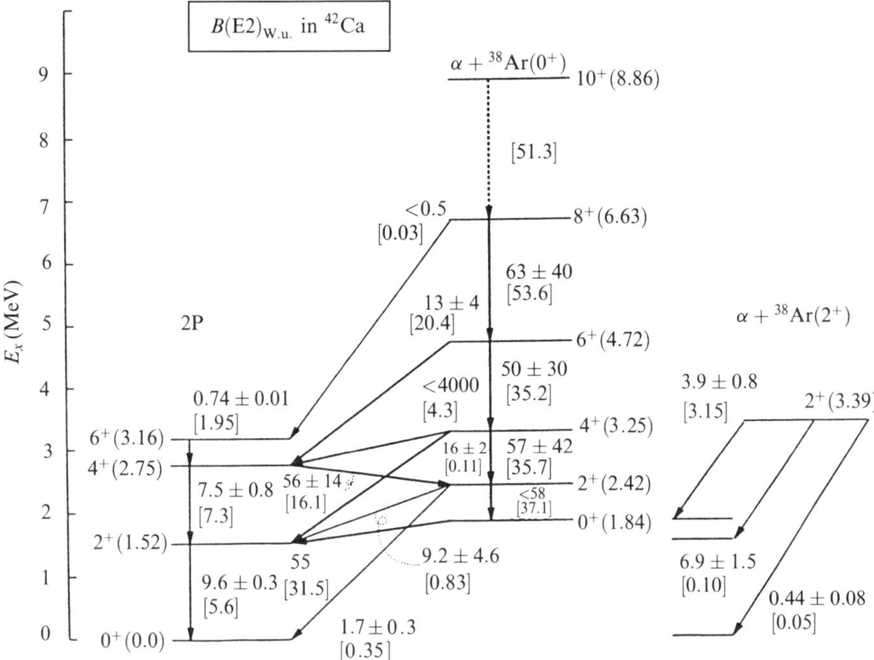

Fig. 6.21 Calculated and experimental E2 transitions of ^{42}Ca (transitions are in the Weisskopf unit). The numbers are the experimental rates and those in parentheses are the calculated rates in the OCM. Classification of the spectra into several bands has been done according to their main components. (Taken from Sakuda and Ohkubo [91].)

is explained. A shell model calculation describes the level structure of the ground state band well [105]. However, although many shell model calculations have been performed to understand this width [106–108], it was difficult to reproduce the observed large α-width of the ground state. A large-space configuration mixing shell model calculation ($13\hbar\omega$) improves this; but the result is still only one-fourteenth of the experimental value [109]. This suggests that other important correlations which are not involved in the shell model are necessary [98]. In fact, by assuming the existence of a genuine α-cluster state at $E_x = 5$ MeV, the resultant α-width of the ground state is greatly enhanced by a coupling to reproduce the experimental result [110]. A microscopic calculation starting from a hybrid of the shell model and the cluster variational wave functions can also reproduce the experimental α-width of the ground state [111]. In the calculation, the α-cluster state was predicted to be as low as 2.5 MeV. These calculations show that the ground state of ^{212}Po has considerable amount of α-clustering. Since not only the ground state but also the excited states of the ground band have large α-widths, it is quite important to know whether the α-cluster model can reproduce the experimental large α-widths. By using phenomenological potentials which were successfully used in a systematic analysis of

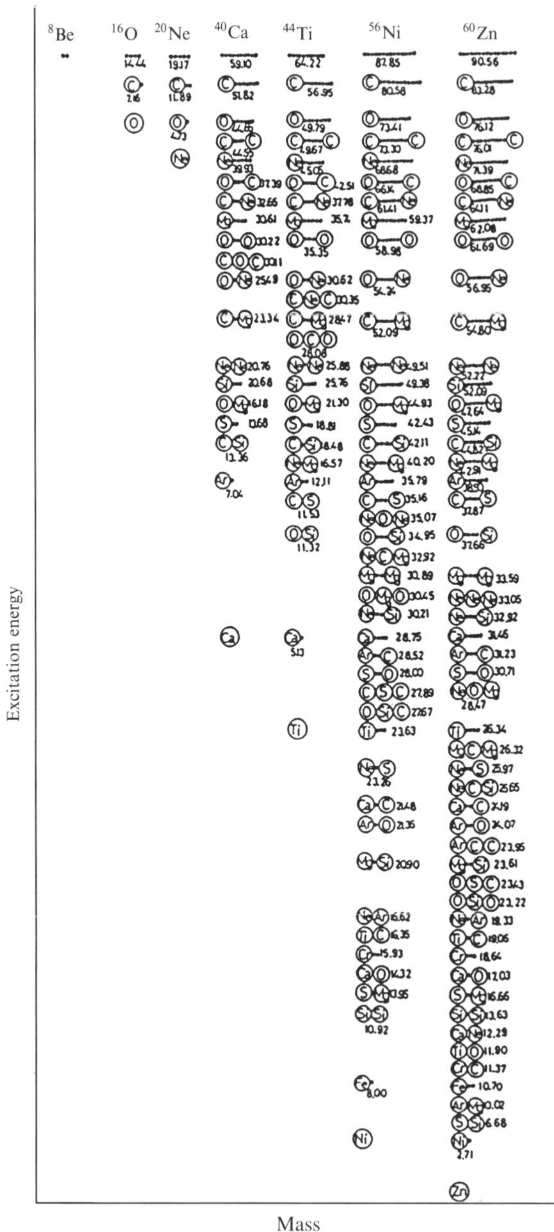

Fig. 6.22   Diagram of molecular viewpoint extended to the fp-shell region. Threshold energies into subunit clusters are given in MeV. (Taken from Ohkubo *et al.* [94].)

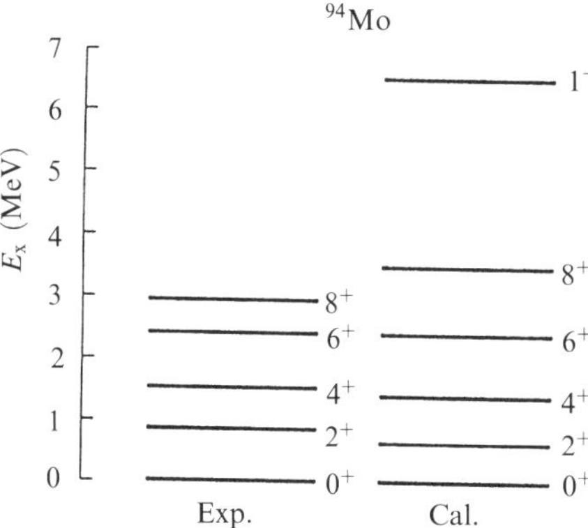

Fig. 6.23 Comparison of the energy levels of ^{94}Mo calculated in the α-cluster model using a folding model with experiment. (Taken from Ohkubo and Sakuda [101].)

α-decays of heavy nuclei, Buck *et al.* [112] have reproduced in their α-cluster model the α-decay widths of the ground band up to 18^+. The semiclassical Bohr–Sommerfeld quantization was used to calculate the α-widths and the preformation probability was assumed to be unity. Their potential was not checked for scattering and the ground state band is assumed to have $N = 20$.

6.5.2 *A unified description of the* $\alpha + {}^{208}$Pb *system and* α*-clustering of* ^{212}Po

A unified description of $\alpha + {}^{208}$Pb scattering and quasi-bound states of ^{212}Po was presented by using the double-folding model [100]. The volume integral of the potential, $328\,\mathrm{MeV\,fm}^3$ ($\lambda = 1.225$), is consistent with the values in other α-particle scattering results from heavy and light nuclei. Remarkably, the calculated bandhead of the quasi-bound state $N = 22$, which is the first Pauli-allowed band, falls above the α-threshold in the range of the experimental ground state band of ^{212}Po. As in the case of the $\alpha + {}^{90}$Zr system, introduction of a very weak L-dependence in the strength of λ ($c/\lambda_0 = 0.0025$) reproduces the spectrum of the ground state band (Fig. 6.24). This dependence shows a trend similar to the one in the $\alpha + {}^{90}$Zr system. In Table 6.4, we show energy levels, rms radii, and the B(E2) values. The volume integral changes little from $J = 0^+$ ($316\,\mathrm{MeV\,fm}^3$) to 10^+ ($309\,\mathrm{MeV\,fm}^3$), which shows that these potentials belong to the same family.

The calculated intercluster distance of the ground state amounts to 90% of the sum of the experimental rms radii for the α-particle and ^{208}Pb, which suggests that the ground state band has a significant amount of α-clustering. The calculated B(E2) values are enhanced very much without effective charges: The agreement with experiment is satisfactory for the $6^+ \rightarrow 4^+$ transition if one takes into account that the shell model calculations give results which, even with effective charges, are one to two orders of magnitude smaller than the experiment. For high-spin states, mixing of symmetry breaking components due to the spin–orbit force would decrease the calculated values [102]. The

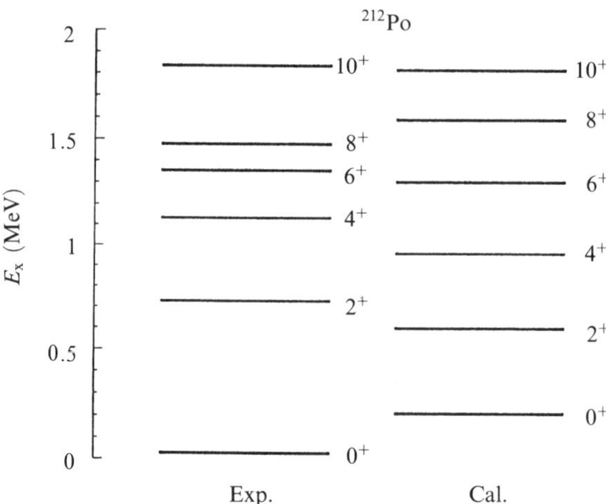

Fig. 6.24 Comparison of the energy levels of ^{212}Po calculated in the α-cluster model using a folding model with experiment. (Taken from Ohkubo and Sakuda [101].)

Table 6.4 Same as Table 6.3 but for ^{212}Po (Taken from Ohkubo [100])

J^π	E_x (MeV)	λ	$\sqrt{\langle R^2 \rangle}$ (fm)	B(E2) (e^2fm^4) Exp.	B(E2) (e^2fm^4) Cal.
0^+	0.0	1.1805	6.60		
2^+	0.727	1.1711	6.62		579
4^+	1.132	1.1655	6.60		814
6^+	1.355	1.1620	6.54	1013	844
8^+	1.476	1.1590	6.46	297	801
10^+	1.834	1.1531	6.38	164	729

calculated wave function of the ground state is very similar to the harmonic-oscillator wave function with $N = 22$ in the inner region; however, at the surface, the wave function is pushed slightly outward, with its outermost peak at 8.2 fm, which brings about the enhancement of B(E2) values and α-widths.

The α-reduced widths can be calculated using the R-matrix theory. The R-matrix reduced width is given by $\gamma_\alpha^2 = (\hbar^2 a/2\mu)\Omega_l^2(a)$, where $\Omega_l(a)$ is the amplitude of the amount of clustering [111] at a channel radius a, and μ is the reduced mass. In the model $\Omega_l(a)$ is taken to be a wave function $\psi_l(a)$ at a, beyond which only the Coulomb potential works. The calculated $\gamma_\alpha^2 = 1.23$ keV for the ground state agrees well with the experimental value, 1.4 keV. For the 2^+ state, the calculated result, 1.36 keV, agrees with the experimental value, 0.48 keV, within a factor of three. All the calculations which reproduce the experimental α-widths, whatever the model or calculational method, involve α-clustering [100,110–113]. The long-standing problem of the α-width of the ground state of ^{212}Po has been solved by recognizing anew the α-cluster model in heavy nuclei.

For the high-spin members, the cluster model overestimates the α-widths, which means that the mixing of symmetry broken shell model components due to the spin–orbit force increases as the spin becomes higher [102] and suggests the importance of the shell model configurations which are orthogonal to the cluster model configurations [111]. The 18^+ isomer, which is in the line of the 8^+ state of ^{20}Ne and the 12^+ state of ^{44}Ti, is understood as a state whose shell model character is extremely intensified. The change of structure from an α-cluster state to a shell model state as the spin becomes higher seems to persist more purely in the ground state band of ^{212}Po.

The model gives the $N = 23$, $K = 0^-$ band at $E_x = 5.4\,\mathrm{MeV}$ and the second $K = 0^+$ band with $N = 24$ at 9.7 MeV. The calculated dimensionless reduced width θ_α^2 for the $N = 24$, 0^+ state is 48%, which is large enough to be observed in experiments. Since the folding potentials do not differ very much for neighbouring nuclei, the persistence of α-clustering around the ^{212}Po region is expected.

6.6 The future

It has been shown that the α-cluster structure persists not only in light nuclei but also in the fp-shell and possibly in heavy nuclei. The existence of parity doublet bands and the higher-nodal band may be universal even in heavy nuclei. This suggests that the α-cluster model is powerful throughout the periodic table. In the heavier-mass region ($A \geq 40$), there have been only a few α-transfer experiments. Systematic α-transfer experiments to observe the members of the parity doublet bands and the higher-nodal band in heavy nuclei are very useful. The α-clustering aspects of the nucleus near the double-shell-closure, such as in ^{56}Ni, ^{60}Zn, ^{100}Sn, and ^{104}Te, are of great interest. Since the targets are unstable, radioactive beams are useful. Exploring the α-clustering aspects of unstable nuclei (neutron-rich and proton-rich) is an important future subject. A detailed study of the α-cluster structure in the fp-shell nuclei is also needed to understand the reaction process in nuclear synthesis. The α-clustering, which occurs both in the surface and the inner regions of the nucleus, may be studied using high-energy beams in the (p, α) reaction. The facility at RCNP is expected to be a powerful tool in future research on nuclear clustering.

Acknowledgements

The authors would like to express their sincere thanks to their colleagues, especially F. Michel, G. Reidemeister, T. Sakuda, D. M. Brink, S. Okabe, T. Yamada, A. C. Merchant, M. Fujiwara, T. Suehiro, S. Kato, and K. Katori for their collaboration in the work leading to the present article. They are also grateful to H. Tanaka, R. Tamagaki, K. Ikeda, A. Arima, H. Ejiri, Y. Abe, H. Horiuchi, B. Buck, and R. Lovas for their interest and encouragement.

Appendix

This appendix summarizes the formalism of the cluster models: (1) Generator Coordinate Method (GCM), (2) Resonating Group Method (RGM), (3) Orthogonality-Condition Model (OCM), and (4) Local-Potential Cluster Model (LPCM). For simplicity and transparency, the model is explained for the α + nucleus system.

6.A.1 *Generator coordinate method*

The cluster wave function for the α + core(C) system in the Brink–Bloch-type formalism is given by

$$\Psi(\boldsymbol{R}_1, \boldsymbol{R}_2) = \sqrt{\binom{A}{4}}\,\mathcal{A}\{\Phi_{\boldsymbol{R}_1}(\alpha)\Phi_{\boldsymbol{R}_2}(\mathrm{C})\}, \tag{A6.1}$$

where $\mathcal{A}$ is the antisymmetrizer operator which exchanges nucleons between the two clusters, and A is the mass number of the total system. The α-cluster wave function $\Phi_{\boldsymbol{R}_1}(\alpha)$ localized around the centre $\boldsymbol{R}_1$ is given by

$$\Phi_{\boldsymbol{R}_1}(\alpha) = \{\phi\}^4. \tag{A6.2}$$

The nucleon spatial wave function at $\boldsymbol{R}_1$ in the GCM space is

$$\phi(\boldsymbol{r}) = (\pi b)^{-3/4}\exp\left[-\frac{1}{2b^2}(\boldsymbol{r} - \boldsymbol{R}_1)^2\right], \tag{A6.3}$$

where b is the size parameter of the α-cluster. We assume that the two clusters have the same size parameter b. The cluster wave function for the core nucleus C, $\Phi_{\boldsymbol{R}_2}(\mathrm{C})$ localized around the centre $\boldsymbol{R}_2$ is given by the shell model wave function. In the case where the core C is $^{40}\mathrm{Ca}$, the $\Phi_{\boldsymbol{R}_2}(^{40}\mathrm{Ca})$ is given by the harmonic-oscillator closed shell around the centre $\boldsymbol{R}_2$,

$$\Phi_{\boldsymbol{R}_2}(^{40}\mathrm{Ca}) = \left\{(000)^4(100)^4(010)^4(001)^4(200)^4(020)^4(002)^4(110)^4(011)^4(101)^4\right\}_{\boldsymbol{R}_2}. \tag{A6.4}$$

The GCM wave function is given by a superposition of $\Psi(\boldsymbol{R}_1, \boldsymbol{R}_2)$ as follows:

$$\Psi^{\mathrm{GCM}} = \int \mathrm{d}\boldsymbol{R}\, f(\boldsymbol{R})\Psi(\boldsymbol{R}_1, \boldsymbol{R}_2), \tag{A6.5}$$

where $\boldsymbol{R} = \boldsymbol{R}_1 - \boldsymbol{R}_2$ is the relative distance parameter, the generator coordinate, between the two clusters. The weight function $f(\boldsymbol{R})$ is determined by solving the Hill–Wheeler equation

$$\int \{\mathcal{H}(\boldsymbol{R}, \boldsymbol{R}') - E\mathcal{N}(\boldsymbol{R}, \boldsymbol{R}')\} f(\boldsymbol{R}')\,\mathrm{d}\boldsymbol{R}' = 0, \tag{A6.6}$$

where E is the total energy of the system, and the energy kernel $\mathcal{H}(\boldsymbol{R}, \boldsymbol{R}')$ and norm kernel $\mathcal{N}(\boldsymbol{R}, \boldsymbol{R}')$ are given by

$$\mathcal{H}(\boldsymbol{R}, \boldsymbol{R}') = \langle \Psi(\boldsymbol{R}_1, \boldsymbol{R}_2)|H|\Psi(\boldsymbol{R}'_1, \boldsymbol{R}'_2)\rangle, \tag{A6.7}$$

$$\mathcal{N}(\boldsymbol{R}, \boldsymbol{R}') = \langle \Psi(\boldsymbol{R}_1, \boldsymbol{R}_2)|\Psi(\boldsymbol{R}'_1, \boldsymbol{R}'_2)\rangle, \tag{A6.8}$$

where H is a nuclear Hamiltonian with two-body interactions. More details, such as parity and angular momentum projections, are given in Ref. [13].

6.A.2 *Resonating group method*

The cluster wave function in the RGM is assumed to be

$$\Psi = \sqrt{\binom{A}{4}}\,\mathcal{A}\{\varphi_\alpha(1,\ldots,4)\varphi_{\mathrm{C}}(5,\ldots,\mathrm{A})\chi(\boldsymbol{r})\}, \tag{A6.9}$$

with

$$\boldsymbol{r} = \frac{1}{4}\sum_{i=1}^{4}\boldsymbol{r}_i - \frac{1}{A-4}\sum_{i=5}^{A}\boldsymbol{r}_i, \tag{A6.10}$$

where φ_α and φ_{C} are the intrinsic wave functions of the clusters α and C, respectively. The relative wave function $\chi(\boldsymbol{r})$ is determined by solving the Schrödinger equation

$$\langle \varphi_\alpha \varphi_{\mathrm{C}}|H - E|\Psi\rangle = 0, \tag{A6.11}$$

where the integration is performed over the internal coordinates and spin–isospin variables. This leads to the following integro-differential equation in the RGM:

$$\{T_r + V_{\mathrm{D}}(\boldsymbol{r})\}\chi(\boldsymbol{r}) + \int M(\boldsymbol{r}, \boldsymbol{r}')\chi(\boldsymbol{r}')\,\mathrm{d}\boldsymbol{r}' = E\left\{\chi(\boldsymbol{r}) - \int K(\boldsymbol{r}, \boldsymbol{r}')\chi(\boldsymbol{r}')\,\mathrm{d}\boldsymbol{r}'\right\}, \tag{A6.12}$$

with

$$E = E - E_\alpha^{(\mathrm{in})} - E_{\mathrm{C}}^{(\mathrm{in})}, \tag{A6.13}$$

where $E_\alpha^{(\mathrm{in})}$ and $E_{\mathrm{C}}^{(\mathrm{in})}$ are the internal energies of the clusters α and C, respectively, T_r is the kinetic energy operator, and the direct (folding) potential $V_{\mathrm{D}}(\boldsymbol{r})$ is given by

$$V_{\mathrm{D}}(\boldsymbol{r}) = \langle \phi_\alpha \phi_{\mathrm{C}}|\sum_{\substack{i\in\alpha \\ j\in\mathrm{C}}} v_{ij}|\phi_\alpha \phi_{\mathrm{C}}\rangle. \tag{A6.14}$$

The exchange kernel $M(\boldsymbol{r}, \boldsymbol{r}')$ is the sum of the kinetic energy kernel and the interaction kernels, which can be calculated by a transformation from the GCM kernels $\mathcal{H}(\boldsymbol{R}, \boldsymbol{R}')$ and $\mathcal{N}(\boldsymbol{R}, \boldsymbol{R}')$. The coupled-channel RGM including the excitation of the core is formulated similarly.

6.A.3 *Orthogonality-condition model*

It is very tedious to calculate the exchange kernels for heavy systems in the RGM. The RGM equation is approximated well by taking into account the Pauli principle only through the norm kernel as follows:

$$\sqrt{1 - K}(T_r + V_D(r) - E)\sqrt{1 - K}\chi(r) = 0. \tag{A6.15}$$

The potential $V_D(r)$ is originally derived from the two-body interactions, but the nuclear potential can be replaced by other local potentials such as a Woods–Saxon squared potential and a double-folding potential.

The norm kernel $K(r, r')$ has the eigenfunctions of the following integral equation:

$$\int K(r, r')\chi_i(r')\, dr' = \mu_i\chi_i(r). \tag{A6.16}$$

When the two clusters have the harmonic-oscillator shell model intrinsic wave functions with the same size parameter, the eigenfunctions are also the harmonic oscillator wave functions. The eigenfunctions whose eigenvalues are unity are forbidden by the Pauli principle and satisfy the following equation:

$$\mathcal{A}\{\phi_\alpha\phi_C\chi_i^{F}(r)\} = 0. \tag{A6.17}$$

The eigenfunctions whose eigenvalues are very close to unity are the almost forbidden states. Equation (A6.15) is further approximated by

$$\Lambda(T_r + V^{(\text{eff})}(r) - E)\Lambda\chi(r) = 0, \tag{A6.18}$$

where Λ is a projection operator given by

$$\Lambda = 1 - \sum_{i=1}^{m} |\chi_i^{F}(r)\rangle\langle\chi_i^{F}(r')|, \tag{A6.19}$$

m being the number of the forbidden states. In eqn (A6.18), it is no longer necessary to calculate the eigenvalues of the norm kernel. The OCM equation can easily be extended to include the excitation of the core.

6.A.4 *Local-potential cluster model*

In the GCM, RGM, and OCM the equations are non-local. When the interaction potential $V^{(\text{eff})}(r)$ can bind the deeply bound states which are very similar to the forbidden states, as is often the case, eqn (A6.18) is approximately equivalent to

$$(T_r + V^{(\text{eff})}(r) - E)\chi(r) = 0, \tag{A6.20}$$

for the physical states whose wave functions satisfy the Wildermuth condition

$$2n + L \geq N_0, \tag{A6.21}$$

where n is the number of the nodes of the wave function with orbital angular momentum L. For the $\alpha + {}^{16}O$ system, $N_0 = 8$ and, for the $\alpha + {}^{40}Ca$ system, $N_0 = 12$.

Equation (A6.20) is a re-interpretation of the OCM and has been very useful for the cluster study of nuclear structure and reactions.

References

1. H. A. Becquerel, *Compt. Rend.* **122** (1896).
2. E. Rutherford, *Phil. Mag.* **21**, 669 (1911).
3. G. Gamov, *Z. Phys.* **53**, 510 (1928).
4. E. U. Condon and R. W. Gurney, *Nature* **122**, 439 (1928).
5. W. Wefelmeier, *Z. Phys.* **107**, 332 (1937).
6. M. G. Mayer, *Phys. Rev.* **75**, 1969 (1949); O. Haxel, J. H. D. Jensen, and H. E. Suess, *Phys. Rev.* **75**, 1766 (1949); M. G. Mayer and J. H. D. Jensen, *Elementary theory of nuclear shell structure*, John Wiley and Sons, New York (1955).
7. A. Bohr, *Kgl. Dan. Mat-Fys. Medd.* **26**(14) (1952); A. Bohr and B. R. Mottelson, *Kgl. Dan. Mat-Fys. Medd.* **27**(16) (1953); **30**(1) (1955).
8. D. M. Brink and J. J. Castro, *Nucl. Phys.* **A216**, 109 (1973).
9. K. Sekine, M. Taketani, T. Toyoda, and K. Inoue, *Soryushiron Kenkyu (Kyoto)* **11**, 195 (1956).
10. K. Wildermuth and Th. Kanelopoulos, *Nucl. Phys.* **7**, 150 (1958); K. Wildermuth and W. McClure, *Cluster representation of nuclei* (Springer-Verlag, Berlin, 1966).
11. B. F. Bayman and A. Bohr, *Nucl. Phys.* **9**, 596 (1958/59).
12. R. Tamagaki and H. Tanaka, *Prog. Theor. Phys. (Kyoto)* **34**, 191 (1965); S. Okai and S. C. Park, *Phys. Rev.* **145**, 787 (1966).
13. D. M. Brink, *Proceedings of the international school of physics*, 'Enrico Fermi' course 36, p. 247 (1966).
14. K. Ikeda, N. Takigawa, and H. Horiuchi, *Prog. Theor. Phys. (Kyoto)* (Suppl.) Extra number, 464 (1968).
15. K. Ikeda, T. Marumori, R. Tamagaki, H. Tanaka, H. Horiuchi, Y. Suzuki, J. Hiura, F. Nemoto, H. Bando, Y. Abe, N. Takigawa, M. Kamimura, K. Takada, Y. Akaishi, and S. Nagata, *Prog. Theor. Phys. (Kyoto)* (Suppl.) **52** (1972).
16. H.Tanaka *et al.*, *Proceedings of the international symposium on cluster structure of nuclei and transfer reactions induced by heavy ions*, Tokyo, H. Kamitsubo *et al.*, eds (IPCR Cyclotron Progress Report, Suppl 4, 1975).
17. K. Ikeda, H. Horiuchi, S. Saito, Y. Fujiwara, M. Kamimura, K. Kato, Y. Suzuki, E. Uegaki, H. Furutani, H. Kanada, T. Kaneko, S. Nagata, S. Okabe, T. Sakuda, M. Seya, Y. Abe, Y. Kondo, T. Matsuse, and A. Tohsaki-Suzuki, *Prog. Theor. Phys. (Kyoto)* (Suppl.) **68** (1980).
18. K. Wildermuth and Y. C. Tang, *A unified theory of the nucleus* (Vieweg, Braunschweig, 1977).
19. A. Arima, in *Heavy ion collisions*, R. Bock, ed. (North Holland, Amsterdam, 1979), Vol. 1, p. 419.
20. D. A. Bromley, *Proceedings of the fourth international conference on clustering aspects of nuclear structure and nuclear reactions*, Chester, UK, J. S. Lilley and M. A. Nagarajan, eds (Reidel, Dordrecht, 1985), p. 1.

21. Cluster conferences (1) *Proceedings of 5th international conference on clustering aspects in nuclear and subnuclear systems* (Kyoto, 1988), *J. Phys. Soc. Japan* **58** (1989); (2) *Proceedings of the international symposium on developments of nuclear cluster dynamics* (Sapporo, 1988), Y. Akaishi *et al.*, eds (World Scientific, 1989); (3) *Clustering phenomena in atoms and nuclei. Proceedings of the international conference on nuclear and atomic clusters* (Turku, Finland, 1991), M. Brenner *et al.*, eds (Springer, 1992); (4) *Atomic and nuclear clusters. Proceedings of the second international conference* (Santorini, Greece, 1993), G. S. Anagnostatos and W. Von Oeretzen, eds (Springer, 1994); (5) *Proceedings of the sixth international conference on clusters in nuclear structure and dynamics* (Strasbourg, 1994), F. Haas, ed. (CRN, Strasbourg, 1995).

22. G. R. Satchler, L. W. Owen, A. J. Elwyn, G. L. Morgan, and R. L. Walter, *Nucl. Phys.* **A112**, 1 (1968).

23. A. Sugie, P. E. Hodgson, and H. H. Robertson, *Proc. Phys. Soc.* **A70**, 1 (1957).

24. B. Buck, H. Friedrich, and C. Wheatley, *Nucl. Phys.* **A275**, 246 (1977); H. Friedrich and L. F. Canto, *Nucl. Phys.* **A291**, 249 (1977).

25. A. Tohsaki, *Phys. Rev.* **C49**, 1814 (1994).

26. T. P. Marvin and P. P. Singh, *Nucl. Phys.* **A180**, 282 (1972).

27. J. J. Simpson, W. R. Dixon, and R. S. Storey, *Phys. Rev. Lett.* **31**, 946 (1973).

28. S. Ali and A. R. Bodmer, *Nucl. Phys.* **80**, 99 (1966).

29. S. Saito, *Prog. Theor. Phys. (Kyoto)* **41**, 705 (1969).

30. V. G. Neudachin, V. I. Kukulin, V. L. Korotkikh, and V. P. Korennoy, *Phys. Lett.* **B34**, 581 (1971).

31. V. I. Kukulin, V. G. Neudachin, and Yu. F. Smirnov, *Nucl. Phys.* **A245**, 429 (1975).

32. I. Shimodaya, R. Tamagaki, and H. Tanaka, *Prog. Theor. Phys. (Kyoto)* **27**, 793 (1962).

33. D. Baye, *Phys. Rev. Lett.* **58**, 2738 (1987); D. Baye and J.-M. Sparenberg, *Phys. Rev. Lett.* **73**, 2789 (1994).

34. G. R. Satchler and W. G. Love, *Phys. Rep.* **55**, 183 (1979).

35. S. A. Sofianos, K. C. Panda, and P. E. Hodgson, *J. Phys. Phys.* **G19**, 1929 (1993).

36. E. Uegaki, S. Okabe, Y. Abe, and H. Tanaka, *Prog. Theor. Phys. (Kyoto)* **57**, 1262 (1977); E. Uegaki, Y. Abe, S. Okabe, and H. Tanaka, *Prog. Theor. Phys. (Kyoto)* **59**, 1031 (1978); **62**, 1621 (1979).

37. Y. Fukushima and M. Kamimura, *J. Phys. Soc. Japan* (Suppl.) **44**, 225 (1978).

38. G. S. Anagnastatos, *Phys. Rev.* **C51**, 152 (1995).

39. Y. Suzuki, *Prog. Theor. Phys. (Kyoto)* **55**, 1751 (1976); **56**, 111(1976).

40. S. Okabe, *Tours symposium on nuclear physics II*, H. Utsunomiya *et al.*, eds (World Scientific, 1995), p. 112.

41. R. A. Baldock, B. Buck, and J. A. Rubio, *Nucl. Phys.* **A426**, 222 (1984).

42. Y. Fujiwara, *Prog. Theor. Phys. (Kyoto)* **62**, 122 (1979).

43. S. Ohkubo, Y. Kondo, and S. Nagata, *Prog. Theor. Phys. (Kyoto)* **57**, 82 (1977).

44. S. Ohkubo and Y. Tokunaga, *Proceedings of the 1989 international nuclear physics conference*, Sao-Paulo, Brazil, Vol. I, p. 306; S. Ohkubo, *Proceedings of the 6th international conference on nuclear reaction mechanisms*, E. Gadioli, ed. (Milan

Univ. Ricerca and Scientifica ed Educazione Permanente Suppl. No. 84, 1991), p. 396.

45. F. Michel, J. Albinski, P. Belerly, Th. Delbar, Gh. Gregoire, B. Tasiaux, and G. Reidemeister, *Phys. Rev.* **C28**, 1904 (1983).

46. F. Michel, G. Reidemeister, and S. Ohkubo, *Phys. Rev.* **C35**, 1961 (1987).

47. B. Buck, C. B. Dover, and J. P. Vary, *Phys. Rev.* **C11**, 1803 (1975).

48. W. J. Gerace and A. M. Green, *Nucl. Phys.* **A93**, 110 (1967); **A123**, 241 (1969).

49. A. Arima, V. Gillet, and J. Ginocchio, *Phys. Rev. Lett.* **25**, 1043 (1970).

50. A. Jaffrin, *Nucl. Phys.* **A196**, 577 (1972).

51. H. Faraggi, A. Jaffrin, M.-C. Lemaire, M. C. Mermaz, J.-C. Faivre, B. G. Harvey, J.-M. Loiseaux, and A. Papineau, *Phys. Rev. Lett.* **24**, 1188 (1970); H. Faraggi, M.-C. Lemaire, J.-M. Loiseaux, M. C. Mermaz, and A. Papinau, *Phys. Rev.* **C4**, 1375 (1971).

52. U. Strohbush, C. L. Fink, B. Zeidman, R. G. Markham, H. W. Fulbright, and R. N. Horoshko, *Phys. Rev.* **C9**, 965 (1974).

53. D. Frekers, H. Eickhoff, H. Löhner, K. Poppensieker, R. Santo, and C. Wiezorek, *Z. Phys.* **A276**, 317 (1976).

54. H. Kihara, M. Kamimura, and A. Tohsaki-Suzuki, *Proceedings of the international conference on nuclear structure*, edited by the Organizing Committee (International Printing Co. Ltd., Tokyo, 1977), p. 235.

55. H. Horiuchi, *Prog. Theor. Phys. (Kyoto)* **73**, 1172 (1985).

56. K. F. Pal and R. G. Lovas, *Phys. Lett.* **96B**, 19 (1980).

57. K. Itonaga, *Prog. Theor. Phys. (Kyoto)* **66**, 2103 (1981).

58. Th. Delbar, Gh. Gregoire, G. Paic, R. Ceuleneer, F. Michel, R. Vanderpoorten, A. Budzanowski, H. Dabrowski, L. Freindl, K. Grotowski, S. Micek, R. Planeta, A. Strazalkowski, and K. A. Eberhald, *Phys. Rev.* **C18**, 1237 (1978).

59. D. M. Brink and N. Takigawa, *Nucl. Phys.* **A279**, 159 (1977).

60. F. Michel, G. Reidemeister, and S. Ohkubo, *Phys. Rev.* **C34**, 1248 (1986).

61. F. Michel, G. Reidemeister, and S. Ohkubo, *Phys. Rev. Lett.* **57**, 1215 (1986); *Phys. Rev.* **C37**, 292 (1988).

62. S. Ohkubo and D. M. Brink, *Phys. Rev.* **C36**, 966 (1987).

63. S. Ohkubo, *Phys. Rev.* **C38**, 2377 (1988).

64. D. Frekers, R. Santo, and K. Langanke, *Nucl. Phys.* **A394**, 189 (1983).

65. A. Tohsaki-Suzuki and K. Naito, *Prog. Theor. Phys. (Kyoto)* **58**, 721 (1977).

66. H. Friedrich and K. Langanke, *Nucl. Phys.* **A252**, 47 (1975).

67. K. Langanke, *Nucl. Phys.* **A377**, 53 (1982).

68. D. Wintgen, H. Friedrich, and K. Langanke, *Nucl. Phys.* **A408**, 239 (1983).

69. A. C. Merchant, K. F. Pal, and P. E. Hodgson, *J. Phys.* **G15**, 601 (1989).

70. P. E. Hodgson and A. C. Merchant, *Proceedings of the 5th international conference on nuclear reaction mechanisms*, E. Gadioli, ed. (Milan Univ. Ricerca and Scientifica ed Educazione Permanente Suppl. No. 66, 1988), p. 375.

71. S. Ohkubo, *Phys. Rev.* **C36**, 551 (1987).

72. P. E. Hodgson, *Proceedings of the 5th international conference on clustering aspects in nuclear and subnuclear systems*, J. Phys. Soc. Japan (Suppl.) **58**, 755 (1988).

73. T. Yamaya, S. Oh-ami, O. Satoh, M. Fujiwara, T. Itahashi, K. Katori, S. Kato, M. Tosaki, S. Hatori, and S. Ohkubo, *Phys. Rev.* **C41**, 2421 (1990).

74. T. Yamaya, S. Oh-ami, M. Fujiwara, T. Itahashi, K. Katori, M. Tosaki, S. Kato, S. Hatori, and S. Ohkubo, *Phys. Rev.* **C42**, 1935 (1990).

75. H. Ikegami, S. Morinobu, I. Katayama, M. Fujiwara, and S. Yamabe, *Nucl. Instr. Meth.* **175**, 335 (1981).

76. J. J. Kolata, J. W. Olness, and E. K. Warburton, *Phys. Rev.* **C10**, 1663 (1974).

77. H. W. Fulbright, C. L. Bennett, R. A. Lindgren, R. G. Markham, S. C. McGuire, G. C. Morrison, U. Strohbush, and J. Toke, *Nucl. Phys.* **A284**, 329 (1977).

78. T. Wada and H. Horiuchi, *Phys. Rev.* **C38**, 2063 (1988).

79. C. Y. Kim and T. Udagawa, *Phys. Rev.* **C46**, 532 (1992).

80. P. Guazzoni, M. Jaskola, L. Zetta, C. Y. Kim, T. Udagawa, and G. Bohlen, *Nucl. Phys.* **A564**, 425 (1993).

81. T. Yamaya, S. Ohkubo, S. Okabe, and M. Fujiwara, *Phys. Rev.* **C47**, 2389 (1993).

82. T. Yamaya, K. Ishigaki, H. Ishiyama, T. Suehiro, S. Kato, M. Fujiwara, K. Katori, M. H. Tanaka, S. Kubono, V. Guimaraes, and S. Ohkubo, *Phys. Rev.* **C53**, 131 (1996).

83. T. Yamada, *Phys. Rev.* **C42**, 1432 (1990).

84. T. Yamada and S. Ohkubo, *Z. Phys.* **A349**, 363 (1994).

85. T. Ogawa, Y. Suzuki, and K. Ikeda, *Prog. Theor. Phys.* (*Kyoto*) **57**, 1072 (1977).

86. S. Ohkubo and K. Umehara, *Prog. Theor. Phys.* (*Kyoto*) **80**, 598 (1988).

87. G. Reidemeister, S. Ohkubo, and F. Michel, *Phys. Rev.* **C41**, 63 (1990).

88. T. Yamaya, M. Saito, M. Fujiwara, T. Itahashi, K. Katori, T. Suehiro, S. Kato, S. Hatori, and S. Ohkubo, *Phys. Lett.* **B306**, 1 (1993).

89. T. Yamaya, M. Saitoh, M. Fujiwara, T. Itahashi, K. Katori, T. Suehiro, S. Kato, S. Hatori, and S. Ohkubo, *Nucl. Phys.* **A573**, 154 (1994).

90. T. Sakuda and S. Ohkubo, *Phys. Rev.* **C49**, 149 (1994).

91. T. Sakuda and S. Ohkubo, *Phys. Rev.* **C51**, 586 (1995).

92. T. Sakuda and S. Ohkubo, *Phys. Rev.* **C57**, 1184 (1998).

93. I. Miyamoto and S. Ohkubo, *Z. Phys.* **A349**, 297 (1994).

94. S. Ohkubo, K. Umehara, and K. Hiraoka, *Proceedings of the international symposium on developments of nuclear cluster dynamics*, Y. Akaishi *et al.*, eds (World Scientific, 1989), p. 114; S. Ohkubo, *Soryushiron Kenkyu* (*Kyoto*) **81**, B3 (1990).

95. E. Uegaki and Y. Abe, *Phys. Lett.* **B340**, 143 (1994).

96. E. Uegaki and Y. Abe, *Prog. Theor. Phys.* (*Kyoto*) **90**, 615 (1993).

97. W. D. M. Rae and A. C. Merchant, *Phys. Lett.* **B279**, 207 (1992).

98. P. E. Hodgson, *Zeit. Phys.* **A349**, 197 (1994).

99. H. Abele and G. Staudt, *Phys. Rev.* **C47**, 742 (1993).

100. S. Ohkubo, *Phys. Rev. Lett.* **74**, 2176 (1995).

101. S. Ohkubo and T. Sakuda, *Nuclear reaction dynamics of nucleon–hadron many body system*, H. Ejiri *et al.*, eds (World Scientific, 1996), p. 363.

102. T. Tomoda and A. Arima, *Nucl. Phys.* **A303**, 217 (1978).

103. E. Adamides, L. D. Skouras, and A. C. Xenoulis, *Phys. Rev.* **C23**, 2016 (1981).

104. B. Buck, A. C. Merchant, and S. M. Perez, *Phys. Rev.* **C51**, 559 (1995).

105. D. Strottman, *Phys. Rev.* **C20**, 1150 (1979).

106. H. J. Mang, *Phys. Rev.* **113**, 1593 (1959).

107. N. K. Glendenning and K. Harada, *Nucl. Phys.* **72**, 481 (1965).

108. T. Fliessbach and H. J. Mang, *Nucl. Phys.* **A263**, 75 (1976).

109. I. Tonozuka and A. Arima, *Nucl. Phys.* **A323**, 45 (1979).

110. S. Okabe, *J. Phys. Soc. Japan* (Suppl.) **58**, 516 (1989).

111. K. Varga, R. G. Lovas, and R. J. Liotta, *Phys. Rev. Lett.* **69**, 37 (1992); *Nucl. Phys.* **A550**, 421 (1992).

112. B. Buck, A. C. Merchant, and S. M. Perez, *Phys. Rev. Lett.* **72**, 1326 (1994).

113. F. Hoyler, P. Mohr, and G. Staudt, *Phys. Rev.* **C50**, 2631 (1994).

7

Physics of snowballs—impurity ions in superfluid helium

N. Takahashi and T. Shimoda

7.1 Introduction

The strange behaviour of liquid helium has been attracting the attention of quite a number of physicists ever since helium was liquified and is still doing so after almost a century. It may well not be an exaggeration to say that new precise measurements of physical quantities concerning liquid helium succeed in reviving discussions about our physical insights into the problems of liquid helium.

The impurity ions in liquid helium are also a subject of extensive research in physics and chemistry, for example. One of the keys to understanding such systems is the electrostriction that takes place in the electric field the charge carriers create. The helium atoms in this electric field are electrically polarized, though slightly, and the Coulomb interaction creates an aggregation around a positive charge. If the carrier is a negatively charged electron, the Pauli repulsion between this and the other electrons in the surrounding helium atoms results in a vacuum, isolating the single electron in liquid helium. The aggregations thus formed around a positive impurity ion are referred to as 'snowballs' and they are $\sim$2 nm in diameter, whereas the vacuum surrounding the negative charges are called 'bubbles' and are as large as 5 nm in diameter. The effective masses are approximately 50 and 200 times the mass of He atom for snowballs and bubbles, respectively.

Snowballs belong to a kind of microcluster which possesses an ordered structure. The bulk helium sustains the hexagonal close packed (hcp) configuration when it solidifies, except in a narrow region in the phase diagram. In a snowball, if this is a solid entity, electrically polarized helium atoms arrange themselves around an impurity ion according to a rule which may not be the same as that for the solidification of the bulk helium. Rather, the structure can be thought to be of higher symmetry, such as a bcc- or an fcc-type configuration or, in some cases, probably an icosahedron may be preferred.

The nuclear spin polarization of light radioactive nuclei may be one of the essential probes to seek the structure of such microclusters. The interaction between the nuclear spin and the internal electric and magnetic fields which the atoms surrounding the impurity nucleus exert may influence the relaxation process and this may provide information on the structure in the cluster.

This novel experimental procedure was used in our previous studies of polarization phenomena in heavy-ion reactions (Sugimoto *et al.*, 1985; Takahashi, 1985), i.e. the degradation of kinetic energy via non-depolarizing passage through matter and implantation of fast radioactive nuclei in a solid to select a specified part of the energy spectrum and the β-ray asymmetry measurement to determine the nuclear polarization. It occurred to the authors during the study of the heavy-ion polarization that the high-energy β-rays from the short-lived light nuclei may penetrate substantially into the material surrounding liquid helium when these ions are introduced in it. Further, during the study of the succeeding radioactive beams (Tanihata *et al.*, 1985*a,b*), it was thought that completely isotope-separated high-energy β-active nuclei may be used to penetrate substantially the necessary windows furnished on the cryostat walls without inducing serious background problems. In this way it has generated a renewed interest in the physics of snowballs with an experimental tool to study individual snowballs through β-ray counting (Takahashi *et al.*, 1991; 1992). The physics described here is based on this idea (for details, see Takahashi *et al.*, 1995; 1996).

The detection of radioactivity proves to be one of the most sensitive methods to trace impurities and identify objects. Energetic β-rays emitted from the short-lived light nuclei, such as ^{12}B, are ideal for tagging the impurities in liquid helium. Such nuclei on the neutron-rich side of the nuclear chart, for which the nuclear polarization can also be appreciable, are more easily produced in heavy-ion reactions. The emitted β-rays are also sufficiently energetic and easily penetrate the impurities. We identify impurities in liquid helium via β-ray and α-particle detections. This has become a turning point of the heavy-ion polarization project at RCNP, Osaka University, and the use of the radioactive beams obtained here brings a flavour different from the methods used in traditional nuclear structure and reaction physics (Takahashi, 1994). Some of the related topics and the results of experiments based on this method, and future scope are described in this article.

7.2 Liquid helium[a]

We begin with a brief review of the basic physical properties of liquid helium, including their macroscopic and microscopic features.

Helium exists in several isotopic configurations, ^{3}He, ^{4}He, ^{6}He, and ^{8}He. The first two isotopes are stable and their natural abundance is 0.000137% and 99.999863%, respectively. The last two isotopes are radioactive and emit β-rays with energies 3.507 and 10.0 MeV, and have half-lives 806.7 and 119.0 ms, respectively. The discussions on helium as a liquid are therefore limited to the isotope ^{4}He and little is said about liquid ^{3}He here, which in itself contains very interesting physics. The other two radioactive helium isotopes are good candidates for the studies considered here. The difference between ^{3}He and ^{4}He is quite large, since the former obeys Bose–Einstein statistics while the latter the Fermi statistics. In the following discussion, condensed systems of helium mean, unless stated otherwise, the most abundant isotope ^{4}He.

---

[a]General references to this section can be found in many introductory textbooks on low-temperature physics. The following texts are already classics in this field: London (1954), Donnelly (1967), McClintock *et al.* (1984), and Wilks (1967). The details necessary for an understanding of the related topics are reviewed briefly in this section. Well-informed readers could, however, skip this section.

Helium atoms are considered as rigid spheres of radius approximately 0.27 nm, interacting via weak attractive forces between them. The electric polarizability of the atom is $0.24 \times 10^{-30} \, m^3$, which is extremely small compared with other atoms and molecules. This also holds for liquid helium above 2.17 K at saturated vapour pressure. Bulk helium in the gas phase starts condensing as the temperature is decreased and at 4.2 K it becomes a liquid. Helium has been the last of the gaseous elements that was successfully liquified at the beginning of this century. Its critical temperature is 5.19 K (Fig. 7.1). This can be compared with the normal boiling temperature of 3.19 K and the critical point at 3.32 K for ^{3}He.

Liquid ^{4}He shows unusual behaviour as the temperature decreases below 2.17 K, a phase transition point usually referred to as the λ-point. There is essentially no latent heat associated with the transition encountered here. Across this temperature, the specific heat shows abrupt changes on both sides of this point; the variation has a shape similar to the shape of the Greek λ and hence the appellation as shown in Fig. 7.2. The point divides the occurrence of the two regimes in liquid helium, He I and He II.

The liquid above the λ-point is composed solely of He I and that below the point is solely He II. He I behaves like an ordinary liquid while He II does not, because of the quantum effects. One of the most conspicuous features of He II is its superfluidity, a part of He II exhibiting apparently frictionless flow. Superfluidity makes it possible to transport helium atoms through extremely fine canals without applying any pressure, which is impossible in normal liquid helium. It was proposed by Landau that there are two components of liquid helium producing the properties of superfluidity below the λ-point. Of the two macroscopic components in liquid helium, the first component comprises He I and is referred to as the normal component. The other component is characteristic of He II only and is referred to as the superfluid component. He II is therefore a mixture of the two components and the mixing rate is dependent on the temperature of the liquid (Fig. 7.3). It is worth mentioning that He II carries no entropy while He I does. The

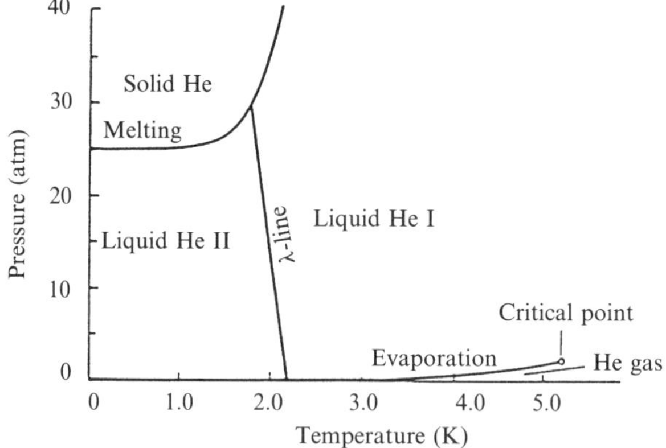

Fig. 7.1   Phase diagram of ^{4}He at low temperature. (Taken from London, 1954.)

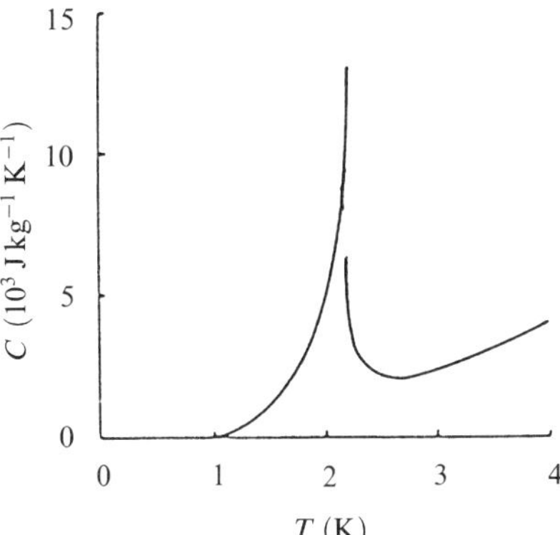

Fig. 7.2   Specific heat of liquid ^{4}He around the λ-transition point at the temperature $T = 2.17$ K. (Taken from Atkins and Rudnik, 1959.)

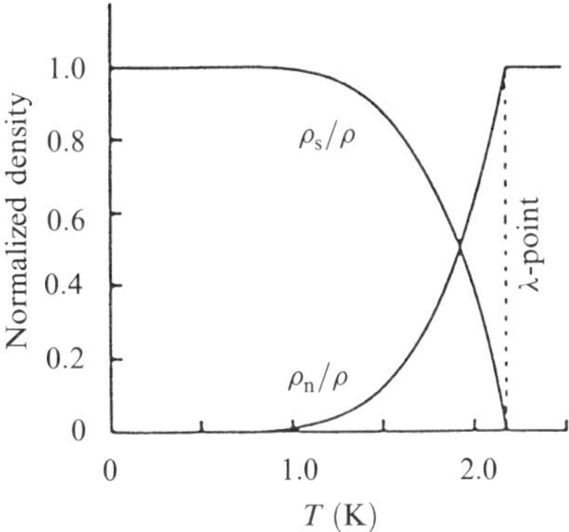

Fig. 7.3   Normalized density in superfluid ^{4}He. Here n and s represent normal and super components, respectively. (Taken from McClintock *et al.*, 1984.)

special properties and phenomena that He II displays are ascribed to the properties of the superfluid component in many cases. Since the energy is minimal, as the temperature is close to zero, and entropy of the superfluid component is zero, it is sensitive to quantum effects. This is the feature which continues to hold physicists' fascination with liquid helium.

An example of the quantum effects can be seen in the solidification at zero temperature. The phase diagram shows that helium does not exist in the solid phase unless the pressure exceeds a certain value even at 0 K. It is necessary to suppress the quantum zero-point motion of the helium atoms and the melting pressure required is usually 25 times the atmospheric pressure.

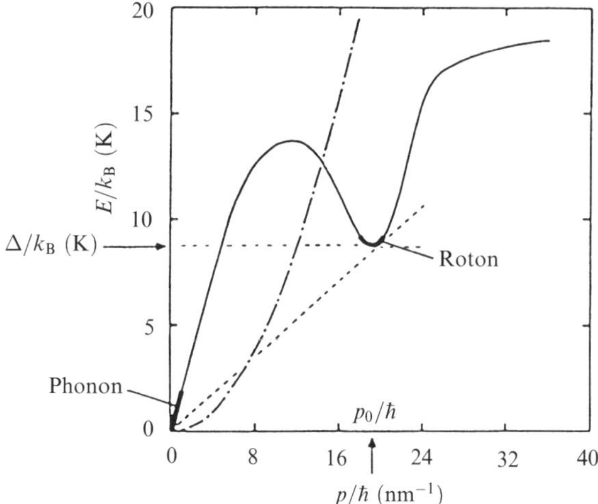

Fig. 7.4 Dispersion relation for basic excitations in liquid helium. The dot-dash curve corresponds to that for a free helium atom. (Taken from McClintock *et al.*, 1984.)

The above essentially concerns the ground states in bulk liquid helium. What are the effects of imparting energy or momentum to the superfluid helium atoms? The answer to this basic question is given by considering two phenomenological excitations: phonons and rotons. Phonons correspond to the motion in the condensed matter as a whole and their energy–momentum relation is

$$E = cp.$$

The interpretation of this type of excitation, which occurs at extremely low temperatures, is straightforward. Roton excitations manifest an energy gap Δ. This is a strange type of excitation and has a characteristic dispersion relation:

$$E = \Delta + \frac{1}{2M_{\text{eff}}}(p - p_0)^2.$$

Here M_{eff} is the effective mass of the roton and p_0 is the mean value for the momentum which an excited group of atoms sustains as a whole. The interpretation of this kind of excitation is by analogy with a moving object and may be compared to a coherent motion of a few helium atoms as a whole. The occurrence of the above two components in the dispersion relation is confirmed experimentally through neutron diffraction. The rotons appear at an energy above $E = \Delta$, where the energy gap Δ is given by $\Delta/k_B = 8.3$ K. The dispersion relation is depicted in Fig. 7.4.

7.3 Impurity ions in liquid helium

Liquid helium behaves quite uniquely as a quantum fluid, and the properties of impurities in this system are subjects of experimental and theoretical condensed-matter study. It is therefore no wonder that it has attracted the attention of many scientists for a long time. In the 1950s and 1960s, the impetus stemmed, for example, from studying the solidification of bulk helium, which takes place even at 0 K under a pressure in excess of 25 times the atmospheric pressure. Does the nucleation around an impurity play the role of an essential initial trigger for the solidification of liquid helium (Dahm and Sanders, 1970)? It may be caused by the ionic impurities in the liquid helium, which may easily be created through the occasional ionization of helium atoms owing to the irradiation, among other things, by the cosmic radiation. Landau indicated, for example, that ^{3}He exists as an impurity and that this impurity may not move with the superfluid component but rather with the normal component (Landau and Pomeranchuk, 1948), although ^{3}He itself possesses the properties of superfluidity.

Liquids of inert elements other than helium were also investigated. Among many such studies, the mobility measurements by Meyer and Reif (1958) were conspicuous by their accuracy and the underlying physics. The results are shown in Fig. 7.5. They give the speed of impurities, helium ions that are produced by the ionization due to the passage of α-particles from the ^{210}Po source immersed in the liquid helium, over a wide range of temperature from normal fluidity to the superfluid regime from 4.2 to 1.0 K. The mobility of helium ions was measured from the time of transmission between two grids arranged halfway between the Po α-source and the collecting electrode.

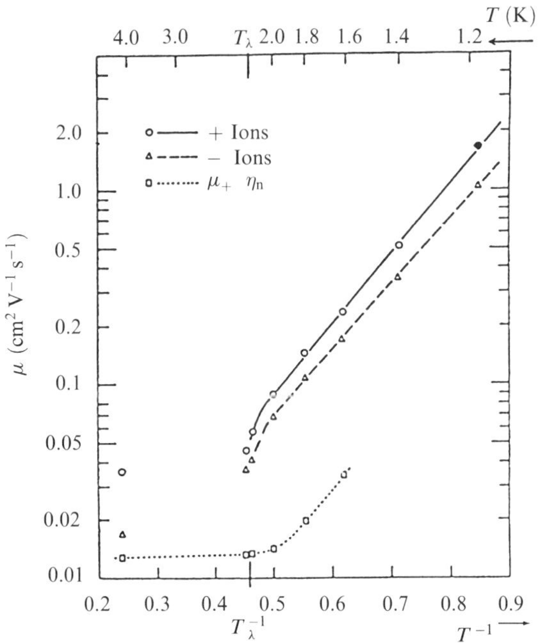

Fig. 7.5 Mobility μ of positive and negative charge carriers in liquid helium at low temperatures, as measured by Meyer and Reif (1958).

The mobility μ, the ratio of velocity v to electric field E, depends on the temperature as

$$\mu = \frac{v}{E} = K \exp\left(\frac{-a}{T}\right)$$

and the value of a found in the experiment was 8.8 K (Reif and Meyer, 1960), essentially identical to the value of the roton energy gap.

It may be of some interest to add that the electrons in liquid helium behave similarly as the helium ions except that the signs of the charges are different and their speeds smaller than those of the positive ions. The slope parameter a' for the electron is 8.1 K. The indication of the exponential behaviour and the similarity of a to the energy gap leads to the consideration that the charge carriers collide with the existing rotons and the mobility indicates an equilibrium between the pull due to the electric force and the drag due to the collision with rotons.

The mobilities of the helium ions are between $0.1\,\mathrm{cm}^2\,\mathrm{V}^{-1}\mathrm{s}^{-1}$ at 2.0 K and $2.0\,\mathrm{cm}^2\,\mathrm{V}^{-1}\mathrm{s}^{-1}$ just below 1.2 K. The speed in an electric field of $200\,\mathrm{V}\,\mathrm{cm}^{-1}$ is accordingly between 20 and $400\,\mathrm{cm}\,\mathrm{s}^{-1}$. These are extremely small values for the motion of an ion. Strangely, electrons move more slowly than He ions. This means that the effective radii of the positive and negative charge carriers are different; for some reason, the latter may be inflated more. This can be understood, according to a semiclassical calculation (Atkins, 1959), on the basis of electrostriction. The helium atoms in this electric field are electrically polarized and the Coulomb interaction creates an aggregation around the positive charge. The ionic structure is different for negative ions; the Pauli repulsive force between the carrier electron and electrons in the surrounding helium atoms exceeds the electrostriction from the negative ions. Thus, negative ions are isolated by a vacuum around them and form 'bubbles' (Kuper, 1961).

The internal pressure within the snowball increases rapidly with the distance from the impurity (Fig. 7.6). For the positive ion, regarded as a point charge, the pressure inside is so high that it easily exceeds the melting pressure of bulk helium at 0 K. For such a tiny, though still macroscopic object, the effect of surface tension is extremely important. Even if the surface tension parameter, as measured by many authors, is included in the calculation, the scenario does not change much. It is almost certain that the microcluster 'snowballs' may exhibit the properties of a solid entity. The bulk helium frozen out at approximately 0 K has the hcp configuration (see Fig. 7.1).

A snowball is therefore a singly charged aggregate of helium atoms created through electrostriction around an alien ion introduced into superfluid helium (Atkins, 1959). While the snowball has been thought of as a highly permanent solid core (Ihas and Sanders, 1970), there is still some doubt as to whether all the existing evidence decisively favours such a picture (Goodstein, 1978). It is indispensable to devise new methods of detection, to test and establish the inherent structure directly.

Many species of impurity ions were introduced and soon it was realized that these impurity ions do create snowballs (Glaberson *et al.*, 1975). The difference in the mobilities of various species of core ions indicates that the internal structure of snowballs

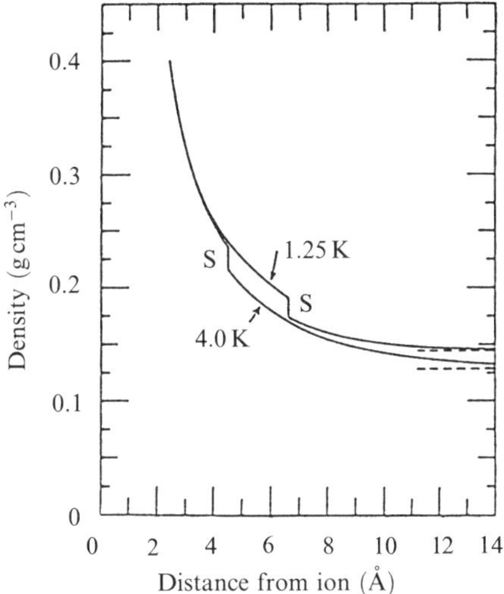

Fig. 7.6 Density around a localized positive point charge in liquid helium, according to a semi-classical calculation (Atkins, 1959), for temperatures $T = 1.25$ and 4 K. S denotes the solid surface around the impurity ion. The dotted lines indicate the normal density of liquid helium.

may not be completely spherical. It is possible that the snowballs may configure themselves as a spherical aggregate rather than as extremely tiny hexagonal structured entities. However, there is no sufficient evidence, experimental or theoretical, that supports this.

Despite the interest in many facets of the physics of snowballs, difficulties persist in detecting them. The snowballs and the core ions are charged equally. Furthermore, the ions are usually singly charged. This means that the minimum number of snowballs to be measured correspond to those forming the measurable electric current of 0.1 pA. In many cases, this is a formidable number, of the order of 10^6–10^7, especially for helium ions that may exist as unstable He$^+$ rather than as simple He$^+$ ion.

Ions may create excitations. What is the precise mechanism of creating basic excitations in superfluid helium? To generate excitations such as rotons, heavy objects such as impurity ions must move faster than a certain threshold velocity v_L, called Landau's critical velocity. A simple algebra suggests that

$$v_L = \left(\frac{E}{p}\right)_{\min} = \frac{\left(2\Delta M_{\mathrm{eff}} + p_0^2\right)^{1/2} - p_0}{M_{\mathrm{eff}}}.$$

Usually it suffices to take $v_L = \Delta/p_0$. The normal ions we treat here move at velocities much lower than v_L. The ion kinetics is in fact dissipative. This is because the ions are trapped in vortices where the kinetic energies are used to increase their internal energy. Vortices are originally quantized rotations of flow in liquid helium. Just as we may squeeze out heat through separation of the superfluid component by passing the liquid helium through a tiny capillary, we may also squeeze out the angular momentum of liquid helium in certain experiments. The origin of rotons lies in the transfer of energy and momentum but their intrinsic nature or our images of them have not yet been revealed. It has also been considered that rotons may be produced in pairs (Allum *et al.*, 1977).

We briefly mention here the trapping time of the ions as envisaged in our experiments. Usually, the trapping time is extremely long for snowballs, unless there is an influx of heat or disturbances. It may be interesting to study the extent to which we may use the snowballs as trapping media and release the core ions as required in an experiment. There is a preliminary test to use snowballs as a means to trap and transport radioactive reaction products before converting them into low-energy beams.

We refer the reader to Günther *et al.* (1995) for more details on the recent topics on the impurities in liquid helium.

7.4 Detection methods and production of polarization

Among numerous attempts to confine atoms and ions in electromagnetic fields, there are a few that try to confine them in the internal electromagnetic fields of atomic and molecular environments. In one of the first decisive efforts, ions were confined in atomic clusters and the nuclear spin polarization of ^{12}B ions (β-radioactive, $T_{1/2} = 20.3$ ms, $I^\pi = 1^+$, $E_{\beta\max} = 13.7$ MeV) was maintained throughout their lifetime as microcluster 'snowballs' in superfluid helium.

A snowball is a singly charged aggregate of helium atoms created around an impurity ion as a result of electrostriction in superfluid helium (Atkins, 1959). It has been observed in a mobility measurement of impurity ions in liquid helium (Meyer and Reif, 1958; Careri *et al.*, 1959). Such a microcluster is presumed to sustain a highly symmetric structure. The relaxation phenomena of nuclear polarization of core ions in snowballs are good probes to seek the internal structure of the aggregate. The result that a change in the mass of the core ion from 4 to 40 u produces only a small change in mobility, is considered to be generally consistent with the electrostrictive model of the ions in liquid helium (Ihas and Sanders, 1970; Johnson and Glaberson, 1972).

While the snowballs are thought to be highly permanent solid entities (Schwarz, 1975), there are still doubts whether all the existing evidence does decisively favour the picture that a solid core is formed (Goodstein, 1978; Bowley, 1972). One of the difficulties encountered in snowball experiments has long been the lack of efficient detection methods other than the electric current measurements. It is indispensable to devise a new method of detection, to elucidate the inherent structure and the interactions of snowballs directly. Recently, laser techniques have widely been applied in this field too (Günther *et al.*, 1995).

The methods used in the study of polarization phenomena in heavy-ion reactions (Sugimoto *et al.*, 1985) enabled us to conceive the possibility of studying individual snowballs by the detection of β-rays and also α-particles from radioactive core ions of snowballs. For example, the lifetime of snowballs trapped in vortices has been measured by incorporating an α-particle detection method with ^{8}Li ions. At 1.8 K, the lifetime was approximately 700 ms. This value suggests that the lifetime of ^{12}B snowballs will be limited by the lifetime of ^{12}B ions themselves (Takahashi *et al.*, 1992). A preliminary report on this work can been found elsewhere (Takahashi *et al.*, 1987).

7.4.1 *Nuclear polarization from heavy-ion reactions*

Large spin polarization up to 35% was observed for reaction products ejected by ^{12}B at an angle of 5° in the reaction of 38.5 MeV/u ^{14}N + ^{9}Be. What is the reaction mechanism leading to such a large polarization? It is considered that the relative motion in the heavy-ion reaction is described, at intermediate energies, by using the concept of classical trajectories. Polarization is expected, if the mean deflection angle along the trajectory is finite. Polarization reflects a competition between the near- and far-side trajectories, since they contribute to polarization with opposite signs. When the fragments are observed at more backward angles, their polarization is larger, since the contribution from only one side is dominant. The angular dependence of the polarization is investigated from this point of view.

Beams of ^{14}N from the ring cyclotron with an energy of 39.3 MeV/u were injected onto a ^{9}Be target of thickness 97.5 mg cm^{-2}. The reaction products emitted by ^{12}B at 0°, 1°, 3°, 5°, 7°, and 9° were introduced into the secondary-beam line. The measured spin polarization is plotted in Fig. 7.7 as a function of outgoing momentum of ^{12}B. The negative polarization in the higher-momentum region suggests that the far-side component dominates over the near-side one (Asahi *et al.*, 1990). At forward angles, 0° and 1°, the polarization is extremely small, as expected from the cancellation of contributions from trajectories on both sides. The polarization from 3° to 9° shows similarity in magnitude- and momentum-dependence independently of the angle. The differential cross sections as momentum spectra are shown in Fig. 7.8. The momentum associated

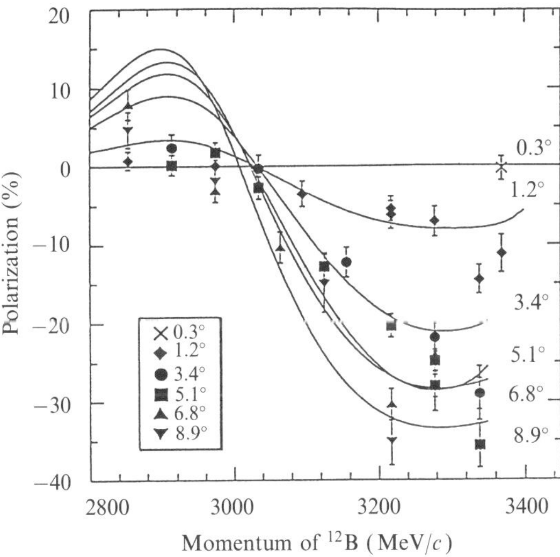

Fig. 7.7 Polarization of the fragments (^{12}B) from the reaction ^{9}Be (^{14}N, ^{12}B) ^{11}C at 38.5 MeV/u as a function of the outgoing ^{12}B momentum for various reaction angles. Solid lines are the results of calculations as explained in the text. (Taken from Shimoda *et al.*, to be published.)

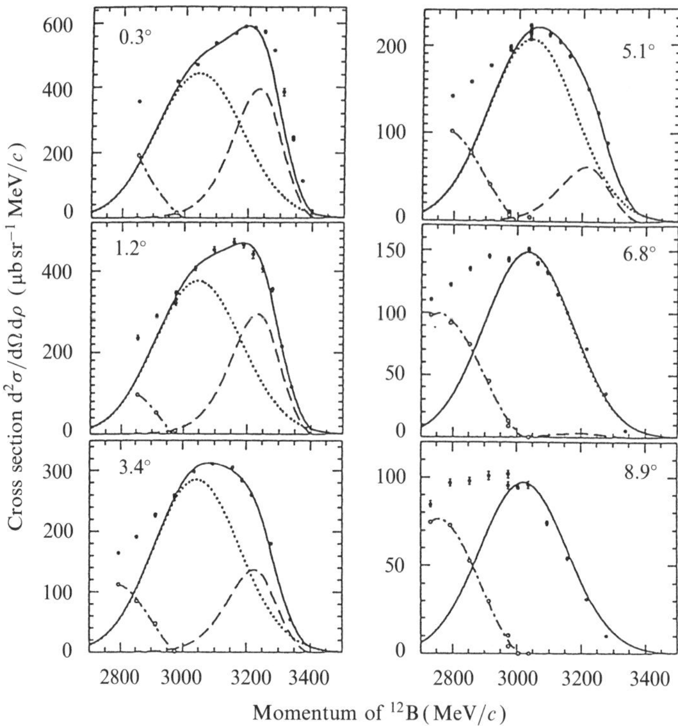

Fig. 7.8 Cross sections for ^{12}B from the reaction ^{9}Be (^{14}N, ^{12}B) ^{11}C at various reaction angles at 37.5 MeV/u as a function of ^{12}B momentum. Dashed and dotted lines are the calculated results for direct-transfer and fragmentation process, respectively, and sums of these cross sections are shown by solid lines. The dot-dashed curves represent the parts of the cross sections unexplained by the sum, most probably from the multistep contributions. (Taken from Shimoda *et al.*, to be published.)

with the maximal cross section at $0°$ is higher by approximately 7% than that corresponding to the beam velocity ($\sim$3050 MeV/c) and the peak momentum shifts rapidly towards lower momenta as the angle increases. The polarization at forward angles shows a non-zero value at the peak momentum. Such observations cannot be explained only by the simple fragmentation mechanism.

In order to examine the contribution of other reaction processes, a Gaussian momentum distribution postulated in the fragmentation model (Goldhaber, 1974) has been fitted to the experimental cross sections. The remainder at higher momenta shows a hump spectrum and the peak momentum decreases gradually with angle. This angular dependence is consistent with the kinematics of the two-body reaction ^{9}Be (^{14}N, ^{12}B) ^{11}C, where the reaction Q-value is about -36 MeV. The cross section for the remainder shows a steep angular distribution. This suggests a direct-reaction mechanism. The present data are therefore analysed with two reaction mechanisms, direct transfer and projectile fragmentation.

As for the direct transfer, Tanaka et $al.$ (1978) account well for the polarization of ^{12}B in (^{14}N, ^{12}B) reaction at lower incident energy of $\sim 6\,\mathrm{MeV/u}$. Following the same prescription, we calculate the cross section and polarization for the present reaction at much higher energy; it is assumed that a proton pair in the $(0p_{1/2}, 0p_{3/2})$ configuration in ^{14}N is transferred to ^{9}Be with the orbital angular momentum $\ell_1 = 2\hbar$ in the initial state. The transferred cluster carries away the angular momentum λ_1, projection of ℓ_1 normal to the reaction plane, with respect to the centre of mass of the projectile. The remaining part (fragment) of the projectile receives the angular momentum $-\lambda_1$. According to Brink (1972), the transition probability is expected to be high when kinematical matching conditions are satisfied, i.e. the linear and angular momenta are conserved. These conditions give rise to selections of Q-value and λ_2 of the final state with respect to λ_1. In the present case the most probable values are estimated as follows:

$$\lambda_2 \sim 4.8\hbar \quad \text{and} \quad Q \sim -33\,\mathrm{MeV} \quad \text{for } \lambda_1 = +2\hbar,$$
$$\lambda_2 \sim 6.5\hbar \quad \text{and} \quad Q \sim -71\,\mathrm{MeV} \quad \text{for } \lambda_1 = 0,$$
$$\lambda_2 \sim 8.3\hbar \quad \text{and} \quad Q \sim -109\,\mathrm{MeV} \quad \text{for } \lambda_1 = -2\hbar.$$

Here, it is shown clearly that the respective λ_1 values prefer significantly different Q-values. The cross section is expressed as a product of the transition probability and the spectroscopic factor. The latter is expressed as the level density of the two-particle and zero-hole states:

$$p_2(E^*, \ell_2) = E^*(2\ell_2 + 1)\,\exp\left[\frac{-\ell_2(\ell_2 + 1)}{4\langle m^2 \rangle}\right],$$

where E^* is the excitation energy and ℓ_2 the relative angular momentum of the final state. The 'spin-cutoff parameter' $\langle m^2 \rangle$ related to the rigid-body moment of inertia of the residual nucleus is estimated to be about $(3/2)\hbar$ in the present system. This small value causes restriction in the ℓ_2 space; the spectroscopic factor for lower ℓ_2 states is much larger than for those with higher ℓ_2. Accordingly, the cross section for $\lambda_1 = +2\hbar$ is more dominant than the others and hence large polarizations can be expected in the region of low $|Q|$ (high momentum)

The calculated cross sections are shown by dashed lines in Fig. 7.8, together with the Gaussians for the fragmentation process (dotted lines). The normalization has been so determined that the sum of the cross sections (solid lines) best reproduces the data in the high-momentum region. The data are reasonably explained by the sum of two cross sections, except for the low-momentum region at backward angles, where other mechanisms also contribute. The mean deflection angle of the trajectories is estimated to be about $-2.5°$ so as to reproduce the experimental angular distribution. The polarization due to individual processes is calculated for each angle by taking into account the components of far- and near-side trajectories. It was assumed that the polarization is zero for the component in the lower-momentum region, where the cross sections are not reproduced. The momentum dependence of the polarization is explained qualitatively, while the calculated magnitudes for transfer process significantly overestimate the quantitative data around 3–7°. The solid lines in Fig. 7.7 are the results obtained by reducing the magnitude of the polarization to 50% for the transfer process only. The saturation of the magnitude at backward angles is reproduced qualitatively. Note that the non-zero polarization at peak momentum is ascribed to the transfer mechanism.

7.5 Measurement of freezing-out of polarization

Polarized radioactive ^{12}B nuclei were produced in the reaction of 38.5 MeV/u ^{14}N on ^{9}Be and separated isotopically through the secondary-beam line of the ring cyclotron facility at RCNP, Osaka University (Shimoda *et al.*, 1992). The maximum polarization of 40% was obtained for the products ^{12}B in the highest outgoing momentum range at a reaction angle of 5°, the sign of the polarization being negative, in terms of Basel convention. The ^{14}N beam was shaped into rectangular pulses of 20 ms duration and 48 ms repetition, and was focused onto a Be target of thickness 294 mg cm^{-2}. The reaction products ^{12}B in the momentum range between 2.8 and 2.9 GeV/c were taken out at 5° with respect to the direction of the incident beam onto the optical axis of the secondary-beam course. An intensity of approximately 10^3 ^{12}B ions s^{-1} was obtained at the measuring port 15 m downstream from the target. The purity of the ^{12}B ion beam was better than 99% and an average initial polarization of -30 ± 2% was evaluated for these ^{12}B products based on the results of the above measurements. Ions undergoing no interactions with the Be nuclei were collected off-axis. The results are reproduced here for the momentum spectrum and polarization in Fig. 7.9.

The ^{12}B ions were implanted into superfluid helium during the in-beam period when the temperature was controlled at 1.43 K within 0.03 K by means of a germanium resistor thermometer attached inside the superfluid chamber of diameter 20 mm and length 400 mm. The low-temperature part of the cryostat, placed at the measuring port, is shown schematically in Fig. 7.10. The ion beam window at the front end of the chamber was made of 100 µm thick stainless steel. Reaction products having kinetic energy, in excess of a specified value, determined from range and energy loss in the energy degrader and the windows, were allowed to enter into liquid helium. The residual range in liquid helium was estimated to be 7.5 ± 3 mm from the window, designated as A in the figure. A substantial number of the implanted ^{12}B ions were neutralized during the stopping process in liquid helium. A considerable fraction of ions, however, survived without being neutralized, until they came to rest as singly charged ions and became core ions of snowballs. Snowballs were detected through the β-rays from the ^{12}B core ions and the detection-sensitive domain was 40 ± 10 mm from the ion beam window, designated as B in the figure, and the estimated stopping area was thus excluded from the sight of the β-ray detector telescopes.

The snowballs produced were dragged approximately 35 mm downstream through a static electric field applied along the axis of the superfluid chamber by means of cylindrical electrodes and a solenoid wound around it.

The β-rays emitted by ^{12}B in the detection domain were detected during the out-beam periods of 24 ms with a pair of plastic detector telescopes deployed parallel to the reaction

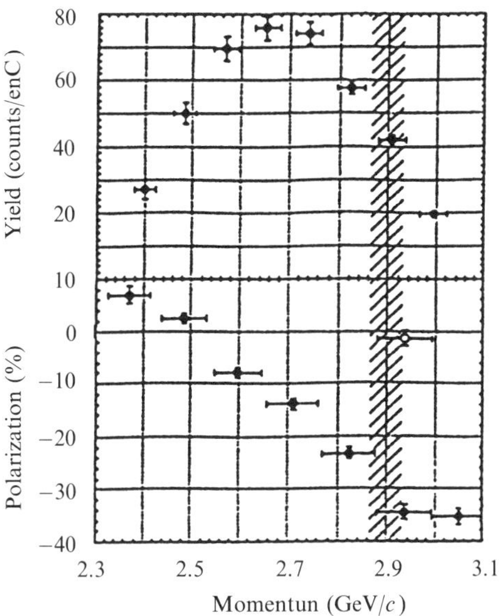

Fig. 7.9 Momentum spectrum (upper) and polarization (lower) of the reaction products ^{12}B at 38.5 MeV/u and 5°. The sign of polarization is according to the Basel convention. The result of a zero-asymmetry check is included in the figure by an open circle. (Taken from Takahashi *et al.*, 1996.)

normal,[b] i.e. in the direction along the expected polarization. The β-rays from the core ions of ^{12}B snowballs penetrated a 500 μm thick FRP wall of the superfluid chamber, a 200 μm thick copper solenoid, the vacuum chamber window of 100 μm thick stainless steel, before impinging on a detector telescope, which consisted of three energy-loss detectors of thickness 1.5, 2, and 5 mm. The energy threshold of the detector telescopes was set at 2.5 MeV.

The angular distribution $W(\theta)$ of β-rays from the polarized ^{12}B is asymmetric, owing to the parity non-conservation in the weak interaction, and is given by

$$W(\theta) = 1 - P\cos\theta$$

with respect to the reaction normal, taken as the axis of ^{12}B polarization. The polarization P is expressed in terms of the square of true up–down asymmetry R as

$$P = \frac{1 - R^{1/2}}{1 + R^{1/2}},$$

---

[b]The reaction normal is defined as $\boldsymbol{n} = \boldsymbol{k}_i \times \boldsymbol{k}_f$, the vector product of the incident and the outgoing wave vectors. The sign of polarization is taken positive when $\boldsymbol{P}$ is parallel to $\boldsymbol{n}$.

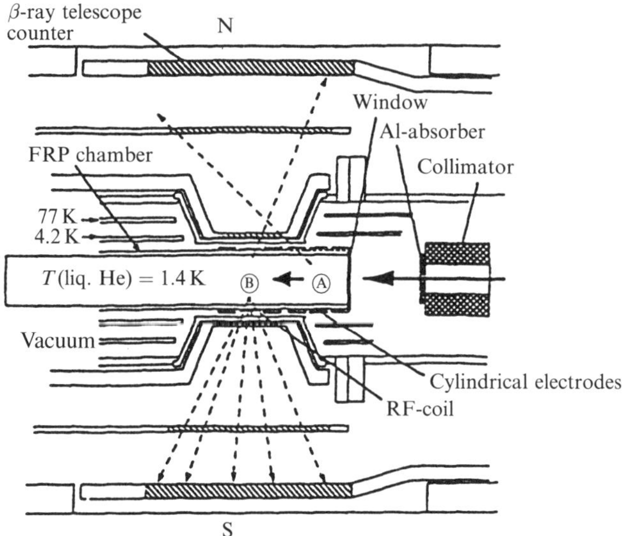

Fig. 7.10 A part of the cryostat and β-ray detector telescopes deployed between the magnetic poles. The ions are implanted into A and snowballs formed are transported to B before β-rays are detected. (Taken from Shimoda *et al.*, 1995.)

where

$$R = \frac{(N_{\text{up}}/N_{\text{down}})_{\text{off}}}{(N_{\text{up}}/N_{\text{down}})_{\text{on}}}.$$

Here N_{up} and N_{down} stand for the counting rates of β-rays measured at $\theta = 0°$ and $180°$ and subscripts 'off' and 'on' refer to the NMR at off- and on-resonance, respectively.

The nuclear polarization of ^{12}B was reversed during every other out-beam period by applying the adiabatic-rapid-passage NMR over the resonance frequency of $534\,\text{kHz}$ corresponding to the value of the bare magnetic moment of ^{12}B nuclei, $1.003\mu_{\text{N}}$, with the range of the frequency modulation of $\pm 3\%$. A static magnetic field was produced in the domain and the strength was $0.7\,\text{kG}$ with an inhomogeneity less than 0.2%. The solenoid produced the rf magnetic field of maximum $23\,\text{G}$ perpendicular to the static magnetic field during the rf period of $3\,\text{ms}$. An interval of $5\,\text{ms}$ was allowed between the in-beam and the rf periods, for collection of all the snowballs. The rf solenoid was activated solely for the rf period and was otherwise kept at the ground potential, serving as one of the electrodes as stated above. In the other out-beam periods, the rf field was not applied and the polarization was kept unaltered. By taking the average of the up–down ratios of the counting rates for the on- and off-rf periods, the β-ray asymmetry was obtained free from the instrumental asymmetries.

The charge carriers were identified here as snowballs from their electric mobility, which was obtained in a separate experiment from the measured transit time to reach the narrowly collimated detection region B after stopping at the region A, with essentially the same experimental configuration of the apparatus as in Fig. 7.10.

After the implantation period of 3 ms was over, increases in the β-ray counting rate at 5 and 15 ms were observed. The decrease in the count-rate for the fast component exhibiting a maximum at around 8 ms, indicated that ^{12}B in this component proceeded further against the repelling electric field, and thus was uncharged. A Monte-Carlo simulation produced a uniform velocity of $5.3\,\mathrm{m\,s}^{-1}$ for the former component and $1.7\,\mathrm{m\,s}^{-1}$ for the latter, which was consistent with the known value for electric mobility of $0.45\,\mathrm{cm}^2\,\mathrm{V}^{-1}\mathrm{s}^{-1}$. The slower component was therefore identified as snowballs formed around the radioactive ions. From a series of measurements with varying implantation depth, it became clear that the charged and the neutralized atoms amounted respectively, to $\sim$20% and $\sim$80% of the implanted ions for superfluid helium, and 40% of the neutralized atoms constituted the fast component, the existence of which was not observed in normal liquid helium.

7.6 Results

The measured polarization is shown in Fig. 7.11, together with the time sequence of the measurements; one half cycle with impression of rf magnetic field and another half without it.

The average polarization was $-19.9 \pm 1.3\%$ and there was essentially no relaxation observed. From this measurement it is seen that the relaxation time is long enough as compared to the lifetime of ^{12}B. This result shows decisively that the nuclear spin

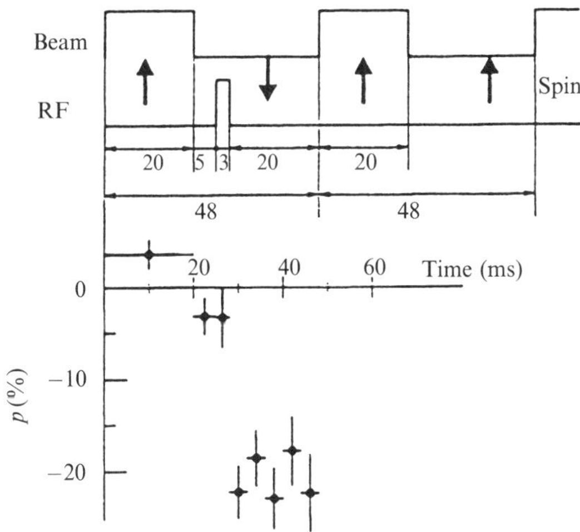

Fig. 7.11 Polarization of ^{12}B in snowballs, as obtained by combining the β-ray asymmetries at the on- and off-rf periods, shown in a typical sequence of measurements. Note that the asymmetry becomes appreciable after the application of rf. The polarization maintains a constant value and essentially no relaxation is observed. (Taken from Takahashi *et al.*, 1995.)

polarization of the core ions is maintained in the snowballs. The measured value is still smaller than the estimated initial polarization of ^{12}B ions, $-30 \pm 2\%$ at the implantation (Shimoda *et al.*, 1995), which can be explained as follows.

The polarization of the ^{12}B was preserved during the flight of ^{12}B ions *in vacuo* from the target to the chamber window, since the ions were energetic enough to be in the fully stripped state and no hyperfine interactions were effective. The collision time is short enough when penetrating the solid material, such as the chamber window, and no deterioration of polarization took place. When the ^{12}B ions were gradually decelerated through ionization and attached with the electrons in liquid helium, the collision time was so short that no appreciable depolarization took place. The singly charged ^{12}B ions had the advantage that four electrons were mainly in the 1s and 2s shells, especially in snowballs where the internal pressure was so high that the possible p state did not contribute much. This means that only slight internal electromagnetic fields, if any, except for those from the surroundings, were effective. On the other hand, the neutralized ^{12}B atoms would see a strong magnetic field due to the last electron and the polarization was apt to be destroyed.

There was small β-ray background, in the absence of an electric field applied to drag the snowballs into the detection region. The β-ray asymmetry was essentially zero for this component. The electric field strength was varied from zero to the applied voltage of 3 kV. The yield and the β-ray asymmetry increased with increase in the applied voltage and saturated at the maximum voltage. The observed polarization of ^{12}B was thus $-23.9 \pm 1.3\%$, with the inclusion of the correction for this component.

7.7 Discussion

A further correction is needed for the discrepancy between the magnitude of the observed polarization of decaying ^{12}B and the value of the initial ^{12}B polarization. This proved that there is a small fraction of snowballs that evade the application of NMR. The snowballs, even under the influence of the applied electric field, drift so slowly that some of the snowballs created close to the end of in-beam period do not reach the detection domain (B in Fig. 7.10) and are not amenable to NMR. The correction increased the value for the polarization of ^{12}B ions in snowballs to 29% and made the measured polarization essentially the same as the initial polarization.

The conclusion, therefore, is that the initial polarization in the ^{12}B core ions was maintained inside the snowballs. The magnetic effects of the surrounding electrons is small, within 3%. On the other hand, the bulk solid helium has hcp structure, which may definitely affect the quadrupole interactions inside the cluster, when the constituent atoms are electrically polarized. In this case, the quadrupole interactions may reduce the spin polarization, as all the helium atoms in the neighbourhood of the impurity core ion are under the effects of electrostriction. In the present case, the polarization lasted for a time longer than the lifetime of ^{12}B and the effect was presumably small. The static magnetic field is essentially strong enough to decouple the electronic effects from the nuclear effect, though precise measurements must be awaited. It may be proper to mention here the spin relaxation time of B implanted in Pt in a pioneering work of Minamisono *et al.* (1973), in which the relaxation time as large as 3.4 s has been reported.

The above facts support strongly the view that if the snowball has a permanent solid core, the structure of such a core is highly symmetric and favours the view of a highly permanent cluster of helium atoms (Schwarz, 1975). In fact, a similar molecular orbital calculation shows that the hydrogen atoms bonding to metal ions constitute a higher symmetry as the number of hydrogen atoms increases (Niu *et al.*, 1992). The electrostriction effects increase the liquid density over a large region surrounding the ion. But it has hitherto been quite uncertain whether any solid-like structure is formed, although there have obviously been good reasons to believe that the density within a few tens of nanometers from the ion is extremely high. If the snowball has a fluid phase (Bowley, 1972) or an asymmetric solid structure (Cole and Toigo, 1978), it will be difficult to assume that the 'freezing-out of polarization' does take place as the experimental result shows. The charge carriers encounter incessantly the electrically polarized atoms, as they are dragged under the electric field. Indeed, Johnson and Glaberson (1972) observed that the mobility of 'snowball' shows a slightly nonlinear dependence on the atomic number of the core ion. They conclude, based on the results of their measurements with various kinds of core ions, that the snowball might be thought of as tiny crystals, while the ion radius would be an oscillating function of the inner-shell radius, and the surface of the ion need not be perfectly spherical.

Another interesting feature is that ions implanted into superfluid helium produce unexpectedly large spatial spreading, which is observed for the first time in this series of measurements (Shimoda *et al.*, 1995). It is worth mentioning here the stopping processes of fast ions in superfluid helium as measured in connection with this experiment. The stopping process of ^{8}Li ions was thus studied; the depth of penetration of ions into liquid helium, the spatial width of stopping region, and the ratio of singly charged ^{8}Li to neutralized ^{8}Li were measured. This was accomplished by using energy degraders of various thicknesses and well-collimated detector telescopes for β-rays from ^{8}Li, with essentially the same apparatus as in Fig. 7.10. It is observed here for the first time that the ions produced unexpectedly long depths of penetration into superfluid helium and the width is also unexpectedly large as compared to the results of estimates on the basis of the normal range and straggling. Specifically, $\sim$20% of neutral ^{12}B, the final stage of the introduced ^{12}B ions, travel with a speed close to the critical velocity, while the rest remain at the depth of a few millimetres from the entrance window (Takahashi, 1997). A similar motion might also have been observed using the laser tracing of some of heavy impurity atoms.

7.8 Conclusions

Summarizing, we have found for the first time, using a novel detection method, that nuclear spin polarization is frozen out, at least for the lifetime of the unstable nucleus ^{12}B, in 'snowballs', microclusters of helium atoms around alien ions equilibrated with the surrounding liquid helium. This fact tempts us to presume that the snowballs have a highly symmetric structure. In general, therefore, once the ions are embedded in the snowballs, nuclear polarization can be preserved for a reasonable length of time. Such spin polarized ions are confined in atomic clusters bound to the electric field, and this trapping produces localization, which is often extremely desirable for the study of isolated systems with specific properties, e.g. polarized radioactive ions far from the stability, etc. This supplies us with an ideal environment for trapping such ions and with an ideal material for freezing-out the nuclear polarization. Also, this gives a new tool for nuclear spectroscopic measurements, such as determination of electromagnetic moments of unstable nuclei, irrespective of their atomic number and their position relative to the stability line. This definitely opens a way to expand the study of nuclei from the hitherto narrow region near the stability line to the large $N-Z$ plane. Other interesting features which will be topics of research in the near future are the issues of behaviour, as seen, for example, in our study of neutral impurities in superfluid helium, and also the inherent structure of snowballs as related to a precursor to solidification. In addition, this novel detection method of snowballs will enable us to observe more closely the transport phenomena of alien ions in superfluid helium, which may enable an examination of new phases in the superfluidity of helium itself. All these avenues open new ways to use snowballs as a versatile tool in nuclear and condensed-matter physics.

Acknowledgements

A part of our project described here was carried out at the RCNP under the programme number E19 and our sincere appreciation goes to the staff of the RCNP for their cooperation throughout the course of the experiments. This work is supported in part by the Grants-in-Aid of Scientific Research from the Ministry of Education, Science and Culture, Tokyo, Toray Science and Technology Grant, and grants from the Mitsubishi Foundation.

References

Allum, D. R., McClintock, P. V. E., Phillips, A., and Bowley, R. M. (1977). *Phil. Trans. R. Soc. London* **A284**, 179.

Asahi, K. *et al.* (1990). *Phys. Lett.* **B251**, 488.

Atkins, K. R. (1959). *Phys. Rev.* **116**, 1339.

Atkins, K. R. and Rudnik, I. (1959). In *Progress in low temperature physics*, J. Gorter, ed., Vol. VI, Chapter 2, North Holland, Amsterdam (1978).

Bowley, R. M. (1972). *J. Low Temp. Phys.* **7**, 185; *ibid.* **8**, 261.

Brink, D. M. (1972). *Phys. Lett.* **B40**, 37.

Careri, G., Scaramuzzi, F., and Thomson, J. O. (1959). *Nuovo Cimento* **8**, 1758.

Cole, M. W. and Toigo, F. (1978). *Phys. Rev.* **B17**, 2054.

Dahm, A. J. and Sanders Jr., T. M. (1970). *J. Low Temp. Phys.* **2**, 199.

Donnelly, R. J. (1967). *Experimental superfluidity*, The University of Chicago Press.

Glaberson, W. *et al.* (1975). *J. Low Temp. Phys.* **20**, 313.

Goldhaber, A. S. (1974). *Phys. Lett.* **B53**, 306.

Goodstein, L. (1978). *J. Low Temp. Phys.* **33**, 137.

Günther, H., zu Putlitz, G., and Tabbert, B. (1995). *Z. Phys.* **B98**(3), 297–446.

Ihas, G. G. and Sanders Jr., T. M. (1970). *Phys. Lett.* **31A**, 502.

Johnson, W. W. and Glaberson, W. I. (1972). *Phys. Rev. Lett.* **29**, 214.

Kuper, C. G. (1961). *Phys. Rev.* **122**, 1007.

Landau, L. and Pomeranchuk, I. (1948). *Dokl. Acad. Nauk USSR* **59**, 669.

London, F. (1954). *Superfluids*, Vol. II, Dover Publications, New York.

McClintock, P. V. E., Meredith, D. J., and Wigmore, J. K. (1984). *Matter at low temperature*, Blackie & Sons Limited, London.

Meyer, L. and Reif, F. (1958). *Phys. Rev.* **110**, 279L.

Minamisono, T., Nojiri, Y., Mizobuchi, A., and Sugimoto, K. (1973). *J. Phys. Soc. Japan* **34**(Suppl.), 156.

Niu, J., Rao, B. K., and Jena, P. (1992). *Phys. Rev. Lett.* **68**, 2277.

Reif, F. and Meyer, L. (1960). *Phys. Rev.* **119**, 1164.

Schwarz, K. W. (1975). *Adv. Chem. Phys.* **33**, 1.

Shimoda, T., Miyatake, H., and Morinobu, S. (1992). *Nucl. Instr. and Meth.* **B70**, 320.

Shimoda, T., Miyatake, H., Mitsuoka, S., Mizoi, Y., Kobayashi, H., Sasaki, M., Shirakura, T., Ueno, H., Izumi, H., Asahi, K., Murakami, T., Morinobu, S., and Takahashi, N. (1995). *Nucl. Phys.* **A588**, 235c.

Shimoda, T. *et al.*, to be published.

Sugimoto, K., Ishihara, M., and Takahashi, N. (1985). *Polarization phenomena in heavy-ion reactions*, in Treatise on Heavy-Ion Science, D. A. Bromley, ed., Vol. 3, pp. 395–536, Plenum Press, New York, and references therein. [See also Takahashi (1985) and Tanaka *et al.* (1986) for more up-to-date description of the results.]

Takahashi, N. (1985). *Hyperfine Interactions* **21**, 173.

Takahashi, N. (1994). *Nucl. Phys.* **A577**, 99c.

Takahashi, N. (1997). *Acta Phys. Polonica* **28**, 41.

Takahashi, N., Shimoda, T., Fujita, Y., Itahashi, T., Ikeda, N., and Hinde, D. J. (1987). *Nuclear structure through static and dynamic electromagnetic moments*, p. 334, University of Melbourne, Melbourne.

Takahashi, N., Shimoda, T., Miyatake, H., Fujita, Y., and Itahashi, T. (1991). *Radioactive nuclear beams*, University of California, p. 82.

Takahashi, N., Shimoda, T., Miyatake, H., Fujita, Y., and Itahashi, T. (1992). *Radioactive nuclear beams*, p. 395, Adam Hilger.

Takahashi, N., Shimoda, T., Fujita, Y., Itahashi, T., and Miyatake, H. (1995). *Z. Phys.* **B98**, 347.

Takahashi, N., Shimoda, T., Miyatake, H., Mizoi, Y., Kobayashi, H., Sasaki, M., Shirakura, T., Mitsuoka, S., Morinobu, S., Asahi, K., and Ueno, H. (1996). *Hyperfine Interactions* **97/98**, 469.

Tanaka, K., Ishihara, M., Kamitsubo, H., Takahashi, N., Mizobuchi, A., Nojiri, Y., Minamisono, T., and Sugimoto, K. (1978). *J. Phys. Soc. Japan* **44**(Suppl.), 825.

Tanaka, K. H., Nojiri, Y., Minamisono, T., Asahi, K., and Takahashi, N. (1986). *Phys. Rev.* **C34**, 580.

Tanihata, I., Hamagaki, H., Hashimoto, O., Nagamiya, S., Shida, Y., Yoshikawa, N., Yamakawa, O., Sugimoto, K., Kobayashi, T., Greiner, D. E., Takahashi, N., and Nojiri, Y. (1985a). *Phys. Lett.* **160B**, 380.

Tanihata, I., Hamagaki, H., Hashimoto, O., Shida, Y., Yoshikawa, N., Sugimoto, K., Yamakawa, O., Kobayashi, T., and Takahashi, N. (1985b). *Phys. Rev. Lett.* **55**, 2676.

Wilks, J. (1967). *The properties of liquid and solid helium*, Clarendon Press, Oxford.

8

Deeply bound pionic atoms

H. Toki, S. Hirenzaki, and N. Matsuoka

8.1 Structure of deeply bound pionic atoms

The pion is a glue that holds nucleons together to form a nucleus. In the nucleus, it is in a virtual state, i.e. it lives only for a short time as allowed by the uncertainty principle, viz. $\Delta t \sim \hbar/m_\pi c^2$. This virtual pion travels a distance $\Delta l \sim c\Delta t \sim 1.4\,\text{fm}$, which determines the range of the Nucleon–Nucleon (NN) interaction. How about a 'real' pion inside a nucleus? We may then experience the case where a pion lives for a long time inside the nucleus and interacts with the nucleons.

This situation might occur for the case of Deeply Bound Pionic Atoms (DBPAs). Suppose a pion is attracted by a point-like Coulomb potential. The Bohr radius is the typical distance of the pion from the centre, which is $a \sim 200Z^{-1}\,\text{fm}$, Z being the nuclear charge number. The Bohr radius a becomes smaller than the nuclear radius $R = 1.2A^{1/3}\,\text{fm}$ above $Z \sim 40$ ($A \sim 90$). Then the real pion stays inside the nucleus! However, the width Γ being too large to be observed, this pion was believed to be absorbed by nucleons instantly. In fact, in the standard pionic atom experiment, the observation of pionic X-rays demonstrates the existence of these atoms. Although shallow bound states are seen by pionic X-rays, DBPAs are not observed for $Z > 12$ for the 1s states and $Z > 30$ for the p states. The possibilities are:

1. they are not observed, because DBPAs do not exist (the problem of structure);
2. DBPAs do exist, but the method is not able to detect them (the problem of formation).

The structure and the formation of DBPAs have been studied theoretically by two groups extensively [1–13]. We follow here the papers by Toki et al. [2,6] to describe the theoretical methods employed in the analysis. We then discuss the experimental methods using direct reactions with hadrons. We would like to mention here that the DBPA states were discovered very recently at GSI using (d, ^{3}He) reactions on ^{208}Pb [14], after some indications of DBPAs were observed at TRIUMF with (n, d) reactions and at the Research Center for Nuclear Physics (RCNP) with (p, ^{2}He) reactions.

The simplest way to start our theoretical discussion is to solve the Klein–Gordon equation incorporating the pion optical potential used for known pionic atoms [15]:

$$\left[-\nabla^2 + m_\pi^2 + 2m_\pi V_{\text{opt}}(r)\right]\phi(r) = [E - V_{\text{Coul}}(r)]^2\,\phi(r). \qquad (8.1)$$

Here, m_π is the pion mass and $V_{\text{Coul}}(r)$ is the Coulomb potential for finite nuclear size,

$$V_{\text{Coul}}(r) = e^2 \int \frac{\rho_p(r')}{|\boldsymbol{r} - \boldsymbol{r}'|}\, \mathrm{d}^3 r', \tag{8.2}$$

where $\rho_p(r')$ is the proton density distribution. We take the Woods–Saxon form for the density distribution and assume the shapes of neutron and proton distributions to be the same. We use the optical potential of Seki–Masutani in the Ericson–Ericson form as [15]

$$2m_\pi V_{\text{opt}}(r) = -4\pi\left[b(r) + \varepsilon_2 B_0 \rho^2(r)\right] + 4\pi\nabla\left[c(r) + \varepsilon_2^{-1} C_0 \rho^2(r)\right]L(r)\nabla \tag{8.3}$$

with

$$b(r) = \varepsilon_1\{b_0\rho(r) + b_1[\rho_n(r) - \rho_p(r)]\},$$

$$c(r) = \varepsilon_1^{-1}\{c_0\rho(r) + c_1[\rho_n(r) - \rho_p(r)]\},$$

$$L(r) = \left\{1 + (4/3)\pi\lambda\left[c(r) + \varepsilon_2^{-1} C_0 \rho^2(r)\right]\right\}^{-1}.$$

The transformation from pion–nucleon to pion–nucleus centre-of-mass provides the kinematical factors $\varepsilon_1 = 1 + m_\pi/M$ and $\varepsilon_2 = 1 + m_\pi/2M$, M being the nucleon mass. The parameters b and c are those obtained from the S- and P-wave pion–nucleon interaction, respectively. λ is the Lorentz–Lorenz correction parameter in the dimesic function $L(r)$. It is definitely questionable to use the optical potential obtained from the analysis of the shallow pionic atoms, which motivated us to study the DBPA states both experimentally and theoretically in order to learn the properties of the pion–nucleus interaction. The first calculation for DBPAs was performed in the momentum space following the method of Kwon and Tabakin [16]. At present, we can calculate the pionic atom states both in momentum and coordinate spaces. (The computer programs are available on request [2].)

The calculated results are surprising and very interesting [1,2]. As shown in Fig. 8.1, the lowest state (1s) in ^{208}Pb is bound around 7 MeV and has a width of about 1 MeV. This state is well separated from the next excited state (2p), which is bound around 5 MeV. Hence, these DBPAs do exist as quasi-stable states. The reason for these relatively small widths is related to the fact that these states are pushed up from those of the finite-size Coulomb potential, shown by dashed lines, due to the pion–nucleus optical potential [2].

This is seen clearly in Fig. 8.2, where the Coulomb and the strong interaction potentials are plotted against the radial coordinate. The strong interaction is repulsive and hence pions are pushed outward from the interior region and, due to the Coulomb attraction, they are localized around the nuclear surface as a pionic halo. Because the pions are pushed out, they avoid overlap with nucleons to be absorbed. Hence, these states have relatively small widths. The pion densities for these DBPA states are shown on the lower side of Fig. 8.2.

8.2 Formation of DBPAs

The theoretical expectation for the existence of DBPAs is positive. Hence, we discuss here the method for detecting these states experimentally. The standard method for observing the pionic X-rays is not suitable for the formation of DBPAs due to strong

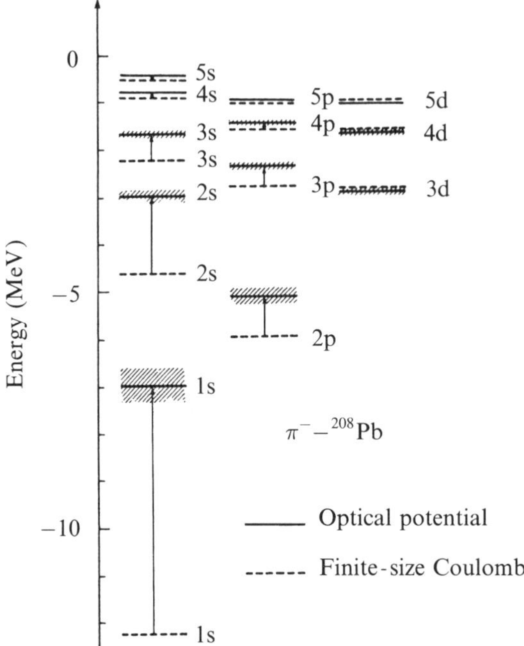

Fig. 8.1 Level scheme of a DBPA in ^{208}Pb. The horizontal solid lines with hatched area indicate levels and widths calculated with the pion–nucleus optical potential, while the dashed lines are those with the finite Coulomb potential. (From Toki and Yamazaki [1].)

absorption. Hence, we have to invent new methods to create and detect these interesting states. We use direct reactions to excite these quasi-stable states. In order to understand the physics of the direct reactions for the formation of pionic atoms, we begin with the (n, p) reactions.

8.2.1 *The* (n, p) *reactions*

In order to understand the physics of DBPA formation, the first process we would like to focus on is the (n, p) pion transfer reaction [1,2]. The incoming neutron is partly a system of a proton and a negative pion, and the pion is transferred to the target nucleus to form a pionic atom. In this case, the formation of DBPAs is signalled in the excitation spectrum of the outgoing proton as a peak around the energy loss of the pionic mass.

In order to get an idea of the theoretical calculation of the (n, p) reaction for DBPA, we work out the cross section within the Plane Wave Born Approximation (PWBA),

$$d\sigma = \frac{1}{v} \frac{M_i}{E_i} \frac{d^3 k_f}{(2\pi)^3} \frac{M_f}{E_f} \frac{1}{2(m_\pi - E_{nl})} (2\pi)\delta(E_f + m_\pi - E_{nl} - E_i)$$

$$\times \sum_{m,s_i,s_f} \left| \int d^3 r\, e^{-ik_f r} \bar{u}(k_f, s_f) ig\gamma_5 \phi_{nlm}^*(r) u(k_i, s_i) e^{ik_i r} \right|^2, \tag{8.4}$$

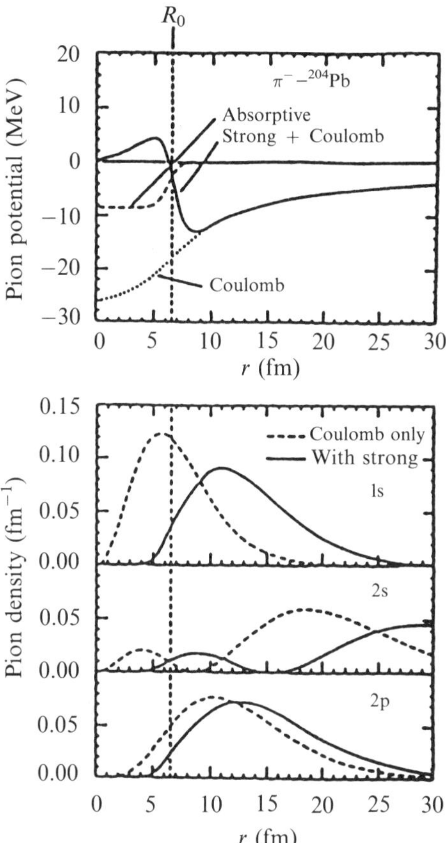

Fig. 8.2 Pion–nucleus optical potential for ^{208}Pb and the pionic atom densities of three lowest states. The broken and the solid curves are obtained with the finite-size Coulomb potential and with the optical potential, respectively. (From Toki *et al.* [2].)

where the P-wave πNN interaction is responsible for the pion production. We then get the following simple expression for the (n, p) cross sections:

$$\left(\frac{d\sigma}{d\Omega}\right)_{PW} = \frac{M^2}{(2\pi)^2} \frac{k_f}{k_i} \left(\frac{f}{m_\pi}\right)^2 \frac{q_L^2}{m_\pi - E_{nl}} \sum_m |\psi_{nlm}(q)|^2. \tag{8.5}$$

Here, $\psi_{nlm}(q)$ is the Fourier transform of the pionic wave function $\phi_{nlm}(r)$ of the (nl) orbit. The Lorentz invariant momentum is $q_L^2 = q^2 - (E_i - E_f)^2$, and k_i and k_f are the initial and the final nucleon momenta. The momentum transfer at zero scattering angle is given by $q \sim m_\pi E_i / p_i$. For an incident energy $T_i \sim 400\,\text{MeV}$, $q \sim 200\,\text{MeV} \sim 1\,\text{fm}^{-1}$, which is very large for pionic atoms. Hence, it is essential to keep the momentum transfer as small as possible to make the cross sections for the formation of pionic atoms reasonably large.

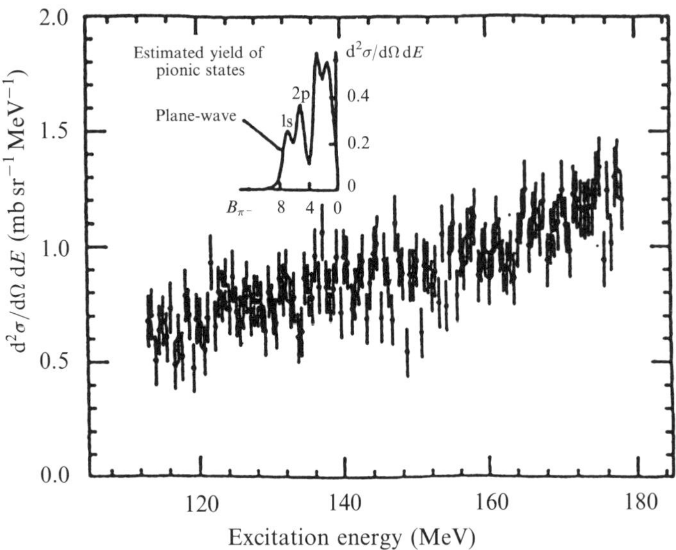

Fig. 8.3 Experimental (n, p) spectrum on ^{208}Pb taken at $T = 418$ MeV as a function of excitation energy in the vicinity of the pion threshold. The PWBA results on pionic atom formation are also shown at the corresponding excitation energy with the same vertical scale above the experimental data points. (Taken from Iwasaki *et al.* [5].)

Within the PWBA, we predicted the (n, p) spectrum on ^{208}Pb for pionic atom formation as shown in Fig. 8.3 around the excitation energy of the pionic mass (~ 140 MeV) [1,2]. Experiments were then performed at TRIUMF with the neutron beam of $T_\mathrm{n} = 418$ MeV, and the proton spectrum in the region of the pionic atom formation is depicted in Fig. 8.3. As seen in the experimental spectrum, no peaks or bumps are found in this experiment, contrary to the theoretical prediction [5]. It was recognized later that the distortion effect is enormous for this reaction with large momentum mismatch and it was concluded that the distortion effect cuts down the cross sections for pionic atom formation by factors of order 10–100. This conclusion was also reached in Ref. [9]. In this sense, it is very interesting to note that if we are able to use Σ as a projectile on ^{208}Pb and measure Λ for pionic atom formation, we can reduce the momentum transfer and hence increase the cross section [10].

8.2.2 *The* (n, d) *reactions*

As we now know the reasons for the small cross sections for DBPA formation in (n, p) reactions, let us consider the (n, d) process allowing the extra neutron play a role in controlling the momentum mismatch [6]. In fact, if we take $T_\mathrm{n} \sim 320$ MeV, then the transferred pion would carry no momentum to fall into the pionic atom states. We formulate this process in terms of the effective number approach as

$$\left[\frac{d\sigma}{d\Omega}\right]_{nA \to d(A-1)\pi^-} = \left[\frac{d\sigma}{d\Omega}\right]^{lab}_{nn \to d\pi^-} N_{eff}, \tag{8.6}$$

where

$$N_{eff} = \sum_{Mm_s} \left| \int \chi_f^*(r) \xi_{1/2m_s}^*(\sigma) \left[\phi_{l_\pi}^*(r) \otimes \varphi_{j_n}(r, \sigma)\right]_{JM} \chi_i(r) \, d^3r \, d\sigma \right|^2. \tag{8.7}$$

Here, the incident and the outgoing waves are distorted waves, which takes into account their absorption in the nuclear medium. The pion production dynamics are included completely in the experimental nn $\to$ dπ cross section.

Our theoretical prediction of the (n, d) spectrum on ^{208}Pb is shown in Fig. 8.4. Pionic atoms are formed with several neutron hole states in ^{208}Pb, and their cross sections are plotted in the vicinity of the pion production threshold down to the pionic binding energy [6]. Those with large cross sections and with large binding energy are the 1s and 2p states with corresponding shallow neutron hole states. With good resolution, we see clearly several states separately, but with, for example, 1 MeV resolution the (n, d) spectrum in the pion bound region is represented by two peaks for the DBPAs and for the shallow states. Above the threshold energy, we expect a continuum pion spectrum, which can also be calculated in the effective number approach.

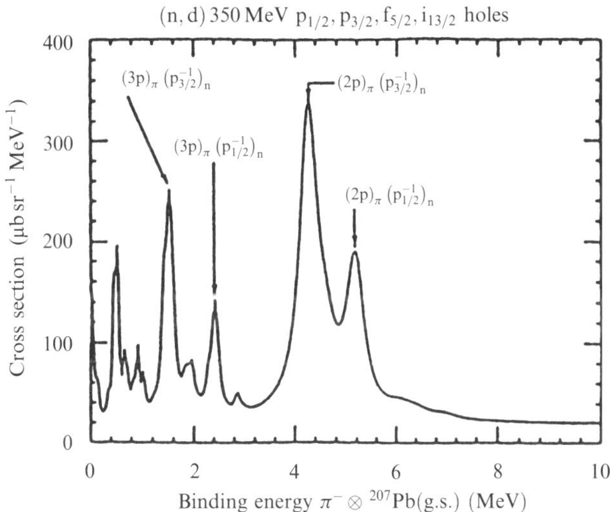

Fig. 8.4 Theoretical (n, d) spectrum at $T = 350$ MeV in the vicinity of the pion threshold down to the pionic binding energy. The pion–neutron hole configurations are denoted for resolved peaks. (From Toki *et al.* [6].)

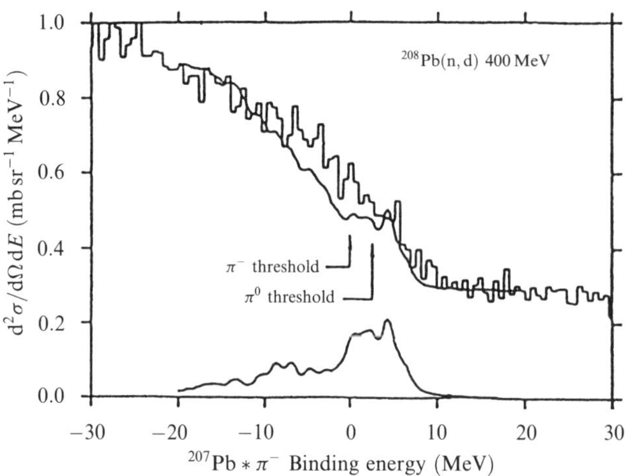

Fig. 8.5 A preliminary result for the ^{208}Pb(n, d) spectrum at $T = 400$ MeV taken at TRIUMF in the vicinity of the negative pion threshold with a mono-energetic neutron beam. The solid curve is the calculated result for the DBPAs and the quasi-free negative and neutral pion production convoluted with 1.5 MeV FWHM resolution for comparison with the experimental data by multiplying by a factor of two. The lower curve shows the cross section due to the pionic atom formation. (From Yamazaki [17].)

The (n, d) experiment was performed with $T_n = 400$ MeV at TRIUMF [17]. The deuteron spectrum in the pionic atom region is shown in Fig. 8.5. Because of small contamination of protons in the target, the deuteron peak due to the np $\rightarrow$ dπ^0 process fixes the pion threshold energy precisely. It is then very interesting to note that there exists appreciable deuteron cross section below the pion threshold. These cross sections should originate from the pionic atom formation.

In fact, we overlay in Fig. 8.5 [6] the calculated results on the pionic atom formation and quasi-free pion production together with the assumption of a constant background of 0.3 mb sr^{-1} MeV^{-1} onto the experimental ones. It is amazing to see that the calculated results with a normalization factor of two, which is expected with this rough estimate, almost coincide with the experimental ones. Hence, we seem to find an indication of DBPA in the (n, d) spectrum. We, however, need a better resolution for experimental identification of DBPAs.

8.3 Search for DBPAs using (p, ^{2}He) reactions

We shall discuss here in detail the ^{208}Pb(p, ^{2}He) reaction to observe DBPAs in ^{207}Pb [18]. The reaction process is considered to be very similar to the (n, d) case [6]. Although the cross section of the elementary process p + n $\rightarrow$ ^{2}He + π^- is estimated to be about 1/40 compared to that of the n + n $\rightarrow$ d + π^- process, we have the advantage of using the primary proton beam for the (p, ^{2}He) reaction. For light nuclei such as ^{12}C, the mean orbital radius of the 1s pion wave function is an order of magnitude larger than the nuclear radius. Hence, the production cross section of bound pionic states using the (p, ^{2}He) reaction on ^{12}C is very small. The ^{12}C(p, ^{2}He) reaction is also measured for comparison with results for the heavy nuclear targets.

The (p, ^{2}He) experiments were carried out at the RCNP, Osaka University, using a 390.6 MeV proton beam accelerated by the ring cyclotron. For the measurements of ^{2}He (two protons coupled to $^1S_0$ state) at 0°, the Large Acceptance Spectrograph (LAS) was used. The LAS is a QD-type spectrograph with angular acceptance of 120 mr horizontally and 200 mr vertically, and momentum acceptance 30%. The central orbit radius and the deflection angle of the dipole magnet are 175 cm and 70°, respectively. The details of the spectrograph parameters and the counter specifications are described in Refs [18,19]. The targets were CH$_2$, CD$_2$, ^{12}C, and ^{208}Pb with thicknesses 8.4, 21.3, 30.8, and 22.3 mg cm^{-2}, respectively. The experimental layout is shown in Fig. 8.6. In

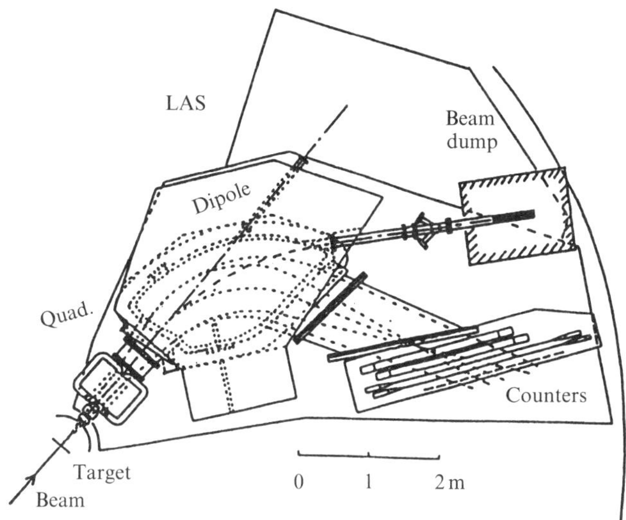

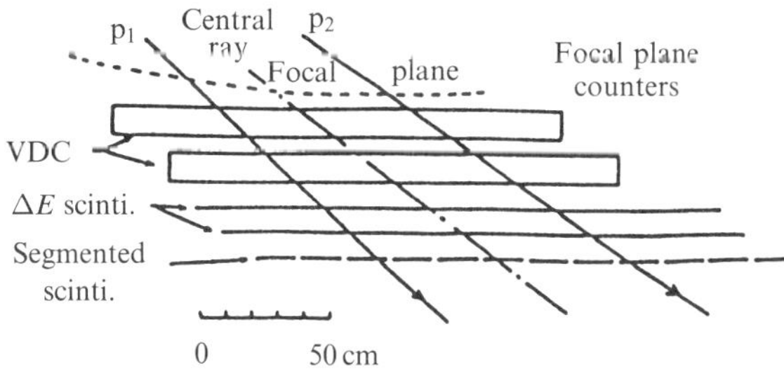

Fig. 8.6 Experimental setup of the LAS spectrograph and the focal plane counter system for the measurements of ^{2}He at 0° after bombarding various nuclear targets with protons. (Taken from Matsuoka *et al.* [18].)

the measurements of ^{2}He at $0°$, the incident beam is transported to an external beam dump through a beam extraction port of LAS. The focal plane counter system consists of two Vertical Drift Chambers (VDC) for ray tracing, two planes of ΔE scintillators, and 12 segmented scintillators (hodoscope). The trigger condition for the detection of ^{2}He is that two protons hit two different segmented scintillators. The segmented scintillators have a width of 15 cm and a height of 40 cm, and are separated from each other by 2 cm to avoid spurious triggering due to inclined incident particles. Throughout the experiment, the beam current was 1.0–1.5 nA and the effects of accidental coincidences were corrected. The true-to-accidental coincidences ratio was about 6.

In order to obtain the absolute cross section for the (p, ^{2}He) reaction, the effective solid angle for the detection of ^{2}He was calculated by a simulation taking into account the counter geometry and the condition of data analysis. The p–p relative energy in ^{2}He was considered up to 5 MeV, and the enhancement factor of the relative wave function due to final state pp ($^1S_0$) interaction as given in Ref. [20] was used. In the simulation, the particle trajectories were determined by ray-tracing calculations. The Q-value dependence of the detection efficiency was obtained. The maximum effective solid angle is 2.4 msr.

In the data analysis, the momentum and the emission angle for each proton of ^{2}He were determined using a trace-back method from the VDC data, and the energy spectrum of ^{2}He was obtained as a function of the sum of kinetic energies of the two protons. For heavy nuclei the spectrum is equivalent to the Q-value spectrum, because the recoil energy is negligibly small. The resolution was about 700 keV (FWHM) in the energy spectrum.

We obtained also the relative-energy spectrum of two coincident protons. An example of the spectrum, together with the simulation calculation, is shown in Fig. 8.7. The spectrum has a peak at around 0.6 MeV and falls off as the relative energy increases, and vanishes at about 5 MeV, consistent with the simulation calculation. In such a case of small relative energy, the contributions of two-proton final states with angular momentum $L > 0$ are small, since the amplitudes of these partial waves in the interaction region are very small compared to the amplitude of S-wave [20].

The elementary processes of pion production in the (p, ^{2}He) reaction are the $p+p \rightarrow$ ^{2}He $+ \pi^0$ and $p + n \rightarrow$ ^{2}He $+ \pi^-$ reactions. We have examined these elementary processes using CH_2, CD_2, and ^{12}C targets. The preliminary results of the cross sections obtained are 8.7 ± 0.4 μb sr^{-1} for $p + p \rightarrow$ ^{2}He $+ \pi^0$ and 137.8 ± 4.0 μb sr^{-1} for the $p + n \rightarrow$ ^{2}He $+ \pi^-$ reaction. The ratio of the cross sections is about 1/16. In the case of $p+n \rightarrow$ ^{2}He$+\pi^-$ process, the initial states with the isospin $T = 1$ and 0 contribute. On the other hand, only the $T = 1$ state contributes to the $p+p \rightarrow$ ^{2}He$+\pi^0$ reaction. From the study of π^- absorption on the pp ($^1S_0$) pair in ^{3}He, the $T = 1$ cross section is found to be much smaller than the $T = 0$ cross section [21]. Our measurement for the ratio of the two cross sections is consistent with these results. In the following discussion, the contribution from quasi-free π^0 production is neglected.

The results for the ^{12}C and ^{208}Pb(p, ^{2}He) reactions are shown in Fig. 8.8. The (p, ^{2}He) reaction leading to π^- production is considered to proceed via the elementary process $p+n \rightarrow$ ^{2}He $+ \pi^-$, with a neutron hole being left in the residual nucleus. The Q-value

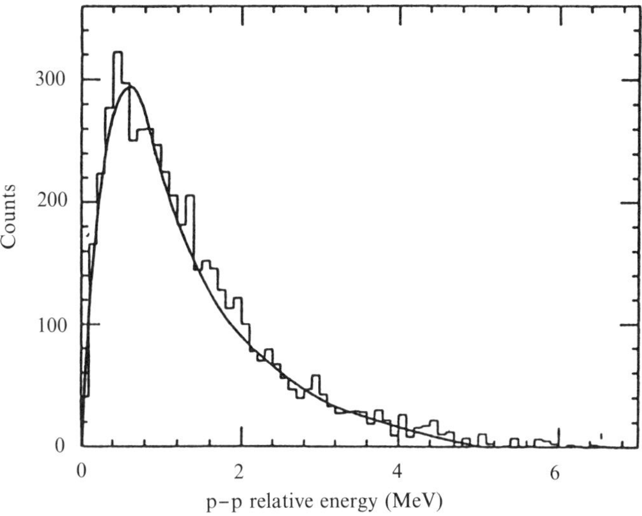

Fig. 8.7   The p–p relative-energy spectrum in the $CH_2(p, {}^2He)$ reaction at $0°$. The smooth curve shows the simulation calculation.

of the reaction is given by $Q = m_n - m_p - B_n - m_\pi + B_\pi$, where B_n and B_π are binding energies of the neutron and π^-, respectively. In Fig. 8.8, the energy scale is defined with respect to the threshold ($B_\pi = 0$) for the free π^- production for the least-bound neutron orbit. The least-bound neutron orbits are $1p_{3/2}$ with $B_n = 18.72\,\mathrm{MeV}$ for ${}^{12}C$, and $3p_{1/2}$ with $B_n = 7.367\,\mathrm{MeV}$ for ${}^{208}Pb$. The Q-values corresponding to the threshold are -157.0 and $-145.65\,\mathrm{MeV}$ for ${}^{12}C$ and ${}^{208}Pb$, respectively. By definition, the spectrum in the positive binding energy region results entirely from contributions from π^- bound states. In the negative energy region, there are contributions from the quasi-free π^- production process and the π^- bound states coupled to neutron deep-hole states.

The spectrum for ${}^{12}C$ shows a nearly flat cross section of about $1.07\,\mu\mathrm{b}\,\mathrm{sr}^{-1}\,\mathrm{MeV}^{-1}$ in the 0–$20\,\mathrm{MeV}$ region, and the cross section increases due to the quasi-free π^- production in the negative binding energy region. Within the statistical error, we have no excess cross section due to pionic bound states near threshold. For ${}^{208}Pb$, the spectrum shows a contribution from π^- bound states, in addition to the quasi-free process. The cross section of the bound states in the 0–$7\,\mathrm{MeV}$ region is $16 \pm 1.4\,\mu\mathrm{b}\,\mathrm{sr}^{-1}$ if we assume the flat continuum to have a cross section of $2.7\,\mu\mathrm{b}\,\mathrm{sr}^{-1}\,\mathrm{MeV}^{-1}$. The shape of the spectrum is similar to the result for the ${}^{208}Pb(n, d)$ reaction at $T_n = 400\,\mathrm{MeV}$ [17], although the bound state cross section in the 0–$7\,\mathrm{MeV}$ region is only about $1/88$ of that in the (n, d) case. The cross section of $n + n \rightarrow d + \pi^-$ reaction is $9.24\,\mathrm{mb}\,\mathrm{sr}^{-1}$ at $400\,\mathrm{MeV}$ and $0°$ [6], so that the ratio of cross sections of elementary processes for the $(p, {}^2He)$ and (n, d) reactions is about $1/67$. The difference of the production cross sections of the pionic

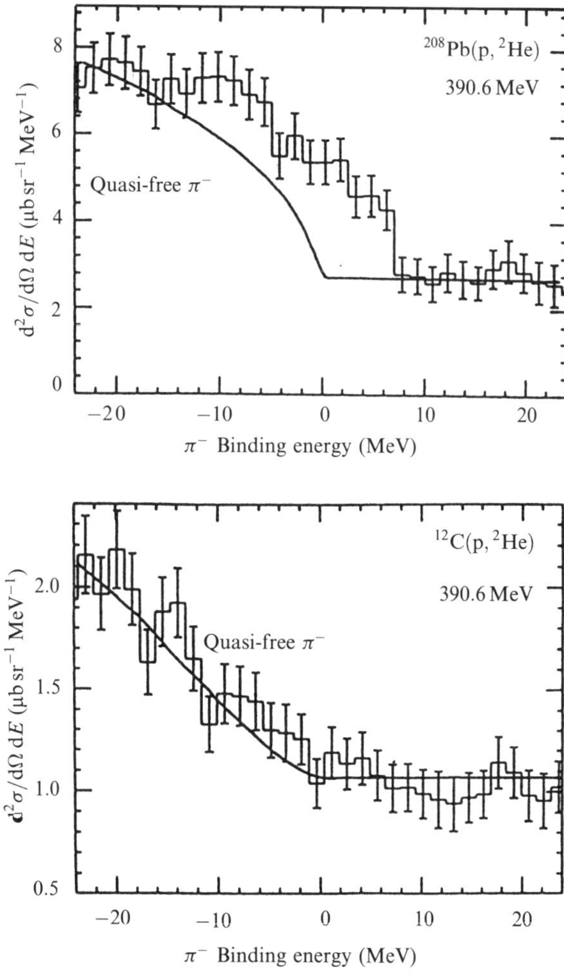

Fig. 8.8　Spectra of the ^{12}C and ^{208}Pb(p, ^{2}He) reactions near the threshold for free π^- production at $0°$ and $T_p\ =\ 390.6\,$MeV. The absolute values of the cross section correspond to the p–p relative-energy cutoff of 5 MeV in ^{2}He. The smooth curves show calculations for the quasi-free π^- production process. (Taken from Matsuoka *et al.* [18].)

bound states between the two reactions is considered to reflect mainly the difference of the cross sections for the elementary processes.

The cross section of the quasi-free π^- production was calculated using a simple model. The free π^- is assumed to be produced via the elementary $p + n \rightarrow {}^2\text{He} + \pi^-$ reaction in the nucleus, and the $(A-1)$ nucleons are assumed to be spectators. The neutron inside the nucleus moves with the momentum $\boldsymbol{p}$ and a momentum distribution $\rho(\boldsymbol{p})$. The neutron is assumed to have the binding energy of the $1p_{3/2}$ orbit for ^{12}C and of the $3p_{1/2}$

orbit for ^{208}Pb. Using energy and momentum conservation, the transferred energy ω and momentum q for the (p, ^{2}He) reaction are $\omega = E_p = E_{^2\text{He}} = E_{\pi^-} - m_n + B_n$ and $q = p_p - p_{^2\text{He}} = p_{\pi^-} - p$.

The following formula is used for the cross section [6,7]:

$$\frac{\mathrm{d}^2\sigma}{\mathrm{d}E\,\mathrm{d}\Omega} = \left[\frac{\mathrm{d}\sigma}{\mathrm{d}\Omega}\right]^{\text{lab}}_{\text{pn}\to{}^2\text{He}\,\pi^-} \bar{N}_{\text{eff}}\, R(\omega, q) \tag{8.8}$$

where

$$R(\omega, q) = \int \mathrm{d}^3 p\, \rho(p)\delta(\omega - E_{\pi^-} + m_n - B_n). \tag{8.9}$$

Here, $[\mathrm{d}\sigma/\mathrm{d}\Omega]^{\text{lab}}_{\text{pn}\to{}^2\text{He}\,\pi^-}$ denotes the elementary cross section for π^- production at $0°$ in the laboratory system, and $\bar{N}_{\text{eff}}$ is the effective number of neutrons which contribute to the reaction. $R(\omega, q)$ is the response function of the target nucleus. For the elementary cross section we use the value of $137.8\ \mu\text{b}\,\text{sr}^{-1}$.

For ^{12}C, the neutron orbits, $1p_{3/2}$ and $1s_{1/2}$, make a major contribution to the quasi-free π^- production. The binding energy of the 1s neutron is about 37 MeV, which is larger than that of the $1p_{3/2}$ orbit by about 20 MeV. So, in the energy region presented in Fig. 8.8, the contribution comes exclusively from the $1p_{3/2}$ orbit. In the calculation, we use the momentum distribution obtained from the $1p_{3/2}$ neutron wave function in a Woods–Saxon potential [22] ($V = 55.35\ \text{MeV}$, $R = 1.36A^{1/3}$ fm, $a = 0.55$ fm, $V_{\text{so}} = 9\ \text{MeV}$) with a correct binding energy. As shown in Fig. 8.8, the shape of the spectrum above the free π^- production threshold is reproduced by the calculation. The value of $\bar{N}_{\text{eff}} = 1.8$ is used to reproduce the experimental cross sections near 20 MeV binding energy.

In the case of ^{208}Pb, we take $\rho(p) = N\exp(-p^2/p_0^2)$, with $p_0 = 177\ \text{MeV}/c$ for simplicity. The value of p_0 was determined so as to reproduce the same $\langle p^2 \rangle$ as the Fermi distribution with $p_F = 280\ \text{MeV}/c$. The result of the calculation is shown in Fig. 8.8. The cross section of the flat continuum is assumed to be $2.7\ \mu\text{b}\,\text{sr}^{-1}\,\text{MeV}^{-1}$. The value of $\bar{N}_{\text{eff}}$ is determined to be 8.1, which reproduces the experimental cross sections near 20 MeV binding energy. The excess cross section above the prediction of the calculation in the negative binding energy region may point to a contribution from pionic bound states coupled to neutron deep-hole states. In order to reproduce the spectrum, we need more refined calculations including the contributions of these bound states and the quasi-free pion production process coming from several neutron orbits—$3p_{1/2}, 2f_{5/2}, 3p_{3/2}, \ldots$, because the differences in the binding energies of these orbits are not large.

8.4 The (d, ^{3}He) reactions for DBPAs

The discovery of DBPAs was made in an experiment performed at GSI with ^{208}Pb(d, ^{3}He) reactions [14]. We describe here the experimental method and its results. The deuteron beam was provided by SIS at GSI with an energy $T_d = 600\ \text{MeV}$. This energy was chosen to be the best suitable for the formation of pionic atoms, due to the predicted small momentum transfer and large elementary cross section [6,7]. Since high resolution and good statistics were required for this experiment, the GSI FRagment Separator (FRS)

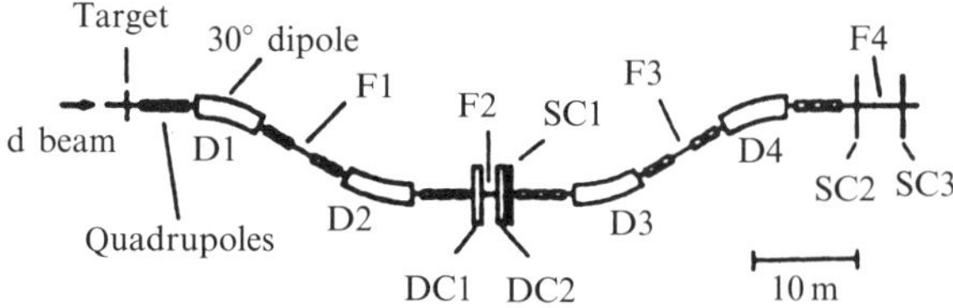

Fig. 8.9   The layout of the experimental setup of FRS used for (d, ^{3}He) reactions. The target is placed at the leftmost part where the deuteron beam is incident from the left. (From Yamazaki *et al.* [14].)

was used. The layout of the experimental setup of FRS is shown in Fig. 8.9. As a target, a 2 mm wide strip, 45.2 mg cm^{-2} thick, enriched ^{208}Pb was used. The resolution of about 0.5 MeV FWHM was achieved in the (d, ^{3}He) reaction for the formation of pionic atoms.

The excitation functions at $0°$ are shown in Fig. 8.10 for the ^{208}Pb target and for ^{27}Al for comparison. The yield of ^{3}He in the range of pionic atom formation (in arbitrary units) is plotted against the Q-value. The Q-value is related to the excitation energy E_x with respect to the ground state of ^{207}Tl as $E_x = -Q - 2.513$ MeV. The π^- binding energy B_π for the ground state of ^{207}Pb is expressed as $B_\pi = Q + 140.147$ MeV. We see clearly a pronounced peak at $Q = -135.6$ MeV with a somewhat skewed shape in the bound π^- region. Above this peak ($Q > -132$ MeV), we see a structureless continuum. In the π^- unbound region ($Q < -140.14$ MeV), we observe a continuum which is associated with the quasi-free production of π^-. Some strength still remains between the peak and the quasi-free region.

In the middle part of Fig. 8.10, we show the theoretical spectrum obtained by Hirenzaki *et al.*, which had been calculated long before the experiment [7]. We see an excellent overall agreement. The experimentally observed skewed shape of the dominant peak is also consistent with the theoretical result. As indicated in the theoretical spectrum, the dominant peak consists of the excitation of the 2p pionic state, and the 1s pionic state coupled with $p_{3/2}^{-1}$ and $p_{1/2}^{-1}$ neutron hole states. These contributions overlap in the energy spectrum due to insufficient experimental resolution. The strength between this peak and the quasi-free pion production is due to the excitation of higher pionic atom states. The pion unbound region is mostly due to the quasi-free pion production. The theoretical results underestimate the strength. This is caused by the use of a small number of neutron hole states. In fact, the use of a larger number of neutron hole states increases the strength to a value compatible with the experiment [23].

The good comparison indicates that the binding energies of the pionic atom states are not very different from the theoretical predictions. A closer look at the comparison of experiment and theory shows a slight shift in energy. The experimentalists provide the binding energy of the 2p pionic state as $B(2p) = 5.4 \pm 0.2$ MeV, a value slightly larger than the theoretical one, $B(2p) = 5.162$ MeV. We also expect that the configuration mixing effect between various pion and neutron hole states is very small. We should be able to extract the pion mass in the nucleus from this information [24].

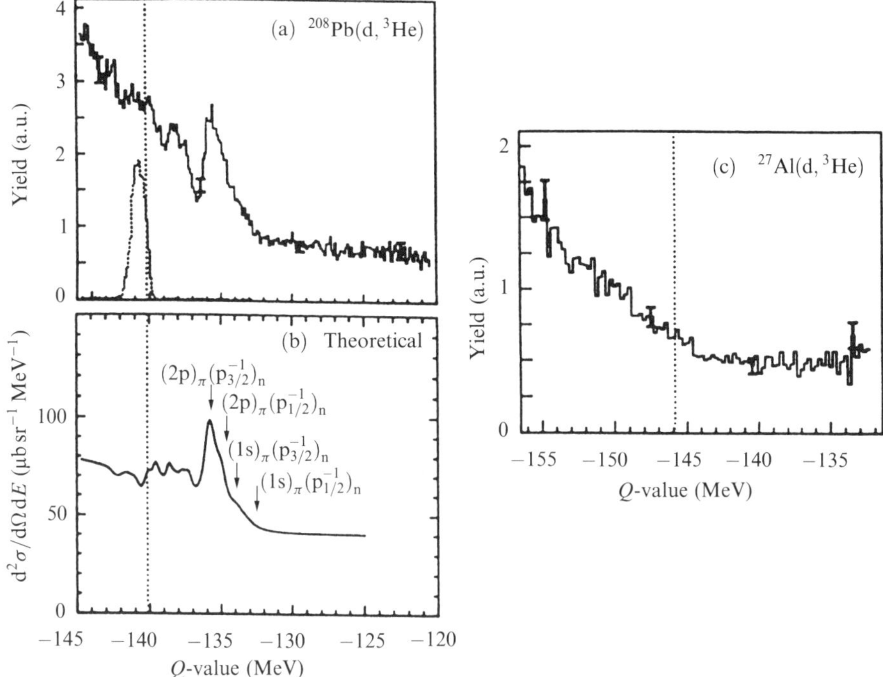

Fig. 8.10 The ^{208}Pb(d, ^{3}He) spectra as a function of the Q-value. The experimental spectrum is shown in (a), while the corresponding theoretical prediction is provided in (b). The p (d, ^{3}He) π^0 reaction peak used as calibration is also shown by the dotted curve in (a). A reference spectrum of ^{27}Al(d, ^{3}He) reaction is shown in (c). The dotted vertical straight line in each figure corresponds to the π^- emission threshold. (From Yamazaki *et al.* [14].)

8.5 Conclusion

We have been studying at RCNP the structure and the formation of DBPAs. The theoretical studies of the structure of DBPAs indicate the existence of these interesting states as quasi-stable states. The discussions have then moved to the formation of DBPAs, first with (n, p) pion transfer reactions and then with (n, d) reactions. There are some indications of the detection of these states in the spectrum of (n, d) reactions [17]. Subsequent work with (p, ^{2}He), which is discussed in detail, seems to indicate a similar feature [18]. For a clear demonstration, however, it is necessary to improve the experimental resolution and statistics.

Recently, Yamazaki *et al.* [14] performed experiments with (d, ^{3}He) reactions on ^{208}Pb. They used the 600 MeV deuteron beam of SIS at GSI and the high-resolution FRS and were able to identify a clear peak in the $0°$ spectrum below the pion production threshold. The position and the shape of the peak agree very well with the theoretical prediction. This comparison enables us to provide the energy shifts and the widths of the 2p and the 1s states, though, at this moment, with large errors. This discovery motivates further theoretical and experimental efforts in determining the changes in the pion properties inside the nucleus.

The use of inverse kinematics is also very interesting [8]. Particularly, we may be able to produce pionic atoms in unstable nuclei, since we can use unstable nuclear beams on deuteron targets. The structures of DBPAs are very sensitive to the neutron surface properties, and we are able to use them for the determination of the latter, even for the unstable nuclei [4]. We hope strongly that DBPAs will be found in various systems experimentally. Then we can extract precious information about the density dependence of the pion–nucleus interaction and also the structure of the neutron density near the nuclear surface, which is otherwise difficult to obtain. In this respect, it is interesting to note that the new data on ^{208}Pb(d, ^{3}He) reactions leading to DBPAs provide the binding energies of 2p and 1s states, which are then used to extract the change in the pionic mass in the nucleus [24].

Acknowledgements

We are grateful to Professors T. Yamazaki and R. Hayano for fruitful collaborations on this subject. There are many of our experimental colleagues who participated in this interesting study. H. T. and S. H. acknowledge fruitful collaborations and discussions with Professor E. Oset on the theoretical aspects of pionic atoms.

References

1. H. Toki and T. Yamazaki, *Phys. Lett.* **B213** (1988) 129.
2. H. Toki, S. Hirenzaki, T. Yamazaki, and R. Hayano, *Nucl. Phys.* **A501** (1989) 653.
3. T. Yamazaki, R. Hayano, H. Toki, and P. Kienle, *NIM* **A929** (1990) 619.
4. H. Toki, S. Hirenzaki, and T. Yamazaki, *Phys. Lett.* **B249** (1990) 391.
5. M. Iwasaki *et al.*, *Phys. Rev.* **C43** (1991) 1099.
6. H. Toki, S. Hirenzaki, and T. Yamazaki, *Nucl. Phys.* **A530** (1991) 679.
7. S. Hirenzaki, H. Toki, and T. Yamazaki, *Phys. Rev.* **C44** (1991) 2472.
8. T. Yamazaki, R. Hayano, and H. Toki, *NIM* **A305** (1991) 406.
9. J. Nieves and E. Oset, *Nucl. Phys.* **A518** (1990) 617.
10. J. Nieves and E. Oset, *Z. Phys.* **A343** (1992) 477.
11. J. Nieves, E. Oset, S. Hirenzaki, H. Toki, and M. J. Vicente-Vacas, *Mod. Phys. Lett.* **A7** (1992) 2991.
12. J. Nieves, E. Oset, and C. Garcia-Recio, *Nucl. Phys.* **A554** (1993) 509.
13. J. Nieves, E. Oset, and R. C. Carrasco, *Nucl. Phys.* **A565** (1993) 785.
14. T. Yamazaki *et al.*, *Z. Phys.* **A355** (1996) 219.
15. R. Seki and K. Masutani, *Phys. Rev.* **C27** (1983) 2799.
16. Y. K. Kwon and F. Tabakin, *Phys. Rev.* **C18** (1978) 932.
17. T. Yamazaki, *Nucl. Phys.* **A553** (1993) 221c.
18. N. Matsuoka *et al.*, *Phys. Lett.* **B359** (1995) 39.
19. N. Matsuoka *et al.*, RCNP annual report (1992) p. 174.
20. R. J. N. Phillips, *Nucl. Phys.* **53** (1964) 650.

21. E. Piasetzky *et al.*, *Phys. Rev. Lett.* **57** (1986) 2135; C. Ponting *et al.*, *Phys. Rev. Lett.* **63** (1989) 1792.

22. L. R. B. Elton and A. Swift, *Nucl. Phys.* **A94** (1967) 52.

23. H. Toki, S. Hirenzaki, and K. Takahashi, *Z. Phys.* **A356** (1997) 359.

24. T. Yamazaki *et al.*, preprint (1997).

9

Atomic physics with cyclotron beams

Ichiro Katayama

9.1 Introduction

Why atomic physics at a cyclotron laboratory? 'The interaction is very clear, so what is left for study?'—this is often what I was asked when we started atomic collision studies at RCNP. I hope this report provides some answers to the question. Firstly, I would like to remark on the so-called 'Massey criterion', which states that between a projectile and an electron, in a certain orbit in a target atom, the strongest interaction is expected when the velocities match each other. As the velocity of a K-shell electron in Pb atom equals that of a 160 MeV proton, this implies that, in order to study the excitation of inner-shell electrons in heavy atoms, one needs very high energy ion beams. Secondly, there is one other aspect, viz. the potential usage of a foil target. All the atomic physics experiments which were done at RCNP utilized thin foils, even though some of them aimed to study a single collision process, e.g. electron capture and loss process. Others were designed to study solid state effects such as channelling, convoy electrons, stopping-power experiments, and so on. (Although they are not topics from pure atomic physics in a strict sense, we include them in this review.) Even for these solid state effect experiments, the use of high energy beams from cyclotron is more effective, since it relaxes us from ultra-high vacuum condition. There is another advantage in studying atomic collisions at RCNP: one can use the high-resolution magnetic spectrograph RAIDEN, which is quite unique if one compares it with facilities available at other institutes.

The study of atomic physics by RAIDEN was active in 1978 when we 'accidentally' observed $^3\mathrm{He}^{1+}$ as an electron capture product in a (^{3}He, t) experiment. The magnetic field was set to a value slightly higher than that for the ground state triton. When the ^{3}He beam was started, we saw a huge peak suddenly growing at the centre of the CRT screen. Soon it was found that the peak was due to $^3\mathrm{He}^{1+}$. This electron capture process was successfully used to calibrate a magnetic-field-dependent ρ–x relation, where ρ is the effective curvature of an ionic orbit in the dipole magnetic field and x is the position of the ions on the focal plane counter. (For more details, see Section 9.5.2.) In 1982 when I came back from Julich where I also started high-energy electron capture experiments using a $^3\mathrm{He}^{2+}$ beam, I proposed a high-resolution study of the ^{3}He electron capture. This was followed by studies of K-shell electron capture by Katayama $et\ al.$, L-shell capture by Ogawa $et\ al.$, electron loss of $^3\mathrm{He}^{1+}$ by Haruyama $et\ al.$ (see Section 9.5.1), charge-state-dependent stopping power in non-equilibrium regions by Ogawa $et\ al.$ (Section 9.6), stopping power for molecular ions in non-equilibrium regions by Susuki *et al.* (Section 9.7), stopping power of metallic elements by Sakamoto *et al.* (Section 9.8), and direct observation of a binding-energy effect by Ishi *et al.* (Section 9.9). Meanwhile, Nakayama *et al.* measured the inner-shell ionization in 1984 (Section 9.2), using the in-beam γ-ray experiment line (F-beam line), and also a large scattering chamber line (E-beam line). The Mannami group studied channelling in 1987 (Section 9.3) and a convoy electron in 1988 (Section 9.4) at E-beam line.

Two experiments which were performed off-line at different institutes after irradiation at RCNP are not covered in this review: search for a line in positron scattering by Sakai *et al.* [1] and material defect study using a positron emitter by Tanikawa *et al.* [2]. A snowball experiment is discussed in Chapter 6.

9.2 Lγ X-ray emission in heavy-ion bombardment of Bi

Nakayama *et al.* performed an 88 MeV ^{20}Ne^{4+} + ^{209}Bi collision experiment to investigate the ionization of inner-shell electrons in a heavy quasi-atom with $Z = 93$ [3]. Their interest was in the multiple inner-shell vacancy formation in a single collision, though the direct Coulomb ionization mechanism plays a role in the ionization process in the collision at a few MeV/amu. The singles measurement of Bi L X-rays observed with a Si(Li) detector revealed the nature of multiple-vacancy formation in heavy-ion–atom collision. Among L X-rays, Lγ X-ray lines consist of the transitions from the N and O shells and hence the degree of multiple ionization of outer shells is reflected strongly in the energy spectrum of Lγ X-rays. Although the emission probability of Lγ X lines is small as compared with other lines, Lγ X-ray spectrum with high statistics was obtained and

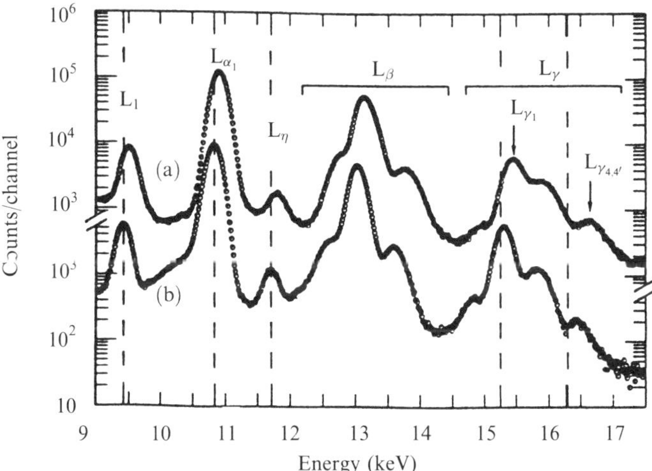

Fig. 9.1 Comparison of the L X-ray spectra from Bi bombarded by (a) 88 MeV Ne ions and (b) 20 MeV O ions. The broken lines indicate the unperturbed (unshifted) transition energies for L$_1$, L$_{\alpha_1}$, L$_{\eta}$, L$_{\gamma_1}$, and L$_{\gamma_{4,4'}}$ lines. (Taken from Ito *et al.* [3].)

228   I. KATAYAMA

a detailed analysis was made. Figure 9.1 shows the Bi L X-ray spectrum observed by 88 MeV Ne ions bombardment and one by 20 MeV O ions from a tandem van de Graaf at Kyoto University [3]. As shown in Fig. 9.2, the Lγ lines in the spectra were decomposed into six transition lines with the assumption that the energy shifts of these lines can be sorted into two groups corresponding to the L–N and L–O transitions and that the shifts are the same in each group. The degree of N-shell multiple ionization has been deduced from the energy shift determined for the L–O transition. They concluded that

(1) approximately 11 spectator vacancies are present in $4f_{7/2}$, $4f_{5/2}$, and $4d_{5/2}$ states for the case of Ne bombardment,

(2) the measured intensity ratios between the L–O and L–N transitons are taken to suggest that the O-shell electrons are less ionized than the N-shell electrons, and

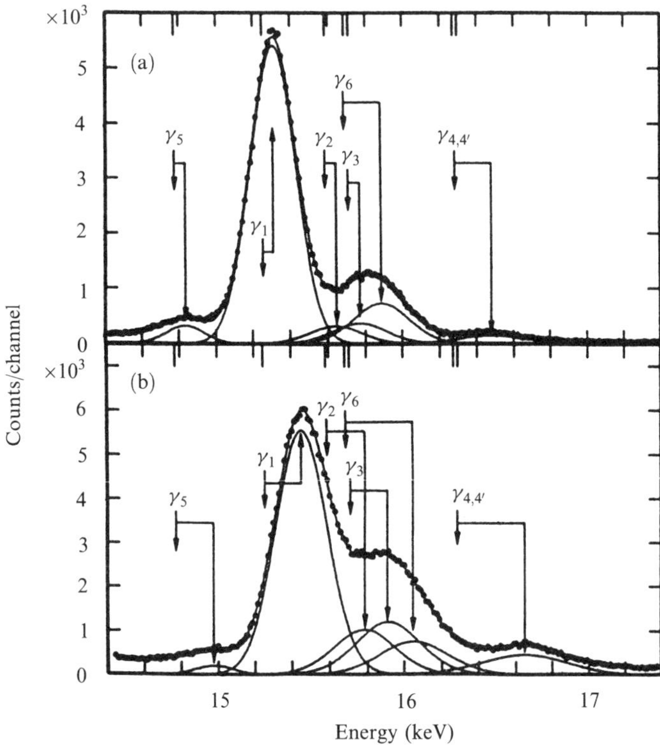

Fig. 9.2   Least-squares analysis for the Lγ X-ray spectra observed by 20 MeV O ions (a) and 88 MeV Ne ions (b). A linear background (not shown) was used. The short vertical arrows below the peak labels indicate the unperturbed (unshifted) energies for the Lγ X-ray lines. The arrows with horizontal lines are the peak centroids of the Gaussians used. The distances between the two arrows corresponding to the same peak label show the energy shifts described in the text. (Taken from Ito *et al.* [3].)

(3) these findings may indicate the rearrangement of outer-shell multiple vacancies prior to L X-ray emission. The theoretical discussion of the result has also been given in Ref. [4].

9.3 Channelling in a bent crystal

Since channelled steering of a beam of fast ions in a bent crystal was first suggested by Tsyganov [5], a number of studies on channelling of relativistic ions in bent crystals have been carried out. Mannami *et al.* proposed a channelling experiment using 80 MeV protons in a bent Si crystal at RCNP [6]. The motivation was to understand the phenomenon more quantitatively, especially the so-called bulk capture process, which was first observed at St. Petersburg. The phenomenon implies the increase of the fraction of channelling protons as they travel further in the crystal. This suggests that the ions which have not been trapped in planar channels at the incident surface are captured into these channels inside the crystal.

Figure 9.3 shows a schematic diagram of the goniometer head on which a thin slab of Si (001) crystal was mounted [7]. One end of the crystal was clamped and the other was pressed by a rod which could be swung, so that the crystal was bent on the goniometer during the experiment.

Proton beams were collimated to have an angular spread of less than 0.1 mrad on the target crystal by using several slits in the upstream beam line. The angular distribution of the scattered protons was measured by a position-sensitive Single Wire Proportional Counter (SWPC) placed 8.9 m downstream from the crystal. Figure 9.4 shows the angular distribution of 80 MeV protons transmitted through (a) straight and (b) bent (011) planes which were at 45° to the (001) surface. The sharp peaks in the middle are due to the protons channelled through the planes. The shift of the peak (see Fig. 9.4(a) and (b)) is caused by the steering of the channelled protons by the (011) planes. Although they observed the channelling in several experiments, the beam positioning and the bending of the crystal could not be reproduced well, which made a systematic analysis difficult. Especially, the radius of curvature of the crystal bend as shown in Fig. 9.3 was found to be non-uniform and uncontrollable. For the bulk capture process, a numerical simulation for the case of (110) planar channelling by a 100 GeV proton has been made. The radius of curvature of the crystal was assumed to change from 20 cm to infinity at 12 μm from the incident surface. This model has clearly shown the bulk capture effect. There do not

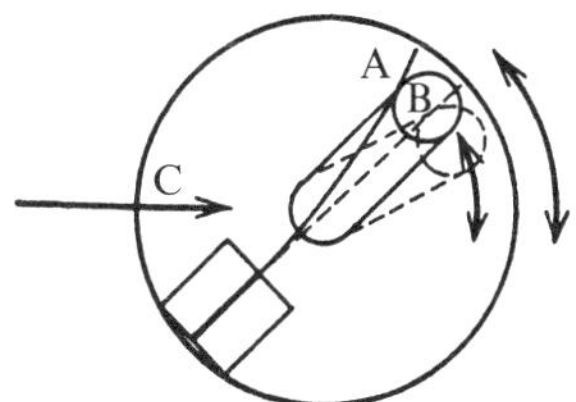

Fig. 9.3 Schematic diagram of goniometer head. A, Si (001); B, bending rod; C, protons. (Sueoka *et al.* [7].)

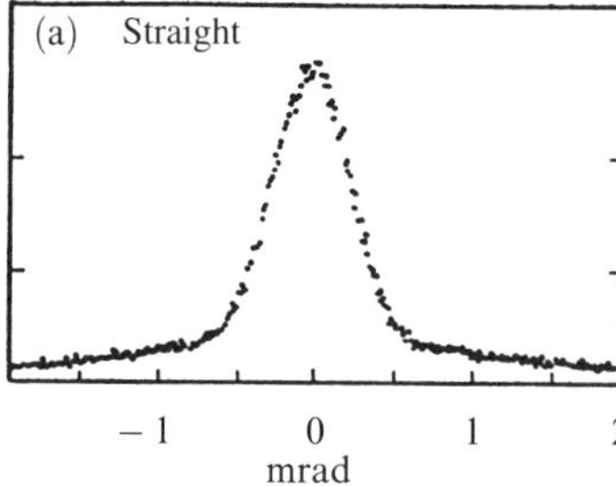

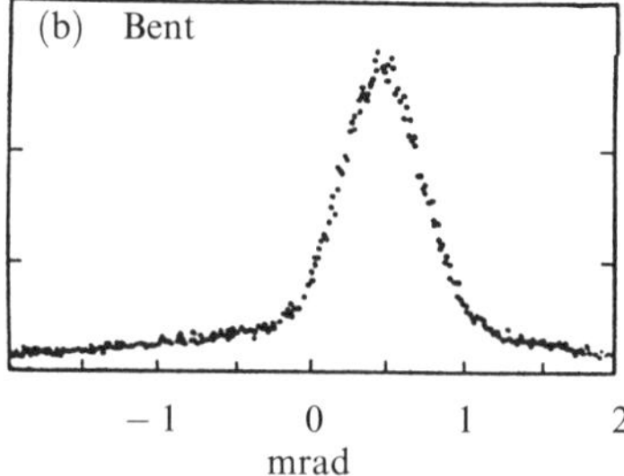

Fig. 9.4   Angular distributions of 80 MeV protons transmitted through straight (a) and bent (b) planar channels of Si. (Sueoka *et al.* [7].)

seem to be any proposals to use a channelled steering of a beam since the SSC project was withdrawn (Mannami, private communication).*

9.4   Convoy electrons

A sharp cusp-like peak appears in the energy spectrum of electrons emitted in the forward direction at the electron velocity equal to the ion velocity when fast ions are transmitted through a thin solid foil. These electrons are called convoy electrons. Similar electrons are observed in ion–gas collisions and are interpreted as due either to Electron Loss (ELC) or Electron Capture (ECC) to low-lying continuum states of the ions. The origin of the convoy electrons, however, is not yet well understood.

Mannami *et al.* performed a series of experiments, from 1988 to 1991, to investigate the mechanism of convoy electrons [8–10]: (1) a target-material dependence of convoy-electron yields using 24 and 50 MeV ^{3}He^{2+} and ^{3}He$^+$ projectiles, (2) a thickness dependence of convoy-electron yields at transmission of 50 MeV ^{3}He^{2+} through carbon foils, (3) convoy-electron yields at 50 MeV ^{3}He^{2+} transmission through two thin carbon foils in order to detect the effects of foil-excited He ions on the production of the convoy electrons, and (4) convoy-electron yields at transmission of 16.7 MeV/amu H$^+$ and H$_2^+$ ions through a thin carbon foil in order to see the effects of spatially correlated two-fragment H$^+$ ions on the formation of convoy electrons. One of the advantages of the use of fast ions for these experiments, as mentioned before, is that suface contamination has a negligible effect both on the motion of convoy electrons and on the charge state of the ions transmitted through the foil even under a conventional vacuum ($\approx 10^{-6}$ Torr), i.e. the mean free paths for charge exchange collisions of 50 MeV ^{3}He ions and for convoy electrons are larger than the thickness of the contamination layers of carefully prepared foils.

A foil or a double foil on a goniometer was irradiated by a parallel collimated beam. The foils were so thin that almost all the scattered ions could transmit through the entrance and exit windows of an electron energy analyser at 6 cm downstream from the foil and were stopped in a Faraday cup for the measurement of the total incident beam current.

---

*A 450 MeV proton beam experiment was performed at CERN [59].

A ring biased at $-4\,$kV was placed at the entrance window of the Faraday cup to prevent the secondary electrons from entering the Faraday cup. The electron energy analyser was a parallel-plate electrostatic type with mean incident angle of $30°$ and the foil was at the source point of the analyser. A ceratron was used for the detection of electrons at the focal point. The energy resolution of the analyser was 0.01 and the entrance window of the analyser limited the acceptance half-angle to $3°$.

9.4.1 *Target material and thickness dependence of convoy-electron yields by* $50\,$MeV ^{3}He^{2+} *and* ^{3}He$^+$ *beams*

Figure 9.5 shows the electron energy spectrum of an Ag target ($3\,$mm thick) [11]. A cusp-like peak at $9.19\,$keV is due to convoy electrons, and a shoulder at $36\,$keV, which is four times the convoy-electron energy, is due to electrons formed by binary collisions (knock-on electrons). At ^{3}He$^+$ incidence, the energy spectrum shows a broad peak at energy smaller than that of the convoy electrons. Due to this broad peak of the foil-stripped electrons, convoy electrons were hardly observed. The yield of convoy electrons was obtained by integrating the spectrum at the cusp after subtracting the broad background. Figure 9.6(a) shows the dependence of the yield of convoy electrons, thus obtained at transmission of $50\,$MeV ^{3}He^{2+} ions [9], on the thickness of the carbon foil. The calculated He$^+$ fraction in the beam transmitted by the foil is shown by a solid curve. The relation $Y = KF_1$ was used, where $K = (4.8 \pm 1.6) \times 10^{-3}$ and F_1 was obtained by extrapolation of the cross sections at 69–$130\,$MeV [12]. Figure 9.6(b) shows the dependence on the atomic number of the target, Z_2. The equilibrium fractions F_1 of ^{3}He$^+$ ions at $50\,$MeV ^{3}He^{2+} transmission through C, Al, Ni, Ag, and Au foils of thickness $1\,\mu$m are connected by a solid line. The relation $Y = KF_1$ mentioned above gives a result shown by the broken curve in the figure. These results were interpreted using the two-step model [13], according to which the yield of the convoy electrons is

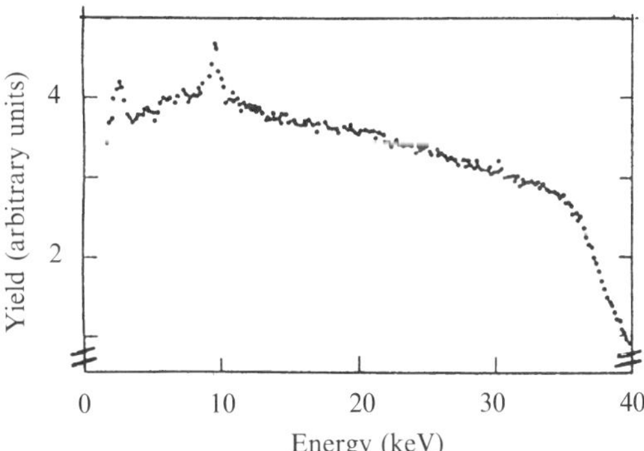

Energy (keV)

Fig. 9.5 Electron energy spectrum at $50\,$MeV ^{3}He^{2+} on Ag ($3\,\mu$m thick). (Taken from Mannami *et al.* [11].)

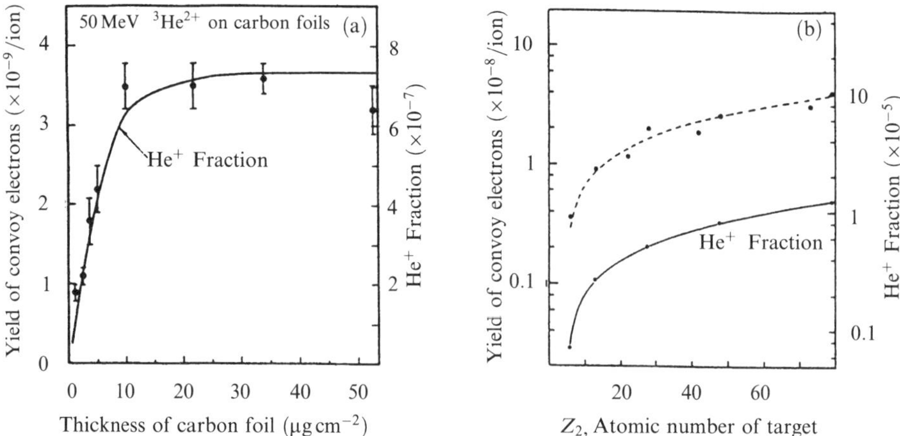

Fig. 9.6 Dependence of the yield of convoy electrons on the thickness of carbon foil (a) and on atomic number of the target Z_2 at transmission of 50 MeV ^{3}He^{2+} ions (b). The calculated He$^+$ fraction in the beam transmitted by the foil is shown by a solid curve in (a). Equilibrium fractions F_1 of ^{3}He$^+$ ions at 50 MeV ^{3}He^{2+} transmission through C, Al, Ni, Ag, and Au foils are connected by a solid line in (b). The relation $Y = K F_1$ is derived, where $K = (4.8 \pm 1.6) \times 10^{-3}$ for both (a) and (b). (Taken from Hasegawa *et al.* [8] and Mannami *et al.* [9].)

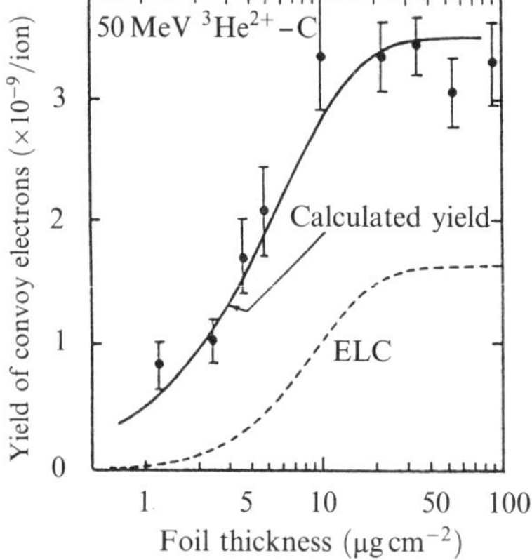

Fig. 9.7 Calculated thickness dependence of the yield of convoy electrons at the transmission of 50 MeV ^{3}He^{2+} ions through a thin carbon foil. The contribution of the ELC process is shown by a broken line. (Taken from Mannami *et al.* [9].)

given by the following relation:

$$Y = N \int_0^D [\sigma_{\mathrm{ELC}} F_1(z) + \sigma_{\mathrm{ECC}} F_2(z)] \exp[-(D - z)/\lambda]\,\mathrm{d}z, \tag{9.1}$$

where N is the density of atoms, σ_{ELC} and σ_{ECC} are ELC and ECC cross sections, respectively, $F_1(z)$ and $F_2(z)$ are, respectively, the fractions of He^+ ions and He^{2+} ions at depth z in the foil, and λ is the mean free path for convoy electrons. Using the relation $Y = K F_1$ together with eqn (9.1), the expressions for σ_{ELC} and σ_{ECC} follow. Using the data for λ and the charge-changing cross section of 50 MeV ^{3}He–C collisions, the σ_{ELC} and σ_{ECC} in carbon were determined. The yields of convoy electrons at 50 MeV $^3\text{He}^{2+}$ transmission through carbon foils, calculated by using eqn (9.1), are shown in Fig. 9.7. The experimental data are well explained. The contribution of the ELC process to the yield is also shown by a broken line in the figure. The ELC contribution increases with the increase of thickness and becomes almost half of the yield from the thickest foil. The observed Z_2 dependence of the yield shown in Fig. 9.6(b) could not be analysed, since no ELC and ECC cross sections for material other than carbon are available. Considering the empirical relation $Y = K F_1$ with a constant K for 50 MeV $^3\text{He}^{2+}$ beam, the convoy electrons can be interpreted, though naively, to be produced from He ions upon emergence from the solid, e.g. the $^3\text{He}^+$ ion loses its electron into one of its continuum states upon the emergence from the solid. It is to be noted, however, that a similar idea proposed in Ref. [14] for ECC and in Ref. [15] for ELC is not accepted in Ref. [16]. So this proportionality has remained unexplained.

9.4.2 *Convoy electrons by foil-excited He ions*

Although there is similarity between convoy electrons and cusp electrons captured at continuum states in ion–gas collisions, several phenomena are specific to the case of a solid; they are a larger population of high-angular-momentum Rydberg states, higher-order multipole content of the convoy-electron velocity distribution, anomalous electron transport properties in the presence of a Coulomb field, and so on. Mannami *et al.* have

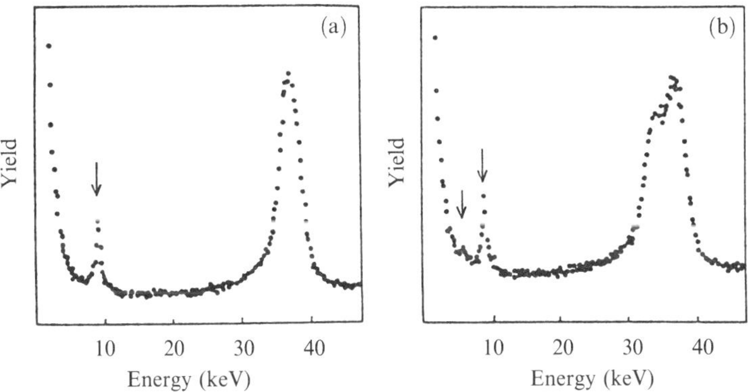

Fig. 9.8 Energy spectra of electrons emitted in the beam direction for 50 MeV $^3\text{He}^{2+}$ on two carbon foils. The peak at about 9 keV is due to a convoy electron. The thicknesses of the upstream and downstream foils are 5.2 and 2.4 $\mu\text{g cm}^{-2}$, respectively. The conditions for the energy spectra were: both foils grounded (a), and the upstream foil biased at 3 kV (b). (Taken from Mannami *et al.* [11].)

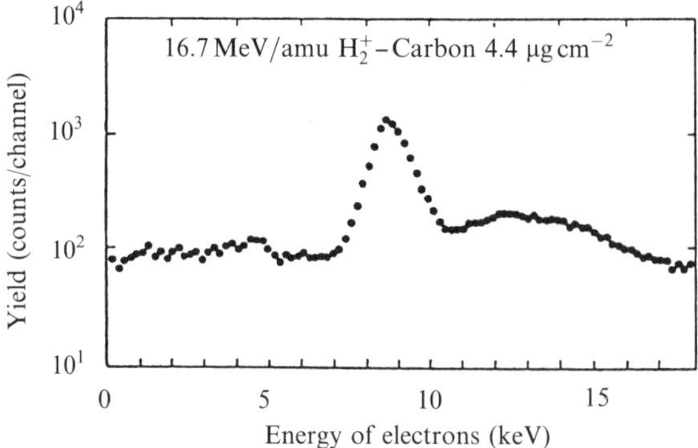

Fig. 9.9   Energy spectrum of secondary electrons at transmission of 16.7 MeV/amu H^{2+} through a carbon foil (4.4 $\mu g\,cm^{-2}$ thick). (Taken from Mannami *et al.* [17].)

studied the production of convoy electrons by foil-excited Rydberg states of fast ^{3}He ions [10]. They utilized two thin carbon foils instead of one. A bias voltage was applied at the upstream foil to decelerate electrons emitted by it. Typical energy spectra of the electrons emitted from two foils are shown in Fig. 9.8. The thicknesses of the upstream and downstream foils were 5.2 and 2.4 $\mu g\,cm^{-2}$, respectively. In Fig. 9.8(b), the cusp-like peak at 6 keV was formed due to deceleration of the convoy electrons from the upstream foil by 3 kV potential. When the downstream foil was replaced by a 5 $\mu g\,cm^{-2}$ foil, this small low-energy cusp peak disappeared. The experiment showed that the yields of convoy electrons from two foils of the 5.2 and 2.4 $\mu g\,cm^{-2}$ combination without a bias voltage are about three times more than the yield from a single 2.4 $\mu g\,cm^{-2}$ foil. With the bias voltage on, the yield decreased to two times the 2.4 $\mu g\,cm^{-2}$ yield. From these results, they have concluded that the 'three times yield' can be due to the ELC of Rydberg states produced at the upstream foil, and that the fraction of the Rydberg states in the beam transmitted through the foil producing convoy electrons in the downstream foil is almost comparable to or larger than that of convoy electrons produced at the upstream foil.

9.4.3   *Secondary electrons produced by fast H^+ and H_2^+ ions*

Figure 9.9 shows the energy spectrum of electrons for 16.7 MeV/amu H_2^+ ions incident on a 4.4 $\mu g\,cm^{-2}$ thick carbon foil [17]. A broad peak was observed at about 8.5 keV, and the energy of the peak decreased with the increasing foil thickness. The yields were about 10^{-2} for incident H_2^+ ions, and were not convoy electrons. For H^+ beams, the foil thickness dependence of the convoy-electron yields was analysed using eqn (9.1). Since the σ_{ELC} and σ_{ECC} were not known, the yield was analysed by assuming pure ELC or ELL process. The result favoured the ELC process. This is different from the ^{3}He result, which has not yet been discussed.

9.5 Electron capture and loss of light projectiles

Electron capture at intermediate energy has been studied for a long time, but it is still not understood completely. Intermediate energy means a region where the projectile velocity is comparable to the velocity of the electron to be captured. At high energies, where the projectile velocity is substantially higher than the electron orbital velocity, a second-order plane wave Born theory is believed to work well. At low energies, on the other hand, the molecular picture has been used to understand the electron capture mechanism. Previous experimental investigations of K-electron capture from the heaviest target at intermediate energy was on $p + Ar$ at $10\,MeV$ [18]. The K-shell was identified by a coincidence method between the K X-ray from the residual atom and the capture product H_0. Katayama *et al.* proposed, in 1985, using RAIDEN (Fig. 9.10) for the detection of the K-electron capture product $^3He^+$.

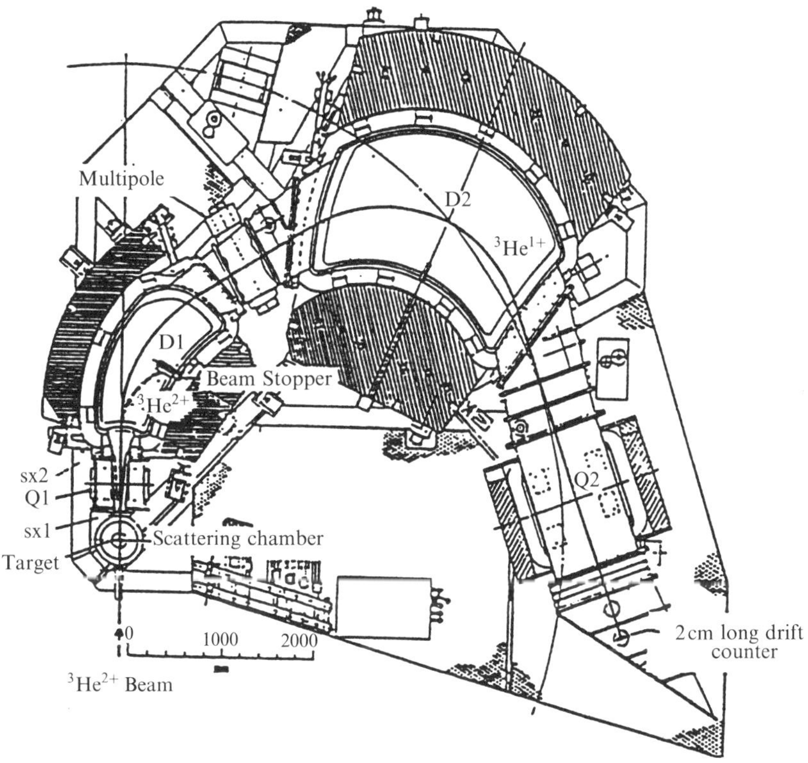

Fig. 9.10 The experimental arrangement using the high-resolution magnetic spectrograph RAIDEN. (Taken from Ogawa *et al.* [23] and Katayama *et al.* [24].)

9.5.1 *The K-shell electron capture of heavy target atoms by ^{3}He^{2+} beams*

The Katayama *et al.* idea was to use the high-resolution property of RAIDEN to resolve the energy loss of the K-electron capture product ^{3}He$^+$ from that resulting from higher-shell capture [19]. The experiment comprised three different measurements. Firstly, the total capture yield was measured as a function of target thickness. The capture product ^{3}He$^+$ was detected by a plastic scintillator on the focal plane of the RAIDEN. Most of the targets were prepared by evaporating the target material onto a $5\,\mu\mathrm{g\,cm}^{-2}$ thick carbon backing. When the target was irradiated by a ^{3}He^{2+} beam coming through the carbon backing, the yield ^{3}He$^+$ is given as

$$Y = N_\mathrm{C} \exp(-\sigma_\mathrm{L} t) + N_0 \left(\frac{\sigma_\mathrm{T}}{\sigma_\mathrm{L}}\right)[1 - \exp(-\sigma_\mathrm{L} t)], \qquad (9.2)$$

where σ_T is the total electron capture cross section for the reaction ^{3}He^{2+} + target $\rightarrow$ ^{3}He$^+$ + target$'$, and σ_L the electron loss cross section for ^{3}He$^+$ + target $\rightarrow$ ^{3}He^{2+} + target$'$, N_C is the ^{3}He yield from the carbon backing, N_0 the beam intensity, and t the target thickness. It was assumed that σ_L is much larger than σ_T and that the neutral fraction of ^{3}He is negligibly small. The strength of N_C was determined experimentally using a C foil of thickness $5\,\mu\mathrm{g\,cm}^{-2}$ as a target. Figure 9.11 shows an example of target thickness dependence of the total ^{3}He$^+$ yields. In principle, a least-squares fitting of eqn (9.2) to the data should provide the σ_L and σ_T values [20]. The cross section σ_L has, however, a relatively larger error compared to the one given by the attenuation method, which is described in the following. Therefore, the data shown in Fig. 9.11 were used effectively only to get the $\sigma_\mathrm{T}/\sigma_\mathrm{L}$ ratio. In the attenuation method, which was the second measurement, intense ^{3}He$^+$ beams were first produced from the electron capture of beams using a thin Au foil placed at the normal target position in the scattering chamber of the RAIDEN. The ^{3}He$^+$ particles were then focused on a target material placed on the focal plane of the spectrograph. The charge state of ^{3}He particles coming out of the target

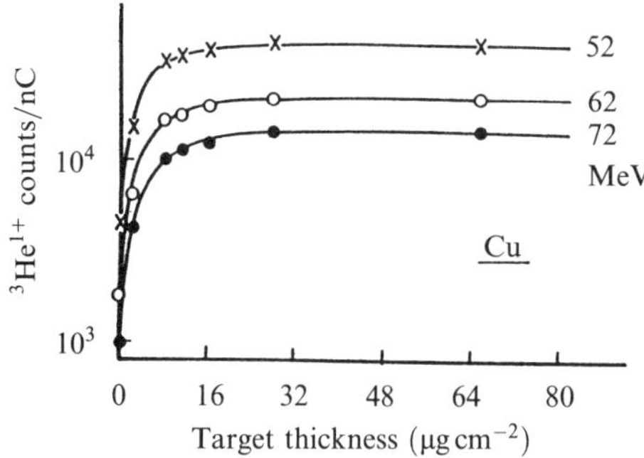

Fig. 9.11 A typical growth curve of the ^{3}He$^+$ yields at energies of 72, 62, and 52 MeV for Cu target. The yields at zero thickness come from the carbon backing. The result of a least-squares fitting to eqn (9.2) is also indicated by solid lines. (Taken from Katayama *et al.* [24].)

material was analysed by a small magnet located downstream of the focal plane. Both the $^3\text{He}^+$ and $^3\text{He}^{2+}$ particles separated by the magnet were detected with a 4 cm long position-sensitive gas proportional counter. This method gave an accurate attenuation of $^3\text{He}^+$ intensity, and also yields a more precise σ_L value than the one given by eqn (9.2) [21]. Surface contamination, including the oxidization effect, on the targets was examined using a van de Graaf beam at Kyoto University, Uji campus. The correction was finally estimated to be 20–30%, depending on the target species. Combining the results on σ_T/σ_L and σ_L, as well as surface contamination effects, σ_T and σ_L strengths were obtained [22]. In the third measurement, high-resolution energy loss spectra of $^3\text{He}^+$ were measured. The result is shown in Fig. 9.12 for targets of C, Cu, Ge, Nb, Ag, Sn, and Au at ^{3}He energy of 72 MeV. The spectrum for a carbon target, which gives a response function for the deconvolution of the spectrum, showed a momentum resolution of 2×10^{-5}. The momentum difference between two lines in a spectrum was found to

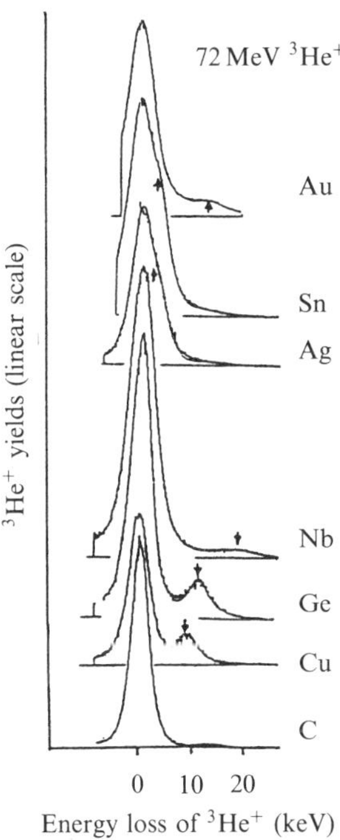

Fig. 9.12 Typical $^3\text{He}^+$ energy spectra measured under the same conditions for C, Cu, Ge, Nb, Ag, Sn, and Au targets at an incident energy of 72 MeV. The arrow pointing downwards indicates the K-shell position of the corresponding target atom, and the upward arrow, the L-shell position. The solid lines are the results of peak fitting using the C-line shape. A small bump observed in case C is due to the beam structure. (Taken from Katayama *et al.* [24].)

give a relation corresponding to

$$\left(\frac{\mathrm{d}p}{p}\right)_1 - \left(\frac{\mathrm{d}p}{p}\right)_2 = \frac{1}{2}\frac{(E_{B1} - E_{B2})}{E(^3\mathrm{He})}, \qquad (9.3)$$

where E_{B1} (E_{B2}) corresponds to the binding energy of the electron in initial state 1 (2), and $E(^3\mathrm{He})$ is the kinetic energy of $^3\mathrm{He}$. The arrows in the figure denote the position of the K–M shell binding energy difference. The good correspondence between the arrow and the peak position of the lines made us assign the lines to the K-shell electron capture of the $^3\mathrm{He}^{2+}$ projectiles [19,22–24]. The shoulder structure at the side of main peaks in Nb, Ag, Sn, and Au targets are similarly assigned to the L-electron capture [D2]. Combining the three measurements mentioned above, the K-shell capture cross sections were obtained from the following relation:

$$\sigma_\mathrm{K} = \sigma_\mathrm{T} \left[\frac{K}{K + L + M + \cdots + C}\right]\left[\frac{K + L + M + \cdots + C}{K + L + M + \cdots}\right], \qquad (9.4)$$

where $K, L, M, \ldots, C$ denote the partial fractions of the $^3\mathrm{He}^+$ components from each target shell (K, L, M, ...) and from the carbon backing, respectively, of each target. The cross sections σ_T are those obtained as mentioned above. The second factor in eqn (9.4) is obtained from the deconvolution of the $^3\mathrm{He}^+$ spectrum. The last factor is obtained as $Y(t_1)/[Y(t_1) - N_c \exp(-\sigma_\mathrm{L} t_1)]$, which is calculable for a specific target thickness t_1 of the target used to measure the $^3\mathrm{He}^+$ energy loss spectrum.

A two-step process of excitation and capture of an electron, which is not discernible experimentally, was considered not to influence the final result [24]. The K-electron capture cross sections at 72 MeV of $^3\mathrm{He}$ vs the atomic number of the targets are shown in Fig. 9.13. The Strong Born Potential (SBP) model [26], which was interesting to test, was found not to explain the experiment. The results for heavy targets are explained well by the two-state coupling model [27].

9.5.2 *Other related studies*

Ogawa *et al.* have tried to implement a new ion optics mode in the RAIDEN, so that a momentum spectrum and an angular distribution could be simultaneously measured in the horizontal and vertical directions. The new mode, which was successfully tested in 1989, can give 0.5 mrad angular resolution, along with the normal energy resolution [28]. This mode was then applied to measure $\mathrm{d}\sigma/\mathrm{d}\Omega$ for electron capture of $^3\mathrm{He}^+$ from a carbon target. The analysis, however, showed that due to a complicated deconvolution procedure, a clear conclusion on theoretical models could not be drawn.

For gas targets, the above method for the $^3\mathrm{He}^{2+}$ beam does not work. Katayama *et al.* have tried to measure electron capture and loss cross sections of hydrogen for $\mathrm{p} + \mathrm{Ar} \to \mathrm{H}_0 + \mathrm{Ar}^+$ collision at a proton energy of 65 MeV. Putting a window foil at the entrance of the recoil mass separator CARP, they filled the upstream section of CARP with Ar gas and measured the pressure dependence of the H_0 yield. The method is reported in Ref. [29].

Just as for $^3\mathrm{He}^+$, electron capture was found as a kind of by-product in a $(^3\mathrm{He}, \mathrm{t})$ experiment, now 450 MeV $^3\mathrm{He}^+$ was used again in GRAND RAIDEN in a similar

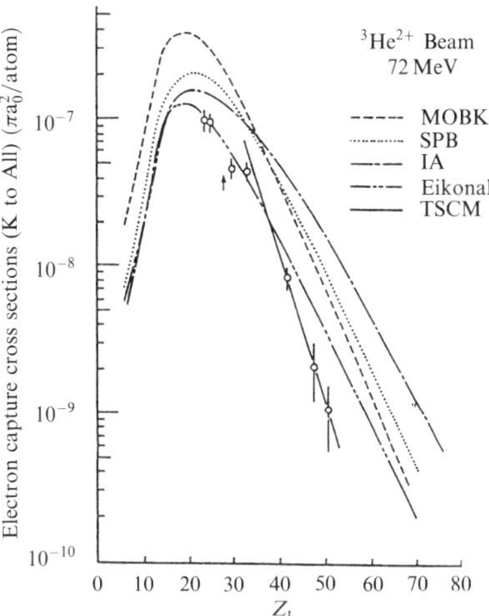

Fig. 9.13   The Z_t dependence of the target K-shell ECC cross sections of the present experimental results together with theoretical calculations for 72 MeV ^{3}He^{2+}. (Taken from Katayama *et al.* [24].)

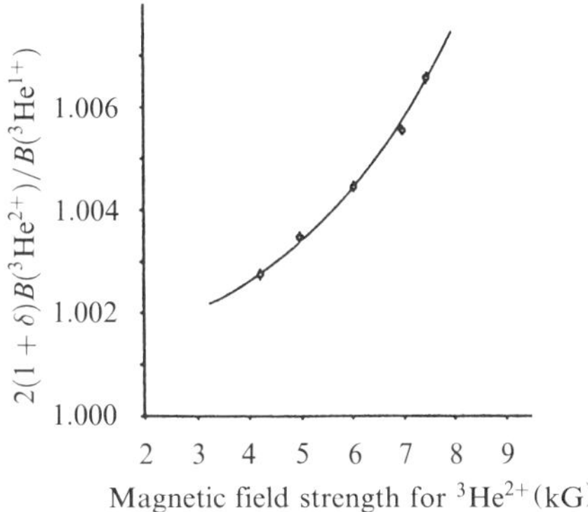

Fig. 9.14   The ratios of the two magnetic fields used for focusing the ^{3}He^{2+} and ^{3}He$^+$ ions at the same position at the centre of the focal line. (Taken from Katayama *et al.* [31].)

experiment. Janecke *et al.* measured the equilibrium product of ^{3}He$^+$ from Al to Au targets using a ring cyclotron beam. They deduced σ_T by using σ_L from the extrapolation of low-energy data [30].

As described in the introduction, the ^{3}He$^+$ electron capture was shown to be useful for calibration of the spectrograph [31]. The result is shown in Fig. 9.14, from which it was found that the effective ion radius in the two dipole magnets in RAIDEN changes by 0.8% when the magnetic field is changed from 4 to 15 kG.

The electron loss of $^3\text{He}^+$, $^{12}\text{C}^{5+}$, and $^{16}\text{O}^{7+}$ at 10.7 MeV/amu was also measured in the course of charge-state-dependent stopping power experiments [32]. The result has been discussed in the context of Bohr [33] and Gillespie [34] theories.

9.6 Charge-state-dependent stopping power of ions in thin carbon foils

Ogawa *et al.* proposed to measure the energy loss of $^3\text{He}^+$ in a carbon foil, thin enough for charge equilibrium not to be attained. This study also utilized a high-resolution property of the RAIDEN. The schematic diagram of the experimental arrangement is shown in Fig. 9.15 for the case of $^3\text{He}^+$ beams. A beam of 9.9 MeV/amu $^3\text{He}^{2+}$ was

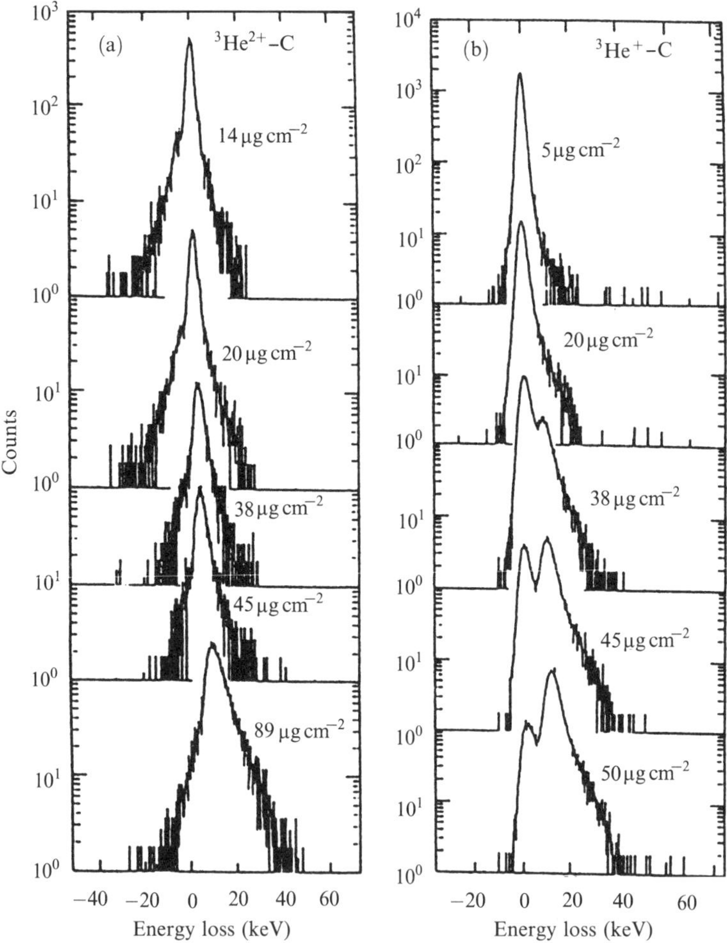

Fig. 9.15 Typical energy loss spectra of 32 MeV $^3\text{He}^{2+}$ (a) and $^3\text{He}^+$ (b) ions after passing through carbon foils. The outgoing ions were in the same charge state as the incident beams. (Taken from Ogawa *et al.* [35].)

extracted from the AVF cyclotron. An Au foil of about $100\,\mu\mathrm{g\,cm^{-2}}$ thickness was used to make the charge exchange at about 3 m upstream of the switching magnet (SWG). A desired charge state was selected by adjusting the magnetic field of the SWG and transported to the target foil in the scattering chamber. A target thickness dependence of the momentum distribution of the emergent ^{3}He was measured using the RAIDEN. Similar measurements were also performed for Li, C, and O ions at the same velocity as ^{3}He.

9.6.1   *A double peak in the ^{3}He$^+$ energy loss spectrum*

The energy loss spectra of ^{3}He, as a function of target thickness, are shown in Fig. 9.15 [35]. The peak position of the line is plotted as a function of the target thickness in Fig. 9.16. For ^{3}He^{2+} ions, the relation was quite linear. On the other hand, for ^{3}He$^+$, though the spectra show a simple structure for thin foils, the spectra for the thicker foils began to show a double-peak structure. The intensity of the higher-energy component is found to give a simple exponential decay, as shown in Fig. 9.17. The lower-energy component, on the other hand, shows a constant strength for the thicker foils. From these observations, the higher-energy component was interpreted as energy loss of the ^{3}He$^+$ without losing the electron during the traverse in the foil. The lower-energy component shows the same energy loss as that for the ^{3}He^{2+}, which allows us to understand that it was a consequence of two charge-exchange collisions, i.e. electron loss near the entrance and electron capture at the exit of a foil [35–37].

The energy loss of the ^{3}He$^+$ ions in carbon was analysed and explained well by Kaneko [38]. The first observation of the double-peak structure shown in Fig. 9.16 was highly appreciated at the ICACS meeting held at Nara in 1991. It is to be noted that the momentum change at electron capture due to a mass increase in ^{3}He$^+$ ions by one

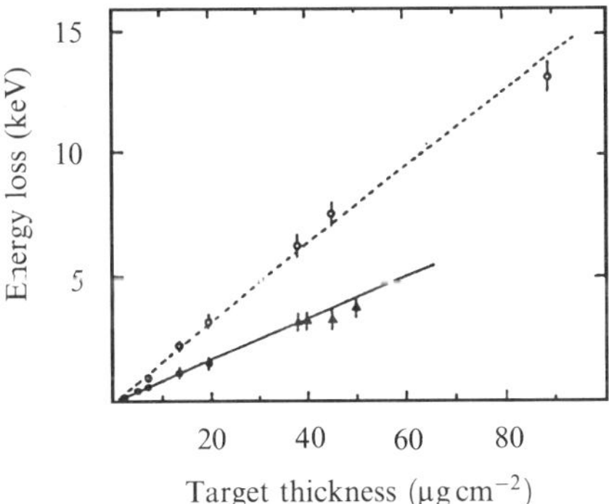

Fig. 9.16   Foil thickness dependence of the energy loss. Open and solid circles are for ^{3}He^{2+} and ^{3}He$^+$, respectively. Solid triangles are for the smaller energy loss component in the two-peak spectrum of ^{3}He$^+$. A dashed line shows the least-squares fitting to data for ^{3}He^{2+}. A solid line shows the least-squares fitting to data from single-peak spectra of ^{3}He$^+$. (Taken from Ogawa *et al.* [35].)

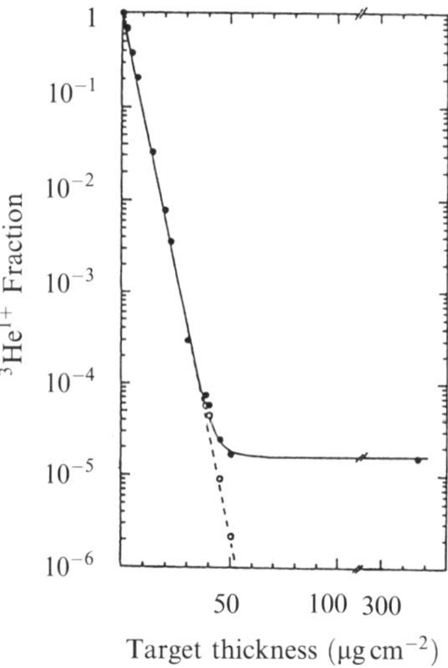

Fig. 9.17 The attenuation of ^{3}He$^+$ beams after passing through carbon foils. The solid circles represent the ratio of outgoing to incident ^{3}He$^+$ ions, and open circles the contributions from the smaller-energy-loss component in the two-peak spectrum. (Taken from Ogawa *et al.* [35].)

electron was also observed for the first time [39]. This momentum change was found to coincide with the amount obtained from the extrapolation, to zero target thickness, of the difference between the two peaks in the ^{3}He$^+$ spectrum.

9.6.2   *Energy loss of* Li, C, *and* O *ions in non-equilibrium charge state*

Similar measurements were performed for Li$^{2+,3+}$ [40], C$^{4+,5+,6+}$ [41], and O$^{5+,6+,7+}$ [42] at an energy of 9.9 MeV/amu for each ion. From the thickness dependence of energy loss for these ions, the stopping power in carbon was obtained for each ion. The result was discussed in view of the screening effect of bound electron and the effective charge of these ions in carbon was deduced [41].

9.6.3   *Other related studies*

The energy losses of neutral particles H and He were similarly measured using a charge state selector and a stripper device placed in the upstream of the RAIDEN scattering chamber. The results have been published in Refs [43,44].

9.7   Stopping power of carbon for H$_2^+$ and H$_3^+$ ions

When molecular ions are incident on a solid, almost all molecular ions dissociate into fragment ions within femtoseconds of entering the solid. Many studies have been performed on the stopping power of carbon for the protons produced by the dissociation of H$_2^+$ ions [45]. The stopping power of the fragment protons was found to be larger than that of a single proton. This anomaly in the stopping power was explained by the

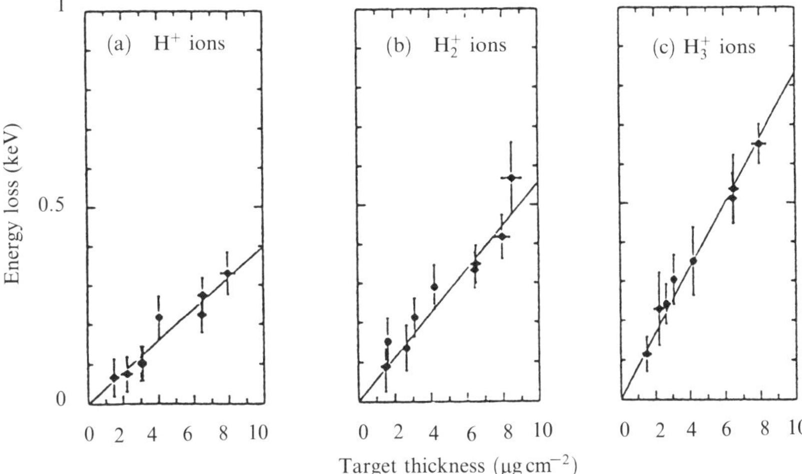

Fig. 9.18 Thickness dependence of the energy loss of ions transmitted through carbon foils: H^+ (a), H_2^+ (b) and H_3^+ (c). The foil of thickness $8.5\,\mu g\,cm^{-2}$ used for H_2^+ ions was made with two half-thickness foils. Solid lines show the least-squares fitting to data. (Taken from Susuki *et al.* [47,48].)

coherent excitation of the target electrons by correlated fragment ions [46]. Before the present study, there had been no reported work on the stopping power of the molecular ions. Susuki *et al.* have measured the stopping power of carbon for 9.6 MeV/amu H_2^+ and H_3^+ ions for the first time using the RAIDEN. The thickness dependence of the mean energy loss of H_2^+ and H_3^+ is shown in Fig. 9.18, together with the result for protons at the same energy [47,48]. The effective charge Z_{eff} of H_2^+ and H_3^+ for the stopping power was obtained as the square root of the energy loss measured relative to the proton energy loss. For H_2^+ and H_3^+ ions the Z_{eff} values in carbon thus deduced are 1.98 ± 0.08 and 1.45 ± 0.09, respectively. The general expression for the stopping power calculated by the first Born approximation is given in Ref. [49] as

$$-\frac{dE}{dx} = \left(\frac{8\pi N e^4}{h^2 v^2}\right) \sum (E_n - E_0) \frac{dq}{q^3} \left|F_{00}^{p}(-q)\right|^2 \left|F_{n0}^{t}(q)\right|^2, \tag{9.5}$$

where N is the density of target atoms, E_n and E_0 are the energies of the excited and ground states of the target atom, q is momentum transfer, $q_{min} = (E_n - E_0)/hv$, $q_{max} = 2mv/h$, m is the electron mass, $F_{00}^{p}(q)$ is the form factor of the projectile, and $F_{n0}^{t}(q)$ is the transition form factor of the target atom. Choosing the wave function of the molecular ions as a linear combination of the ground state wave function of hydrogen atom, $\phi_{1s}(r)$, and neglecting the cross-terms of $\phi_{1s}(r)$'s in the calculation of the form factor of the projectiles, eqn (9.5) reduces to

$$-\frac{\mathrm{d}E}{\mathrm{d}x} = \left(\frac{8\pi N Z e^4}{h^2 v^2}\right) \int_{q_{\min}}^{q_{\max}} \frac{\mathrm{d}q}{q} \left[1 + \frac{(\sin qR)}{qR}\right]\left[1 - \frac{1}{2(1 + a_{\mathrm{B}}^2 q^2/4)^2}\right] \quad (\mathrm{H}_2^+),$$

$$(9.6)$$

$$-\frac{\mathrm{d}E}{\mathrm{d}x} = \left(\frac{12\pi N Z e^4}{h^2 v^2}\right) \int_{q_{\min}}^{q_{\max}} \frac{\mathrm{d}q}{q} \left[1 + \frac{(2\sin qR)}{qR}\right]\left[1 - \frac{2}{3(1 + a_{\mathrm{B}}^2 q^2/4)^2}\right]^2 \quad (\mathrm{H}_3^+),$$

$$(9.7)$$

where R is the internuclear distance for the ground state molecular ions, a_{B} is the Bohr radius, Z and I are the atomic number and the mean excitation energy of target atom, respectively. Susuki *et al.* have calculated the stopping powers taking the most probable values of R to be 1.17 and 0.97 Å and $I = 79\,\mathrm{eV}$ (carbon). Figure 9.19 shows the calculated stopping powers and squares of the effective charge number, Z_{eff}^2, of the ions, in the energy range from 0.5 to 100 MeV/amu. Similar stopping powers have been obtained by Kaneko using a dielectric function method and the wave packet model for H_2^+ and H_3^+ ions [50] and the result agrees well with the present calculated values. The calculated stopping powers are slightly larger than the experimental values. In the H_3^+ experiment, the energy spectrum of the dissociative fragments, H_2^+, was also measured. The spectrum was found to be almost independent of the carbon thickness.

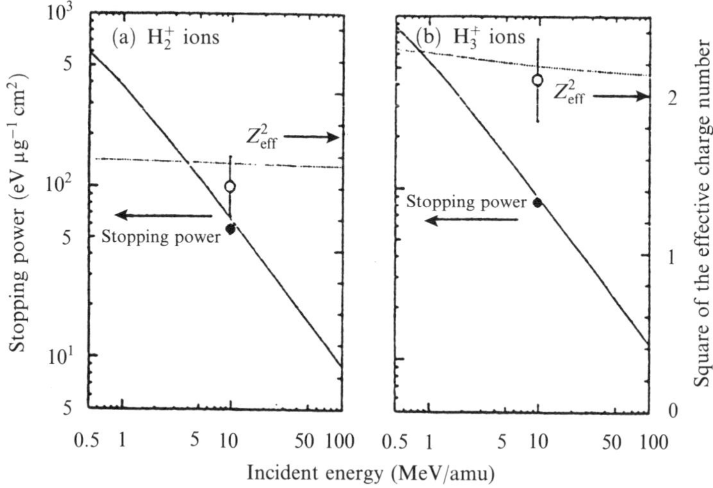

Fig. 9.19 Theoretical stopping powers and Z_{eff}^2 for 0.5–100 MeV/amu H_2^+ (a) and H_3^+ (b) ions in carbon. The experimental values are shown for comparison. (Taken from Susuki *et al.* [47,48].)

9.8 Stopping power of metallic elements for light projectiles

The stopping powers of ten metallic elements ranging from Al to Pb have been measured for 15, 25, 35, 45, 55, 65, and 73 MeV protons and for 13 MeV/amu ^{4}He and ^{12}C using the RAIDEN by Sakamoto *et al.* [51]. They measured ions scattered by a sample target at 4° instead of those transmitted in the beam direction. Sample targets were obtained commercially. To monitor the energy of the incident beam, they measured ions scattered by an Au foil (100 µg cm^{-2}) at 4° before and after each sample measurement (see Fig. 9.20) [52]. The results are as follows: (1) The stopping powers of ten metallic elements for 55, 65, and 73 MeV protons are shown in Fig. 9.21. Comparisons

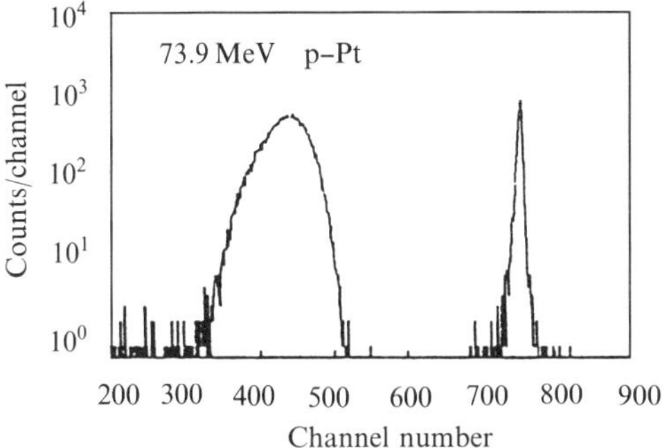

Fig. 9.20 A typical momentum spectrum of protons measured with RAIDEN. (Taken from Sakamoto *et al.* [52].)

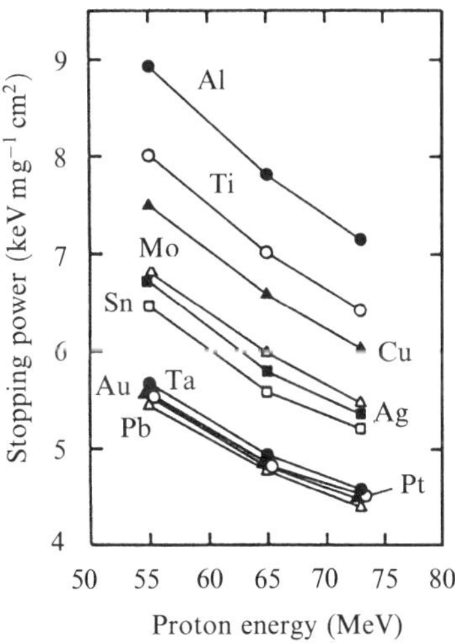

Fig. 9.21 The experimental stopping powers of ten metallic elements are shown for 55, 65, and 73 MeV protons. The solid lines are given only to connect the data points. The experimental uncertainties are smaller than the data points. (Taken from Sakamoto *et al.* [51].)

between experimental and calculated stopping powers were made using 73 MeV protons. The calculated values were taken from compilations of Anderson and Ziegler [53] and others. The experimental values were smaller than the calculated values of Anderson and Ziegler by 1.18%. They deduced the mean excitation energies (I-values) by analysing data using the Bethe–Bloch formula. (2) From the I-values obtained, they discussed the problem of the I-value for metallic Al. In order to clarify the discrepancy between the reported I-values for Al, stopping-power measurements for protons at higher energies are desirable, where correction terms make very little contribution to the I-value. (3) They also discussed the problem of the relation between I-values and shell correction. They found that Bichsel's shell correction could give consistent I-values regardless of proton energy, while Bonderup's shell correction gave I-values dependent on the proton energy. (4) From the difference between stopping powers for ^{4}He and ^{12}C ions, they extracted the magnitude of the Barkas correction, which was found to be consistent with that reported in Ref. [54]. The final result will be reported in the near future [55].

9.9 Direct measurement of the binding energy effect

The excitation function of K X-ray yields vs incident projectile energy at low energy, where particle velocities are smaller than the mean orbital velocity in the target K-shell, is described well by the Perturbed Stationary State (PSS) theory [56]. The PSS considers (1) the deflection of a projectile in the target Coulomb field and (2) the increase in binding energy of electrons due to the formation of a temporary $Z_1 + Z_2$ pseudo-atom, where Z_1 and Z_2 are the atomic numbers of the projectile and the target, respectively. Although the necessity of the binding-energy effect has been established well, there has been no direct test of the effect experimentally. Ishii *et al.* proposed to observe the effect by using the RAIDEN. After some trials, they concluded that a coincidence measurement between the scattered protons and the K X-rays from the residual atom is necessary. They considered the combination of a Cu target and a 10 MeV proton beam. A new target box was prepared in the beam line 50 cm upstream from the scattering chamber of the RAIDEN. Two gas proportional counters were used to detect the Cu K X-rays. Scattered protons in the direction of $0.7°$ were detected by the RAIDEN at $3°$. They observed, after several trials, the inelastic scattering peak due to the K-shell ionization of Cu atoms. The result, however, showed that the energy shift is much larger than that expected from the atomic number of the united atom, i.e. $Z_1 + Z_2 = 30$. The physical interpretation of the result is under discussion [57]. Meanwhile, a precise line shape of the proton elastic line from carbon was disscused in terms of the Vavylov model [58].

9.10 Conclusion

The facility at RCNP has provided a place where unique studies on high-energy atomic physics have been performed. Due to the general policy, however, which has been adopted recently, to terminate the experiments using direct AVF cyclotron beams, there is no new proposal in atomic physics at the RAIDEN any more. The experiments which were already completed have perhaps been published, though the results on the binding-energy effect are yet to be understood.

A ring cyclotron facility, for electron capture using a high-energy $^3\text{He}^{2+}$ beam has been established. Some unique subjects in atomic physics could be studied using a high-resolution ring cyclotron beam and a high-resolution magnetic spectrograph. Especially, the atomic physics associated with pion production, by 400 MeV high-resolution proton beam, would be a challenging field.

Acknowledgements

Comments on the manuscript by Professors M. Mannami, N. Sakamoto, H. Ogawa, and S. Itoh are gratefully acknowledged.

References

1. M. Sakai *et al.*, RCNP Ann. Rep. (1991), p. 71.
2. S. Tanigawa *et al.*, RCNP Ann. Rep. (1988), p. 114.
3. S. Ito, M. Shoji, N. Maeda, R. Katano, T. Mukoyama, R. Ono, and Y. Nakayama, *J. Phys.* **B20** (1987) L597.
4. T. Mukoyama and S. Ito, *Bull. Inst. Chem. Res. Kyoto Univ.* **65** (1987) 163.
5. E. N. Tsyganov, Fermilab TM-684, Batavia (1976), p. 66.
6. M. Mannami, K. Kimura, Y. Susuki, Y. Fujii, N. Sakamoto, H. Ogawa, T. Noro, I. Katayama, and H. Ikegami, *Nucl. Instr. Meth.* **B33** (1988) 62.
7. K. Sueoka *et al.*, RCNP Ann. Rep. (1991), p. 100.
8. M. Hasegawa, K. Sueoka, Y. Susuki, K. Kimura, M. Mannami, H. Ogawa, N. Sakamoto, T. Noro, I. Katayama, and H. Ikegami, *Phys. Lett.* **A144** (1990) 357.
9. M. Mannami, K. Kimura, M. Hasegawa, Y. Susuki, Y. Mizuno, N. Sakamoto, H. Ogawa, I. Katayama, T. Noro, and H. Ikegami, *Nucl. Instr. Meth.* **B56/57** (1991) 180.
10. M. Mannami, K. Kimura, Y. Susuki, M. Hasegawa, Y. Mizuno, M. Tsuji, N. Sakamoto, H. Ogawa, I. Katayama, T. Noro, and H. Ikegami, *Nucl. Instr. Meth.* **B67** (1992) 114.
11. M. Mannami *et al.*, RCNP Ann. Rep. (1988), p. 103.
12. I. Katayama, G. P. A. Berg, S. A. Martin, J. Meissberger, W. Oelert, A. Retz, M. Rogge, J. G. M. Roemer, G. Gaul, H. Hasai, J. Tain, and L. Zemlo, *Z. Phys.* **D3** (1986) 73.
13. P. Koschar, A. Clovias, O. Heil, M. Burkhard, J. Kemmler, and K. O. Groenveld, *Nucl. Instr. Meth.* **B24/25** (1987) 153.
14. K. Dettmann, K. G. Harrison, and M. W. Lucas, *J. Phys.* **B7** (1974) 269.
15. R. Latz, J. Schader, H. J. Frischkorn, P. Kosschar, D. Hofmann, K. O. Groenveld, and W. Mechbach, *Nucl. Instr. Meth.* **B2** (1984) 265.
16. W. Mechbach and P. Focke, *Nucl. Instr. Meth.* **B33** (1988) 255.
17. M. Mannami *et al.*, RCNP Ann. Rep. (1991), p. 80.
18. E. Horsdal-Pedersen, C. L. Cocke, and M. Stockli, *Phys. Rev. Lett.* **46** (1981) 170.
19. I. Katayama, M. Fujiwara, S. Morinobu, T. Noro, H. Ikegami, F. Fukuzawa, and I. Sugai, *Proc. second Asia-Pasific physics conf.* (Bangalore, 1986), S. Chandrasekhar, ed. (World Scientific, 1987), p. 622.

20. Y. Haruyama, I. Katayama, H. Ogawa, T. Noro, H. Ikegami, F. Fukuzawa, K. Yoshida, A. Aoki, and I. Sugai, *Nucl. Instr. Meth.* **B33** (1988) 220.

21. Y. Haruyama, H. Ogawa, I. Katayama, T. Noro, H. Ikegami, F. Fukuzawa, K. Yoshida, M. Tosaki, A. Aoki, and I. Sugai, *Nucl. Instr. Meth.* **B48** (1990) 130.

22. I. Katayama, Y. Haruyama, and H. Ogawa, *Butsuri* **43** (1988) 611 (in Japanese).

23. H. Ogawa, I. Katayama, Y. Haruyama, T. Noro, H. Ikegami, F. Fukuzawa, K. Yoshida, A. Aoki, and I. Sugai, *Nucl. Instr. Meth.* **A262** (1987) 23.

24. I. Katayama, H. Ogawa, Y. Haruyama, H. Ikegami, F. Fukuzawa, K. Yoshida, A. Aoki, and I. Sugai, *Phys. Rev.* **A52** (1996) 23.

25. H. Ogawa, I. Katayama, Y. Haruyama, T. Noro, H. Ikegami, F. Fukuzawa, K. Yoshida, A. Aoki, and I. Sugai, *Proc. third workshop on high-energy ion–atom collision process* (Debrecen, 1988), D. Berenyi and G. Hock, eds (Lecture Note in Physics, Vol. 294, Springer-Verlag, 1988), p. 145.

26. J. Macek and S. Alston, *Phys. Rev.* **A26** (1982) 815.

27. C. D. Lin, S. C. Soong, and L. N. Tunnell, *Phys. Rev.* **A17** (1978) 1646.

28. H. Ogawa *et al.*, RCNP Ann. Rep. (1988), p. 107.

29. I. Katayama, *AIP conference proceedings on "nuclear physics with stored beams"*, Vol. 128 (1985), p. 275.

30. K. Dennis, H. Akimune, G. P. A. Berg, S. Chang, B. Davis, M. Fujiwara, M. N. Harakeh, J. Janecke, J. Liu, K. Pham, D. A. Roberts, and E. J. Stephenson, *Phys. Rev.* **A50** (1994) 3992.

31. I. Katayama, Y. Fujita, M. Fujiwara, S. Morinobu, T. Yamazaki, and H. Ikegami, *Nucl. Instr. Meth.* **171** (1980) 195.

32. H. Ogawa, I. Katayama, I. Sugai, Y. Haruyama, M. Saito, K. Yoshida, M. Tosaki, and H. Ikegami, *Nucl. Instr. Meth.* **B88** (1994) 350.

33. N. Bohr and K. Dan Vidensk, *Selsk. Mat. Fys. Medd.* **18**(8) (1948).

34. G. H. Gillespie, *Phys. Rev.* **A18** (1978) 1967.

35. H. Ogawa, I. Katayama, H. Ikegami, Y. Haruyama, A. Aoki, M. Tosaki, F. Fukuzawa, K. Yoshida, I. Sugai, and T. Kaneko, *Phys. Rev.* **A43** (1991) 11 370.

36. H. Ogawa, I. Katayama, H. Ikegami, Y. Haruyama, M. Tosaki, A. Aoki, F. Fukuzawa, K. Yoshida, and I. Sugai, *Rad. Effects and Defects in Solids* **117** (1991) 213.

37. H. Ogawa, I. Katayama, I. Sugai, Y. Haruyama, A. Aoki, M. Tosaki, F. Fukuzawa, K. Yoshida, and H. Ikegami, *Nucl. Instr. Meth.* **B69** (1992) 108.

38. T. Kaneko, *Phys. Rev.* **A43** (1991) 4780.

39. H. Ogawa, I. Katayama, H. Ikegami, Y. Haruyama, A. Aoki, M. Tosaki, F. Fukuzawa, K. Yoshida, and I. Sugai, *Phys. Lett.* **A160** (1991) 77.

40. H. Ogawa, I. Katayama, I. Sugai, Y. Haruyama, M. Saito, K. Yoshida, M. Tosaki, Y. Susuki, M. Fritz, K. Kimura, and M. Mannami, *Nucl. Instr. Meth.* **B115** (1996) 66.

41. H. Ogawa, I. Katayama, I. Sugai, Y. Haruyama, M. Tosaki, A. Aoki, K. Yoshida, and H. Ikegami, *Phys. Lett.* **A167** (1992) 487.

42. H. Ogawa, I. Katayama, I. Sugai, Y. Haruyama, M. Saito, K. Yoshida, M. Tosaki, and H. Ikegami, *Nucl. Instr. Meth.* **B82** (1993) 80.

43. H. Ogawa, N. Sakamoto, I. Katayam, Y. Haruyama, M. Saito, K. Yoshida, M. Tosaki, Y. Susuki, and K. Kimura, *Phys. Rev.* **A54** (1996) 5027.
44. H. Ogawa, N. Sakamoto, I. Katayam, Y. Haruyama, M. Saito, K. Yoshida, M. Tosaki, Y. Susuki, and K. Kimura, *Nucl. Instr. Meth.* **B132** (1997) 36.
45. D. S. Gemmel and Z. Vager, *Treaties on heavy-ions science*, A. Bromley, ed. (Plenum Press, New York, 1985).
46. W. Brandt *et al.*, *Phys. Rev. Lett.* **33** (1974) 1325.
47. Y. Susuki, M. Fritz, K. Kimura, M. Mannami, N. Sakamoto, H. Ogawa, I. Katayama, T. Noro, and H. Ikegami, *Phys. Rev.* **A50** (1994) 3533.
48. Y. Susuki, M. Fritz, K. Kimura, M. Mannami, N. Sakamoto, H. Ogawa, I. Katayama, T. Noro, and H. Ikegami, *Phys. Rev.* **A51** (1995) 3868.
49. Y. K. Kim and K. T. Cheng, *Phys. Rev.* **A22** (1980) 61.
50. T. Kaneko, *Phys. Rev.* **A51** (1995) 535.
51. N. Sakamoto, H. Ogawa, M. Mannami, K. Kimura, Y. Susuki, M. Hasegawa, I. Katayama, T. Noro, and H. Ikegami, *Rad. Effects and Defects in Solids* **117** (1991) 193.
52. N. Sakamoto *et al.*, RCNP Ann. Rep. (1988), p. 117.
53. H. H. Anderson and J. F. Ziegler, *Hydrogen: stopping powers and ranges in all elements* (Pergamon, New York, 1977).
54. N. Sakamoto, N. Shiomi-Tsuda, H. Ogawa, and R. Ishiwari, *Nucl. Instr. Meth.* **B33** (1988) 158.
55. N. Sakamoto *et al.* (to be published).
56. G. Basbas, W. Brandt, and R. Laubert, *Phys. Rev.* **A5** (1978) 1655.
57. K. Ishi, I. Katayama, A. Ando, Y. Haruyama, H. Ogawa, and H. Orihara (to be published).
58. P. V. Vavylov, *Sov. Phys. JETP* **5** (1957) 749.
59. B. N. Jensen *et al.*, *Nucl. Instr. Meth.* **B71** (1992) 155.

10

Nuclear physics with strange flavour

Tadafumi Kishimoto and Toshio Motoba

10.1 Introduction

It is well known that quarks are real constituents of hadrons. The interactions among quarks are described by Quantum ChromoDynamics (QCD), where the interaction between three colour charges is mediated by gluons. Nuclei are made up of nucleons held by the strong interaction, though it has been quite rare to require explicit quark degrees of freedom to understand many aspects of their low-energy phenomena. Since hadrons are colour singlets, the interactions among nucleons are the residual interactions of quarks, which is responsible for the success of the meson theory.

It is desirable to describe nuclei in terms of quarks by which a consistent picture of the low- and high-energy phenomena can be given. The strong interaction among quarks is independent of their flavour, which leads to the flavour symmetry of the strong interaction among hadrons. The first recognized flavour symmetry is the isospin symmetry. It holds well in nucleon–nucleon (NN) interactions, and the small observed breaking is attributed to the tiny mass difference between the proton and the neutron (up and down quarks). The next flavour we can introduce relatively easily is strangeness. Strange hyperons are a few tens of a percent heavier than nucleons, although one can still use $\mathrm{SU_F}(3)$ symmetry to describe the hyperon–nucleon (YN) interaction where breaking of the symmetry can again be attributed to the mass differences of nucleons and hyperons.

The strong interaction between baryons can be studied by scattering experiments. There have been a large number of NN scattering experiments through which the NN interaction has been studied in detail [1]. However, few data on YN scattering and almost no data on hyperon–hyperon (YY) scattering are available. Thus, the study of the strong interaction among octet baryons has been carried out with the help of a theoretical model that is based on the meson theory and the $\mathrm{SU_F}(3)$ symmetry by which a vast amount of NN data were related to the scarce YN data [2].

Since the strong interactions between the octet baryons are similar, this makes possible the existence of a nucleus where a nucleon is replaced by a hyperon; this is called a hypernucleus. Knowledge of the strong YN interaction from YN scattering is so limited that the hypernuclear structure has been a source of complementary information. So far, these limited data have shown that the YN interaction is described well by the meson exchange model where $\mathrm{SU_F}(3)$ symmetry with explicit breaking due to physical mass difference of hadrons is assumed. These pictures are based on the flavour independence of the strong interaction among quarks, although the quarks are buried in hadrons.

The weak interaction can play a special role since it changes the flavour of quarks; thus, one can see explicit quark degrees of freedom even for the low-energy phenomena. The weak baryon–baryon interaction has not been so well explored as the strong interaction. Experimental study of the weak NN interaction can only be made through parity violation because the parity-conserving part is completely masked by the strong interaction.

Parity violation causes the cross section to depend on the helicity of the particles participating in the reaction. However, since the parity-conserving strong interaction is overwhelming, one has to struggle with a very small effect (10^{-7}–10^{-8}) in an NN scattering experiment [3]. Recently, large parity-violating effects have been observed in low-energy neutron scattering from specific nuclear levels [4]. In this experiment, one can obtain even a $\sim 10\%$ effect with fairly good accuracy. However, since the enhancement is due to a complex mixture of the adjacent levels with opposite parity, the connection of such data to the weak NN interaction is indirect.

In a hypernucleus, one can observe weak hyperon decay in a nucleus. The so-called non-mesonic decay is due to the weak YN interaction. Recently, polarized hypernuclei were produced and asymmetric emission of decay protons from the weak non-mesonic decay has been measured. The large asymmetry is the first demonstration of parity violation in the weak YN interaction. The asymmetric decay of the polarized hypernuclei introduced a new method for the study of the weak decay of hyperons in a nucleus. The weak non-mesonic decay has a large momentum transfer, which means that the short-range effect has to be considered to understand the process. This is the region where the quark degree of freedom has to be considered in addition to the meson exchange. By extending this study, we wish to pursue the unified picture of the weak baryon–baryon interaction.

In the following, we will discuss production of polarized hypernuclei (Section 10.2), basic aspects of hypernuclear weak decays (Section 10.3), weak non-mesonic decay and weak baryon–baryon interactions (Section 10.4), weak decays of polarized hypernuclei (Section 10.5), and strangeness nuclear physics at RCNP and SPring-8 (Section 10.6).

10.2 Production of polarized hypernuclei

In order to produce a hypernucleus, one has to use reactions to introduce strangeness to a nucleus, e.g. the (K^-, π^-) reaction has been used since the early stages of the study. The elementary (K^-, π^-) reaction has a large cross section since a strange quark in K^- is easily transferred to a neutron which results in a Λ. The small momentum transfer of the reaction makes the cross section of the so-called substitutional state quite large. On the contrary, the K^- beam intensity is quite limited. The (π^+, K^+) reaction can also produce the hypernucleus. Although, in order for the reaction to be an alternative way to produce the hypernucleus, intense pion beams are available, the characteristics of the reaction must be studied first.

The feasibility of the (π^+, K^+) reaction has been investigated by Dover *et al.* [5], who predicted that the natural-parity 'stretched' states $[(0\ell_N j_N)^{-1}(0\ell_\Lambda j_\Lambda)]_{J=\ell_N+\ell_\Lambda}$ are most strongly populated at forward angles. The optimal matching of the form factor $F_J(q)$ occurs kinematically at $J = (bq)^2/2 \simeq 1.6A^{1/3}$, where the momentum transfer

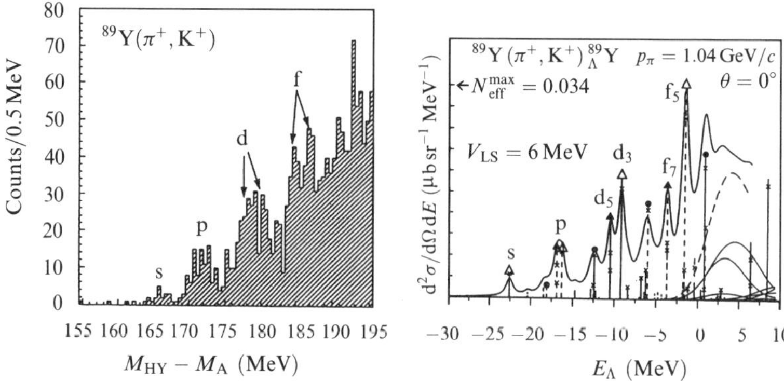

Fig. 10.1 The experimental spectrum (left) for the ^{89}Y $(\pi^+, K^+)\,^{89}_{\Lambda}$Y reaction at $p_\pi \simeq$ 1.06 GeV/c compared with the theoretical prediction for $p_\pi = 1.04$ GeV/c (right). The corresponding Λ single-particle orbits are indicated by s, p, etc.

$q \simeq 350$ MeV/$c = 1.78$ fm^{-1} at $p_\pi = 1.05$ GeV/c, and the oscillator size parameter $b \simeq A^{1/6}$ are relevant. As a direct extension, it was shown by Bandō and Motoba [6] that the (π^+, K^+) reaction has promising characteristics for probing a series of hyperon states, particularly in medium and heavy hypernuclei. In fact, the (π^+, K^+) experiments performed at Brookhaven [7,8] and KEK [9,10] proved its usefulness by disclosing a beautiful series of Λ single-particle states in heavy hypernuclei. Figure 10.1 compares a typical experimental spectrum of ^{89}Y $(\pi^+, K^+)\,^{89}_{\Lambda}$Y [10] with the theoretical strength function [10], where the $[g_{9/2}^{-1}\ell_\Lambda]$ series with $\ell_\Lambda = 0s, 0p, 0d, 0f$ are clearly observed. This comparison demonstrates that the Distorted-Wave Impulse Approximation (DWIA) framework works quite well. The DWIA theory also works satisfactorily in describing the (K^-, π^-) process [11].

In this chapter, however, we confine ourselves to the study of polarized hypernuclear production on the basis of DWIA, since the polarization provides a novel chance to investigate the weak interactions of Λ hyperons. In the following sections, the ability to produce polarization is discussed for the (π^+, K^+), (K^-, π^-), and (γ, K^+) reactions.

10.2.1 *Hypernuclear Λ-spin polarizations in the (π^+, K^+) and (K^-, π^-) reactions*

The hypernuclear polarization was studied theoretically for the first time by Bandō [12] and Ejiri *et al.* [13]; however, their actual calculations were limited to the absorption-based polarization without any spin-flip interaction. After a reanalysis [14] of the π^-p $\rightarrow$ K$^0\Lambda$ elementary data, the spin-non-flip and spin-flip amplitudes became available in a convenient form. In this subsection, polarized hypernuclear productions in the (π^+, K^+) [15,16] and (K^-, π^-) [17] reactions at the incident momenta larger than 1 GeV/c are discussed.

Denoting (π^+, K^+) and (K^-, π^-) by (a, b), the differential cross section for the hypernuclear production reaction, $a + A(J_i) \rightarrow H(J_f) + b$, is expressed, in the laboratory frame, by

$$\left.\frac{d\sigma}{d\Omega}\right|_{\theta_b}^{\text{lab}} = \frac{(2\pi)^4 p_b^2 E_b E_a E_H}{p_a\{p_b(E_H + E_b) - p_a E_b \cos\theta_b\}} \overline{|T_{if}^{\text{lab}}|^2}, \tag{10.1}$$

where E (p) are laboratory energies (momenta) and θ_b is the laboratory angle of outgoing b particle, all evaluated in the frame appropriate to the A-body nuclear target. In the impulse approximation, the T-matrix is expressed in terms of the elementary t-matrix for a reaction on a single nucleon. The squared T-matrix element is given in the 'frozen-nucleon' approximation by

$$\overline{|T_{if}^{\text{lab}}|^2} = \sum_{M_f} R(J_i J_f; M_f), \tag{10.2}$$

$$R(J_i J_f; M_f) = \frac{1}{2J_i + 1} \sum_{M_i} \left|\langle J_f M_f T_f \tau_f | \hat{O}_{ab}(\theta) | J_i M_i T_i \tau_i \rangle\right|^2. \tag{10.3}$$

The transition operator $\hat{O}_{ab}(\theta)$ here is defined with the interaction consisting of both spin-non-flip (f) and spin-flip (g) components:

$$\hat{O}_{ab}(\theta) = \int d^3 r \, \chi_{p_b}^{(-)}(r)^* \chi_{p_a}^{(+)}(r) \sum_{\nu=1}^{A} \hat{X}(\nu) \, \delta\left(r - \frac{M_c}{M_A} r_\nu\right) \lambda[f + ig(\sigma_\nu \cdot n)]. \tag{10.4}$$

The factor λ accounts for the laboratory–c.m. transformation. The unit vector $n = (\hat{p}_a \times \hat{p}_b)/|\hat{p}_a \times \hat{p}_b|$ is perpendicular to the reaction plane. The operator $\hat{X}(\nu)$ converts the νth nucleon into a hyperon.

The polarization of the hypernuclear state $|J_f\rangle$ is given by

$$\mathcal{P}_H(J_f) = \sum_{M_f} \frac{M_f}{J_f} \mu_{fi}(M_f), \quad \mu_{fi}(M_f) = \frac{R(J_i J_f; M_f)}{\sum_{M_f} R(J_i J_f; M_f)}, \tag{10.5}$$

where $\mu_{fi}(M_f)$ is the magnetic subspace population defined by the matrix element of eqn (10.3). Note that a frame with $\{z \parallel n, \, y \parallel q, \, x = y \times z\}$ $(q = p_a - p_b)$ is used throughout and, therefore, M is quantized along the direction n. We can express the quantity $R(J_i J_f; M_f)$ of eqn (10.3) in terms of the 'reduced effective numbers' $\rho(J_i J_f; M_f)$ as

$$R(J_i J_f; M_f) = \lambda^2 \left\{|f|^2 \rho^{ff}(J_i J_f; M_f) + |g|^2 \rho^{gg}(J_i J_f; M_f)\right.$$
$$\left. + 2\,\text{Im}[fg^*]\rho^{fg}(J_i J_f; M_f)\right\}. \tag{10.6}$$

Detailed expressions for $\rho(J_i J_f; M_f)$ are given by Bandō *et al.* [15]. Correspondingly, $\overline{|T_{if}^{\text{lab}}|^2}$ of eqn (10.2) consists of three contributions,

$$N^{ff}(J_i J_f) = \sum_{M_f} \rho^{ff}(J_i J_f; M_f), \quad N^{gg}(J_i J_f) \text{ and } N^{fg}(J_i J_f). \tag{10.7}$$

The N's here are generalization of the 'effective nucleon number' which has been used often. There are two different origins of polarization: One is the spin polarization ($\mathcal{P}^s$) due to the f–g interference and the other is the orbital polarization ($\mathcal{P}^\ell$) due to the absorption (of particles a, b). The latter arises from a difference between the absorptions of near-side and far-side passing particles and/or between the absorptions of up-stream (outgoing) and down-stream (incoming) particles. The far–near mechanism works for the momentum transfer q in the longitudinal direction, while the up–down mechanism does so for the transverse component of q. The far–near mechanism dominates in the large-momentum transfer reactions such as (π^+, K^+), while the up–down mechanism dominates for the low-momentum transfer reactions such as (K^-, π^-). Both origins are incorporated in the above formalism. It is noted that the cross section is related to the $|f|^2$ and $|g|^2$ terms, while the polarization originates from the f–g interference term.

The elementary process $\pi^- p \rightarrow K^0 \Lambda$ (and, therefore, $\pi^+ n \rightarrow K^+ \Lambda$ also) is experimentally known to give a large positive polarization [18,14] for Λ hyperon at $p_\pi \sim 1\,\text{GeV}/c$, which is in the optimal momentum range for Λ-hypernuclear production. Figure 10.2 (left) shows the dependence of the elementary polarization $\mathcal{P}^{\text{elem}}$ and

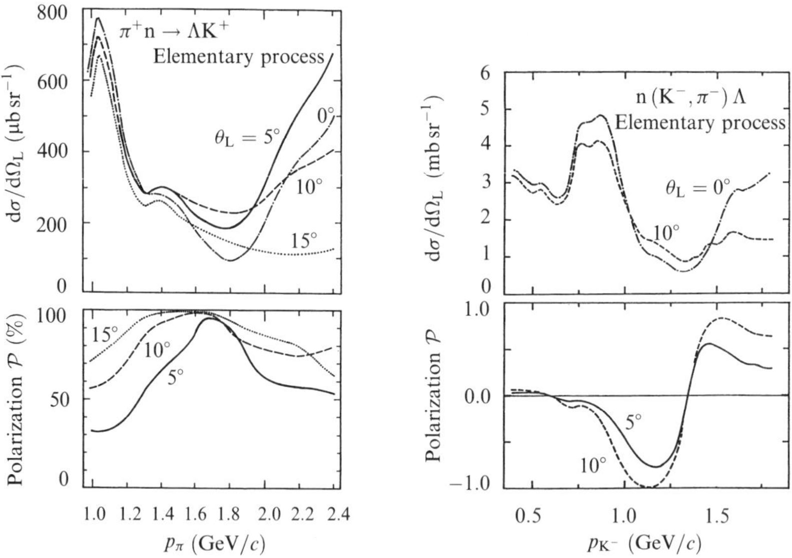

Fig. 10.2   Left: The elementary cross sections and polarizations for the $\pi^+ n \rightarrow \Lambda K^+$ process as a function of projectile momentum for various laboratory scattering angles. Right: Those for the $K^- n \rightarrow \Lambda \pi^-$ process.

cross section on the incident pion momentum p_π over a wide range. The polarization easily exceeds 50% at 5–10° depending on p_π.

The (K^-, π^-) reaction in the low-momentum region, which has been used as a selective method to produce substitutional hypernuclear states, yields essentially no polarization. However, the same (K^-, π^-) reaction in the higher-momentum region with K^- above $1\,\text{GeV}/c$ has a possibility of yielding large polarizations, like in the elementary $K^- n \rightarrow \Lambda\pi^-$ process, as is clearly suggested by the amplitudes due to Gopal *et al.* [19]. Figure 10.2 (right) shows the laboratory differential cross sections and polarizations of the $K^- n \rightarrow \Lambda\pi^-$ elementary process as a function of the K^- momentum p_{K^-} [17], which are evaluated by using the amplitudes obtained by Gopal *et al.* [19]. It is remarkable that the polarization reaches almost 100% around $p_K = 1.1$ and $1.5\,\text{GeV}/c$ (with opposite sign), where the cross sections are still reasonably large (1–$2\,\text{mb}\,\text{sr}^{-1}$). It is notable that the recoil momentum q_Λ at $\theta_L = 0°$ remains small (90–$120\,\text{MeV}/c$) at $p_K = 1.1$–$1.5\,\text{GeV}/c$, which is about one-third of $q_\Lambda \sim 350\,\text{MeV}/c$ in the (π^+, K^+) reaction at $p_\pi = 1.04\,\text{GeV}/c$. Thus, the (K^-, π^-) reaction at $p_K = 1.5\,\text{GeV}/c$ may populate low-lying Λ-shell states with small ℓ_Λ through $\Delta\ell \simeq \ell_n - \ell_\Lambda$, in contrast to the substitutional states through $\Delta\ell = 0$ in the (K^-, π^-) reaction at $p_K \sim 0.6\,\text{GeV}/c$, and to the high-spin states through $\Delta\ell = \ell_n + \ell_\Lambda$ in the (π^+, K^+) reaction.

In calculating the cross sections and polarizations of hypernuclei, the π^+ and K^+ distorted waves have been obtained by employing the eikonal approximation together with the absorptive potentials given in terms of the empirical averaged $\pi^+ N$ and $K^+ N$ reaction cross sections. (For example, $\bar{\sigma}_{\pi^+ N} = 41\,\text{mb}$ and $\bar{\sigma}_{K^+ N} = 14\,\text{mb}$ at $p_\pi = 1.04\,\text{GeV}/c$.) The Λ and nucleon single-particle wave functions employed here are given by the density-dependent Hartree–Fock solutions.

The configuration-mixed shell model calculations have been carried out to obtain target and hypernuclear wave functions, in which the Λ kinetic energy and the YNG(ΛN) and NN interactions are employed. In order to describe the radial behaviour of weakly bound Λ states reasonably well and also to simulate approximately the unbound Λ states in the low-energy continuum region, we include the Harmonic Oscillator (HO) higher-nodal Λ-orbits of 1ℓ ($\ell = $ s, p, d, f) as well as the 0ℓ orbits in the model space. The correct treatment of the continuum final states is discussed by Motoba *et al.* [20].

Table 10.1 summarizes the typical results of polarizations and production cross sections for $^{12}_\Lambda$C. One sees that some states such as 1^-_1 and 2^+_2 have both a sizable polarization and a large cross section. In the first (π^+, K^+) experiment at KEK [21–23], the optimum angle has been chosen to be $\theta_K \simeq 14°$, where $P_H^2(d\sigma/d\Omega)$ was predicted to be the maximum. It is remarkable that the (K^-, π^-) reaction at $p_K = 1.1\,\text{GeV}/c$ gives rise to larger polarizations even at smaller angles when compared with the (π^+, K^+) reaction. This suggests a new possibility of using the (K^-, π^-) reaction at larger p_K. In fact, $p_K \simeq 1.5\,\text{GeV}/c$ also yields large polarization [17].

The calculated ^{12}C(π^+, K^+) strength function is displayed in Fig. 10.3 (left) for a scattering angle $\theta_K = 5°$. This pattern itself is not strongly dependent on θ_K and it is in good agreement with several experiments done at Brookhaven [8,24] and KEK [9].

Table 10.1 Hypernuclear polarizations $P_{\rm H}(J_f)$ and differential cross sections $d\sigma/d\Omega$ in $(\mu{\rm b}\,{\rm sr}^{-1})$ calculated for typical $^{12}_{\Lambda}{\rm C}$ states

Energy (MeV)	$(\pi^+, {\rm K}^+)$ $p_\pi = 1.04\,{\rm GeV}/c$				$({\rm K}^-, \pi^-)$ $p_{\rm K} = 1.1\,{\rm GeV}/c$		
$E_\Lambda^{\rm cal}$ $(J^\pi; E_{\rm x}^{\rm cal})$	$\theta_{\rm K} = 5°$	$10°$	$15°$	$[d\sigma/d\Omega(10°)]$	$\theta_\pi = 5°$	$15°$	$[d\sigma/d\Omega(10°)]$
-10.76^* $(1_1^-; 0.0)$	-0.12	-0.25	-0.38	(7.73)	0.42	0.74	(48.1)
-9.01 $(1_2^-; 1.8)$	0.38	0.65	0.81	(1.84)	-0.76	-0.99	(17.5)
-5.86 $(1_3^-; 4.9)$	-0.14	-0.29	-0.43	(1.00)	0.47	0.80	(7.0)
-0.76 $(2_1^+; 10.0)$	0.04	0.07	0.05	(5.52)	-0.13	-0.09	(30.1)
-0.16 $(2_2^+; 10.6)$	-0.24	-0.45	-0.63	(4.58)	0.46	0.77	(37.7)
1.02 $(2_3^+; 11.8)$	0.27	0.48	0.60	(2.03)	-0.56	-0.76	(15.6)

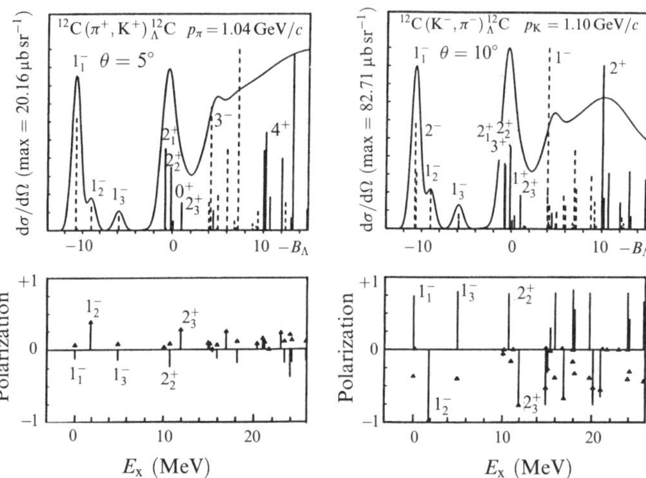

Fig. 10.3 Left: The calculated excitation functions (top) and hypernuclear polarizations $\mathcal{P}_{\rm H}$ (bottom, solid line) and Λ-polarizations $\mathcal{P}_\Lambda$ (bottom, triangle) for the $^{12}{\rm C}\,(\pi^+, {\rm K}^+)\,^{12}_{\Lambda}{\rm C}$ reaction at $p_\pi = 1.04\,{\rm GeV}/c$. Right: Those for the $^{12}{\rm C}\,({\rm K}^-, \pi^-)\,^{12}_{\Lambda}{\rm C}$ reaction at $p_{\rm K} = 1.10\,{\rm GeV}/c$. The configuration-mixed wave functions [26] are used.

The cross sections for the ground state peak are calculated to be

$$d\sigma/d\Omega(1_1^- + 1_2^-; \theta_{\rm L})^{\rm cal} = 15.4\ (5°),\ 10.3\ (10°),\ 4.9\ (15°) \quad \text{in } \mu{\rm b}\,{\rm sr}^{-1},$$

for which the experimental value is $10.36 \pm 0.61\,\mu{\rm b}\,{\rm sr}^{-1}$ $(10°)$ [8]. For another pronounced peak around $B_\Lambda = 0\,{\rm MeV}$, we list the theoretical cross sections summed over the neighbouring $2_1^+, 3_1^+, 2_2^+$, and 0_1^+ states. The experimental value is $17.0\,\mu{\rm b}\,{\rm sr}^{-1}$ $(5.6°)$ [24], while the calculated values are $d\sigma/d\Omega(B_\Lambda \simeq 0;\ \theta_{\rm L})^{\rm cal} = 17.6\ (5°),\ 12.8\ (10°),$ $5.5\ (15°)$ in $\mu{\rm b}\,{\rm sr}^{-1}$. The agreement is remarkably good.

It is interesting to make a detailed comparison of the new experimental spectrum by Hasegawa *et al.* [9] with two small peaks predicted theoretically [25] in the valley (Fig. 10.3). Their structures are attributed to the $[^{11}{\rm C}((1/2)^-; 2.0\,{\rm MeV}) \otimes s_{1/2}^\Lambda]_{1_2^-}$ and

$[^{11}\mathrm{C}((3/2)^-; 4.8\,\mathrm{MeV}) \otimes s^{\Lambda}_{1/2}]_{1^-_3}$ states, respectively. The 1^-_2 state seems to be confirmed in the experiment as the shoulder of the ground state peak. On the other hand, the 1^-_3 peak is calculated at $E_{\mathrm{x}} \simeq 5\,\mathrm{MeV}$ which seems considerably lower in excitation energy than the other small peak obtained experimentally [9] below the pronounced 2^+ peak. Figure 10.3 (right) shows the predicted $^{12}\mathrm{C}(\mathrm{K}^-, \pi^-)$ strength function and hypernuclear polarization.

10.2.2 *Effects of depolarization processes before the weak decay*

The hypernuclear ground state polarization $P_{\mathrm{H}}(J_{\mathrm{G}})$ produced directly in the reaction (production stage: e.g. $P_{\mathrm{H}}(1^-_1)$ in Table 10.2) should be modified by the γ-cascades or particle emission from the excited states. Thus, the final polarization $\mathcal{P}_{\mathrm{H}}(J_{\mathrm{G}})$ just before the weak decay can be expressed as

$$\mathcal{P}_{\mathrm{H}}(J_{\mathrm{G}}) = \frac{1}{\mathcal{S}}\left[P_{\mathrm{H}}(J_{\mathrm{G}})\frac{\mathrm{d}\sigma(J_{\mathrm{G}})}{\mathrm{d}\Omega} + \sum_{J'} P_{\mathrm{H}}(J')\frac{\mathrm{d}\sigma(J')}{\mathrm{d}\Omega}C(J' \to J_{\mathrm{G}}) \right], \qquad (10.8)$$

where $C(J' \to J)$ denotes the attenuation coefficient for the γ-ray or particle emission [27] and J' represents symbolically all the intermediate states, and $\mathcal{S}$ is the sum of the relevant cross sections involved.

In order to evaluate the modified polarization by eqn (10.8), we have to calculate the γ-transition rates (branching ratios) between bound states. As an example, we show in Fig. 10.4 the decay scheme of $^{12}_{\Lambda}\mathrm{C}$ calculated with the shell model wave functions [25]. The lowest particle emission threshold is the $^{11}_{\Lambda}\mathrm{B} + \mathrm{p}$ channel at $E_{\mathrm{x}} = 9.21\,\mathrm{MeV}$. Therefore, the large 2^+ peak leads to the $^{11}_{\Lambda}\mathrm{B}$ states. In Fig. 10.4 one sees that the branching ratios within the bound states are determined by the M1 transitions, as the M1 rates are calculated to be two or three orders of magnitude larger than the E2 transition rates. For example, we note that

$$T(\mathrm{M1};\ 1^-_2 \to 1^-_1) = 1.40 \times 10^{13}\,\mathrm{s}^{-1}, \qquad T(\mathrm{E2};\ 1^-_2 \to 1^-_1) = 1.80 \times 10^{11}\,\mathrm{s}^{-1}.$$

The M1 transitions occur mostly between the $^{11}\mathrm{C}$ excited states. The calculated γ-energies and the branching ratios, together with the (π^+, K^+) cross sections, are consistent with the recent experimental observation of γ-rays from $^{12}_{\Lambda}\mathrm{C}$ as reported from KEK [28].

In $^{13}_{\Lambda}\mathrm{C}$, three (π^+, K^+) peaks are obtained below the lowest particle emission threshold of $^{12}\mathrm{C}+\Lambda$ at $E_{\mathrm{x}} = 11.69\,\mathrm{MeV}$. They all have large polarizations and are, respectively, attributed to the following structures:

$$\left[^{12}\mathrm{C}(0^+_{\mathrm{G}}) \otimes s^{\Lambda}_{1/2}\right](1/2)^+_{\mathrm{G}}, \quad \left[^{12}\mathrm{C}(2^+_1) \otimes s^{\Lambda}_{1/2}\right](3/2)^+, \quad \left[^{12}\mathrm{C}(0^+_{\mathrm{G}}) \otimes p^{\Lambda}_{3/2}\right](3/2)^-.$$

It is interesting to have the possibility of observing the strong $\mathrm{E1}((3/2)^- \to (1/2)^+_{\mathrm{G}})$ γ-ray transition with the theoretical rate $T(\mathrm{E1}) = 1.01 \times 10^{17}\,\mathrm{s}^{-1}$, whose energy provides the precise splitting between the $p^{\Lambda}_{3/2}$ and $s^{\Lambda}_{1/2}$ orbits. It is notable, however, that the $(1/2)^-$ state is only very weakly excited in the (π^+, K^+) reaction due to the $L = 0$

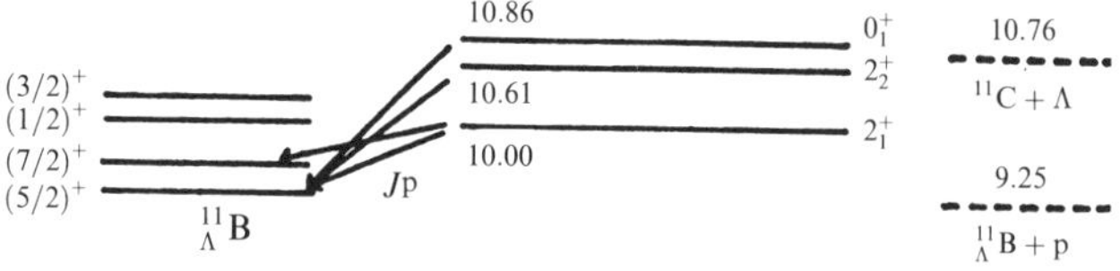

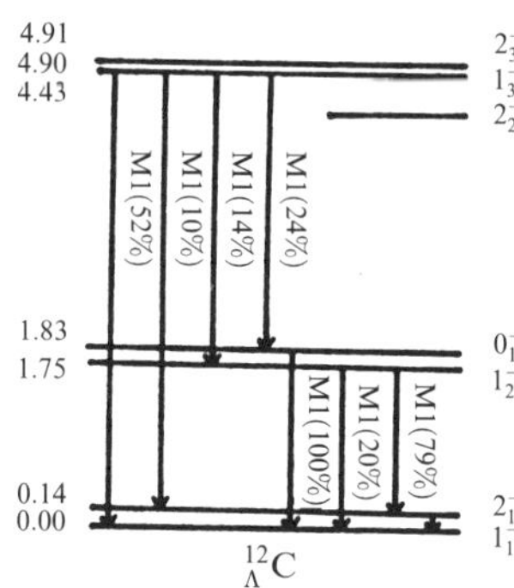

Fig. 10.4 The calculated decay scheme of the excited states in $^{12}_\Lambda$C.

transition, so that we can hardly extract the position of the $\mathrm{p}^\Lambda_{1/2}$ orbit. In the (K^-, π^-) reaction, on the other hand, this $(1/2)^-$ state should be preferentially excited.

In general, the Λ-particle itself has a polarization different from P_H, and the Λ-spin polarization $P_\Lambda(J_f)$ in a hypernuclear state is given as the weighted sum of the expectation values of the hyperon spin with respect to the many-body wave function $|J_f M_f\rangle$:

$$P_\Lambda(J_f) = \sum_{M_f} \langle J_f M_f | \vec{\sigma}_\Lambda \cdot \hat{n} | J_f M_f \rangle \mu_{fi}(M_f). \tag{10.9}$$

If we use $|J_\mathrm{G} M_\mathrm{G}\rangle$ and eqn (10.8), we obtain the modified ground state Λ-spin polarization $\mathcal{P}_\Lambda$ under the effect of depolarization processes, and it is this $\mathcal{P}_\Lambda$ that should be relevant in the weak decay stage:

$$\mathcal{P}_\Lambda(J_\mathrm{G}) = \sqrt{J_\mathrm{G}/[(J_\mathrm{G} + 1)(2J_\mathrm{G} + 1)]}\langle J_\mathrm{G} \| \vec{\sigma}_\Lambda \| J_\mathrm{G} \rangle \mathcal{P}_\mathrm{H}(J_\mathrm{G}). \tag{10.10}$$

Table 10.2 compares $\mathcal{P}_\mathrm{H}(J_\mathrm{G})$ with the original $P_\mathrm{H}(J_\mathrm{G})$, and one sees the depolarization effect due to γ-cascade is considerable. The reduction is about 20% in $^{12}_\Lambda$C and 40% in $^{13}_\Lambda$C. The (K^-, π^-) reaction at $p_\mathrm{K} = 1.1\,\mathrm{GeV}/c$ clearly provides larger polarization than the (π^+, K^+) reaction. The calculation suggests that the ^{13}C target should be a suitable candidate for future experiments involving large hypernuclear polarization.

Table 10.2 Hypernuclear polarization $P_H(J_G)$, the modified one and the Λ-spin polarization

| | $(\pi^+, K^+)\, p_\pi = 1.04\,\text{GeV}/c$ | | | | $(K^-, \pi^-)\, p_\pi = 1.1\,\text{GeV}/c$ | | | |
| | ${}^{12}_{\Lambda}\text{C}(1^-_1)$ | | ${}^{13}_{\Lambda}\text{C}((1/2)^+_1)$ | | ${}^{12}_{\Lambda}\text{C}(1^-_1)$ | | ${}^{13}_{\Lambda}\text{C}((1/2)^+_1)$ | |
	$\theta_K = 10°$	$15°$	$\theta_K = 10°$	$15°$	$\theta_K = 10°$	$15°$	$\theta_K = 10°$	$15°$
P_H	-0.25	-0.38	0.61	0.77	0.74	0.96	-0.99	-0.90
$\mathcal{P}_H$	-0.21	-0.32	0.38	0.47	0.61	0.74	-0.57	-0.53
$\mathcal{P}_\Lambda$	0.11	0.16	0.38	0.47	-0.31	-0.37	-0.57	-0.53

Experiments to study the production and weak decay of polarized ${}^{12}_{\Lambda}\text{C}$ [21–23] and ${}^{5}_{\Lambda}\text{He}$ [29] by the (π^+, K^+) reaction are described in Section 10.5.

10.2.3 *The (γ, K^+) reaction and its characteristics*

Hypernuclear production by (γ, K^+) or $(e, e'K^+)$ is an interesting alternative to the 'standard' reactions described in the previous subsections, since both incoming and outgoing waves are only weakly distorted. This simplifies the description of the influence of the nuclear medium on the process, but the relatively weak interaction also means a smaller cross section for hypernuclear production. In fact, the photoproduction cross section of $\sim 100\,\text{nb\,sr}^{-1}$ is three orders of magnitudes smaller than that of (K^-, π^-), and the electroproduction cross section typically a further two orders less in efficiency in the most pronounced states. Nevertheless, the small cross sections should be compensated by the availability of high intensities at CEBAF and SPring-8. A few extensive reviews of electroproductions have been published recently [30–32], so only a brief summary of the subject is given here.

The understanding of the structure of the elementary amplitudes and the role of kaon–hyperon–nucleon coupling constants and its extension to the process in the nuclear target is far from complete. In addition to the early work [33], several attempts have been made for a description based on the diagrammatic techniques [34–37]. Here, following Schorsch *et al.* [38], we adopt a phenomenological approach for the elementary amplitudes.

The elementary amplitude for $p\,(\gamma, K^+)\,\Lambda$ in the 2-body Centre-of-Mass (2-CM) system is written as

$$M \equiv \frac{(2\pi)^2}{\sqrt{s}} \sqrt{e_\gamma e_p e_\Lambda e_K}\,\langle q, -q|t|\kappa, -\kappa\rangle$$
$$= \mathrm{i}F_1(\sigma \cdot \hat{\epsilon}) + F_2(\sigma \cdot \hat{q})[\sigma \cdot (\hat{\kappa} \times \hat{\epsilon})] + \mathrm{i}F_3(\sigma \cdot \hat{\kappa})(\hat{q} \cdot \hat{\epsilon}) + \mathrm{i}F_4(\sigma \cdot \hat{q})(\hat{q} \cdot \hat{\epsilon}).$$

$$(10.11)$$

Here, σ is the baryon Pauli matrix, e's are 2-CM energies, and $\hat{\epsilon}$ and $\hat{\kappa}$ $(\hat{q})$ are unit vectors for the photon polarization, 2-CM in- (out-) momentum, respectively. By transforming the t-matrix from the 2-CM system to the two-body laboratory system, the laboratory cross section becomes

$$\left.\frac{d\sigma}{d\Omega}\right|_{\text{lab}}^{\text{two-body}} = \frac{(2\pi)^4 p_K^2 E_K E_\gamma E_\Lambda}{k_\gamma\{p_K E_\Lambda + E_K(p_K - k_\gamma \cos\theta_L)\}} \left|\langle k_\gamma - p_K, p_K |t| k_\gamma, 0\rangle\right|^2, \quad (10.12)$$

where E's are the laboratory energies. For the reaction on a nuclear target, the many-body T-matrix is calculated using the t-matrix and the K^+ distorted wave within the impulse approximation. Then one obtains many-body laboratory cross section in analogy with eqn (10.12), where all the relevant quantities should be calculated from the many-body kinematics.

Using the Schorsch parameterization for the elementary amplitudes [38], the cross sections and the polarizations for $p(\gamma, K^+)\Lambda$ are shown in Fig. 10.5; note that the optimal incident energies are around $E_\gamma = 1.1$–$1.3\,\text{GeV}$ [39]. There, the cross sections reach $350\,\text{nb}\,\text{sr}^{-1}$ in collinear geometry with appreciable polarization ability at $\theta_K > 10°$.

In analogy with description of other hypernuclear productions, the above elementary amplitudes can be used for the reactions on nuclear targets. To illustrate this, the calculated spectrum for $^{12}\text{C}\,(\gamma, K^+)\,^{12}_\Lambda\text{B}$ is displayed in Fig. 10.6 for $p_\gamma = 1.2\,\text{GeV}/c$ [40]. It is noteworthy that, among others, the non-normal parity states such as 3^+ and 2^- are produced most strongly with a cross section of 70–$90\,\text{nb}\,\text{sr}^{-1}$, while the 1^- ground state does not reach $30\,\text{nb}\,\text{sr}^{-1}$. This characteristic is attributed to the dominance of the spin-flip terms in the transition interaction and also to the large-momentum transfer $q \simeq 350\,\text{MeV}/c$.

The main advantage of the present process as seen from Figs 10.6 and 10.7 is that unnatural parity states are strongly excited. This characteristic, which is different from

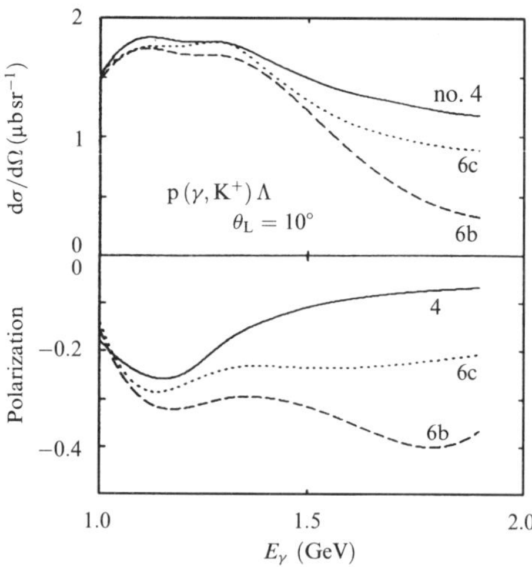

Fig. 10.5   Cross section and polarization of the elementary $\gamma p \to \Lambda K^+$ process as a function of laboratory photon momentum at $\theta_L = 10°$. The curves denoted by 4, 6b and 6c correspond to the amplitudes given by Schorch *et al.* [38].

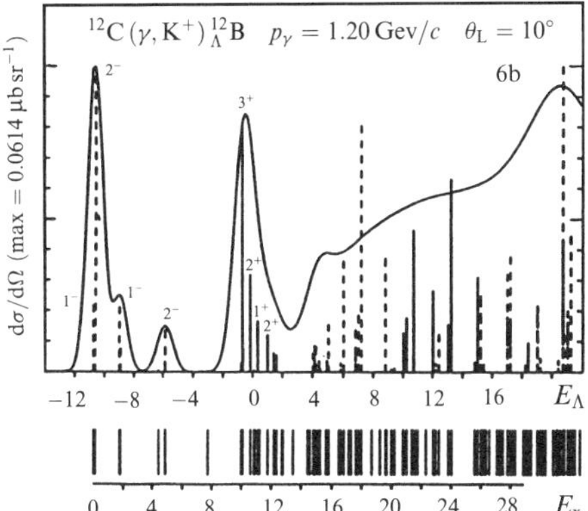

Fig. 10.6 Calculated (γ, K^+) excitation function for the ${}^{12}C\,(\gamma, K^+)\,{}^{12}_{\Lambda}B$ reaction at $E_\gamma = 1.2\,\text{GeV}$ at $\theta_K = 10°$.

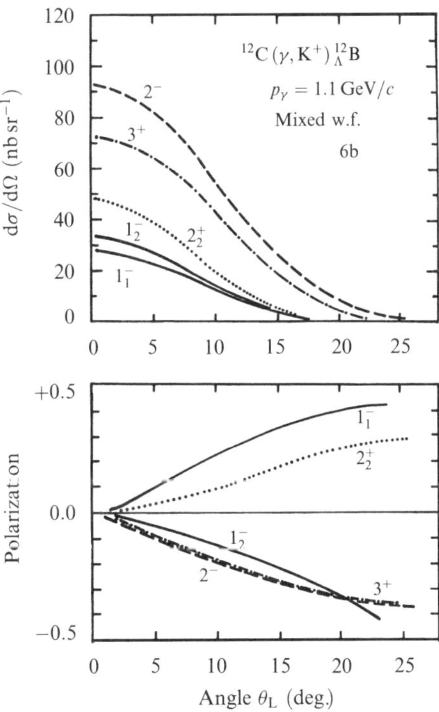

Fig. 10.7 Differential cross sections (top) and hypernuclear polarization $P_H(J_f)$ for the ${}^{12}C\,(\gamma, K^+)\,{}^{12}_{\Lambda}B$ reaction plotted as a function of the kaon laboratory scattering angle.

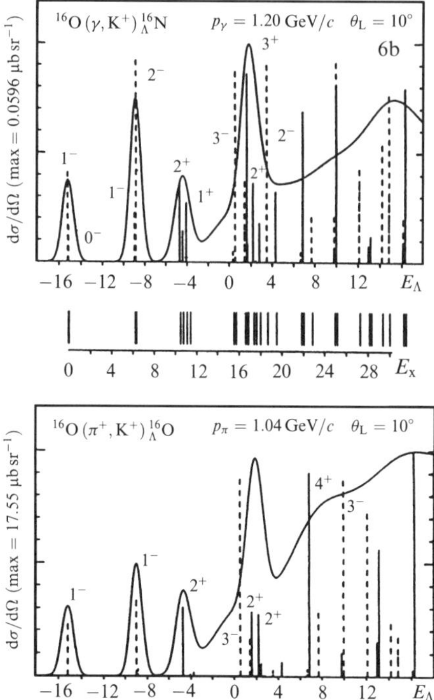

Fig. 10.8 Comparison of theoretical (γ, K^+) and (π^+, K^+) excitation functions for the $^{16}\mathrm{O}$ target. The solid (dashed) line indicates the positive (negative) parity state. The excitation energy is given in units of MeV with respect to the Λ-threshold.

those of (K^-, π^-) and (π^+, K^+) reactions, makes the photo- and electroproductions useful complementary tools for hypernuclear spectroscopic studies. Both electro- and photoproductions can lead to polarization of the resulting hypernucleus through the interference of spin-flip and spin-non-flip amplitudes [41]. It can also be noted that the strongly excited 3^+ state has large polarization $P(3^+) = -0.35$ to -0.50 at $\theta_{\gamma\mathrm{K}} = 10$–$20°$.

We also include the predictions for the $^{16}\mathrm{O}\,(\gamma, \mathrm{K})\,^{16}_{\Lambda}\mathrm{N}$ reaction, as this hypernucleus is well known to have a simple structure: Because of the large separation of the nucleon-hole states, $\epsilon(\mathrm{p}_{3/2}^{-1}) - \epsilon(\mathrm{p}_{1/2}^{-1}) = 6.3\,\mathrm{MeV}$, the hypernuclear spectrum is clearly divided into four groups, corresponding to the $[\mathrm{p}_{1/2}^{-1}, \mathrm{s}_\Lambda]$, $[\mathrm{p}_{3/2}^{-1}, \mathrm{s}_\Lambda]$, $[\mathrm{p}_{1/2}^{-1}, \mathrm{p}_\Lambda]$, and $[\mathrm{p}_{3/2}^{-1}, \mathrm{p}_\Lambda]$ dominant configurations. In fact, the configuration mixing is practically negligible except for $J = 1^+$ states. In the shell model diagonalization, we include the $0\mathrm{p}^\Lambda$ and $1\mathrm{p}^\Lambda$ oscillator basis states so as to meet the extended p-state radial behaviour expected from the shallow Λ binding energy $B_\Lambda \simeq 2.5\,\mathrm{MeV}$.

In Fig. 10.8 the calculated excitation function of $^{16}\mathrm{O}\,(\gamma, \mathrm{K}^+)\,^{16}_{\Lambda}\mathrm{N}$ ($E_\gamma = 1.2\,\mathrm{GeV}$ and $\theta_\mathrm{L} = 10°$) shows four pronounced peaks for $J = 1_1^-$ (g.s.), $J = 2^-$ ($E_\mathrm{x} =$

6.4 MeV), $J = 2_1^+$ ($E_x = 10.3$ MeV), and $J = 3^+$ ($E_x = 16.6$ MeV) states [40]. Compared with the (π^+, K^+) excitation function in Fig. 10.8 (bottom) [42], we see that the patterns are quite similar to each other. It should be remarked, however, that the second and fourth peaks in the top figure are attributed to the unnatural parity states $[p_{3/2}^{-1}, s_{1/2}^\Lambda]J = 2^-$ and $[p_{3/2}^{-1}, p_{3/2}^\Lambda]J = 3^+$, respectively. Both states are excited only through the spin-flip process, which is dominant in the (γ, K^+) reaction. In addition to the pronounced four peaks in both excitation functions, the non-negligible transition strengths are also distributed over the high excitation region $E_x > 20$ MeV. This is due to the sizable momentum transfer ($q \simeq 350$ MeV$/c$) involved similarly in both reactions. In drawing a smooth excitation function, however, we use a large smearing width (6 MeV Gaussian) for such high-lying states so as to simulate large escape widths, resulting in the structureless continuum background in that energy region.

The calculated values of the hypernuclear polarizations and differential cross sections are summarized in Table 10.3, where the results for the (γ, K^+), (π^+, K^+), and (K^-, π^-) reactions are compared. The three kinds of the Schorsch amplitudes (no. 4, 6b, and 6c) do not cause much difference within the region of $0° \le \theta_L < 20°$. Therefore, in this section, we show the results with amplitude no. 6b. The (γ, K^+) cross sections for the major peaks are predicted to be several tens of nb sr^{-1} at $\theta_L = 10°$ and, therefore, they amount to about 1/100 of the (π^+, K^+) cross sections (about 1/1000 of the (K^-, π^-) ones). The amount of hypernuclear polarization is about half as large as the

Table 10.3 Hypernuclear polarizations and differential cross sections (in square brackets) calculated in DWIA for the (γ, K^+), (π^+, K^+), and (K^-, π^-) reactions on the ^{16}O target. Each incident momentum is indicated in GeV/c. The lowest two lines show the polarizations after modification due to the effect of electromagnetic cascades and/or particle emission

Level energy E_x (MeV)			Polarization $\mathcal{P}_H(J_i)$ at $\theta_L = 10°$		
Exp.	Exp.	Cal.	(γ, K^+)	(π^+, K^+)	(K^-, π^-)
KEK[a]	CERN[b]	$(J^\pi; E_x)^{\mathrm{cal}}$	1.20 GeV/c (nb sr^{-1})	1.04 GeV/c (μb sr^{-1})	1.10 GeV/c (μb sr^{-1})
$(-13.0)_{\mathrm{g.s.}}$	$(-13.0)_{\mathrm{g.s.}}$	$(1_1^-; \mathrm{g.s.})$	-0.20 [23.6]	0.59 [3.81]	-0.99 [38.2]
6.3	6.0	$(1_2^-; 6.23)$	0.26 [13.0]	-0.28 [5.63]	0.79 [47.7]
		$(2_1^-; 6.36)$	0.20 [52.7]	0.02 [5.63]	0.01 [26.5]
10.6		$(2_1^+; 10.30)$	-0.21 [20.8]	0.46 [5.13]	-0.75 [47.1]
	10.5	$(0_1^+; 11.24)$	0.0 [0.2]	0.0 [0.34]	0.0 [0.5]
16.6		$(2_2^+; 16.58)$	0.02 [1.8]	0.02 [4.80]	0.04 [31.5]
		$(3_1^+; 16.63)$	-0.20 [49.2]	0.01 [0.59]	0.00 [34.9]
		$(2_3^+; 17.20)$	0.12 [20.8]	-0.44 [4.60]	0.77 [45.6]
	17.1	$(0_2^+; 17.42)$	0.0 [0.50]	0.0 [1.0]	0.0 [1.5]
	$\tilde{\mathcal{P}}_H(^{16}_\Lambda\mathrm{N};\ 1_{\mathrm{g.s.}}^-)$		-0.18	0.32	-0.39
	$\tilde{\mathcal{P}}_H(^{15}_\Lambda\mathrm{C};\ (3/2)_{\mathrm{g.s.}}^+)$		-0.20	0.46	-0.61

[a] E_x^{exp} with respect to $-B_\Lambda(\mathrm{g.s.})$, taken from the KEK (π^+, K^+) data (E336)
[b] Taken from the CERN (K^-, π^-) data ($p_K = 0.79$ GeV/c)

(π^+, K^+) reaction for the normal parity states, while non-normal parity states such as 2^- and 3^+ acquire sizable polarization only in the (γ, K^+) reaction. Therefore, the high-resolution experimental observation of such peaks will provide the energy positions of such multiplet members. In general, as discussed elsewhere [25], the (K^-, π^-) reaction at $p_K = 1.10$ and $1.50\,\text{MeV}/c$ turns out to be the best tool to produce polarized hypernuclear states. In Table 10.3 we also list the modified polarization of the ground state after taking into account the depolarization effect arising from electromagnetic cascades from upper states and/or particle emissions.

10.3 Basic aspects of hypernuclear weak decays

The π-mesonic decay is the unique decay mode of the free Λ-hyperon and it has two channels:

$$\Lambda \to p + \pi^- + 37.8\,\text{MeV} \quad (64.2\%), \tag{10.13}$$

$$\Lambda \to n + \pi^0 + 41.1\,\text{MeV} \quad (35.8\%). \tag{10.14}$$

The experimental decay branching ratio ($\Gamma_{\pi^-}^{\text{free}} : \Gamma_{\pi^0}^{\text{free}} \simeq 2 : 1$) obeys the $\Delta I = 1/2$ rule, and the total decay rate ($\Gamma_\Lambda = \Gamma_{\pi^-}^{\text{free}} + \Gamma_{\pi^0}^{\text{free}}$) determines the lifetime: ($\tau_\Lambda = \hbar/\Gamma_\Lambda = 2.63 \times 10^{-10}\,\text{s}$). This decay mode is characterized by a small pion (nucleon) momentum $q_0 \simeq 100\,\text{MeV}/c$, which is much smaller than the nuclear Fermi momentum $k_F \simeq 280\,\text{MeV}/c$. In the hypernucleus, therefore, the π-mesonic decay mode generally tends to be suppressed by the Pauli exclusion principle acting on the created nucleon. In spite of the suppression in nuclear medium, the π-mesonic decay of hypernuclei remains important and useful because of its unique features.

In the nuclear medium, a new decay process (the non-mesonic decay mode) of Λ can take place, on the basis of the two-body weak interaction with nucleons:

$$\begin{aligned} \Lambda + n &\to n + n + 176\,\text{MeV}, \\ \Lambda + p &\to n + p + 176\,\text{MeV}. \end{aligned} \tag{10.15}$$

The $176\,\text{MeV}$ energy release corresponds to much larger momentum $q_{\text{nm}} \simeq 400\,\text{MeV}/c$ than k_F. Therefore, the neutron-stimulated (Γ_n) and proton-stimulated (Γ_p) non-mesonic decay rates do not undergo Pauli suppression so strongly and, hence, this decay rate dominates over the π-mesonic decay rate except in very light hypernuclei. In spite of the large experimental error bars, this fact is known from the observed ratios of non-mesonic decay rate ($\Gamma_{\text{nm}} = \Gamma_n + \Gamma_p$) to π^- decay rate, $Q^- \equiv \Gamma_{\text{nm}}/\Gamma_{\pi^-}$,

$$\left(\frac{\Gamma_{\text{nm}}}{\Gamma_{\pi^-}}\right)^{\text{exp}} \simeq 1 \text{ (for } A \simeq 5\text{)}, \quad 5\text{–}50 \ (A \simeq 10), \quad 100\text{–}200 \ (40 < A < 100). \tag{10.16}$$

The total weak decay rate of a hypernucleus thus consists of four contributions:

$$\Gamma_{\text{tot}}(^A_\Lambda Z) = \Gamma_{\pi^-} + \Gamma_{\pi^0} + \Gamma_p + \Gamma_n. \tag{10.17}$$

Figure 10.9 summarizes the observed hypernuclear lifetimes $\tau(^A_\Lambda Z) = \hbar/\Gamma_{\text{tot}}(^A_\Lambda Z)$. These data indicate nearly constant total weak decay rates across a wide range of A [43].

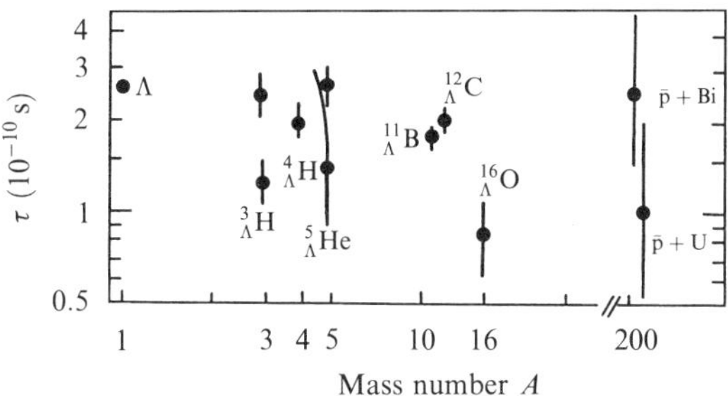

Fig. 10.9 The observed lifetimes of Λ-hypernuclei.

However, the branching ratios, e.g. Γ_{nm} vs Γ_{π}, vary widely so that the π-mesonic decay and non-mesonic decay modes roughly anti-correlate to produce a uniform lifetime. It is thus quite interesting to know how $\Gamma_{\pi}(^{A}_{\Lambda}Z)$ is suppressed, and how $\Gamma_{nm}(^{A}_{\Lambda}Z)$ behaves as a function of A.

10.3.1 *Calculational framework of the mesonic decay rate*

As the π-decay of Λ takes place in the deep interior of the nucleus with only small momentum transfer ($q_0 \simeq 100\,\mathrm{MeV}/c$), it can sensitively probe the many-body structure of the nucleus [44], and also reflects the behaviour of the pion in the nuclear interior [45,46]. In addition, the selective character of the π-decay interaction makes the π-mesonic decay a good spectroscopic tool [44,47–49].

The π-mesonic decay interaction for Λ hyperon is expressed [47] as

$$\mathcal{H}_{\pi} = \mathrm{i}g_{\mathrm{w}}\bar{\psi}_{\mathrm{N}}(1 + \lambda\gamma_5)\tau\Psi_{\Lambda}\phi_{\pi}, \tag{10.18}$$

where $\Psi_{\Lambda} = \begin{pmatrix} 0 \\ \psi_{\Lambda} \end{pmatrix}$ is the spurion with ψ_{Λ} being the Λ field, which is introduced to guarantee the $\Delta I = 1/2$ rule. By using the data on lifetime and asymmetry of the pion from the polarized Λ, the coupling constants are empirically known to be $g_{\mathrm{w}} = -0.23 \times 10^{-6}$ and $\lambda = -6.9$. The non-relativistic Hamiltonian for the π^{-}- and π^{0}-decays of Λ in nuclear medium is given by

$$H_{\pi} = \left[s_{\pi}X^{(s)}(r) + \mathrm{i}p_{\pi}X^{(p)}(r)\frac{(\sigma \cdot \nabla)}{q_0} \right] \chi_{\pi}^{(-)*}(q;r), \tag{10.19}$$

$$s_{\pi^{-}} = -\frac{\sqrt{2}}{\sqrt{4\pi}}g_{\mathrm{w}} = 0.96 \times 10^{-7}, \qquad s_{\pi^{0}} = \frac{-s_{\pi^{-}}}{\sqrt{2}}, \tag{10.20}$$

$$p_{\pi^{-}} = \frac{q_0}{2\sqrt{M_{\Lambda}M_{\mathrm{N}}}}\lambda s_{\pi^{-}} = -0.35 \times 10^{-7}, \qquad p_{\pi^{0}} = \frac{-p_{\pi^{-}}}{\sqrt{2}}, \tag{10.21}$$

where σ is the spin operator for the baryon, $\chi_{\pi}^{(-)*}(q;r)$ is the distorted wave of the pion with momentum q, which tends to a plane wave $\mathrm{e}^{\mathrm{i}q \cdot r}$ if there is no distortion, and

$X^{(s,p)}(r)$ is the vertex renormalization which should be taken in a manner consistent with the distorted wave χ_π adopted, so that, in principle, the result does not depend on the off-shell extension of the πN amplitude. This renormalization is inspired by the Begg–Agassi–Gal theorem [50] and developed further [51,52]. Note that the spin-non-flip parity-violating interaction (s_π) is much stronger than the spin-flip parity-conserving interaction (p_π); $p_\pi^2/s_\pi^2 \simeq 1/9$, which is due to the smallness of q_0 in the π-decay. The situation is different in non-mesonic decay where the relevant momentum is four times larger.

To generate the pion distorted wave χ_π in eqn (10.19), the standard pion optical potential ('full form') is employed, which reproduces the pion–nucleus elastic scattering data,

$$2\omega U_\pi(r)^{\text{full}} = -4\pi[b(r) + B(r)] + 4\pi \nabla \cdot L(r)[c(r) + C(r)]\nabla$$
$$- 4\pi\left[\frac{p_1 - 1}{2}\nabla^2 c(r) + \frac{p_2 - 1}{2}\nabla^2 C(r)\right], \tag{10.22}$$

where $L(r)$ is the Lorentz–Lorentz–Ericson–Ericson (LLEE) term [53], and other quantities depending on the nuclear density $\rho(r)$ are given by Carr *et al.* [54] (Set-E version).

When the above Kisslinger form for the optical potential is adopted, the consistent renormalization factors $X^{(s,p)}(r)$ in eqn (10.19) should be

$$X^{(s)}(r) = 1 \quad \text{and} \quad X^{(p)}(r) = L(r) = \frac{1}{1 + (4\pi/3)\xi[c(r) + C(r)]}. \tag{10.23}$$

The combined use of eqns (10.22) and (10.23) essentially removes the dependence on the particular πN off-shell extension used in the optical potential construction [53] and, hence, stabilizes the theoretical predictions of the π-mesonic decay rates.

The total π-mesonic decay rate of a hypernucleus $(^A_\Lambda Z)$ is obtained by summing the partial decay rates over all final nuclear states as

$$\Gamma_\pi\left(^A_\Lambda Z; \, J_i T_i \tau_i\right) = \sum_f \Gamma_\pi(J_i T_i \tau_i \rightarrow J_f T_f \tau_f), \quad \pi = \pi^0, \, \pi^-. \tag{10.24}$$

The partial decay rate for the transition from a hypernuclear state $|^A_\Lambda Z; \, J_i T_i \tau_i\rangle$ to a nuclear state $|^A Z'; \, J_f T_f \tau_f\rangle$ is expressed as [44]

$$\Gamma_\pi\left(^A_\Lambda Z; \, J_i T_i \tau_i \rightarrow J_f T_f \tau_f\right)$$
$$= \frac{2q}{1 + (\omega/M_A)}\left[s_\pi^2 \, S_\pi^{(s)}(i, f; q) + p_\pi^2\left(\frac{q}{q_0}\right)^2 S_\pi^{(p)}(i, f; q)\right], \tag{10.25}$$

where the 'suppression factors' $S_\pi^{(s)}$ and $S_\pi^{(p)}$, which reflect the hypernuclear and nuclear structures and the pion wave distortion, are given by

$$S_\pi^{(s,p)}(i, f; q) = \frac{4\pi}{2J_i + 1}\sum_K\left|\sqrt{A}\left\langle^A Z'; \, J_f T_f \tau_f\left|F_K^{(s,p)}\theta_\pi\right|^A_\Lambda Z; \, J_i T_i \tau_i\right\rangle\right|^2. \tag{10.26}$$

The transition operators $F_K^{(s)}$ and $F_K^{(p)}$ come from the multipole expansion of the s_π and p_π parts of H_π of eqn (10.18), respectively, with $\tilde{j}_K(q; r)$ being the kth partial wave of $\chi_\pi^{(-)*}$:

$$F_K^{(s)} = X^{(s)}(r)\, \tilde{j}_K(q; r)\, Y_K(\hat{r}), \tag{10.27}$$

$$F_K^{(p)} = -i \sum_{k=K\pm 1} (10K0|k0)\, X^{(p)}(r) f_k^K(q; r)\, [\sigma \times Y_k(\hat{r})]_K \tag{10.28}$$

with

$$f_k^K(q; r) = \frac{\partial \tilde{j}_K(q; r)}{q\, \partial r} + \left\{ 1 - \frac{k(k+1) - K(K+1)}{2} \right\} \frac{\tilde{j}_K(q; r)}{qr}. \tag{10.29}$$

In the no-distortion limit, $\tilde{j}_k(q; r)$ and $f_k^K(q; r)$ tend, respectively, to $j_k(qr)$ and $\pm j_k(qr)$ for $K = k \pm 1$. The operator θ_π transforms a Λ hyperon into a nucleon. In the above expressions, the pion momentum q and energy $\omega_q = \sqrt{m_\pi^2 + q^2}$ are determined by the relation

$$M_\Lambda + E\left(^A_\Lambda Z;\, J_i T_i\right) = M_{\mathrm{N}} + E\left(^A Z';\, J_f T_f\right) + \frac{q^2}{2M_A} + \omega_q. \tag{10.30}$$

For the decay of free Λ, eqn (10.25) is reduced to

$$\Gamma_\Lambda = \Gamma_{\pi^0}^{\mathrm{free}} + \Gamma_{\pi^-}^{\mathrm{free}}, \quad \text{where } \Gamma_\pi^{\mathrm{free}} = \frac{2q_0}{1 + (\omega_{q_0}/M_{\mathrm{N}})}(s_\pi^2 + p_\pi^2). \tag{10.31}$$

If the polarized hypernucleus is produced and then decays, the emitted pion exhibits an asymmetric angular distribution which is attributed to the interference between the s_π- and p_π-terms of the decay interaction, eqns (10.19) and (10.21). For a particular initial state $|J_i M_i T_i \tau_i\rangle$ with M_i fixed, the pion angular distribution $\Gamma_\pi(J_i M_i T_i \tau_i \to J_f T_f \tau_f;\, \boldsymbol{q})$ is given by

$$\Gamma_\pi(J_i M_i T_i \tau_i \to J_f T_f \tau_f;\, \boldsymbol{q})$$

$$= \sum_Q \Gamma_\pi^Q(J_i M_i T_i \tau_i \to J_f T_f \tau_f;\, q) P_Q(\hat{\boldsymbol{q}})$$

$$= \Gamma_\pi^{Q=0}(J_i M_i T_i \tau_i \to J_f T_f \tau_f;\, q)\left\{1 + \alpha_1 P_1(\hat{\boldsymbol{q}}) + \alpha_2 P_2(\hat{\boldsymbol{q}}) + \cdots\right\}, \tag{10.32}$$

with $\alpha_Q = \Gamma_\pi^Q / \Gamma_\pi^{Q=0}$. Here, non-zero odd-$Q$ terms lead to asymmetric pion angular distributions. The asymmetry parameter α_1 for the free $\Lambda\pi^-$ decay is [55]

$$\alpha_1^\pi(\text{free}) = -2p_\pi/s_\pi \simeq -0.642 \pm 0.013. \tag{10.33}$$

10.3.2 *Mesonic decays of* p-*shell hypernuclei*

The detailed calculations of π-decay rates for all p-shell and sd-shell hypernuclei reveal an interesting shell structure dependence of $\Gamma_\pi(^A_\Lambda Z)$. Here, we confine ourselves to the p-shell nuclear species for which wave functions are obtained by diagonalizing the

Cohen–Kurath type Hamiltonian with an appropriate ΛN effective interaction, while the sd-shell ones are approximated by single shell model configurations.

Figure 10.10 shows the calculated results for the p-shell hypernuclei. One can see quite drastic variations depending on the species [44]. For $^{8}_{\Lambda}$Be and $^{12}_{\Lambda}$C, the π^{-}-decay rate is even weaker than the π^{0}-decay, while for $^{8}_{\Lambda}$Li and $^{12}_{\Lambda}$B the π^{-}-decay is four times as strong as the π^{0}-decay. Recall that $\Gamma_{\pi^{-}}/\Gamma_{\pi^{0}} = 2$ in the free-Λ decay (see eqn (10.21)). Such a drastic variation of the π-decay rates comes from the shell effects connected with the Pauli blocking for the available final nuclear states. For example, the proton from the π^{-}-decay of $^{12}_{\Lambda}$C has to go only to high-lying $T = 1$ states of ^{12}N, while the neutron in the π^{0}-decay is allowed to go to $T = 0$ states (^{12}C) as well as $T = 1$ states. The final-state energy difference ($E_{T=1} - E_{T=0} > 15\,\mathrm{MeV}$) gives rise to the considerable difference between available π^{-} and π^{0} momenta ($q_{\pi^{-}} < 83\,\mathrm{MeV}$ vs $q_{\pi^{0}} < 121\,\mathrm{MeV}$) on which the decay rate depends rather strongly. In the decay of $^{12}_{\Lambda}$B, the situation is completely opposite.

Up to now, the total and partial decay rates of $^{12}_{\Lambda}$C and $^{5}_{\Lambda}$He were measured at Brookhaven [56,57], and the π^{0}-decay branching ratio was measured at KEK [58] for $^{11}_{\Lambda}$B and $^{12}_{\Lambda}$C. These available data are compared with the various theoretical estimates in Fig. 10.11. The predictions are consistent with the observed ratio $\Gamma_{\pi^{-}}/\Gamma_{\pi^{0}}$ of Brookhaven and $\Gamma_{\pi^{0}}^{\mathrm{exp}}$ for ^{12}C observed at KEK, although the two data of $\Gamma_{\pi^{0}}(^{12}_{\Lambda}$C) differ considerably from each other.

Another interesting aspect of π-decay is the angular distribution of the pions from the polarized hypernucleus. If the parent hypernucleus is polarized, the decay pions have asymmetric angular distributions [44]. The asymmetry which arises from the interference between the parity-conserving and parity-violating interactions also reflects the structures of the parent and daughter nuclear states.

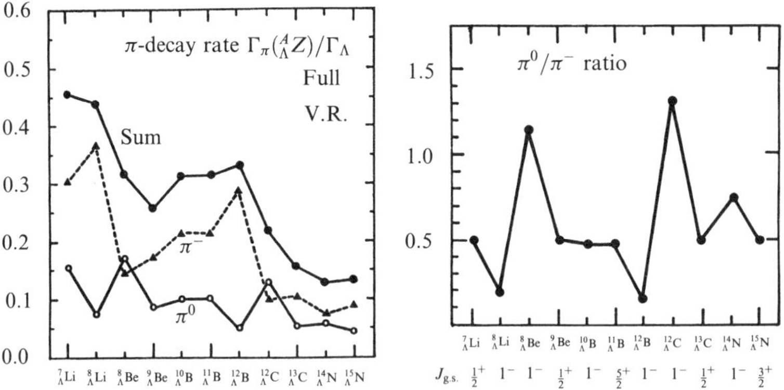

Fig. 10.10 Left: calculated π-mesonic decay rates, Γ_{Λ}, of typical p-shell hypernuclei expressed in units of the free-Λ decay rate. Right: the theoretical ratios $\Gamma_{\pi^{0}}(^{A}_{\Lambda}Z)/\Gamma_{\pi^{-}}(^{A}_{\Lambda}Z)$.

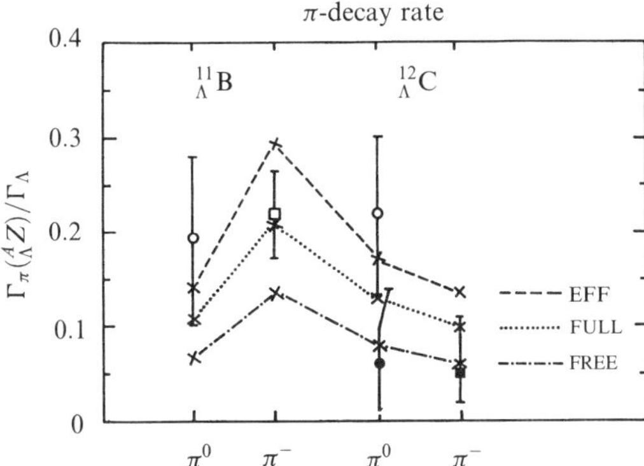

Fig. 10.11 Calculated π-mesonic decay rates in units of Γ_Λ compared with the existing data. FULL corresponds to the results with the full form of the optical potential given by eqn (10.23), while EFF to the effective form. FREE denotes the case of no pion distortion.

10.3.3 Asymmetry in the pionic weak decays of $^5_\Lambda$He

In the decay of very light hypernuclei, in general, the final nuclear states consist predominantly of the continuum states and include a few (if any) bound states. An example is the pionic decay of $^4_\Lambda$H leading to $^4\text{He} + \pi^-$, $^3\text{H} + \text{p} + \pi^-$ $(^3\text{H} + \text{n} + \pi^0)$, $\text{d} + \text{d} + \pi^-$, etc. where only the first channel involves the discrete pion energy. Here, by employing the Kapur–Peierls method [59] in treating the nuclear continuum states, a realistic calculation has been performed. For the present purpose, we consider the $^5_\Lambda$He decay, for which a representative three-body mode, $^5_\Lambda\text{He} \rightarrow {}^4\text{He} + \text{p} + \pi^-$ $(^4\text{He} + \text{n} + \pi^0)$, is adopted, since experimentally this π^--decay channel (1012 events) dominates exclusively over the $^3\text{He} + \text{d} + \pi^-$ (12 events), $^3\text{H} + \text{p} + \text{p} + \pi^-$ (1 event), $^3\text{He} + \text{n} + \text{p} + \pi^-$ (no event), and $\text{d} + \text{d} + \text{p} + \pi^-$ (no event) channels [60].

In Fig. 10.12, the calculated π^- spectrum is shown as a function of p–α energy [63] (or the outgoing pion energy T_π) in comparison with the observed pion distribution from the emulsion data [61,62,64]. The theoretical spectrum has a distinct peak at

$$E^{\text{cal}}_{\text{p}\alpha} = 1.5\,\text{MeV}\ (q_{\pi^-} = 99.9\,\text{MeV}/c,\ \ T_{\pi^-} = 32.1\,\text{MeV};\ \ \ \Delta E \simeq 1.0\,\text{MeV}),$$

$$(10.34)$$

which is in good agreement with the experiment. The peak reflects the proton $\text{p}_{3/2}$ resonance in ^5He (width: $\simeq 1.5\,\text{MeV}$) observed at 1.97 MeV above the p–α threshold [65]. It should be noted that the peak of width $\Delta E_{\text{FWHM}} \simeq 1.0\,\text{MeV}$ $(\Delta q \simeq 1.7\,\text{MeV}/c)$ obtained in the continuum π^- spectrum is sharp enough to be utilized in experiment as a characteristic (nearly mono-energetic) pion. The proton $\text{p}_{1/2}$ 'resonance' is too broad (width: $5 \pm 2\,\text{MeV}$) at $E_{\text{p}\alpha} = 5\text{–}10\,\text{MeV}$ to be identifiable in the π^- spectrum.

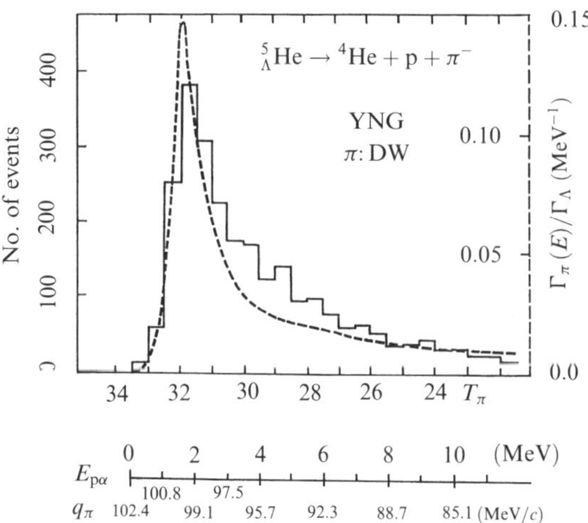

Fig. 10.12 Calculated π-mesonic decay spectrum (dashed line) of $^5_\Lambda$He as a function of the pion kinetic energy T_π (MeV). The experimental spectrum (solid line) [60–62] is also shown for comparison.

The total π^--decay rate of $\Gamma_{\pi^-}(^5_\Lambda$He$)$ is obtained by

$$\Gamma_\pi = \int \Gamma_\pi(E_N)\, dE_N. \tag{10.35}$$

It reveals that the YNG theoretical value is closer to the recent experiment [66] than the ORG one:

$$\Gamma_{\pi^-}(^5_\Lambda\text{He})^{\text{cal}}/\Gamma_\Lambda = 0.321\ (\text{ORG}),\ 0.393\ (\text{YNG})\ \text{vs}\ 0.44 \pm 0.11\ (\text{Exp}). \tag{10.36}$$

From the emulsion data, Bohm *et al.* [67] deduced a somewhat smaller but comparable decay rate $(0.26$–$0.42)\Gamma_\Lambda$ from the observed lifetime $\tau(^5_\Lambda$He$)$ combined with the ratio $Q^-(^5_\Lambda$He$)$ [68].

In the π^--decay, the three-body ^{4}He + p + π^- final state dominates, and the continuum pion spectrum as well as the absolute π-decay rates have been explained well by applying the Kapur–Peierls method [69]. As shown in Fig. 10.12, the pion spectrum has a sharp peak at $q_{\pi^-} \simeq 99.9$ MeV/c ($\Delta q \simeq 1.7$ MeV/c) corresponding to the dominant p$_{3/2}$ proton–^{4}He resonance.

Starting with the non-relativistic expression of the mesonic decay interaction [44], we obtain the partial transition rate for the decay from the fully polarized $^5_\Lambda$He($J_i = 1/2$,

$M_i = 1/2$) to the final state of $^5\mathrm{Li}((3/2)^-) = {}^4\mathrm{He} + \mathrm{p}(\mathrm{p}_{3/2})$,

$$\Gamma_{\pi^-}((3/2)^-) = \frac{8\pi q \langle \ell = 1||Y_1||\ell = 0\rangle^2}{1 + \omega_q/M}$$

$$\times \frac{2}{3} \left\{ s_{\pi^-}^2 a^2 + p_{\pi^-}^2 \left(\frac{q}{q_0}\right)^2 c^2 \right\} [1 + \alpha^\pi P_1(\cos\theta)], \qquad (10.37)$$

$$\alpha^\pi = \frac{2 s_{\pi^-} p_{\pi^-} (q/q_0)\,(c/a)}{s_{\pi^-}^2 + p_{\pi^-}^2 (q/q_0)^2\,(c/a)^2}, \qquad (10.38)$$

$$a \equiv \langle u_p^{\mathrm{N}} | \tilde{j}_1(q;r) | u_s^\Lambda \rangle, \qquad c \equiv \langle u_p^{\mathrm{N}} | f_1^{(2)}(q;r) | u_s^\Lambda \rangle.$$

Here, $\tilde{j}_L(q;r)$ is a partial wave of the pion distorted wave $\chi_\pi^{(-)}(q;r)^*$ and $f_L^{(K)}(q;r)$ the linear combination of a partial wave and its derivative. In the case of no pion distortion, $\tilde{j}_L(q;r)$ tends to the spherical Bessel function $j_L(qr)$, and $f_L^{(K)}(q;r)$, with $K = L \pm 1$, tend to $\pm j_L(qr)$. Then the ratio c/a tends to 1 and, hence, $\alpha^\pi\,(^5_\Lambda\mathrm{He})$ approaches the expression for the free-Λ π^--decay: $\alpha^\pi_{\mathrm{free}} = 2 s_{\pi^-} p_{\pi^-} / \{s_{\pi^-}^2 + p_{\pi^-}^2\} = -0.642$.

In the present case, we note that $q \simeq q_0$, and the magnitude of the ratio c/a is essential for estimating α^π. In general, the pion distortion effect on the mesonic decay rate is appreciable and sometimes drastic in heavy hypernuclear decay. Although the effect is not negligible even for light systems, the ratio c/a is not affected so much as the decay rate itself. Thus, we can expect that the pion asymmetry α^π remains persistently near $\alpha^\pi_{\mathrm{free}}$. For reference, we list $\alpha^\pi = -0.62, -0.59$, and -0.57 for $c/a = 0.95, 0.90$, and 0.85, respectively. If the actual Λ-polarization of $^5_\Lambda\mathrm{He}(J_i = (1/2)^+)$ is $\mathcal{P}_\Lambda$, then α^π in eqn (10.7) should be replaced by the asymmetry parameter $A^\pi \equiv \mathcal{P}_\Lambda \alpha^\pi$.

10.4 Non-mesonic weak decay and baryon–baryon interactions

Recent measurements [66] of the weak decay branching ratios for $^5_\Lambda\mathrm{He}$ and $^{12}_\Lambda\mathrm{C}$ have attracted renewed theoretical interest for the mechanism of the hypernuclear non-mesonic decay. It provides information on the spin–isospin dependence of the decay interaction, which has not yet been established theoretically. Moreover, in this $\Lambda\mathrm{N}$ weak process, one can see both parity-conserving and parity-non-conserving interactions. This is in contrast to the NN weak interaction, where it is actually impossible to see the parity-conserving component, as it is masked by the strong NN interaction.

Table 10.4 lists the six allowed channels [70] if the non-mesonic decay is assumed to start from the $\Lambda\mathrm{N}$ relative S-states. Among these six, only three channels (a, b, f) having isospin $I_{\mathrm{NN}} = 1$ in the final NN system are allowed for the decay $\Lambda\mathrm{n} \to \mathrm{nn}$, while all six contribute to $\Lambda\mathrm{p} \to \mathrm{np}$. The neutron-stimulated and the proton-stimulated decay rates are given, respectively, by

$$\Lambda\mathrm{n} \to \mathrm{nn}: \quad \Gamma_n = 2\Gamma_{I=1} = 2(a^2 + b^2 + f^2), \qquad (10.39)$$

$$\Lambda\mathrm{p} \to \mathrm{np}: \quad \Gamma_p = \Gamma_{I=0} + \Gamma_{I=1} = (c^2 + d^2 + e^2) + (a^2 + b^2 + f^2). \qquad (10.40)$$

Table 10.4 Six interaction channels contributing to the non-mesonic decays starting from the ΛN relative $L = 0$ state. Here, σ_1 denotes the spin for Λ or the final neutron, and $\boldsymbol{q}$ the final neutron momentum. The expressions for the a, b, and f channels are appropriate for $\Lambda n \rightarrow (nn)_{I=1}$, so a factor $1/\sqrt{2}$ should be multiplied for $\Lambda p \rightarrow (np)_{I=1}$, following the $\Delta I = 1/2$ rule

ΛN	NN			I_{NN}	S_{NN}	Parity change
$^1S_0$	$\rightarrow {}^1S_0$	a		1	0	No
	$\rightarrow {}^3P_0$	b	$(1/2)(\sigma_1 - \sigma_2)\cdot\boldsymbol{q}$	1	1	Yes
$^3S_1$	$\rightarrow {}^3S_1$	c		0	1	No
	$\rightarrow {}^3D_1$	d	$(\sqrt{2}/4)S_{12}(\boldsymbol{q})$	0	1	No
	$\rightarrow {}^1P_1$	e	$(\sqrt{3}/2)(\sigma_1 - \sigma_2)\cdot\boldsymbol{q}$	0	0	Yes
	$\rightarrow {}^3P_1$	f	$(\sqrt{6}/4)(\sigma_1 + \sigma_2)\cdot\boldsymbol{q}$	1	1	Yes

Here, the factor 2 for Γ_n comes from the $\Delta I = 1/2$ rule, which is assumed to hold in the non-mesonic decay. The basic quantities to be discussed in the comparison of theory and experiment are Γ_{nm}, Γ_n, Γ_p, and the ratio $\xi_{np} = \Gamma_n/\Gamma_p$.

The ratio of the parity-violating to the parity-conserving rates is also an interesting quantity and is given by

$$\zeta \equiv \frac{\Gamma_{PV}}{\Gamma_{PC}} = \frac{b^2 + e^2 + f^2}{a^2 + c^2 + d^2}. \tag{10.41}$$

Based on the old emulsion data [71,72] on Γ_{nm} and $\xi_{np} \equiv \Gamma_n/\Gamma_p$ of $^4_\Lambda$He and Γ_{nm} of $^4_\Lambda$H, Dalitz and his collaborators [48,70] found the f-dominance ($I_{NN} = 1$) character of the non-mesonic decay. Other data for heavy hypernuclei [68,73] also suggest this character: $\xi_{np} = 1.5$–9 and 5.7.

The modern counter experiments performed at Brookhaven [56,57,66,74] have recently yielded the partial decay rates Γ_n, Γ_p, Γ_{π^-}, and the lifetime τ (hence the total decay rate Γ) for $^5_\Lambda$He and $^{12}_\Lambda$C. Including these data as well, Dover [75] concluded that the $S = 1$, $I = 1$ channel is the most important but the others are not negligible. As the observed ratio $\xi_{np}(^5_\Lambda$He) is finally reported [66] to be $0.93^{+0.55}_{-0.55}$ (modified from the previously reported value [56] $1.30^{+0.65}_{-1.60}$), it differs considerably from the other data $\xi_{np}(^{12}_\Lambda$C) $= 1.33^{+1.12}_{-0.81}$. It is interesting to note that, empirically, for light systems such as $^5_\Lambda$He and $^{12}_\Lambda$C, Γ_{nm} is approximately proportional to $A - 1$. However, it should be noted that individually Γ_n and Γ_p behave differently even in such light systems, as $\Gamma_n/N = 0.10\Gamma_\Lambda(^5_\Lambda$He) vs $0.13\Gamma_\Lambda(^{12}_\Lambda$C), and $\Gamma_p/Z = 0.11\Gamma_\Lambda(^5_\Lambda$He) vs $0.08\Gamma_\Lambda(^{12}_\Lambda$C), although the experimental errors are still quite large. For heavy systems, a saturation effect has been pointed out [11,75], using the data [76] for $^{238}_\Lambda$U, as $\Gamma_{nm}(^{238}_\Lambda$U)/237 $\simeq$ $(1/10)\Gamma_{nm}(^5_\Lambda$He)/4.

There have been two theoretical treatments: one is the meson-exchange description which incorporates a weak and a strong meson–baryon vertex in a manner consistent with the $\Delta I = 1/2$ rule [46,77–80]. The other is the hybrid quark model calculation in which the One-Pion-Exchange (OPE) $\Delta I = 1/2$ description is employed for larger Λ–N distances and the quark–quark interaction with both $\Delta I = 1/2$ and 3/2 components is introduced for shorter Λ–N distances [81].

The π- and ρ-exchange contributions were calculated [79] along the lines of the treatment given in Ref. [78]. The two choices $\pi \pm \rho$ are due to the theoretical uncertainty of the relative phase. The $\pi + \rho$ case reproduces the experimental decay rates $\Gamma_{nm}/\Gamma_\Lambda$ for both $^5_\Lambda$He and $^{12}_\Lambda$C, but fails completely for the ξ_{np} ratios. The extremely small ξ_{np} ratio reflects the $I_{NN} = 0$ channel (c, d, e) dominance in the meson-exchange weak interaction. The inclusion of heavier mesons (η, K, ω, and K*) [80], in addition to π and ρ, seems to improve Γ_{nm} but the ξ_{np} ratio is still too small compared to the experimental values. The hybrid quark model [81] also gives a small ξ_{np} ratio in the $\Delta I = 1/2$ limit, only when the decay rate can be close to the observed magnitude.

Thus, no models have succeeded in reproducing even qualitative features of the non-mesonic decay mechanism and the Γ_n/Γ_p ratio is indeed a puzzling problem.

Given this situation, in order to understand the non-mesonic decay mechanism, it is desirable to consider experiments to obtain different types of information. One such example would be the angular distribution, in particular the asymmetry, of protons from the non-mesonic decay of polarized hypernuclei.

When the Λ hyperon with a spin polarization $\mathcal{P}_\Lambda$ interacts with spin-up and spin-down nucleons, the non-mesonic decay interactions give rise to the following angular distribution of protons;

$$W_p(\theta) = 1 + A_1 \mathcal{P}_\Lambda P_1(\cos\theta), \quad A_1 = \frac{2\sqrt{3}(\sqrt{2}c + d)f}{a^2 + b^2 + 3(c^2 + d^2 + e^2 + f^2)}, \quad (10.42)$$

where the angle θ is measured from the polarization axis. The asymmetry parameter A_1 comes from interference between the parity-conserving (a, c, d) and parity-violating (b, e, f). The asymmetry thus provides new information which is independent of that from the Γ_{nm} and ξ_{np} data. The implications in the actual measurement will be discussed in the next section.

Here, for instructive purposes, we point out the reason why the meson-exchange models result in Γ_p (hence $\Gamma_{I=0}$) dominance. The OPE diagram for $\Lambda N \to NN$ consists of the weak $\Lambda N\pi$ and the strong $NN\pi$ vertices. They are given in charge space by

$$H^{\text{weak}}_{\Lambda N\pi} = ig_w \bar{\psi}_n(1 + \lambda\gamma_5)\pi^0\psi_\Lambda - i\sqrt{2}g_w \bar{\psi}_p(1 + \lambda\gamma_5)\pi^+\psi_\Lambda, \quad (10.43)$$

$$g_w = -0.23 \times 10^{-6}, \qquad \lambda = -6.9 \quad (10.44)$$

and

$$H^{\text{strong}}_{NN\pi} = ig_s\{\bar{\psi}_p\gamma_5\pi^0\psi_p - \bar{\psi}_n\gamma_5\pi^0 + \sqrt{2}\bar{\psi}_p\gamma_5\pi^+\psi_n + \sqrt{2}\bar{\psi}_n\gamma_5\pi\}, \quad (10.45)$$

$$g_s = 13.26.$$

It is noticed that the product of H^{strong} and the γ_5-terms (parity-conserving) of H^{weak} has exactly the same structure as the OPE contribution to the NN force. In fact, the resulting two-body weak interaction is divided into the parity-conserving (V^{PC}_π) and parity-violating (V_π^{PV}) components: $V_\pi = V_\pi^{\mathrm{PC}} + V_\pi^{\mathrm{PV}}$:

$$V_\pi^{\mathrm{PC}}(\Lambda \mathrm{n} \to \mathrm{nn}) = -\lambda \frac{g_{\mathrm{w}} g_{\mathrm{s}}}{4\pi} \frac{m_\pi^2}{4 M_{\mathrm{N}} \bar{M}} \, m_\pi Y(x) \left\{ \tfrac{1}{3}(\sigma_1 \cdot \sigma_2) + T(x) S_{12} \right\}, \tag{10.46}$$

$$V_\pi^{\mathrm{PV}}(\Lambda \mathrm{n} \to \mathrm{nn}) = \mathrm{i} \frac{g_{\mathrm{w}} g_{\mathrm{s}}}{4\pi} \frac{m_\pi}{2 M_{\mathrm{N}}} m_\pi Y(x) V(x) (\sigma_2 \cdot \hat{r}), \tag{10.47}$$

and, for $\Lambda \mathrm{n} \to (\mathrm{np})_I$,

$$V_\pi(\Lambda \mathrm{p} \to (\mathrm{np})_I) = (-)^{I+1} \frac{3}{\sqrt{2}(2I+1)} V_\pi(\Lambda \mathrm{n} \to \mathrm{nn}), \tag{10.48}$$

where $\bar{M} = \sqrt{M_{\mathrm{N}} M_\Lambda}$, $x = m_\pi r$, and

$$Y(x) = \frac{\mathrm{e}^{-x}}{x}, \quad V(x) = 1 + \frac{1}{x}, \quad T(x) = \frac{1}{3} + \frac{1}{x} + \frac{1}{x^2}. \tag{10.49}$$

The relative ratios among $\Lambda \mathrm{n} \to \mathrm{nn}$, $\Lambda \mathrm{p} \to (\mathrm{np})_{T=1}$, and $\Lambda \mathrm{p} \to (\mathrm{np})_{T=0}$ interactions follow from the $\Delta I = 1/2$ rule.

The potential V_π^{PC} is of the well-known OPE form. Recall that in the π-mesonic decay the parity-violating term of $H_{\Lambda \mathrm{N}\pi}^{\mathrm{weak}}$ is more important than the other, because of the small momentum release $q_0 \simeq 100\,\mathrm{MeV}/c$. The situation is different in the non-mesonic decay, where the relevant momentum is four times larger, $q_{\mathrm{nm}} \simeq 400\,\mathrm{MeV}/c$, which makes the parity-conserving term relatively more important. This fact is reflected in V_π of eqns (3.28) and (3.29), where the strengths of V_π^{PC} and V_π^{PV} are essentially comparable. The explicit potentials V_π for the six channels are also listed in Table 10.4.

It is known that the NN OPE potential is dominated by its strong tensor component. Similarly, in the ΛN weak interaction, the tensor component makes a much larger contribution (d channel in Table 10.4). Thus, the OPE model results in the $I_{\mathrm{NN}} = 0$ (hence Γ_{p}) dominance, which is contradictory to the experimental indication. The $^3\mathrm{D}_1$-wave admixture into the $^3\mathrm{S}_1$-wave is most significant in the final-state NN correlations. Thus, the transition from $\Lambda \mathrm{N}(^3\mathrm{S}_1)$ to $\mathrm{NN}(^3\mathrm{D}_1)$ by $V_\pi^{(d)}$ and further to $\mathrm{NN}(^3\mathrm{S}_1)$ by the final-state correlation makes the major contribution to Γ_{p}. Such a feature seems to remain unchanged as long as the meson-exchange mechanism is maintained.

10.5 Weak decays of polarized hypernuclei

Until recently, the study of weak decays has been restricted to the partial decay rates. The rates give spin–isospin structure of the decay for both mesonic and non-mesonic decays. For instance, study of partial decay rates of hyperons disclosed the $\Delta I = 1/2$ rule. For the analysis of non-mesonic decay, the $\Delta I = 1/2$ rule has been assumed since the rule holds well for almost all strangeness-changing hadronic decays and the experimental data themselves are not good enough to verify the rule. Recently, the validity of the rule in the non-mesonic decay has been questioned [82]. This will be decided only when further experimental data are obtained [83].

There is another quantity that one can investigate to study the hyperon weak decay. It is known that mesonic decay has an asymmetric angular distribution with respect to the hyperon spin polarization ($\mathcal{P}_\Lambda$) given by eqn (10.32). It is also represented by

$$W_\pi(\theta) = 1 + \alpha_1^\pi \mathcal{P}_\Lambda \cos\theta, \tag{10.50}$$

where θ is the polar angle with respect to the polarization axis. For instance, we know the asymmetry parameter of Λ-decay precisely [55], so that the measurement of the pion asymmetry can be used to determine the Λ-polarization. The asymmetry parameter for the mesonic decay of hypernuclei has been calculated theoretically for typical hypernuclei [84]. It is found that the parameter depends largely on the initial hypernuclear and final nuclear spins. The theoretical prediction of the parameter is rather unambiguous, because the quantities needed for the calculation, the amplitude of Λ-decay and initial hypernuclear and final nuclear wave functions, are rather well known. Thus, the asymmetry parameter can be used to give polarization of hypernuclei. Such an experiment to measure the weak decay of $^5_\Lambda$He is under way and will be described in Section 10.5.2.

What is particularly interesting is the asymmetry parameter A_1 of the non-mesonic decay [85] defined in the previous section, because none was known for the elementary ($\Lambda n \to pn$) process experimentally. The asymmetry comes from the interference between the amplitudes with different isospin and parity. It is thus necessary to specify the isospin for the two-nucleon system in the final state. For instance, no asymmetry can be obviously expected for the nn or pp systems which have definite isospin 1. The initial state has to be $^3S_1$ since no spin orientation exists in $^1S_0$. The study of the asymmetry parameter gives a complementary constraint on the non-mesonic decay amplitude, which cannot be obtained by measurement of the branching ratios. The amplitude f is found to be dominant by the phenomenological analysis of the branching ratios [70] and is known to represent the dominant kaon contribution [80,86]. The amplitude d is due mostly to pion exchange. The observation of the asymmetry parameter gives the relative phase of these dominant amplitudes. Thus, it elucidates the dominant part of the non-mesonic decay mechanism. Furthermore, if it turns out to be large, it should be quite useful for the measurement of magnetic moments [85,87,88].

10.5.1 *Study of the $^{12}C(\pi^+, K^+)$ reaction*

In this section, we describe the experiment carried out at the KEK 12 GeV Proton Synchrotron (KEK-PS), where production of polarized hypernuclei and observation of their asymmetric weak decay were carried out for the first time. In the experiment, polarized hypernuclei were produced by the (π^+, K^+) reaction on the ^{12}C target, and the asymmetry of the weak-decay particles was measured with respect to the polarization axis [89]. Currently, KEK and BNL are the only laboratories that can provide intense meson beams in the GeV region for the study of hypernuclei. The K2 beam line was used to provide 1.05 GeV/c pion beams, the typical beam flux being around 2–3 $\times$ 10^6/spill for 2–3 $\times$ 10^{12} primary protons/spill, where a spill consisted of 2.0 s of continuous beam every 4.0 s. The momenta of the incident pions and the outgoing kaons were measured by the PIK spectrometer system [90]. The Λ-hypernuclei were polarized by detecting kaons at $\pm14°$, where the best figure-of-merit (cross section $\times$ (polarization)^2) is obtained. The scattered kaons were identified clearly with negligible background by the mass spectrum obtained by combining Time-Of-Flight (TOF) and momenta.

A plastic scintillator was used as a carbon target which covered an area of 60(wide) $\times$ 20(high) mm^2 with a thickness of 64 mm and was divided into two halves ('upper' and 'lower'), each consisting of 32 segments of plastic scintillator to obtain better spatial resolution of the interaction vertex. The active segmented target helped in selecting the hypernuclear events by the energy-loss pattern in the target [91]. The counter system ROYAL was designed to detect 10–50 MeV pions and 30–200 MeV protons emitted from the target. These are the energies expected from the mesonic and non-mesonic decays of hypernuclei. Further details of the experimental setup can be found elsewhere [22,23].

The asymmetry (A) was obtained from the experimental coincidence yields as

$$\left(\frac{N_1^\uparrow(14)N_2^\uparrow(14)N_1^\downarrow(-14)N_2^\downarrow(-14)}{N_2^\downarrow(14)N_1^\downarrow(14)N_2^\uparrow(-14)N_1^\uparrow(-14)}\right)^{1/4} = \frac{1+A}{1-A}. \tag{10.51}$$

Here, $N_1^\uparrow(\theta)$ stands for number of counts in the ROYAL counter system 1 when it was in the 'up' position and the kaon spectrometer was set at $\theta = 14°$ for the present setup. First-order systematic errors related to normalization of beam intensity, fluctuation of beam intensity, asymmetry of solid angle, and misalignment of beam centre—all cancel in this ratio. In order to determine any spurious instrumental asymmetry, the $(\pi, \pi x)$ and (π^+, pX) reactions were measured simultaneously with the (π^+, K^+) reaction, since no up–down asymmetry is expected in the former reactions because no parity-violating weak interaction is involved. These measurements indicated that the instrumental asymmetry was less than 2%. Details of the experimental conditions and results are presented elsewhere [22,23].

The level scheme of $^{12}_\Lambda$C excited by the (π^+, K^+) reaction is shown in Fig. 10.13. The weak decay takes place from the ground state which is populated directly by the reaction as well as through deexcitation due to γ and/or nucleon emission. The ground state and a few low-lying states (region B) with $p_n^{-1}s_\Lambda$ yield weak decay from the ground state of $^{12}_\Lambda$C. The so-called substitutional states (region AB) with $p_n^{-1}p_\Lambda$ are open for proton emission but closed for Λ emission so that $^{11}_\Lambda$B is exclusively produced in this region [92]. The region QB consists of states with mostly $p_n^{-1}d_\Lambda$ and $s_n^{-1}s_\Lambda$. Here, Λ has to be trapped in a nucleus in order to produce hypernuclei through Λ–nucleus interaction.

The measured excitation energy spectrum in coincidence with the protons is shown in Fig. 10.14 for the low excitation region [23]. The fitted spectrum, which is based upon the theoretically calculated cross section [93] of the (π^+, K^+) reaction smeared by the experimental resolution of $\sim$6 MeV, is also shown. One can see two peaks corresponding to the ground state and a few excited states (region B), and the substitutional states (region AB). The protons in coincidence with these peaks are from the non-mesonic decay of the $^{12}_\Lambda$C and $^{11}_\Lambda$B ground states, as discussed above. Other light hypernuclei, which are produced in the region LH, also emit protons.

The proton asymmetry A^p, where the attenuation factor due to solid angle of the ROYAL decay counter is corrected, is obtained as follows. The non-mesonic decay protons from the $^{12}_\Lambda$C(1^-) ground state show a negligible asymmetry, $A^p = -0.01 \pm 0.11$,

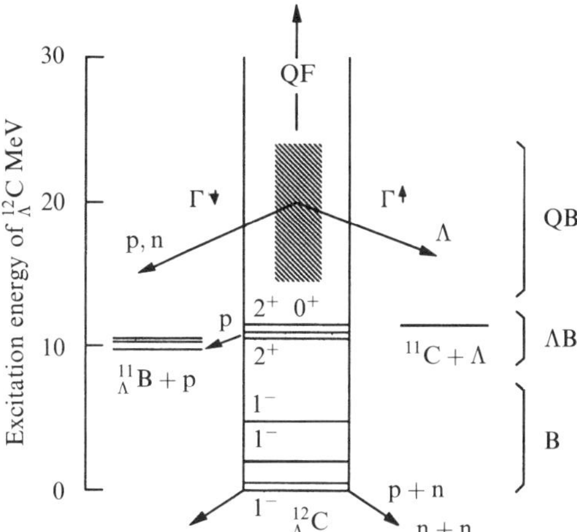

Fig. 10.13 Level scheme of low-lying $^{12}_\Lambda$C. Shaded region is above the Λ-emission threshold, which is mostly excited by the Quasi-Free (QF) process. $\Gamma^\uparrow$ and $\Gamma^\downarrow$ are escape (Λ-emission) width and spreading (proton or neutron emission) width, respectively.

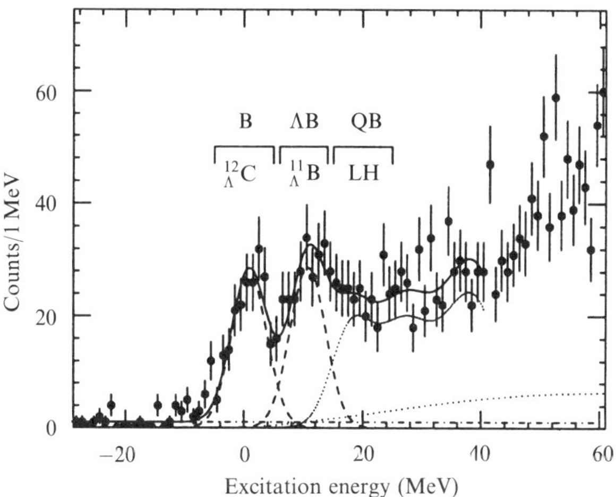

Fig. 10.14 A kaon spectrum gated by decay protons for the ^{12}C(π^+, K$^+$) reaction at p_π =1.05 GeV/c. Protons whose energies are higher than 45 MeV are selected. The horizontal axis gives the excitation energy of ^{12}C. The solid line represents fitted spectrum based on the theoretical cross section (see text).

while that from the $^{11}_\Lambda$B($(5/2)^+$) ground state shows a finite asymmetry of $A^{\mathrm{p}} = -0.19\pm$ 0.10. The asymmetry of the QB region appears to be even higher, $A^{\mathrm{p}} = -0.24 \pm 0.9$.

The non-mesonic decay can be taken as a two-body reaction, since the momentum transfer involved in the decay is much larger than the Fermi momentum. The observed asymmetry A^p is then related to the intrinsic asymmetry parameter α^p of the non-mesonic decay and the Λ-spin polarization P_Λ. Here, we are interested in the two-body non-mesonic decay and Λ can interact with all nucleons in a nucleus. Each nucleon carries little spin polarization on average, even when the polarization of the core-nuclear spin is large. Therefore, we can neglect the effect of the hypernuclear-spin polarization P_J. The weak decay takes place from the ground state of the hypernucleus and the Λ is in the 1s-shell. The hypernuclear spin J is given as $J = J_c + 1/2$ or $J_c - 1/2$, where J_c is the spin of the core nucleus. The spin polarization P_Λ is given by $P_\Lambda = P(J)$ for $J = J_c + 1/2$, and $P_\Lambda = -P(J) \cdot J/(J+1)$ for $J = J_c - 1/2$ [27].

The observed asymmetry of the non-mesonic decay proton is attenuated by intranuclear cascade processes and the Fermi motion of Λ and nucleon. Consequently, the proton asymmetry (A^p) is given by $A^p = k\alpha^p P_\Lambda$. Here, k is the attenuation factor due to the intranuclear cascade process and the Fermi motion, and it is estimated that $k = 0.8$ for non-mesonic decay protons whose energy is above 45 MeV.

The asymmetry parameter α^p can be derived from the observed value of A^p/k providing one knows the P_Λ for each region. Here, we have to rely on the theoretical calculation to obtain P_Λ, since practically no experiment can measure it. The calculation is divided into two steps. First, the polarization of the hypernuclear states produced by the (π^+, K^+) reaction was evaluated [15,93]. Second, the depolarization in the particle and γ-deexcitation process was calculated [27]. The polarizations of hypernuclear states populated by the (π^+, K^+) reaction vary largely depending on their configuration [15]. However, one expects large and similar polarization for the ground states even though they are populated through deexcitation of particles and γ-rays [27]. Although one needs detailed calculation for each case, this general tendency can be explained as follows. The variation in polarization of states populated by the (π^+, K^+) reaction is due mostly to the coupling of Λ-spin to core nuclear spin [15]. The polarization of Λ-spin, which is mainly determined by the elementary reaction, is rather stable. The (π^+, K^+) reaction excites high-spin states, so the deexcitation process is frequently the stretched transition $(J_f = J_i - L)$, where no depolarization takes place [27]. A spin-flip transition, which causes large depolarization, takes place mostly in the low-excitation region.

A large polarization was predicted for the ground state of $^{12}_{\Lambda}\text{C}$ for the simple particle hole configuration where the strength of $p_n^{-1}s_\Lambda$ configuration is concentrated on the ground state [15]. However, the calculation, which includes nuclear core excitation [93], predicts strong scattering to the higher 1^-_2 and 1^-_3 states, which populate the ground state (1^-_1) through γ-transitions. The polarization of the $^{12}_{\Lambda}\text{C}(1^-)$ ground state, $P(1^-)$ is estimated to be between -0.12 and -0.18, depending on the γ-decay scheme employed. The corresponding polarization of the Λ-spin is $P_\Lambda(1^-) = 0.06$–0.09. Similarly, the $^{11}_{\Lambda}\text{B}((5/2)^+)$ polarization is estimated to be between -0.22 and -0.28 and its corresponding Λ-spin polarization is $P_\Lambda((5/2)^+) = 0.16$–$0.21$. The weak-decay proton, associated with the QB region, is considered to be due to the non-mesonic decay proton of Light Hypernuclei (LH) produced through emissions of p, n, d, ^{3}He, or α. These effectively open channels in the QB energy region lead to the production of such LH as $^{11}_{\Lambda}\text{B}$, $^{11}_{\Lambda}\text{C}$, $^{10}_{\Lambda}\text{B}$, $^{9}_{\Lambda}\text{Be}$, or $^{8}_{\Lambda}\text{Be}$. The average polarization of the Λ-spin in the ground states of these LH is found to be $P_\Lambda(\text{LH}) = 0.26$–$0.15$, depending on the polarization due to the (π^+, K^+) reaction and particle emission processes involved in populating the LH.

The experimental values of A^{p}/k are plotted vs the above theoretical values of P_Λ in Fig. 10.15. One obtains the asymmetry parameter $\alpha^{\text{p}} = -1.3 \pm 0.4$ for the p-shell nuclei. The asymmetry parameter cannot be larger than unity in magnitude, so one can say that it is less than -0.6. The theoretical calculation based on the one-boson-exchange model was carried out where pion and kaon exchanges were included [94] and the asymmetry parameter was given for each hypernucleus. It shows strong dependence on a hypernucleus. However, it is the polarization of Λ that is relevant to the non-mesonic decay, as discussed above. The asymmetry parameter of the two-body non-mesonic decay extracted from the theoretical calculation turns out to be independent of the hypernuclei and is close to -0.6. This reflects the fact that the non-mesonic decay is almost a two-body reaction and the calculation reproduces the asymmetry parameter well—not only in sign but also in magnitude.

An appreciably large asymmetry parameter (-1.3 ± 0.4) was observed for the non-mesonic decay of the polarized $^{12}_\Lambda\text{C}$ hypernucleus by using the $^{12}\text{C}(\pi^+, \text{K}^+)$ reaction [23]. It suggests that the amplitudes for isospin 0 and 1 are equally important. Theoretical models based on the meson exchange predict dominance of isospin 0, whereas a phenomenological analysis of experimental branching ratio prefers isospin 1. The present asymmetry parameter, however, appears to have a value between these two.

So far, we have an error of 40% for the asymmetry parameter. The accuracy is far inferior to what is needed for the detailed comparison with theoretical calculations. In the analysis of the asymmetry parameter, only an s-wave Λ–N system is considered. However, nucleons are in p and s orbits in the $^{12}_\Lambda\text{C}$. Thus, one has to consider the p-wave

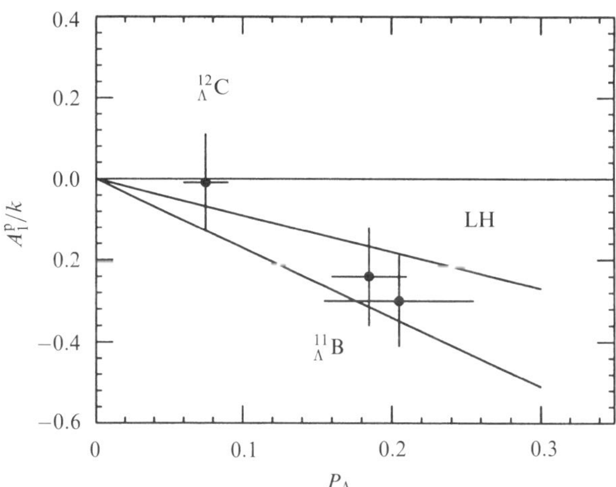

Fig. 10.15 Asymmetries plotted as a function of calculated polarization (P_Λ). Here, k represents the attenuation factor due to the intranuclear cascade process. The two solid lines show the range obtained for the asymmetry parameter.

Λ–N system which makes the comparison with the theoretical calculation complex. We, thus, have carried out an experiment where asymmetric non-mesonic decay was observed from the weak non-mesonic decay of polarized $^5_\Lambda$He [95] instead of repeating the study of $^{12}_\Lambda$C with improved statistics. The experiment is particularly suitable for the study of the asymmetry parameter because: (1) the expected polarization is 1.5–2 times larger; (2) polarization of Λ in $^5_\Lambda$He is unambiguously equal to that of $^5_\Lambda$He; (3) polarization of the $^5_\Lambda$He can be obtained by observing the asymmetry of pions from the mesonic decay of $^5_\Lambda$He (the theoretical calculation was employed to derive the Λ-polarization in $^{12}_\Lambda$C); (4) final state interaction is small and can be estimated reliably; (5) only relative s-wave contributes to the decay. These points are described in detail below.

10.5.2 Study of $^5_\Lambda$He by the ^{6}Li $(\pi^+, \mathrm{K}^+\mathrm{p})\,^5_\Lambda$He reaction

The polarized $^5_\Lambda$He was produced by the ^{6}Li $(\pi^+, \mathrm{K}^+\mathrm{p})\,^5_\Lambda$He reaction. The level scheme of the $^6_\Lambda$Li is shown schematically in Fig. 10.16. The ground state of $^6_\Lambda$Li lies 0.6 MeV above the proton-emission threshold although the Λ is bound by about 4.5 MeV [66,74]. Therefore, the ground state produces the $^5_\Lambda$He by emitting a proton. Since there is no low-lying excited state in $^5_\Lambda$He, the ground state $((1/2)^+)$ is produced exclusively.

The cross section and polarization have been calculated for the ^{6}Li(π^+, K^+) reaction at 1.05 GeV/c [96]. Some of the values relevant to the present discussion are shown in Table 10.5. The ground state of $^6_\Lambda$Li is a doublet of $1^-_{\mathrm{g.s.}}$ and 2^-_1 with the configuration $(\mathrm{p}_{3/2})_n^{-1}(\mathrm{s}_{1/2})_\Lambda$. The 2^-_1 state has the largest cross section among the ground state region. The state emits only protons with $\mathrm{p}_{3/2}$ partial wave. No depolarization takes place for such stretched transitions [27]. The polarization of the $1^-_{\mathrm{g.s.}}$ state is opposite to that of the 2^-_1 state. This is due to the fact that the spin polarization of Λ is the same for both states and the coupling is parallel for the 2^-_1 state and antiparallel for the $1^-_{\mathrm{g.s.}}$ state. The partial

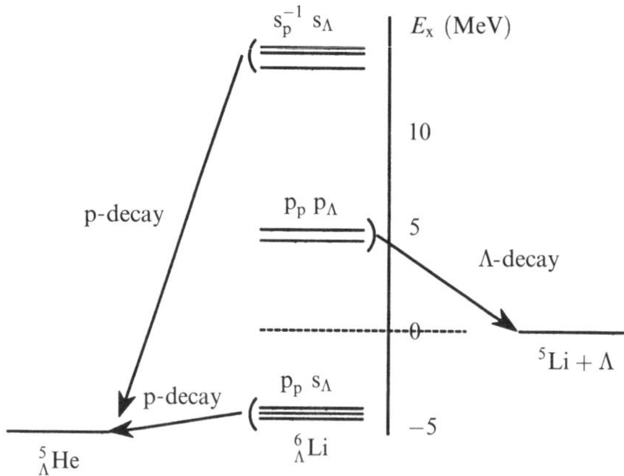

Fig. 10.16 The level scheme of $^6_\Lambda$Li. A proton is emitted from states where a Λ is in an s-orbit even though the Λ emission is energetically possible.

Table 10.5 Predicted cross sections and polarizations of low-lying $^6_\Lambda$Li hypernuclear states. The $1^-_{\text{g.s.}}$ and 2^-_1 states have primarily a $(p_{3/2})_p(s_{1/2})_\Lambda$ configuration, whereas the 1^-_2 and 0^-_1 states mostly a $(p_{1/2})_p(s_{1/2})_\Lambda$ configuration. State $(^6_\Lambda$Li); $\theta = 6°, \theta = 12° \theta = 18°$

$1^-_{\text{g.s.}}$	$(d\sigma/d\omega)(\mu b\,sr^{-1})$	0.78	0.64	0.34
	P	-0.443	-0.586	-0.570
	$P(^5_\Lambda He)$	0.295	0.391	0.380
2^-_1	$(d\sigma/d\omega)(\mu b\,sr^{-1})$	6.90	4.19	1.64
(0.081 MeV)	P	0.249	0.441	0.568
	$P(^5_\Lambda He)$	0.249	0.441	0.568
1^-_2	$(d\sigma/d\omega)(\mu b\,sr^{-1})$	0.97	0.81	0.44
(0.687 MeV)	P	0.127	0.163	0.154
	$P(^5_\Lambda He)$	0.127	0.163	0.154
0^-_1	$(d\sigma/d\omega)(\mu b\,sr^{-1})$	3.43	1.92	0.68
(0.721 MeV)				
$(1/2)^+ (^5_\Lambda He)$	$(d\sigma/d\omega)(\mu b\,sr^{-1})$	12.08	7.56	3.10
(Total)	P	0.17	0.29	0.36
	$P^2(d\sigma/d\omega)$	0.35	0.64	0.40
$(1/2)^+ (^5_\Lambda He)$	$(d\sigma/d\omega)(\mu b\,sr^{-1})$	7.68	4.83	1.98
$(1^-_{\text{g.s.}} + 2^-_1)$	P	0.25	0.43	0.54
	$P^2(d\sigma/d\omega)$	0.48	0.89	0.58

wave of the emitted proton is still dominantly $p_{3/2}$, so that the final polarization of the Λ is in the same direction. The expected polarization P_Λ after proton emission from all states is expected to be populated by the amounts $0.25-0.39$ depending on the assumption. This value is 1.5–2 times larger than that for the $^{12}C(\pi^+, K^+)$ case [96].

In the study of the asymmetry parameter of the non-mesonic decay, the polarization of the initial ΛN system is relevant, since the non-mesonic decay is almost a two-baryon process. The $^5_\Lambda$He has a particularly simple structure. The polarization of $^5_\Lambda$He is equal to that of Λ in a nucleus since the spin of the core ^{4}He is 0. Thus, the polarization of the initial ΛN system is given by the polarization of $^5_\Lambda$He.

For $^5_\Lambda$He, the mesonic decay can be used to obtain the hypernuclear polarization. On the contrary, the theoretical calculation was used to obtain the Λ-polarization (P_Λ) for the $^{12}C(\pi^+, K^+)$ reaction, since detection of asymmetry for the mesonic decay is practically impossible. The asymmetry parameter of the mesonic decay of $^5_\Lambda$He is predicted to be almost the same as that of free Λ-decay with little ambiguity [97]. The large branching ratio of the mesonic decay $(\Gamma_{\pi^-}/\Gamma_{\text{tot}} \sim 0.4)$ [66] together with the large asymmetry parameter gives a polarization of $^5_\Lambda$He with enough precision to give the asymmetry parameter of the non-mesonic decay being studied.

For the discussion of the non-mesonic decay, we have been assuming a relative s-wave for the initial ΛN system. However, the p-wave can contribute to the decay for the p-shell hypernuclei. Actually, it is pointed out that its contribution is appreciable

[86,94]. The initial ΛN system is only a relative s-wave for $^5_\Lambda$He, which simplifies the comparison with the theoretical calculation.

$^5_\Lambda$He is the lightest hypernucleus that can be used for the polarization measurement. The correction to the asymmetry parameter due to Fermi motion and intranuclear cascade is estimated to be 4%, which is much smaller than 20% for the $^{12}_\Lambda$C case.

The experiment was carried out at the K6 beam line of KEK-PS. The SKS spectrometer, which is used to measure the (π^+, K^+) reaction, is quite appropriate for the measurement since it has good energy resolution (2 MeV FWHM) and large acceptance ($\sim$100 msr) [98]. The large solid angles make the simultaneous measurement of positive and negative scattering angles possible. Combining the decay counter systems placed below and above the target, we can eliminate the lowest-order systematic errors even though the counter system is not rotatable. In the present experiment, it is quite important to measure pions from the mesonic decay of $^5_\Lambda$He, since its asymmetry gives the polarization experimentally. Use of the (π^+, K^+) reaction and the SKS spectrometer with energy resolution of around 2 MeV is particularly important for the measurement. In the previous experiment, which used the (K^-, π^-) reaction [66], pions from the mesonic decay of $^5_\Lambda$He were not observed clearly because of the large cross section of the free Λ-production due to substitutional transition and limited energy resolution.

The decay counter system is shown in Fig. 10.17. In the present experiment, we measure the neutral as well as the charged pions to obtain the polarization of $^5_\Lambda$He. Two sets of decay counter systems are set below and above the target. Since we measure kaons scattered at positive and negative angles simultaneously, spurious asymmetry can be eliminated even though the counter system cannot be rotated. Each detector system consists of a Si microstrip detector of 1 mm pitch, an MWPC, a range shower counter, and 36 NaI detectors. Figure 10.18 shows the range shower counter, which we developed for the present experiment. It works as a range counter for protons below 100 MeV and for charged pions below 40 MeV. It also works as a γ-ray shower counter with fast timing response and moderate energy resolution. The counter consists of 32 layers of thin (0.2 mm) lead and 1 mm plastic sheets. A multi-anode PMT views each plastic scintillator sheet that gives information on the range. A PMT gives information on the total energy. As a γ-ray counter, it achieves an energy resolution about half that of the usual 1 mm lead and 5 mm plastic shower counter for γ-ray energy around 100 MeV because of much finer sampling rate [99]. The energy resolution is twice as bad as that of NaI detector. However, the γ-ray energy deposited in the shower detector is less than half of that in a NaI detector, which makes the energy resolution sufficient for the present purpose. It is safe to use NaI counters under the low counting rate condition because of the range shower counter.

In order to obtain final results, we still have work to do to complete the data analysis. However, we can already give the branching ratio by using the current status of the analysis. Detailed discussion of the preliminary results can be found elsewhere [100]. Briefly, the results suggest that the Γ_n / Γ_p is close to 2, which suggests dominance of isospin 1. It is consistent with the Block and Dalitz analysis [70] but contradicts the meson-exchange predictions.

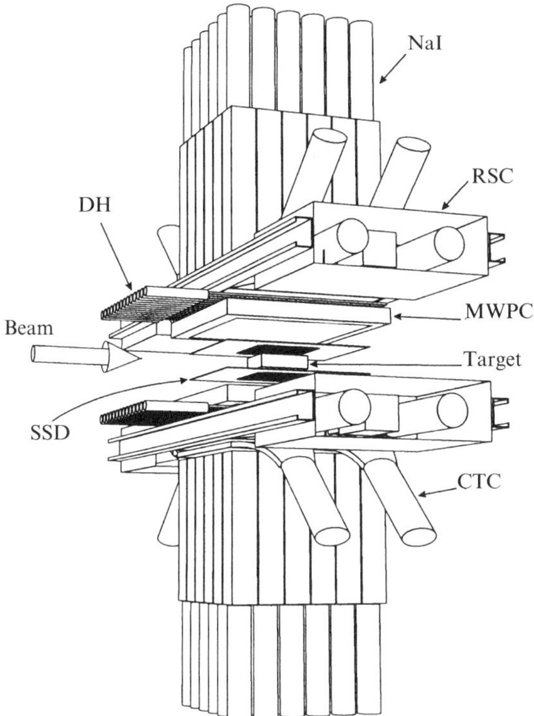

Fig. 10.17 Perspective view of the decay counter system to detect decay particles from the ^{6}Li $(\pi^{+}, K^{+}p)\,^{5}_{\Lambda}$He reaction. Each NaI detector has a dimension of $2.5'' \times 2.5'' \times 12''$ and a PhotoMultiplier Tube (PMT), which gives an idea of the size of the whole system.

The branching ratio indicates the dominance of isospin 1 that corresponds to kaon exchange. The quark cluster model predicts such a dominance. However, the model has difficulty in reproducing the asymmetry parameter [101]. On the other hand, the meson exchange model predicts the asymmetry parameter fairly well, though it fails to explain the branching ratio [94]. The process has a short-range part and a long-range part, which are treated well by the quark model and the meson-exchange model, respectively. A realistic theoretical model needs to include both aspects. A consistent combination of these two models is not straightforward, though efforts have indeed been made in this direction [84,101]. It has been suggested recently that non-mesonic decay involving three nucleons in the final state may make a substantial contribution to the decay rate [102]. Pions virtually emitted by weak decay of Λ are absorbed by the two nucleons which, naturally, are a proton–neutron pair with isospin 0. In this case, the measured branching ratio (Γ_{n}/Γ_{p}) becomes 2, though it does not represent the isospin structure of the two-body reaction. The asymmetry parameter of $^{5}_{\Lambda}$He with good precision should be quite important to clarify the situation. The results will be obtained soon.

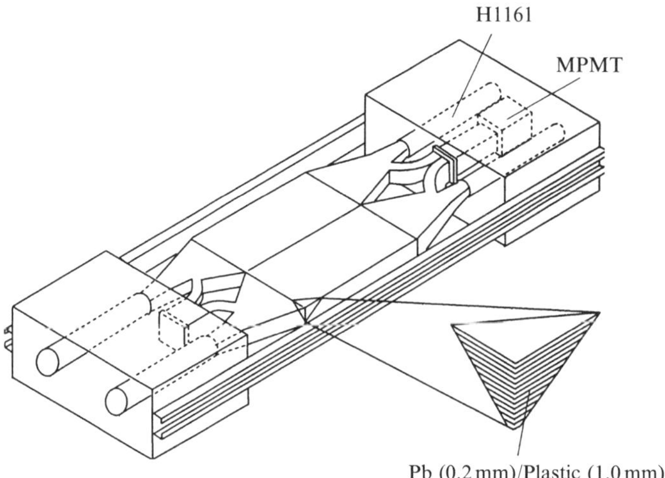

Fig. 10.18 The range shower counter (E278). It gives the energy and the range of charged particles. It has a radiation length (L_R) of $1.2L_R$ and acts as a γ-ray counter with fast timing response. H1161 is the usual 2″ PMT, and MPMT is a multi-anode-type PMT used to observe the range information.

10.6 Strangeness nuclear physics at RCNP and SPring-8

10.6.1 *Strangeness production by the weak* pn → pΛ *process*

The weak non-mesonic decay of Λ-hypernuclei gives information on the weak YN interaction. However, the initial angular momentum and relative momentum states are limited to those given by the hypernuclear wave function. Furthermore, since non-mesonic decay takes place in a nucleus, it is always possible that other nucleons cannot be spectators. The three-nucleon mechanism may be quite important as discussed in the previous section, which obscures the weak YN interaction. If one can measure the cross section of two-body Λp → pn scattering, which is obviously impossible practically, the two-body weak interaction can be derived directly.

On the other hand, the inverse reaction (pn → Λp), that represents a direct generation of the strangeness via the weak interaction, can occur. Since it is pure two-body reaction, the cross section gives directly the transition amplitude of the process. Theoretically, the transition amplitude is calculated first and the non-mesonic decay rate is then obtained by folding the calculated amplitude in the nuclear wave function to obtain the decay rate for comparison with experimental data. Such a procedure is unnecessary for the pn → Λp reaction. The transition rate of non-mesonic decay may depend on the relative wave of the initial ΛN state, dominantly the s- and p-waves. No clear separation is possible for the study of the non-mesonic decay of hypernuclei. The energy dependence of the cross section of the inverse reaction allows one to make a partial wave decomposition of the process if one can measure the cross section.

The asymmetry parameter of the non-mesonic decay is the result of the interference between parity-conserving and party-violating amplitudes. The nuclear effects on the initial Λ-polarization and final state interaction can be estimated only to limited accuracy. Further, the Λ-polarization that we can obtain in the Λ-hypernuclei cannot always be large. Actually, the $^5_\Lambda$He experiment that we carried out seems to give the largest polarization of 0.3–0.4.

This situation can be solved by observing the inverse pn $\rightarrow$ Λp reaction. The analysing power of the reaction cross section with longitudinally polarized proton beam,

$$A = \frac{\sigma(h=1) - \sigma(h=-1)}{\sigma(h=1) + \sigma(h=-1)}, \tag{10.52}$$

where $h = J_\mathrm{p} \cdot P_\mathrm{p}/|J_\mathrm{p}||P_\mathrm{p}|$, gives the interference of parity-conserving and parity-violating amplitude. It is the same quantity as the asymmetry parameter. The polarization of the proton beam can be as large as 0.8. The analysing power has been measured in proton–proton scattering to investigate the parity-violating weak NN interaction. The effects are very small (10^{-7}–10^{-8}), which makes the estimation of the systematic error of the experiment difficult. Since quite large asymmetries were observed in the non-mesonic decay of Λ-hypernuclei, we can expect an effect of the order of unity for the pn $\rightarrow$ Λp reaction. The large asymmetry parameter combined with large longitudinal polarization makes the systematic error easy to handle, though in this particular experiment one has to overcome the drawback of a very small cross section. The overall difficulty of the experiment is similar to the experiments on parity violation in proton–proton scattering.

The Λ produced in the final state has an analysing power for the polarization. This can be used to search for the time-reversal (T) violation in the process, by observing the T-odd correlation $(J_\mathrm{p} \times k_\mathrm{p}) \cdot J_\Lambda$. This process gives an interesting new testing ground for the search. So far, many tests have been carried out in nuclear reactions, β-decays, and decay of mesons, especially kaons. Obviously, the T violation can be expected in the kaon decay, where CP violation has been known for a long time.

It is believed that baryon asymmetry in the present universe is evidence for the T violation in baryogenesis. It has been argued, however, that the CP violation given by the Kobayashi–Maskawa matrix is not large enough to explain the baryon asymmetry of the present universe [103]. Therefore, it is quite probable that a new process is yet to be discovered which has the large T violation. The pn $\rightarrow$ Λp reaction is a strangeness-changing weak process which is similar to the kaon decay, though it is different in some aspects. The kaon is a $q\bar{q}$ system with spin parity 0^-. The pn $\rightarrow$ Λp reaction involves only quarks, and the spin of the intermediate state is predominantly 1 and many other integer spins are possible. The correlation for the present investigation is P-odd and T-odd. There have been relatively accurate measurements for the P-even and T-odd correlation in nuclear β-decay and meson decay, especially for kaons. However, it is difficult to construct a theory that is P-even and T-odd [104], which makes the study of the present correlation interesting. Such a new test has to be made even if the precision is inferior to the existing standard.

It has been proved that there is no null test for the T violation experiment [105]. Thus, one cannot measure the T violation with unlimited accuracy. One always has to investigate the accuracy achievable by the final state interaction. In that sense, neutron scattering at $0°$ is promising [4]. However, one has to practically prepare the analyser of the polarization that brings inefficiency to the measurement. The Λ itself is a good probe for the polarization. The inverse reaction pn $\rightarrow$ Λp presents a good opportunity to investigate the elementary weak YN interaction in terms of cross section, parity violation, and T violation. The question is whether it is feasible and can give enough statistics to investigate the process, which will be discussed in the following section.

10.6.2 *Properties of the* pn $\rightarrow$ Λp *reaction*

The kinematics of the reaction are quite simple. It is a two-body reaction and the Q-value, the mass difference of Λ and neutron, is 176 MeV. The energy of the proton beam has to be larger than 370 MeV for a free neutron target. This energy is achievable in medium-energy accelerators like RCNP at Osaka University, which has a ring cyclotron that provides 400 MeV protons. Since the cross section is very small, as shown in the following section, the proton is a practically unique choice for the beam. One has to use nuclei to provide the neutron target. In this situation, the cross section depends on the Fermi motion of the neutron in a nucleus. The proton and Λ in the final state have small relative momenta. Both particles move in the forward direction, and the Λ decays into pions and nucleons after moving a few centimeters. This decay vertex is characteristic of the production of Λ, and there is no other process that produces pions in the region after the target.

One has to estimate the cross section to test the feasibility of the experiment. The cross section around threshold region can be estimated by using the weak non-mesonic decay of Λ-hypernuclei. The weak non-mesonic decay rate is related to the cross section by

$$\frac{1}{\tau_{\Lambda p \rightarrow pn}} = \langle v\sigma(\Lambda p \rightarrow pn)\rangle_{av} \int \rho_N |u_\Lambda|^2 \, d^3r, \tag{10.53}$$

where $\tau_{\Lambda p \rightarrow pn}$ is the partial-life of hypernuclear non-mesonic decay, ρ the density distribution of a nucleon, and u_Λ the wave function of a Λ. The density distribution of a nucleon is deduced from the charge distribution of a proton. The wave function of Λ can be obtained simply by adjusting the binding energy of the Λ-hypernucleus. Here, the cross section was estimated to be $\sim 10^{-39}$ cm^2, which is quite small compared to the cross section usually seen in nuclear physics, though it is fairly large when compared to usual neutrino cross sections ($\sim 10^{-44}$ cm^2) that typically represent the weak interaction. The cross section due to the weak interaction usually increases as the Q-value increases in the energy region much below the mass of W and Z. Thus, one can expect an increase of the cross section by increasing the beam energy. No Λ is produced by the strong interaction which inevitably accompanies kaon production up to 1.6 GeV. Thus, it should be quite interesting to observe the energy dependence of the cross section up to this energy. Typically, we can have a target with thickness 10^{24} nuclei cm^{-2} and a beam intensity of 10^{12} protons s^{-1}. Since the detection efficiency is of the order of 10%, we can have roughly ten events/day at RCNP. This is sufficient to investigate the cross section and parity violation in a reasonable time. However, for the experiment to search for T violation, much better statistics are necessary. It is interesting to measure the energy dependence of the cross section to investigate the process, and if one can find the energy region where the cross section is much larger than the present estimate, the experiment can be realized. Also, a much higher beam intensity makes the experiment feasible, for which only a couple of laboratories are available in the world.

There are several background processes produced by the strong interaction. The pion production, for example, is found to be of the order of 10^{-27} cm^2, which means that one should overcome a cross section difference of 12 orders of magnitude [106]. For this purpose, we considered a design where the target region is shielded by a heavy metal, thereby shielding the detector from charged particles from the target. However, for such experimental conditions, devices that can tolerate a high counting rate have to be developed. The detector design is currently in progress.

10.6.3 *New possibilities of the (γ, K^+) reaction at SPring-8*

So far, we have discussed the weak YN interaction. It is shown that the 400 MeV ring cyclotron at RCNP can be used for the study. In this section, we would like to mention briefly the strangeness production by the (γ, K^+) reaction, since RCNP is promoting a new beam line to obtain 1–4 GeV photon beams by laser electron Compton scattering. The facility will be constructed at SPring-8, which is a new 8 GeV electron storage ring for the study of synchrotron radiation. This energy is particularly suitable for the production of strangeness by the (γ, K^+) reaction. As we demonstrated earlier, the reaction is characterized by the large spin-flip amplitude which guarantees efficient production of polarized hypernuclei. The beam line and detector design is in progress. Despite this effort, the energy resolution of the γ-ray beam achievable would be a couple of tens of MeV, which is not particularly good to resolve the low-energy structure of the hypernuclear states. It is practically unavoidable because of the energy resolution of the high-energy primary beam. Under this circumstance, the investigation of the elementary process $(\gamma + p \rightarrow K^+ Y)$, where the polarization and asymmetry parameter are vital for the study, is of primary importance [107]. Because of the limited energy resolution in the production process, precise measurement of the decay particles is one way to investigate hypernuclei. For such purposes, the large polarization of the reaction combined with the large asymmetry parameter are useful.

As an example, we would like to discuss an experiment which is based along these lines. One can produce polarized Λ abundantly in the (γ, K^+) reaction. The weak radiative decay of hyperons (e.g. $\Lambda \rightarrow n\gamma$) is less studied compared to the dominant hadronic decay. Again, the branching ratio and the asymmetry parameter are two physical quantities to be studied. Especially, the asymmetry parameter of the radiative decay has been measured fairly precisely only for Σ^+. The observed asymmetry parameter is surprisingly large since a naive model predicts a null result [108]. Since then, many theories have been proposed, and their accuracy can be checked only by conducting new experiments. Recently, the branching ratio of the $\Lambda \rightarrow n\gamma$ decay was measured, though many theories predict similar values [109]. The measurement of the asymmetry parameter is quite important for which the (γ, K^+) reaction could play an important role.

10.7   Conclusions

Strangeness in nuclear physics introduced a new flavour for the study of nuclei. Our discussion has focussed on the weak baryon–baryon interaction from the viewpoint of understanding the consequences of this flavour. Study of the weak ΛN interaction is a vital step towards a unified understanding of the weak baryon–baryon interaction based on the $SU_F(3)$ symmetry. Dynamically, the short-range nature of this process naturally requires the quark degrees of freedom in addition to the meson exchange. Recent progress in this field is largely due to the strong collaboration between theory and experiment, and we can expect that interesting new findings will emerge continuously in the near future.

Acknowledgements

The authors are grateful to Professors H. Ejiri and H. Toki for continuous encouragement during the course of the work discussed here. One of the authors (T. K.) thanks all the members of the KEK experiments E160 and E278 for collaboration. T. M. would like to express his sincere thanks to Professors K. Itonaga and M. Sotona for collaboration that led to several published articles, which formed a major source of the theoretical estimates used in this review. This work has been supported by the Grant-in-Aid for Scientific Research in Priority Areas (Strangeness Nuclear Physics) from the Ministry of Education, Science, Sports and Culture of Japan.

References

1. J. J. de Swart, T. A. Rijken, P. M. Maessen, and R. G. Timmermans, *Il Nuovo Ciment* **A102** (1989) 203 and references therein.
2. Th. A. Rijken, P. M. M. Maessen, and J. J. deSwart, *Nucl. Phys.* **A547** (1992) 245c.
3. E. G. Adelberger and W. C. Haxton, *Ann. Rev. Nucl. Sci.* **35** (1985) 501.
4. A. Masaike, *Proc. IV int. symp. on weak and electromagnetic interactions in nuclei, WEIN'95*, June 1995, Osaka, World Scientific (1995), p. 20.
5. C. B. Dover, L. Ludeking, and G. E. Walker, *Phys. Rev.* **C22** (1980) 2073.
6. H. Bandō and T. Motoba, *Prog. Theor. Phys.* **76** (1986) 1321.
7. R. E. Chrien, *Nucl. Phys.* **A478** (1988) 705c.
8. P. H. Pile *et al.*, *Phys. Rev. Lett.* **66** (1991) 2585.
9. T. Hasegawa *et al.*, *Phys. Rev. Lett.* **74** (1995) 224; T. Hasegawa, Ph.D. Thesis, University of Tokyo (1994).
10. T. Nagae, *Proc. 23rd INS int. symp. on nuclear and particle physics with meson beams in the* 1 GeV$/c$ *region*, 1995, Tokyo; see also T. Motoba, *ibid.*
11. C. B. Dover and E. Walker, *Phys. Rep.* **89** (1982) 1.
12. H. Bandō, *Genshikaku-Kenkyu* (*INS, Tokyo*) **31** (1986) 99.
13. H. Ejiri, T. Fukuda, T. Shibata, H. Bandō, and K.-I. Kubo, *Phys. Rev.* **C36** (1987) 1435.
14. M. Sotona and J. Žofka, *Prog. Theor. Phys.* **81** (1989) 160.
15. H. Bandō, T. Motoba, M. Sotona, and J. Žofka, *Phys. Rev.* **C39** (1989) 587.
16. M. Sotona, J. Žofka, H. Bandō, and V. N. Fetisov, *Czech. J. Phys.* **B39** (1989) 153.

17. T. Kishimoto, H. Ejiri, and H. Bandō, *Phys. Lett.* **B232** (1989) 24.

18. R. D. Baker *et al.*, *Nucl. Phys.* **B141** (1987) 29; *ibid.*, *Nucl. Phys.* **B126** (1977) 365; D. H. Saxon *et al.*, *Nucl. Phys.* **B162** (1980) 522; K. W. Bell *et al.*, *Nucl. Phys.* **B222** (1983) 389.

19. G. P. Gopal *et al.*, *Nucl. Phys.* **B119** (1977) 362.

20. T. Motoba, H. Bandō, R. Wünsch, and J. Žofka, *Phys. Rev.* **C38** (1988) 1322.

21. S. Ajimura *et al.*, *Nucl. Phys.* **A547** (1992) 47c.

22. S. Ajimura *et al.*, *Phys. Rev. Lett.* **68** (1992) 2137.

23. S. Ajimura *et al.*, *Phys. Lett.* **B282** (1992) 293.

24. C. Milner *et al.*, *Phys. Rev. Lett.* **54** (1985) 1237.

25. K. Itonaga, T. Motoba, and M. Sotona, *Prog. Theor. Phys.* (Suppl.) **117** (1994) 17.

26. K. Itonaga, T. Motoba, and H. Bandō, *Prog. Theor. Phys.* **84** (1990) 291.

27. H. Ejiri, T. Kishimoto, and H. Noumi, *Phys. Lett.* **B225** (1989) 35.

28. A. Higashi *et al.*, *Abstract booklet of the int. conf. on spin–isospin responses and weak processes in hadrons and nuclei*, 1994, Osaka, p. 10.

29. T. Kishimoto *et al.*, KEK-PS proposal E278; see also *Nucl. Phys.* **A577** (1994) 263c.

30. J. Cohen, *Int. J. Mod. Phys.* **A4** (1989) 1.

31. R. A. Adelseck and B. Saghai, *Phys. Rev.* **C42** (1990) 108.

32. C. B. Dover and D. J. Millener, in *Modern topics in electron scattering*, B. Frois and I. Sick, eds (World Scientific, Singapore, 1992), Vol. 184, No. 1 (1989), p. 1.

33. V. N. Fetisov, M. I. Kozlov, and A. I. Lebedev, *Phys. Lett.* **B38** (1972) 129.

34. A. M. Bernstein, T. W. Donnelly, and G. N. Epstein, *Nucl. Phys.* **A358** (1981) 195c.

35. S. S. Hsiao and S. R. Cotanch, *Phys. Rev.* **C28** (1983) 1668; *Nucl. Phys.* **A450** (1986) 419c.

36. J. Cohen, M. W. Price, and G. E. Walker, *Phys. Lett.* **B188** (1987) 393.

37. H. Tanabe, M. Kohno, and C. Bennhold, *Phys. Rev.* **C39** (1989) 741.

38. W. Schorsch, J. Tietge, and W. Weilnböck, *Nucl. Phys.* **B25** (1970) 179.

39. M. Sotona, K. Itonaga, T. Motoba, O. Richter, and J. Žofka, *Nucl. Phys.* **A547** (1992) 63c.

40. T. Motoba, M. Sotona, and K. Itonaga, *Prog. Theor. Phys.* (Suppl.) **117** (1994) 123.

41. M. Sotona and J. Žofka, *Czech. J. Phys.* **B40** (1990) 1091.

42. K. Itonaga, T. Motoba, O. Richter, and M. Sotona, *Phys. Rev.* **C49** (1994) 1045.

43. H. Bandō, T. Motoba, and J. Žofka, *Int. J. Mod. Phys.* **5** (1990) 4021.

44. T. Motoba, K. Itonaga, and H. Bandō, *Nucl. Phys.* **A489** (1988) 683.

45. H. Bandō, *Nuclear weak process and nuclear structure* (World Scientific, 1989), p. 450.

46. E. Oset and L. L. Salcedo, *Nucl. Phys.* **A443** (1985) 704; E. Oset, P. Fernandez de Cordoba, L. L. Salcedo, and R. Brockmann, *Phys. Rep.* **188**(2) (1990) 79.

47. R. H. Dalitz, *Phys. Rev.* **112** (1958) 605.

48. R. H. Dalitz and L. Liu, *Phys. Rev.* **116** (1959) 1312.

49. D. Ziemeńska, *Nucl. Phys.* **A242** (1975) 461.

50. M. A. B. Beg, *Ann. Phys.* **13** (1961) 110.

51. J. Delorme, M. Ericson, and G. Fäldt, *Nucl. Phys.* **A240** (1975) 493.

52. M. Ericson and H. Bandō, *Phys. Lett.* **B273** (1990) 169.

53. M. Ericson and T. E. O. Ericson, *Ann. Phys.* (*NY*) **36** (1966) 323.

54. J. A. Carr, H. McManus, and K. Stricker-Bauer, *Phys. Rev.* **C25** (1982) 952.

55. Review of particle properties, *Phys. Rev.* **D50** (1994) 1173.

56. P. D. Barnes, *Nucl. Phys.* **A450** (1986) 43c; **A478** (1988) 127c; **A479** (1988) 89c.

57. P. D. Barnes and J. J. Szymanski, *Proc. 1986 INS int. symp. on hypernuclear physics* H. Bandō, O. Hashimoto, and K. Ogawa, eds 1986, Tokyo, p. 136.

58. A. Sakaguchi *et al.*, *Phys. Rev.* **C43** (1991) 73.

59. P. L. Kapur and R. E. Peierls, *Proc. Roy. Soc. London* **A166** (1938) 277; A. M. Lane and D. Robson, *Phys. Rev.* **185** (1969) 1403; C. Bloch, *Nucl. Phys.* **4** (1957) 503.

60. G. Bohm *et al.*, *Nucl. Phys.* **B4** (1968) 511.

61. W. Gajewski, J. Naisse, J. Sacton, P. Vilain, and G. Wilquet, *Nucl. Phys.* **B14** (1969) 11.

62. G. Bohm and U. Krecker, *Sov. J. Part. Nucl.* **3** (1972) 162.

63. T. Motoba, *Proc. IV int symp. on weak and electromagnetic interactions in nuclei, WEIN'95*, June 1995, Osaka, World Scientific (1995), p. 504.

64. D. H. Davis, private communication; Bulletin de I.U.L.B. 26, 27 (1966).

65. F. Ajzenberg-Selove, *Nucl. Phys.* **A413** (1984) 1, and references therein.

66. J. Szymanski *et al.*, *Phys. Rev.* **C43** (1991) 849.

67. G. Bohm *et al.*, *Nucl. Phys.* **B23** (1970) 93; *Nucl. Phys.* **B24** (1970) 248.

68. J. P. Lagnaux *et al.*, *Nucl. Phys.* **60** (1964) 97.

69. T. Motoba, H. Bandō, T. Fukuda, and J. Žofka, *Nucl. Phys.* **A534** (1991) 597.

70. M. M. Block and R. H. Dalitz, *Phys. Rev. Lett.* **11** (1963) 96.

71. M. M. Block *et al.*, *Proc. int. conf. on hyperfragments*, St. Cergue, 1963, CERN Report 64-1 (Geneva, 1964), p. 63.

72. R. J. Prem and P. H. Steinberg, *Phys. Rev.* **136** (1964) 1803.

73. J. Cuevas *et al.*, *Nucl. Phys.* **B1** (1967) 411.

74. R. Grace *et al.*, *Phys. Rev. Lett.* **55** (1985) 1055.

75. C. B. Dover, *Few-Body Systems* (Suppl.) **2** (1987) 77.

76. J. P. Bocquet *et al.*, *Phys. Lett.* **182B** (1986) 146; **192B** (1987) 312.

77. J. B. Adams, *Phys. Rev.* **156** (1967) 1611.

78. B. H. J. McKellar and B. F. Gibson, *Phys. Rev.* **C30** (1984) 322.

79. K. Takeuchi, H. Takaki, and H. Bandō, *Prog. Theor. Phys.* **73** (1985) 841.

80. J. Dubach, *Nucl. Phys.* **A450** (1986) 71c.

81. C. Y. Cheung, D. P. Heddle, and L. S. Kisslinger, *Phys. Rev.* **C27** (1983) 335.

82. R. A. Schumacher, *Nucl. Phys.* **A547** (1992) 143c.

83. R. A. Schumacher, *Proc. U.S.–Japan Seminar on 'properties and interactions of hyperons'*, Hawai, October 1993, World Scientific (1994), p. 85.

84. T. Motoba, K. Itonaga, and H. Bandō, *Nucl. Phys.* **A849** (1988) 683.

85. T. Kishimoto, KEK 83-6 (1983) 51.

86. A. Ramos, C. Bennhold, E. van Meijgaard, and B. K. Jenning, *Phys. Lett.* **B264** (1991) 233.

87. T. Fukuda *et al.*, *Proc. 1986 INS int. symp. on hypernuclear physics*, Tokyo, 1986, p. 170.

88. A. Masaike and H. Ejiri, *Proc. 3rd LAMPF II workshop*, Los Alamos, 1983, LA-9933-C P823.

89. H. Ejiri *et al.*, KEK proposal E160.

90. M. Akei *et al.*, *Nucl. Inst. Meth.* **A283** (1989), 46.

91. L. Lee *et al.*, *Nucl. Inst. Meth.* **A327** (1993) 287.

92. R. H. Dalitz, D. H. Davis, and D. N. Tovee, *Nucl. Phys.* **A450** (1986) 311c.

93. K. Itonaga, T. Motoba, O. Richter, M. Sotona, and J. Žofka, *Proc. 5th int. symp. mesons & light nuclei*, Prague, 1991.

94. A. Ramos, E. van Meijgaard, C. Bennhold, and B. K. Jannings, *Nucl. Phys.* **A544** (1992) 703.

95. T. Kishimoto *et al.*, KEK-PS proposal E278.

96. T. Motoba and K. Itonaga, paper in preparation.

97. T. Kishimoto, T. Motoba, and K. Itonaga, contributed paper to the PANIC, 1993, p. 629.

98. T. Fukuda *et al. Nucl. Inst. Meth.* **A361** (1995) 485.

99. T. Kishimoto *et al.* INS annual report 1993.

100. H. Noumi *et al.*, *Proc. IV int. symp. on weak and electromagnetic inteactions in nuclei, WEIN'95*, June 1995, Osaka, World Scientific (1995), p. 550.

101. M. Oka, *Proc. IV int. symp. on weak and electromagnetic inteactions in nuclei, WEIN'95*, June 1995, Osaka, World Scientific (1995), p. 540.

102. W. M. Alberico, A. De Pace, M. Ericson, and A. Molinari, *Phys. Lett.* **B256** (1991) 134; A. Ramos, E. Oset and L. L. Salcedo, Phys. Rev. **C50** (1994) 2314; S. Shinmura, *Proc. IV int symp. on weak and electromagnetic inteactions in nuclei, WEIN'95*, June 1995, Osaka, World Scientific (1995), p. 528.

103. A. G. Cohen, D. B. Kaplan, and A. E. Nelson, *Ann. Rev. Nucl. Part. Sci.* **43** (1993) 27.

104. V. P. Gudokov, *Phys. Rep.* **212**(2) (1992) 77.

105. H. E. Conzett, *Proc. IV int symp. on weak and electromagnetic inteactions in nuclei, WEIN'95*, June 1995, Osaka, World Scientific (1995), p. 68.

106. W. W. Daehnick *et al.*, *Phys. Rev. Lett.* **74** (1995) 2913.

107. K. Maeda, private communication, 1995.

108. M. Kobayashi *et al.*, *Phys. Rev. Lett.* **59** (1987) 868.

109. B. Bassalleck, *Nucl. Phys.* **A585** (1995) 255c.

11

Quark nuclear physics

Hiroshi Toki and Hideo Suganuma

11.1 Introduction

In traditional nuclear physics, protons and neutrons, the constituents of a nucleus, are assumed to be point particles and are described by a Hamiltonian with a kinetic term and an interaction term, possibly obtained from the Nucleon–Nucleon (NN) scattering data. The purpose is then to describe the binding energies of nuclei, their structures, decay properties, reaction rates, etc. The fundamental model used is the shell model [1].

About 20 years ago, other degrees of freedom were introduced to nuclear physics. The sensational one is pion condensation, in which the energy of pion becomes zero in nuclear matter, due to the interaction of pions with the nuclear medium, which includes also the delta isobar [2,3]. More recently, deep inelastic scattering has been used to probe the nucleons in nuclear medium and to study differences from their counterparts in free space, which is called the EMC effect [4]. A more drastic process is the production of high-density and high-temperature nuclear matter by relativistic heavy-ion collisions [5]. Such processes could even reduce nuclear matter from hadronic to a quark–gluon plasma phase.

There are several hints that we treat nucleons in nuclei differently from those in free space. This leads us to study the nucleons (hadrons) and nuclei in terms of quarks and gluons. We term this field of study as Quark Nuclear Physics (QNP). Here, the building blocks of matter are the new elementary particles of the standard model: quarks interacting through gluons. We believe that their fundamental theory is Quantum ChromoDynamics (QCD), a non-abelian gauge theory of SU(3) colour.

In QNP, the most essential phenomena are confinement of quarks and gluons, and chiral symmetry breaking. Quarks are not found in free space, but are observed in deep inelastic scattering. They are present in hadrons and hence an understanding of confinement is essential to determine the sizes of hadrons.

The chiral symmetry is found in the QCD Lagrangian,

$$L_{\mathrm{QCD}} = \bar{q}(i\gamma_\mu D^\mu - m)q + \tfrac{1}{4}G_{a\mu\nu}G^{a\mu\nu}, \tag{11.1}$$

where q denotes the quark field with the current mass m, γ_μ the Dirac matrix, and D^μ the covariant derivative. The second term provides the gluon dynamics with self-interaction terms due to its non-abelian nature. In the u–d sector, the current mass m is considered to be negligible ($\sim$5 MeV) compared with the hadron mass ($\sim$1 GeV), and hence the chiral symmetry is realized with high accuracy. We then expect the chiral partners, i.e.

parity doublets, to be degenerate in the u–d meson spectrum. Nature, on the other hand, shows that the pion (0^-) has a mass of 139 MeV with no 0^+ partner. The ρ meson (1^-) is 500 MeV apart from its chiral partner, the a_1 meson (1^+). The same feature is also found for the baryon spectrum. This property of the hadron spectra tells us that the chiral symmetry is broken spontaneously and the pion appears as its Nambu–Goldstone particle. The chiral symmetry breaking should be understood for a full description of QNP. This would provide the constituent quark masses of about 0.3 GeV, which then give the hadron masses, and even the small-mass pions, which are responsible for NN interaction.

How do these phenomena occur? To answer this question, let us consider the QCD quark–gluon coupling constant α_s, which varies with relevant momentum scale obeying the renormalization group of QCD [6]. At large momenta, α_s diminishes and the theory becomes asymptotically free, and hence the momentum dependence of deep inelastic scattering can be calculated perturbatively. At small momenta, where confinement and chiral symmetry breaking are expected, α_s blows up and hence we face highly non-perturbative processes. We ought to find the essential degrees of freedom to get an insight into these phenomena.

In this chapter, we discuss how these most basic phenomena are described in terms of the Dual Ginzburg–Landau (DGL) theory and the ideas needed to arrive at the DGL theory. The DGL theory and its outcome are described in Section 11.2. In Section 11.3 we shall discuss the relation of the DGL theory, in particular the monopoles, with the other important topological object of QCD, i.e. the instantons. We mention briefly in Section 11.4 the experimental efforts using the laser Compton scattered energetic photons at the SPring-8 light source facility to clarify various important quantities of the DGL predictions. In Section 11.5, we provide the conclusions of this paper and discuss future perspectives of QNP. For interested readers, we provide details of various concepts in the appendix.

11.2 The dual Ginzburg–Landau theory

11.2.1 *Colour confinement modelling*

It was Nambu who introduced an interesting view of colour confinement in 1974 [7]. Now, suppose we insert a superconductor, which does not allow the magnetic field to penetrate it, into a magnetic field. If it were to allow the magnetic field, as in a superconductor of the second kind, the magnetic field should be confined in a small vortex-like configuration by breaking the Cooper pairs slightly. This is known as the Meissner effect. Nambu takes its dual version for quark confinement. If the vacuum is normal, the colour electric field should look like the Coulomb potential between a positive and a negative colour charge. If the vacuum is superconductor-like (dual superconductor), then the vacuum inhibits the electric field from passing through and, hence, the colour electric flux ought to be confined in a vortex-like configuration. This is called the dual Meissner effect. This picture, however, did not become popular, because it required colour magnetic monopoles. In the superconductor, the charged object, a Cooper pair, condenses, while in the QCD vacuum, the magnetically charged object, a colour magnetic monopole, also condenses.

't Hooft demonstrated the natural appearance of colour magnetic monopoles in QCD [8]. In a non-abelian gauge theory like QCD, he introduced a particular gauge named abelian gauge, to reduce it to the abelian gauge theory like QED. From a topological argument, colour magnetic monopoles appear—when some condition is met—in the abelian space. (For details, see the appendix.) This finding then supports the idea of Nambu for confinement. Hence, QCD reduces naturally to QED with magnetic monopoles, which is described by Maxwell equations with magnetic charges and currents, as studied by Dirac [9]. These Maxwell equations have the duality symmetry, which arises naturally in the special gauge of QCD.

11.2.2 *The dual Ginzburg–Landau theory*

It took further 10 years before the above idea was formulated in the form of the DGL Lagrangian [10]. The DGL theory is expressed by the following Lagrangian:

$$L_{\text{DGL}} = L_{\text{dual}} + \bar{q}\big(i\gamma_\mu \partial^\mu - m + e\gamma_\mu A^\mu H\big)q$$

$$+ \sum_{\alpha=1}^{3} \big|(i\partial_\mu + g\boldsymbol{\varepsilon}_\alpha \boldsymbol{B}_\mu)\chi_\alpha\big|^2 - \lambda \sum_{\alpha=1}^{3} \big(|\chi_\alpha|^2 - v^2\big)^2. \tag{11.2}$$

Here, L_{dual} denotes the dual version of the gauge field tensor (see the explicit form in Appendix A.5), q is the quark field with mass m and χ_α is the monopole field with the monopole charge $g\boldsymbol{\varepsilon}_\alpha$, where $\boldsymbol{\varepsilon}_\alpha$ is the root vector of SU(3) and g the dual coupling constant, which satisfies the Dirac condition, $eg = 4\pi$; $\boldsymbol{B}_\mu$ is the dual gauge field and A^μ is the gauge field with third and eighth components; $\boldsymbol{H} = (\lambda_3/2, \lambda_8/2)$ is the diagonal part of the SU(3) generators. The last term is the Higgs term which causes monopole condensation, where λ and v are the parameters of the DGL Lagrangian. It is important to note that this DGL Lagrangian is derived from the QCD Lagrangian by assuming the existence of the monopole field and the abelian dominance [11]. It is also supported by the recent lattice QCD calculations [12]. This is the Lagrangian Dirac anticipated, which ought to appear in some non-abelian gauge theory like QCD.

At the mean field level, the energy minimum appears at $\langle \chi_\mu \rangle = v$, which indicates monopole condensation. Writing the monopole fields in terms of the vacuum expectation value and their fluctuation,

$$\chi_\alpha = (v + \tilde{\chi}_\alpha)\mathrm{e}^{i\xi_\alpha}, \tag{11.3}$$

we get the DGL Lagrangian as

$$L = L_{\text{dual}} + \bar{q}\big(i\gamma_\mu \partial^\mu - m + e\gamma_\mu A^\mu H\big)q + \frac{1}{2}m_B^2 B_\mu^2 + \sum_{\alpha=1}^{3} \big[\big(\partial_\mu \tilde{\chi}_\alpha\big)^2 - m_\chi^2 \tilde{\chi}_\alpha^2\big]$$

$$+ \sum_{\alpha=1}^{3} \big[g^2\big(\boldsymbol{\varepsilon}_\alpha \cdot \boldsymbol{B}_\mu\big)^2\big(\tilde{\chi}_\alpha^2 + 2\tilde{\chi}_\alpha v\big)^2 - \lambda\big(4v\tilde{\chi}_\alpha^3 + \tilde{\chi}_\alpha^4\big)\big]. \tag{11.4}$$

Here, the dual gauge field has mass $m_B = \sqrt{3}gv$, and the monopole fluctuation field also has mass $m_\chi = 2\sqrt{\lambda}v$. The monopole condensation provides the mass m_B to the dual gauge field and the two degrees of freedom of the phase of the monopole field are changed into the longitudinal components of the dual gauge fields. We note that the monopole condensation breaks the internal gauge invariance, $U(1)_m$, but not the original gauge invariance, $U(1)_e$. The finite mass of the dual gauge fields indicates that the colour electric fields are able to penetrate the QCD vacuum only by a distance $1/m_B$, which is the penetration length. Hence, we get the dual Meissner effect, by which the colour electric fields are to be confined.

11.2.3 *The static confining potential*

The first application is the $q\bar{q}$ static potential obtained by putting a $q\bar{q}$ pair at some distance r [13]. We treat the $q\bar{q}$ pair as the external source j_μ and calculate the current–current correlation,

$$L_j = -\tfrac{1}{2}j_\mu D^{\mu\nu}j_\nu. \tag{11.5}$$

Here, the gluon propagator in the monopole condensed vacuum is written in the Lorenz gauge,

$$L_{\text{G.F.}} = -\frac{1}{2\alpha}\left(\partial_\mu A^\mu\right)^2,$$

as

$$D_{\mu\nu} = \frac{1}{\partial^2}\left(g_{\mu\nu} + (\alpha - 1)\frac{\partial_\mu \partial_\nu}{\partial^2}\right) - \frac{1}{\partial^2}\frac{m_B^2}{\partial^2 + m_B^2}\frac{n^2}{(n \cdot \partial)^2}X_{\mu\nu}, \tag{11.6}$$

where

$$X_{\mu\nu} = \frac{1}{n^2}\varepsilon^\lambda_{\mu\alpha\beta}\varepsilon_{\lambda\nu\gamma\delta}n^\alpha n^\gamma \partial^\beta \partial^\delta. \tag{11.7}$$

We can arrive at the above expressions by integrating the partition functional first with respect to B_μ and then with respect to A_μ. Putting the $q\bar{q}$ source term explicitly as

$$j_\mu(x) = -\left(Q_a g_{\mu 0}\delta^3(x - a) + Q_b g_{\mu 0}\delta^3(x - b)\right), \tag{11.8}$$

which is written in the momentum space as

$$j_\mu(k) = -\left(Q_a g_{\mu 0}e^{-ika} + Q_b g_{\mu 0}e^{-ikb}\right), \tag{11.9}$$

we get the action

$$S_j = -\frac{1}{2}\int \frac{d^4k}{(2\pi)^4}j_\mu(-k)D^{\mu\nu}j_\nu(k) = -V(x)\int dt. \tag{11.10}$$

Hence, we get the potential as $V(r; n) \equiv V_{\text{Yukawa}}(r) + V_{\text{linear}}(r)$ and the Yukawa part is given as

$$V_{\text{Yukawa}}(r) = \frac{Q_a Q_b}{4\pi}\frac{e^{-m_B r}}{r}. \tag{11.11}$$

The linear potential is given as

$$V_{\text{linear}}(r) = \frac{e^2 m_B^2}{24\pi} \ln\left(\frac{m_\chi^2 + m_B^2}{m_B^2}\right) \cdot r. \qquad (11.12)$$

Here we have used the charge product $Q_a Q_b = -e^2/3$ for a colour singlet $q\bar{q}$ pair. Hence, we can write the string tension as

$$k = \frac{e^2 m_B^2}{24\pi} \ln\left(\frac{m_\chi^2 + m_B^2}{m_B^2}\right). \qquad (11.13)$$

This form corresponds to the energy per unit length of the vortex in the superconductor of the second kind

$$\varepsilon = \frac{\phi_0^2 m_A^2}{32\pi^2} \ln\left(\frac{m_\phi^2}{m_A^2}\right). \qquad (11.14)$$

This expression does not contain m_A^2 in the logarithm, since it has been neglected in comparison to m_ϕ^2. This correspondence suggests that our way of deriving the potential is correct.

The result is shown in Fig. 11.1 together with the phenomenological potential. This procedure fixes the parameters of the DGL Lagrangian. Due to the existing ambiguity, there exist various sets of parameters. In terms of the glueball mass, which appears in the DGL theory and has a strong connection with the QCD vacuum and confinement, it turns out that $M(0^{++}) \sim 1.5\,\text{GeV}$ and $M(1^+) \sim 0.5\,\text{GeV}$. These masses, however, should be looked at with caution. These objects exist in abelian space, and we have to check if they are $SU(3)_c$ invariant. This condition is discussed in Appendix A.3. The 0^{++} glueball is found to be a colour singlet, but not the 1^+ particle. Hence, it is very

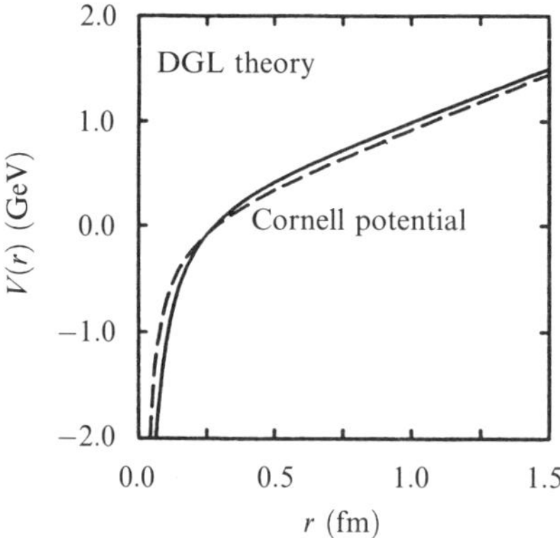

Fig. 11.1   The static potential between a quark and antiquark pair calculated within the DGL theory compared with the phenomenological potential of the Cornell group. (From Suganuma *et al.* [13].)

interesting to search for the 0^{++} glueball experimentally. The appearance of the linear potential is not surprising, since it is modelled in the DGL theory. It is worthwhile to stress, however, that there are no other models which are able to realize confinement of colours and at the same time have a strong link with QCD. The real challenge now is the chiral symmetry breaking, which is discussed next.

11.2.4 *Spontaneous chiral symmetry breaking*

Spontaneous chiral symmetry breaking is directly related to the quark mass generation in the QCD vacuum [13]. How do quarks then behave in a monopole condensed vacuum? This requires solving the Schwinger–Dyson (SD) equation, where quarks get the self-energy corrections due to the non-perturbative interaction with gluons. This seems, however, unphysical, because quarks are confined. It means that whenever a quark is present, there should be an antiquark or a diquark to make the system colour singlet. In principle, therefore, we ought to solve the many-body system to discuss single quarks. Suppose we have the SD equation schematically written as

$$S^{-1}(p) = S_0^{-1}(p) + \int_0^\infty S(p-q)D(q)\,\mathrm{d}q. \tag{11.15}$$

The quark, which is confined, should be seen from the position of the antiquark. Then the probability to find the quark is finite only within the distance of the hadronic scale, i.e. ~ 1 fm. Therefore, gluons cannot travel any distance freely. Rather, they are confined also within the hadronic distance. This indicates that there should be an infrared cut-off, which is of the order of the inverse of the confining distance R as $q > q_c = 1/R$. Hence, the SD equation is modified simply to

$$S^{-1}(p) = S_0^{-1}(p) + \int_{q_c}^\infty S(p-q)D(q)\,\mathrm{d}q. \tag{11.16}$$

We shall write the expression more precisely using the rainbow approximation:

$$S_q^{-1}(q) = \mathrm{i}p\!\!\!/ + \int \frac{\mathrm{d}^4 k}{(2\pi)^4} Q^2 \gamma^\mu S_q(k)\gamma^\nu D_{\mu\nu}(p-k). \tag{11.17}$$

We assume the following simple form for the quark propagator:

$$S_q^{-1}(q) = \mathrm{i}p\!\!\!/ - M(p^2). \tag{11.18}$$

We assume that the n dependence is not included in this expression, since we take the angle average of the Dirac string in the gluon propagator, as the light quarks move around in the confined region. The angle average is expressed as

$$\left\langle \frac{1}{(n\cdot k)^2 + \varepsilon^2} \right\rangle_{\mathrm{ave}} = \frac{1}{2\pi^2} \int \mathrm{d}\Omega_n \frac{1}{(n\cdot k)^2 + \varepsilon^2} = \frac{2}{\varepsilon} \frac{1}{\varepsilon + \sqrt{k^2 + \varepsilon^2}}. \tag{11.19}$$

Here, ε is introduced as a cut-off mass in a covariant way in order to represent the confining nature of quarks and gluons in hadrons instead of the low-momentum threshold,

q_c, as discussed above. Taking the trace of the SD equation, we find the follwing equation for the mass function:

$$M(p^2) = \int \frac{\mathrm{d}^4 k}{(2\pi)^4} Q^2 \frac{M(k^2)}{k^2 + M(k^2)^2} \left[\frac{2}{\tilde{k}^2 + m_B^2} + \frac{1 + \alpha_e}{\tilde{k}^2} \right.$$
$$\left. + \frac{4}{\varepsilon} \frac{1}{\varepsilon + \sqrt{k^2 + \varepsilon^2}} \left(\frac{m_B^2 - \varepsilon^2}{\tilde{k}^2 + m_B^2} + \frac{\varepsilon^2}{\tilde{k}^2} \right) \right].$$

$$(11.20)$$

We show in Fig. 11.2 the result of chiral symmetry breaking, expressed in terms of the quark mass $M(p)$ [13]. It becomes finite with increasing strength of the monopole condensation, which is indicated in the figure by m_B, the mass of the dual gluon field. We find also the pion decay constant and the quark condensate to be close to their semi-empirical values. This calculation shows that monopole condensation is the source of both the confinement and the chiral symmetry breaking. It is interesting to note that the condition for the appearance of the finite quark mass could be worked out by studying the equation for the mass function near the threshold of symmetry breaking and is written as [13]

$$e^2 m_B \geq 2\pi\varepsilon.$$

$$(11.21)$$

This relation implies that the non-trivial solution appears only for finite m_B, which is related to the strength of the monopole condensation.

It is interesting to mention that lattice QCD supports the findings of the DGL theory. The principal assumption of the DGL theory is the abelian dominance for the non-perturbative phenomena. The $q\bar{q}$ potential calculated with full lattice QCD agrees with the result of only its abelian part [12]. This agreement indicates that the confinement

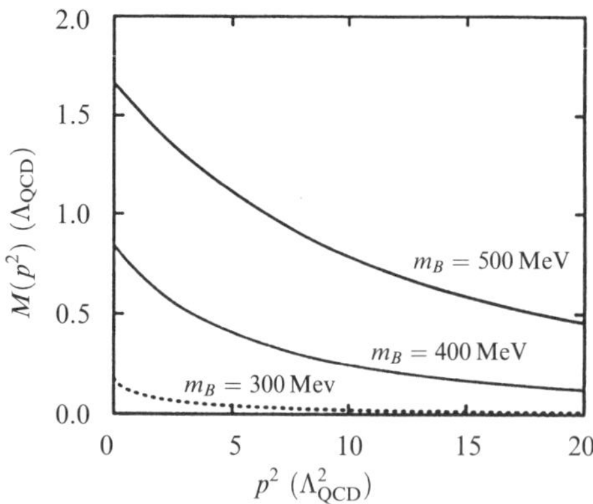

Fig. 11.2 The constituent quark mass calculated within the DGL theory with various values of the dual glueball mass, which indicates the strength of the monopole condensation as a function of the Euclidean momentum squared. The unit Λ_{QCD} is 200 MeV. (From Suganuma *et al.* [13].)

physics is describable in terms of only the abelian gluons. Stimulated by the finding of the DGL theory on the chiral symmetry breaking, the lattice QCD study was carried out also for chiral symmetry breaking with full QCD gluons and with only the abelian gluons. Again, the abelian gluons were found to play a dominant role in chiral symmetry breaking [14]. Other investigations in this direction are being done by several groups [15–17]. All these results indicate that the low-momentum phenomena could be described by the abelian gauge gluons (abelian dominance), provided the gauge is chosen suitably.

11.2.5 *Recovery of chiral symmetry at finite temperature*

We now discuss the recovery of the chiral symmetry at finite temperature [18], which we can be formulated in the imaginary time prescription:

$$p_4 \rightarrow \omega_n = (2n+1)\pi T, \tag{11.22}$$

$$\int \frac{\mathrm{d}^4 k}{(2\pi)^4} \rightarrow T \sum_{n=-\infty}^{n=\infty} \int \frac{\mathrm{d}^3 k}{(2\pi)^3}, \tag{11.23}$$

$$M(p^2) \rightarrow M_T(\omega_n, \boldsymbol{p}). \tag{11.24}$$

Since the SD equation obtained by this replacement is not manageable, we make 'covariant-like ansatz' for the quark mass at various energies as

$$M_T(\omega_n, p) = M_T(\hat{p}^2); \quad \hat{p}^2 = p^2 + \omega_n^2. \tag{11.25}$$

In the limit $T \rightarrow 0$ this approximation reproduces the SD equation at $T = 0$. The correctness of this ansatz is also checked at low temperatures.

We show the calculated results in Fig. 11.3. We, in fact, find the recovery of the chiral symmetry as indicated in Fig. 11.3 by the ratio of quark condensate at finite and zero temperatures, $\langle \bar{q}q \rangle_T$, which decreases with temperature and eventually drops to zero, indicating the recovery of chiral symmetry. We note that the temperature of the phase

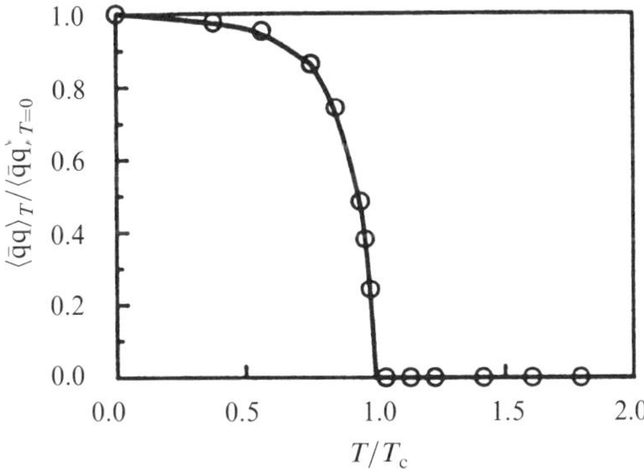

Fig. 11.3 The ratio of quark condensate at finite and zero temperatures within the DGL theory as a function of temperature. The critical temperature is about 0.1 GeV. (From Sasaki *et al.* [13].)

transition, $T_c \sim 0.1$ GeV, seems too low compared to the result of the lattice QCD. This difference is caused by the use of constants in the parameters of the Higgs term. Since this term is introduced at zero temperature, it is likely that it depends on temperature as in the case of a superconductor. In addition, the hadronic scale should also depend on temperature. Here, our intention is merely to demonstrate that the DGL theory provides phase transition to the normal phase at finite temperature.

11.2.6 *Deconfinement phase transition at finite temperature*

We can also discuss the confinement–deconfinement phase transition at finite temperature [19]. In the quenched approximation, we can write the DGL Lagrangian in terms of the dual gauge fields by integrating out the gauge fields A_μ. It amounts to calculating the partition functional

$$Z[J] = \int D\chi_\alpha \, DB_\mu \exp\left(i \int d^4x \left\{ L_{\text{DGL}} - J \sum_{\alpha=1}^{3} |\chi_\alpha|^2 \right\}\right), \tag{11.26}$$

where

$$L_{\text{DGL}} = -\frac{1}{4}(\partial_\mu B_\nu - \partial_\nu B_\mu)^2 + \sum_{\alpha=1}^{3} |(i\partial_\mu + g\boldsymbol{\varepsilon}_\alpha B_\mu)\chi_\alpha|^2 - \lambda \sum_{\alpha=1}^{3}(|\chi_\alpha|^2 - v^2)^2. \tag{11.27}$$

Limiting ourselves up to the harmonic terms, after the separation of the monopole field into the mean field value and the fluctuation as $\chi_\alpha = (\bar\chi + \tilde\chi_\alpha)\exp(i\xi_\alpha)$, we find the partition function as

$$Z[J] = \exp\left(i \int d^4\{-3\lambda(\bar\chi^2 - v^2)^2 - 3J\bar\chi^2\}\right) \left[\det\left(iD_B^{-1}\right)\right]^{-1} \left[\det\left(iD_\chi^{-1}\right)\right]^{-3/2}. \tag{11.28}$$

Here, the exponents -1 and $-(3/2)$ originate from the numbers of the internal degrees of freedom. D_B and D_χ are the propagators of the dual gluon and the monopole fields in the QCD monopole condensed vacuum. The effective potential or the thermodynamical potential is then obtained as

$$V_{\text{eff}} = 3\lambda(\bar\chi^2 - v^2)^2 + 3\frac{T}{\pi^2} \int_0^\infty dk\, k^2 \ln(1 - e^{-E_B/T})$$

$$+ \frac{3}{2}\frac{T}{\pi^2} \int_0^\infty dk\, k^2 \ln(1 - e^{-E_\chi/T}) \tag{11.29}$$

in the one-loop approximation after introducing the temperature in the standard way using the Matsubara frequency.

The result is shown in Fig. 11.4, where the effective potential is plotted as a function of the monopole condensate, $\bar\chi$. The absolute minimum appears at finite value of $\bar\chi$ in the lower-temperature side and jumps to $\bar\chi = 0$. This indicates a deconfinement phase transition of first order. We can calculate the string tension as a function of temperature.

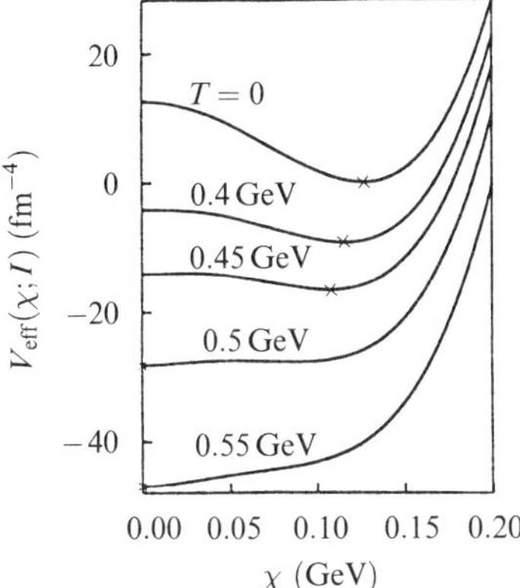

Fig. 11.4 The effective potential (thermodynamical potential) at various temperatures within the DGL theory as a function of the monopole condensate. The absolute minimum is indicated by $\times$ for each curve, indicating the jump (phase transition of first order) around $T = 0.5$ GeV. (From Ichie *et al.* [19].)

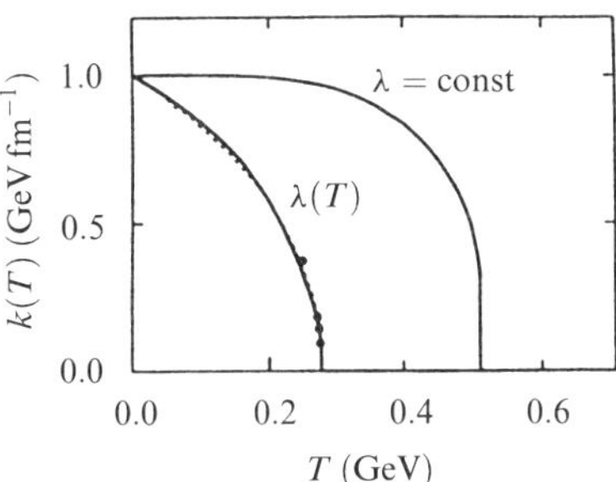

Fig. 11.5 The string tension between a quark and antiquark pair for constant and variable λ's within the DGL theory as a function of temperature. The dots are the results of the pure lattice QCD [20]. (From Ichie *et al.* [19].)

Adjusting λ as a temperature-dependent parameter, so as to reproduce the critical temperature at 0.2 GeV, we find the string tension to reproduce the lattice QCD results [20]. This is shown in Fig. 11.5.

Since the effective potential can be calculated at the critical temperature, we can provide the surface tension σ between the confinement and deconfinement phases at the critical temperature. The surface tension is defined as [21]

$$\sigma = \int_{-\infty}^{\infty} dz \left\{ 3 \left(\frac{d\bar{\chi}(z)}{dz} \right)^2 + V_{\text{eff}}(\bar{\chi}(z); T_c) \right\}, \tag{11.30}$$

where the boundary conditions for $\bar{\chi}$ are given as $\bar{\chi}(z = -\infty) = 0$ and $\bar{\chi}(z = \infty) = \bar{\chi}_c$ in the mixed phase with the boundary surface being taken in the xy-plane ($z = 0$).

Approximating V_{eff} by a sine curve, we have

$$V_{\text{eff}}(\bar{\chi}; T_c) = \frac{h}{2}\left[1 - \cos\left(\frac{2\pi\bar{\chi}}{\bar{\chi}_c}\right)\right],\tag{11.31}$$

where h and $\bar{\chi}_c$ correspond to the height and the width of the effective potential. The field equation, which is obtained by minimizing the free energy, yields the analytic solution

$$\bar{\chi}(z) = \frac{2\sqrt{6}}{3}\bar{\chi}_c \tan^{-1} e^{z/\delta} \quad \text{with } \delta = \frac{\sqrt{3}\bar{\chi}_c}{\pi\sqrt{h}}.\tag{11.32}$$

Hence, the surface tension is expressed as

$$\sigma = \frac{4\sqrt{3}}{\pi}\sqrt{h}\bar{\chi}_c.\tag{11.33}$$

In our case, we find $\sigma^{1/3} = 114\,\text{MeV}$. This value is to be compared with the recent lattice QCD data in the quenched level, e.g. $\sigma^{1/3} = 80\,\text{MeV}$, which is considered to be the size of the surface tension, since the scaling behaviour of this value is not yet achieved [22].

11.3 Instantons and monopoles

11.3.1 *Instantons*

We have already stated that QCD monopoles are the essential degrees of freedom for non-perturbative QCD phenomena in the DGL theory. What is the physical origin of QCD monopoles? Are they merely mathematical objects related to the abelian gauge fixing? In order to answer these questions, we go back to the condition of QCD monopoles by choosing some SU(3) variables, $X(x)$, to be diagonalized in order to find the monopole trajectories. We can show that the QCD monopole appears at the point where X becomes a hedgehog configuration [8,13] (see Appendices A.1 and A.2). What then causes $X(x)$ to have the hedgehog configurations?

Instantons are the classical solutions of the non-abelian gauge theory in Euclidean R^4 space. It is also well known that the instantons provide the $U_A(1)$ anomaly and explain the η' mass problem. We observe instantons clearly after the cooling procedure in lattice QCD simulations of the QCD configurations. In this connection, it is very interesting to see if the instantons are connected with the QCD monopoles [23].

The multi-instanton solution is written as

$$A_\mu(x) = i\bar{\eta}^a_{\mu\nu}\frac{\tau^a}{2}\partial^\nu \ln\phi(x),\tag{11.34}$$

with

$$\phi(x) = 1 + \sum_k \frac{\rho_k^2}{|x - x_k|^2}.\tag{11.35}$$

Here, x_k and ρ_k denote the centre and the size of the kth instanton. The 't Hooft symbol is defined as

$$\bar{\eta}^{a\mu\nu} = -\bar{\eta}^{a\nu\mu} = \begin{cases} \varepsilon^{a\mu\nu} & \text{for } \mu, \nu = 1, 2, 3, \\ -\delta^{a\mu} & \text{for } \nu = 4. \end{cases} \tag{11.36}$$

We first find a simple connection of the monopole trajectories with the instanton configurations. For this purpose, we choose A_4 as $X(x)$ to define the abelian gauge fixing, which is called the Polyakov-like gauge. For the case of multi-instanton solution, we find

$$A_4(x) = \mathrm{i}\frac{\tau^a}{\phi(x)} \sum_k \frac{a_k^2(x - x_k)^a}{|x - x_k|^4} \xrightarrow[x \sim x_k]{} \mathrm{i}\frac{\tau^a(x - x_k)^a}{|x - x_k|^2}. \tag{11.37}$$

Near the centre of each instanton, this gauge provides monopole trajectories at $x = x_k$, which goes through the centre of each instanton, which has the hedgehog configuration. Thus, the instanton centre is inevitably penetrated by the monopole trajectory along the temporal direction in the Polyakov-like gauge. Far from instantons, we need to diagonalize A_4 in order to find the monopole trajectories.

11.3.2 *Multi-instantons and monopole condensation*

It would be extremely interesting to study a multi-instanton system. For this purpose, we make an abelian projection on the lattice by distributing instantons randomly in the four-dimensional space [24]. The multi-instanton configuration is approximated by the sum of instanton and anti-instanton configurations, which is called the sum-ansatz,

$$A_\mu(x) = \sum_k A_\mu^{\mathrm{I}}(x; z_k, \rho_k, O_k) + \sum_k A_\mu^{\bar{\mathrm{I}}}(x; z_k, \rho_k, O_k). \tag{11.38}$$

Here, the instanton configuration is written in the singular gauge as

$$A_\mu^{\mathrm{I}}(x; z, \rho, O) = \mathrm{i}\frac{\tau^a O^{ab} \bar{\eta}^b_{\mu\nu}(x - z)_\nu \rho}{(x - z)^2\left[(x - z)^2 + \rho^2\right]}, \tag{11.39}$$

where the collective coordinates of the instanton are: the instanton size, ρ_k; the instanton centre, z_k, and the colour orientation, O_k^{ab}. The anti-instanton configuration is obtained by replacing the 't Hooft symbol $\bar{\eta}^{a\mu\nu}$ by $\eta^{a\mu\nu}$. As for the distribution function, we take

$$f(\rho) = \left[(\rho/\rho_{\mathrm{IR}})^\nu + (\rho/\rho_{\mathrm{UV}})^{5-11N_c/3}\right]^{-1}. \tag{11.40}$$

Here, the ultraviolet behaviour is obtained by the perturbation calculation, while the infrared behaviour is assumed to be represented by $\nu = 5$. The fall-off at large instanton size is difficult to obtain, but we take the value indicated by recent lattice calculations.

We distribute randomly the instantons in a four-dimensional box and extract the link variable at each lattice site s and the direction μ, which is defined as

$$U_\mu(s) = \mathrm{e}^{\mathrm{i}a A_\mu^a(s)\tau^a}. \tag{11.41}$$

The abelian gauge fixing is performed by the condition of maximizing the quantity

$$R = \sum \text{Tr}\left[U_\mu(s)\tau_3 U_\mu^{-1}(s)\tau_3\right]. \tag{11.42}$$

This is called the Maximally Abelian (MA) gauge and many lattice studies have been performed in this gauge. The details of the MA gauge are given in Appendix A.5. As discussed there, the maximization of R corresponds to the maximization of the abelian components in the link variables and hence the gluon physics is represented maximally with the abelian gluons. The monopole current is defined in the DeGrand–Toussaint form. We then calculate the monopole loop length for many lattice configurations.

The monopole loop length distributions are calculated for various instanton densities. As the density is increased, there appear longer monopole loops. When the instanton density is small, we find only short monopole trajectories. On the other hand, as the instanton density is increased, which corresponds to the closed-packing configuration used for chiral symmetry breaking, we see the appearance of very long monopole trajectories. This is used as an indication of monopole condensation in lattice QCD simulations. Hence, we compare in Fig. 11.6 the monopole distributions due to the instanton distributions and the ones with the QCD simulations [24].

11.3.3 *Multi-instantons and the Wilson loop*

As discussed above, the numerical study of multi-instanton configurations indicates the possibility of confinement due to instantons through monopole condensation. If this is the case, it is interesting to calculate directly the Wilson loop using the multi-instanton configuration [25]. This loop is defined as

$$W[c] = \text{Tr}\left[\exp\left(i\oint_c A_\mu\,dx_\mu\right)\right] = \text{Tr}\left[\prod_c U_\mu(s)\right]. \tag{11.43}$$

If the ensemble average of the Wilson loop is calculated in the limit of large time T and distance R, it shows the area-law behaviour

$$\langle W[R, T]\rangle \to \exp\{-\sigma TR\}. \tag{11.44}$$

We can claim directly the role of instantons for confinement and that the coefficient σ is identified as the string tension. A recent study in fact shows the area-law behaviour. With a reasonable instanton density, the value of the string tension is also close to the empirical value. These studies indicate that the instantons, which are the classical solutions of the non-abelian gauge theory, seem to be playing an important role in non-perturbative phenomena.

11.4 Experimental facilities for quark nuclear physics

11.4.1 *SPring-8 light source facility*

It is important to make measurements on hadron properties carefully using accelerators in the GeV energy range in order to establish the QNP theories. Such new-generation machines are available at KFA Juelich and Utrecht for high-quality hadron beams. As for electron beams, CEBAF has just been constructed in the US, where many interesting

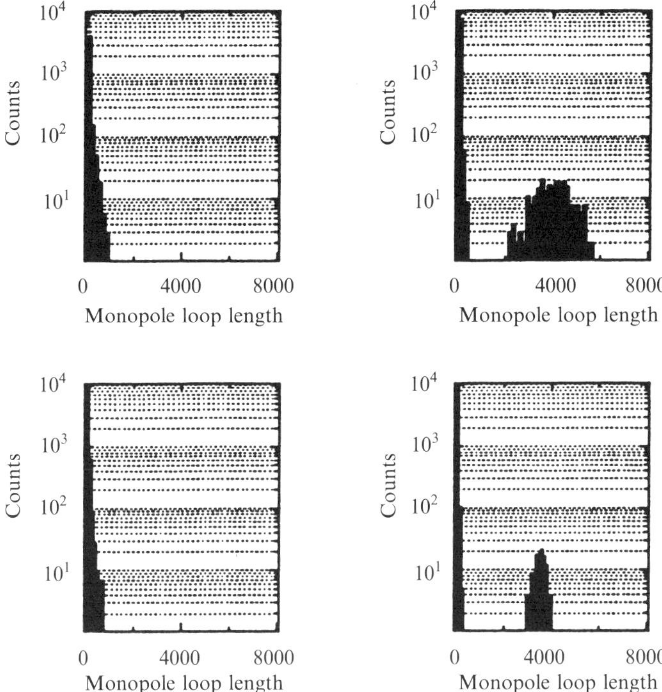

Fig. 11.6 The monopole length distribution for low (left) and high (right) instanton densities (upper figures). The monopole length distribution in the case of pure SU(2) gauge theory on $16^3 \times 4$ at high (left) and low (right) temperatures, which corresponds to the confined phase (lower figures). (From Fukushima *et al.* [24].)

experiments have been proposed to be performed. In Japan, a powerful light facility is being constructed where the electron beam with 8 GeV circulates in a storage ring at SPring-8.

The Japanese nuclear physics community has proposed to extract a high-quality photon beam at about 3 GeV by shooting a laser photon beam into the storage ring of 8 GeV electrons to make complete backward Compton scattering. With the present experimental technique, we can make the accuracy of the photon beam of the order of 10 MeV. The photon beam is almost completely polarized. This is extremely important in extracting various properties of hadrons accessible by the energetic photons.

The concept of the laser photon beam facility is sketched in Fig. 11.7. The laser beam is shot into the electron storage ring completely from the front direction. Here, we can polarize the laser beam either linearly or circularly in order to extract the outcoming energetic photon to have the desired polarization. When the Compton scattering occurs, the energetic photon is scattered in the backward direction relative to the laser beam. The energy of the photon could be identified by measuring the outcoming electrons by position-sensitive counters placed in the inner side of the electron storage ring. We thus

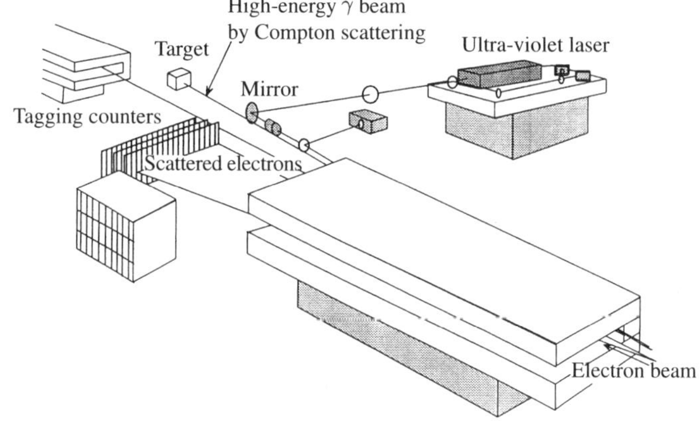

Fig. 11.7   A sketch of the laser photon beam facility at SPring-8.

achieve a photon beam with intensity of 10^7 particles per second. This high-quality GeV photon beam is then brought to the target area, which is surrounded by various detectors.

11.4.2   *Physics with the GeV photon facility*

The photon beam of 2–3 GeV has the wavelength of about 0.1 fm. Hence, when the photon is absorbed by a nucleon, it is absorbed by a quark. This energy is then transformed to other intrinsic degrees of freedom to form higher resonances. The outcoming particles are various mesons, baryons, and/or photons. By measuring these particles, we can study the properties of the known resonance or find new resonances in the baryon and meson sectors. There is much exciting physics to be extracted from these experiments. Gluonic objects, such as glueballs and gluonic Reggeons, are a strong focus of our study. Since we have accumulated much information on the quark sector, we should be able to understand the non-perturbative physics by experimentalists and theoreticians working closely together.

The GeV photon facility at SPring-8 is promising for QNP experiments. We believe it is extremely important to construct a GeV hadron facility for QNP; R&D for such a facility as a possible future machine has already begun.

11.5   Summary and perspectives

We believe QCD is the theory for hadrons. The non-abelian property makes it highly non-perturbative and, hence, the QCD phenomena are extremely interesting. Experimentally, we measure hadron properties and their decays and interactions, and compare them with the QCD predictions. In addition, lattice QCD can provide 'experimental data' on the critical temperature of confinement and chiral phase transitions or on the hadron properties at finite temperature. They are to be interpreted with effective theories like the DGL theory.

In QNP one describes nucleons, mesons, and nuclei in terms of quarks and gluons. The fundamental theory of QNP is QCD. Non-perturbative phenomena are, however, difficult to describe directly in terms of QCD. Hence, we introduce an effective theory as the workable fundamental theory of QNP. The QCD theory is a non-abelian gauge theory. The abelian gauge fixing and the abelian dominance assumption, which are supported by the recent lattice QCD, lead naturally to the DGL theory, where monopole fields are the essential degrees of freedom and their condensation in the QCD vacuum is essential. Here, the Lagrangian is written in terms of the abelian gluons and the colour magnetic monopoles. The monopole condensation can describe the confinement of quarks and even chiral symmetry breaking. Since they are the most essential phenomena in QNP, the DGL theory could be said to be a workable fundamental theory as the shell model is for nuclear physics. Now, it is a matter of time before theoreticians can study mesons and baryons and their properties at finite temperature and confront the exciting experimental phenomena of QNP.

In this respect, we would like to mention the activity at RCNP in developing a GeV photon facility at SPring-8, the photon factory with 8 GeV stored electron ring. Shooting a laser beam of, say, 3.5 eV would lead to $\sim$3 GeV photon beam by Compton backward scattering. We can use this beam for the study of QNP. In addition, RCNP is engaged in R&D work for a high-quality hadron beam as a future project.

Appendix

11.A.1 *Abelian gauge fixing*

We take some gauge-dependent variable $X(x)$, which belongs to su(N), and diagonalize it by a gauge transformation:

$$X(x) \in \mathrm{su}(N),$$

$$X(x) \rightarrow X'(x) = \Omega(x)X(x)\Omega^+(x) = X_{\mathrm{d}}(x),$$

$$X_{\mathrm{d}}(x) = \begin{pmatrix} \lambda_1(x) & & 0 \\ & \lambda_2(x) & \\ 0 & & \lambda_3(x) \end{pmatrix},$$

$$\Omega(x) = \mathrm{e}^{-i\theta^a T^a} \in \mathrm{SU}(N).$$

There remains, however, a residual gauge unfixed, which we write $\omega(x)$. For illustration, we write the case of SU(3) in the matrix form.

1. U(1)$^{N-1}$ gauge transformation

$$\omega(x) = \begin{pmatrix} \mathrm{e}^{i\varphi_1(x)} & & 0 \\ & \mathrm{e}^{i\varphi_2(x)} & \\ 0 & & \mathrm{e}^{i\varphi_3(x)} \end{pmatrix} \in \mathrm{U}(1)^{N-1},$$

$$X'(x) = (\omega(x)\Omega(x))X(x)(\omega(x)\Omega(x))^+ = \omega(x)X_{\mathrm{d}}(x)\omega^+(x) = X_{\mathrm{d}}(x).$$

2. Weyl transformation

$$\omega(x) = \begin{pmatrix} 0 & i & 0 \\ i & 0 & 0 \\ 0 & 0 & 1 \end{pmatrix} \in P(N),$$

$$X'(x) = (\omega(x)\Omega(x))X(x)(\omega(x)\Omega(x))^{+} = \omega(x)\begin{pmatrix} \lambda_1(x) & & 0 \\ & \lambda_2(x) & \\ 0 & & \lambda_3(x) \end{pmatrix}\omega^{+}(x)$$

$$= \begin{pmatrix} -\lambda_2(x) & & 0 \\ & -\lambda_1(x) & \\ 0 & & \lambda_3(x) \end{pmatrix}.$$

Hence, we can diagonalize $X(x)$ with these transformations. It means that the abelian gauge fixing amounts to the gauge fixing of SU(N) $\to$ U(1)$^{N-1}$ $\times$ Weyl.

11.A.2 *Appearance of the QCD monopole*

After the diagonalization of an su(N) matrix, suppose two eigenvalues coincide, which are denoted as $\lambda_i(x) = \lambda_{i+1}(x)$. We focus on the u(2) subspace, where the eigenvalues coincide. The variable $X(x)$ in the u(2) subspace is written as

$$X(x) = X_0(x) + X_a(x)\frac{\tau_a}{2}.$$

Diagonalizing the matrix, we find the eigenvalues as

$$\lambda_i = X_0 \pm \frac{1}{2}\sqrt{X_1^2 + X_2^2 + X_3^2}.$$

$\lambda_i(x) = \lambda_{i+1}(x)$ means $X_i(x) = 0$ for $i = 1, 2, 3$, corresponding to a point in R^3, for a fixed time. Next, we show that this point becomes the QCD monopole.

We now consider the variable X around this point, denoted x_0. Expanding $X(x)$ about this point, we get

$$X(x) = X(x_0) + \partial_i X(x_0)(x - x_0)^i + O(x - x_0)^2$$

$$\approx \lambda(x_0) + \partial_i X_0(x_0)(x - x_0)^i + \partial_i X_a(x_0)(x - x_0)^i \frac{\tau_a}{2}.$$

Writing $C_{ai} = \partial_i X_a(x_0)$, we introduce a variable w_a as $w_a = C_{ai}(x - x_0)^i$. This is a change of variable from x to w by a linear transformation and a rotation. With this variable, the relevant part is written as

$$\tilde{X}(w_a) = w_a\frac{\tau_a}{2}.$$

This is a hedgehog configuration in the w space. In spherical coordinates, this gives

$$w = (w_i, w_2, w_3) = (r\sin\theta\cos\varphi, r\sin\theta\sin\varphi, r\cos\theta),$$

$$\tilde{X} = \frac{1}{2}\begin{pmatrix} w_3 & w_1 - iw_2 \\ w_1 + iw_2 & -w_3 \end{pmatrix} = \frac{r}{2}\begin{pmatrix} \cos\theta & e^{-i\varphi}\sin\theta \\ e^{i\varphi}\sin\theta & -\cos\theta \end{pmatrix}.$$

The matrix diagonalizing $\tilde{X}$ is

$$\Omega(w) = \begin{pmatrix} e^{i\varphi}\cos(\theta/2) & \sin(\theta/2) \\ -\sin(\theta/2) & e^{-i\varphi}\cos(\theta/2) \end{pmatrix}.$$

The gauge field after this transformation is written as

$$A_\mu^{'3} = \mathrm{Tr}\left(\tau_3 \Omega \left(A_\mu - \frac{i}{e}\partial_\mu \right) \Omega^+ \right) = \mathrm{Tr}(\tau_3 \Omega A_\mu \Omega^+) - \frac{1}{e}(1 + \cos\theta)\partial_\mu\varphi.$$

We calculate then the magnetic flux due to the second term going through an area surrounded by C:

$$\Phi = \oint_C A_i^{'3}\,dx^i = -\frac{1}{e}\oint_C (1 + \cos\theta)\partial_i\varphi\,dx^i = -\frac{2\pi}{e}(1 + \cos\theta).$$

This means $\Phi(\theta = 0) = -4\pi/e$ and a Dirac string comes in through the z-axis to the centre of the hedgehog, which corresponds to the QCD monopole with the magnetic charge $4\pi/e$.

11.A.3 SU(N) *invariance condition in the abelian gauge*

Since we work with the abelian gauge, which is the subspace of the full SU(N) space, we have to know the condition if some quantity under consideration is gauge invariant or not in the full SU(N) space. Here, we would like to verify the statement that the quantity which is invariant under the residual symmetry, U(1)$^{N-1}$ × Weyl, in the abelian gauge is invariant in the full SU(N) space. In the following proof, we assume, for simplicity, the adjoint transformation for X and O by the gauge transformation. However, this proof can be generalized in a straightforward way. First, take $X(x)$, which is diagonalized by choosing a suitable Ω, to define the abelian gauge, $\Omega X \Omega^+ = X_{\mathrm{d}}$. Now, let us consider the transformation property of Ω under an arbitrary gauge transformation W. Here, Ω^W is defined so as to diagonalize X^W, which is obtained by the gauge transformation W,

$$X(x) \xrightarrow{W} X^W = WX(x)W^+.$$

X^W is diagonalized by Ω^W as $\Omega^W X^W \Omega^{W\,!} = X_{\mathrm{d}}^W$. We can show that $\Omega^W = d\Omega W^!$, where $d = \mathrm{U}(1)^{N-1} \times$ Weyl.

Now consider any variable O, which becomes $\tilde{O} = \Omega O \Omega^+$ in the original abelian gauge. We find how it transforms in the W gauge:

$$O \xrightarrow{W} O^W = WOW^+$$

and

$$\tilde{O}^W = \Omega^W O^W \Omega^{W+} = d\Omega W^+ WOW^+ W\Omega^+ d^+ = d\Omega O \Omega^+ d^+ = d\tilde{O}d^+.$$

Hence, if $d\tilde{O}d^+ = \tilde{O}$, i.e. U(1)$^{N-1}$ × Weyl invariant, then $\tilde{O}^W = \tilde{O}$.

11.A.4 *Maxwell equations with magnetic monopole*

We follow the work of Zwanziger. The Maxwell equations are written as

$$\partial_\mu F^{\mu\nu} = j_e^\nu, \qquad \partial_\mu {}^*F^{\mu\nu} = j_g^\nu,$$

where ${}^*F^{\mu\nu} = (1/2)\varepsilon^{\mu\nu}{}_{\alpha\beta} F^{\alpha\beta}$ and $\varepsilon^{0123} = 1$ with the metric $g^{\mu\nu} = (1, -1, -1, -1)$ due to the presence of the magnetic current j_g^ν.

The solutions of the above equations are written as

$${}^*F = {}^*(\partial \wedge A) + (n \cdot \partial)^{-1}(n \wedge j_g).$$

We can check this by putting this solution into the second differential Maxwell equation. Then for the dual of this solution, we get

$$F = (\partial \wedge A) - (n \cdot \partial)^{-1}\,{}^*(n \wedge j_g).$$

We can do the same thing by introducing the dual potential B^μ in favour of the second equation. We find the solution of the first equation as

$$F = -{}^*(\partial \wedge B) + (n \cdot \partial)^{-1}(n \wedge j_e)$$

and its dual is written as

$${}^*F = (\partial \wedge B) - (n \cdot \partial)^{-1}\,{}^*(n \wedge j_e).$$

Here, n denotes the direction of the Dirac string.

Now, we have an identity for any antisymmetric tensor:

$$G = \frac{1}{n^2}\{[n \wedge (n \cdot G)] - {}^*[n \wedge (n \cdot {}^*G)]\}.$$

Since we have the relations

$$n \cdot F = n \cdot (\partial \wedge A), \qquad n \cdot {}^*F = n \cdot (\partial \wedge B),$$

F can be written in a simpler form by using the gauge as well as the dual gauge fields as

$$F = \frac{1}{n^2}\{[n \wedge (n \cdot (\partial \wedge A))] - {}^*[n \wedge (n \cdot {}^*(\partial \wedge B))]\}.$$

Using this F in the original Maxwell equations, we find equations of motion for A and B in a complicated form. The remaining task is to obtain the Lagrangian which provides this coupled differential equation. This Lagrangian is found to be

$$L_{\text{dual}} = -\frac{1}{2n^2}[n \cdot (\partial \wedge A)][n \cdot {}^*(\partial \wedge B)] + \frac{1}{2n^2}[n \cdot (\partial \wedge B)][n \cdot {}^*(\partial \wedge A)]$$

$$- \frac{1}{2n^2}[n \cdot (\partial \wedge A)]^2 - \frac{1}{2n^2}[n \cdot (\partial \wedge B)]^2 - j_e A - j_g B.$$

We call this the dual gauge field Lagrangian.

11.A.5 *Maximally abelian gauge*

In order to describe the non-perturbative physics in QCD with only the abelian variables, it is very important to choose a suitable gauge. The most successful gauge, found to date, is the MA gauge and is used frequently in lattice simulations. The MA gauge is defined by the gauge fixing of maximizing the quantity R,

$$R = \sum_{s,\mu} \text{Tr}[U_\mu(s)\tau_3 U_\mu^+(s)\tau_3].$$

Here, $U_\mu(s)$ is the link variable at the site s. In the SU(2) case, it is written as

$$U_\mu(s) = U_0(s,\mu) + i\tau_a U_a(s,\mu).$$

Inserting this expression into the definition of R, we find

$$R = 2\sum_{s\mu}\left[U_0^2 + U_3^2 - U_1^2 - U_2^2\right].$$

It means that the maximization of R corresponds to the maximization of the role of the abelian gluons using the knowledge of the full gluon dynamics.

We can identify the variable X to be diagonalized in defining the abelian gauge. This can be achieved by making an infinitesimal gauge transformation at each site as

$$U_\mu(s) \to V(s)U_\mu(s)V^+(s+\mu) \quad \text{with } V(s) = \mathrm{e}^{i\boldsymbol{\alpha}(s)\cdot\boldsymbol{\tau}}.$$

With this transformation, the quantity R becomes $R' = R + \delta R$ with

$$\delta R = \sum_{s,\mu} \text{Tr}\{i\boldsymbol{\alpha}(s)\boldsymbol{\tau}([X(s,\mu), \tau_3])\},$$

where

$$X(s,\mu) = U_\mu(s)\tau_3 U_\mu^+(s) + U_\mu^+(s-\mu)\tau_3 U_\mu(s-\mu).$$

The maximization of R corresponds to the condition of diagonalization of the variable $X(s,\mu)$ at each site.

We can find the condition of the MA gauge on the gauge fields in the continuum limit. Since in the continuum limit the link variable is written as

$$U_\mu(s) = \exp(iaeA_\mu(s)) \approx 1 + iaeA_\mu(s),$$

we substitute this expression into $X(s,\mu)$ and impose the condition that the non-diagonal components of $X(s,\mu)$ are zero. This leads to the condition necessary for the gauge fields to satisfy in the MA gauge,

$$(\partial_\mu \mp ieA_\mu^3(x))A^{\mu\pm}(x) = 0.$$

References

1. A. Bohr and B. Mottelson, *Nuclear structure*, Benjamin (1975).
2. A. B. Migdal, *Rev. Mod. Phys.* **50** (1978) 107.
3. E. Oset, H. Toki, and W. Weise, *Phys. Rep.* **83** (1982) 281.
4. A. W. Thomas, Proc. of Yukawa international seminar on "mesons and quarks in nuclei", *Prog. Theo. Phys.* **91**(Suppl.) (1987) 204.
5. *Proceedings of "quark matter"*, E. Stenlund *et al.*, eds, *Nucl. Phys.* **A566** (1994).
6. T. Chen and L. Li, *Gauge theory of elementary particle theory*, Oxford University Press (1984).
7. Y. Nambu, *Phys. Rev.* **D10** (1974) 4262.
8. G. 't Hooft, *Nucl. Phys.* **B190** (1981) 455.
9. P. A. M. Dirac, *Proc. R. Soc.* **A133** (1931) 60.
10. T. Suzuki, *Prog. Theor. Phys.* **80** (1988) 929; **81** (1989) 752; S. Maedan and T. Suzuki, *Prog. Theor. Phys.* **81** (1989) 229.
11. Z. F. Ezawa and A. Iwazaki, *Phys. Rev.* **D25** (1982) 2681; **D26** (1982) 631.
12. A. S. Kronfeld *et al.*, *Phys. Lett.* **B198** (1987) 516; T. Suzuki and I. Yotsuyanagi, *Phys. Rev.* **D42** (1990) 4257; S. Hioki *et al.*, *Phys. Lett.* **B272** (1991) 326.
13. H. Suganuma, S. Sasaki, and H. Toki, *Nucl. Phys.* **B435** (1995) 207; H. Toki, H. Suganuma, and S. Sasaki, *Nucl. Phys.* **A577** (1994) 353c; H. Suganuma, S. Sasaki, H. Toki, and H. Ichie, *Prog. Theor. Phys.* **120**(Suppl.) (1995) 57.
14. O. Miyamura and S. Origuchi, *Proc. of international RCNP workshop on "color confinement and hadrons"*, H. Toki *et al.*, eds, World Scientific (1995), p. 235.
15. H. Shiba and T. Suzuki, *Phys. Lett.* **B343** (1995) 315.
16. S. Kitahara, Y. Matsubara, and T. Suzuki, *Prog. Theor. Phys.* **93** (1995) 1.
17. H. Suganuma, A. Tanaka, S. Sasaki, and O. Miyamura, *Proc. of Lattice'95*, Melbourne, July 1995, *Nucl. Phys. B* **47**(Suppl.) (1996) 302.
18. S. Sasaki, H. Suganuma, and H. Toki, *Prog. Theor. Phys.* **94** (1995) 373; S. Sasaki, Doctoral Thesis, Osaka University (1997).
19. H. Ichie, H. Suganuma, and H. Toki, *Phys. Rev.* **D52** (1995) 2944.
20. M. Gao, *Nucl. Phys.* **B9**(Suppl.) (1989) 368.
21. H. Monden, H. Ichie, H. Suganuma, and H. Toki, RCNP preprint, hepph-9701271 (1997).
22. Y. Iwasaki, K. Kanaya, and L. Karkkainen, *Phys. Rev.* **D49** (1994) 3540.
23. H. Suganuma, K. Itakura, H. Toki, and O. Miyamura, *Proc. of international workshop on "non-perturbative approach to QCD"*, Trento, Italy, D. Diakonov, ed., PNPI Press (1995), p. 224; H. Suganuma, K. Itakura, and H. Toki, RCNP preprint, hep-th/9512141 (1995); H. Suganuma, H. Ichie, S. Sasaki, and H. Toki, *Proc. of international RCNP workshop on "color confinement and hadrons"*, H. Toki *et al.*, eds, World Scientific (1995), p. 65; F. Araki, H. Suganuma, and H. Toki, RCNP report (1996) 81.
24. M. Fukushima, A. Tanaka, S. Sasaki, H. Suganuma, H. Toki, and D. Diakonov, *Phys. Lett.* **B399** (1997) 141.
25. M. Fukushima, Masters Thesis, Osaka University (1997).

12

Lepton nuclear physics

H. Ejiri

12.1 Lepton nuclear physics

Fundamental properties of leptons and weak interactions are important topics in physics today. Their studies allow for crucial tests of the standard theory and may provide evidence for unified theories beyond the standard model.

The atomic nucleus is a system of nucleons in an eigenstate of energy E, spin and parity J^π, isospin T, and so on. Thus, it is a suitable 'microlaboratory' (detector) for studying fundamental properties of particles and their interactions. Nuclear levels are used to select particular modes of interactions.

The lepton nuclear physics studies the fundamental properties of leptons, weak interactions, and related subjects in the nuclear microlaboratory. Thus, being complementary to the quark nuclear physics as discussed in Chapter 11, it is one of the important new fields in nuclear physics.

Several interesting topics of the lepton nuclear physics are currently under study at RCNP, Osaka. The subjects discussed in the present chapter include neutrinos (ν's) studied by double beta ($\beta\beta$) decays, spin-coupled dark matter (DM) studied by nuclear elastic and inelastic scatterings, nucleon instabilities and nucleon decays studied by exotic nuclear transitions, and so on. It is interesting to note that charge non-conservation is also studied by searching for exotic X-rays and γ-rays.

It should be noted that these problems in lepton nuclear physics are related to symmetries and conservation laws in nuclear and astroparticle physics. Thus, their precision studies are very useful for sensitive tests of the standard $SU(2)_L \times U(1)$ theory. Actually, finite violations of the symmetry/conservation laws lead to evidence for unified theories beyond the standard model.

Experimental studies of the lepton nuclear physics necessarily involve rare nuclear decays/transitions because of weak nuclear processes in these systems. Consequently, one has to use high-sensitivity detectors at low-background laboratories, namely the underground laboratories.

Rare nuclear decays have so far been studied experimentally at the Kamioka underground laboratory located 350 km north-east of RCNP, Osaka. Fresh measurements are being made at a new underground laboratory, the Oto Cosmo Observatory, located 100 km south of RCNP, Osaka.

Nuclear weak processes for vector interactions are associated with the nuclear isospin responses and those for axial vector interactions with the spin–isospin responses. Such nuclear spin–isospin responses are studied by investigating charge-exchange spin-flip nuclear reactions such as (p, n), (^{3}He, t), (d, ^{2}He), and (t, ^{3}He). In particular, proton and ^{3}He beams with 0.2–0.5 GeV energies, which are provided by the RCNP cyclotron, are used for preferential excitations of nuclear spin–isospin modes.

12.2 Neutrinos and weak interactions studied by double beta decays

12.2.1 *Double beta decays*

Double beta decays are very interesting from both astroparticle physics and nuclear physics viewpoints. Astroparticle physics is concerned with the fundamental properties of neutrinos and weak interactions, and nuclear physics with the nuclear spin–isospin responses and spin–isospin giant resonances.

The normal $\beta\beta$ decay in the framework of the minimal $SU(2)_L \times U(1)$ theory is expressed as

$$2\nu\beta\beta: \quad A \to B + \beta + \beta + \nu + \nu, \tag{12.1}$$

where ν is the electron neutrino (anti-neutrino $\bar\nu$) in case of β^+ (β^-) decay. This represents a two-neutrino $\beta\beta$ decay ($2\nu\beta\beta$). Noting that lepton numbers $L = 1$ for β^- and ν, and $L = -1$ for β^+ and $\bar\nu$, the $2\nu\beta\beta$ process conserves the lepton number.

The $\beta\beta$ decays of particular interest in particle physics are the neutrinoless $\beta\beta$ decays ($0\nu\beta\beta$) beyond the standard theory:

$$0\nu\beta\beta: \quad A \to B + \beta + \beta. \tag{12.2}$$

The $0\nu\beta\beta$ process, which gives $L = \pm 2$, violates the lepton number conservation law, $\Delta L = 0$. Thus, $0\nu\beta\beta$ is a very sensitive probe for studying such quantities as the Majorana neutrino mass $\langle m_\nu \rangle$, the right-handed weak current $\langle RHC \rangle$, the Majoron–ν coupling, the R-parity-violating coupling with SUper SYmmetry (SUSY) particles, and others, which are all beyond the standard $SU(2)_L \times U(1)$ theory.

The $2\nu\beta\beta$ and $0\nu\beta\beta$ processes are shown schematically in Fig. 12.1. Double beta decays have been discussed extensively. (For details, see review articles [1–6] and references therein.)

The transition rate $T^{0\nu}$ for $0\nu\beta\beta$ (Fig. 12.1B) is given in terms of $\langle m_\nu \rangle$ and $\langle RHC \rangle$ as follows [1–6]:

$$T^{0\nu} = G^{0\nu} |M^{0\nu}|^2 \big[\langle m_\nu \rangle^2 + C_{\lambda\lambda} \langle \lambda \rangle^2 + C_{\eta\eta} \langle \eta \rangle^2$$
$$+ C_{m\lambda} \langle m_\nu \rangle \langle \lambda \rangle + C_{m\eta} \langle m_\nu \rangle \langle \eta \rangle + C_{\lambda\eta} \langle \lambda \rangle \langle \eta \rangle \big], \tag{12.3}$$

where $\langle \lambda \rangle$ and $\langle \eta \rangle$ are the $\langle RHC \rangle$ terms, λ is written as $(M_W^L / M_W^R)^2$ with M_W^L (M_W^R) being the mass of the left-handed (right-handed) weak boson, and η is the coupling of left and right weak bosons.

The transition rate for the gluino ($\tilde{g}$) exchange process (Fig. 12.1C) is given as [1–7]

$$T_S(\tilde{g}) = G^S f^2 m_{\tilde{g}} A(m_{\tilde{g}}) / m_{\tilde{u}}^4, \tag{12.4}$$

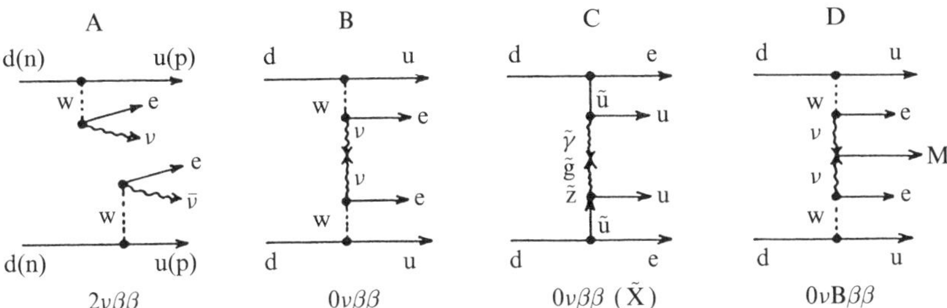

Fig. 12.1 Schematic diagrams of $\beta\beta$ decay processes: A, two-neutrino process; B, neutrinoless process with the Majorana neutrino exchange; C, neutrinoless process in the SUSY model; and D, neutrinoless process followed by the Goldstone boson (Majoron). d and u are down and up quarks, and n and p are neutron and proton, respectively.

where G^{S} and $A(m_{\tilde{g}})$ are the phase space factor and the nuclear form factor, respectively. $m_{\tilde{g}}$ ($m_{\tilde{u}}$) and f are the gluino (squark) mass and the R-parity-violating interaction. The transition rate $T^{0\nu}$ for the Majoron-emitting process (0νββB, Fig. 12.1D) is written as [1–6]

$$T^{0\nu B} = G^{B}|M^{B}|^{2}\langle g_{B}\rangle^{2}, \tag{12.5}$$

where G^{B}, M^{B}, and $\langle g_{B}\rangle$ are the phase space factor, the matrix element, and the Majoron–ν coupling, respectively. The double Majoron emitting process (0νββBB) is discussed in Ref. [8].

The 0νββ, 0νββB, and 2νββ processes are studied by measuring the summed-energy, $E_{\beta} + E_{\beta'}$ spectra, as shown in Fig. 12.2. The 0νββ spectrum shows a sharp peak at the $\beta\beta$ decay Q-value $Q_{\beta\beta}$, while the 0νββ and 2νββ spectra show continuum bumps in the energy region below $Q_{\beta\beta}$ because νB and 2ν escape from the detectors. Individual terms of $\langle m_{\nu}\rangle$, $\langle\lambda\rangle$, and $\langle\eta\rangle$ of 0νββ are determined by studying the energy and angular correlations of the two β-rays [2].

Nuclear diagrams for 0νββ are shown schematically in Fig. 12.3: two-nucleon modes (Fig. 12.3A and B) are dominant; isobar (Δ) and meson modes are also possible.

The transition rate $T^{2\nu}$ for 2νββ (Fig. 12.1A) is written as

$$T^{2\nu} = G^{2\nu}|M^{2\nu}|^{2}, \tag{12.6}$$

where $G^{2\nu}$ and $M^{2\nu}$ are the phase space factor and the matrix element, respectively. Thus, the observed half-lives give the 2νββ matrix elements $M^{2\nu}$. They are used to verify nuclear structure calculations for $M^{2\nu}$ and the parameters for spin–isospin interactions. They are also used for calculating $M^{0\nu}$. The 2νββ process proceeds mostly via 1^{+} intermediate states. The 0νββ process associated with the ν-exchange between two nucleons in a nucleus involves large momentum transfers ($q = 0.2$–$0.4\,\mathrm{GeV}/c$) and, accordingly, many intermediate J^{π} states up to $J^{\pi} \approx 6^{\pm}$.

The neutrino study by the $\beta\beta$ decays is a good example of lepton nuclear physics, where the nucleus acts as a microlaboratory for studying fundamental interactions. The

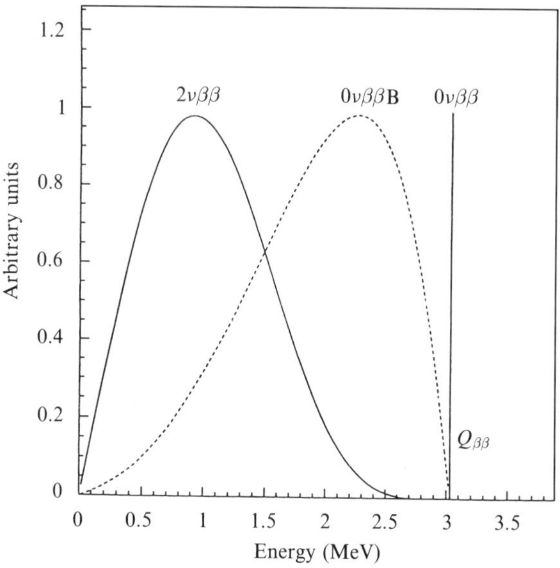

Fig. 12.2 Summed-energy, $E = \varepsilon_1 + \varepsilon_2$, spectra for $\beta\beta$ decays from ^{100}Mo(0^+) to the 0^+ ground state of ^{100}Ru. The solid lines show the calculated shapes on an arbitrary scale.

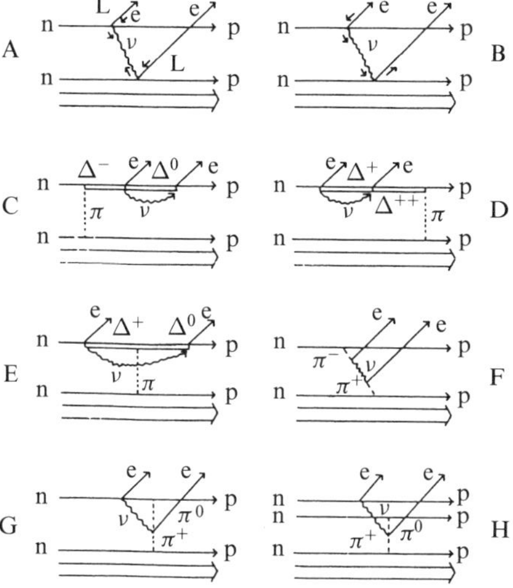

Fig. 12.3 Neutrinoless $\beta\beta$ decay schemes: A, 2n mode with the mass term $\langle m_\nu \rangle$; B, 2n with the right-handed current term $\langle$RHC$\rangle$; C, Δ^- in ψ_i; D, Δ in ψ_f; E, $\Delta^+\Delta^0$ mode; F, 2n double-pion-exchange current; G, 2n single-pion exchange; and H, 3n single-pion exchange. The arrows in A and B give the relevant helicity. (From Ref. [4].)

$0\nu\beta\beta$ decays, which are relevant to the neutrino mass studies, are enhanced by a factor of $\sim 10^7$ in a nucleus, since they involve essentially a ν-exchange between two nucleons (quarks) in a nucleus (baryon). The phase space factor $G^{0\nu}$ and the nuclear matrix element $M^{0\nu}$ become much larger than those for $2\nu\beta\beta$ decays. Studies of $\beta\beta$ decay in nuclei, where the intermediate nuclei are heavier than the initial nuclei, become feasible as strong single-β decays to intermediate nuclei are energetically forbidden. Thus, the nucleus, being used for studying rare $0\nu\beta\beta$ process, acts as an excellent telescope with a large enlargement for $0\nu\beta\beta$ and with a large rejecting power for single-β rays and other backgrounds.

12.2.2 *Double beta decay experiments*

Double beta decays are second-order weak processes. Therefore, their decay rates are extremely small. The half-lives for $2\nu\beta\beta$ decays are of the order of 10^{19}–10^{24} yr, and those for the $0\nu\beta\beta$ decays with $\langle m_\nu \rangle \approx 1$ eV are of the order of 10^{22}–10^{26} yr, both of which depend greatly on their Q-values ($Q_{\beta\beta}$) as well as on their nuclear matrix elements (M). Actually, the leading terms in the phase space factors $G^{0\nu}$ and $G^{2\nu}$ are proportional to $Q_{\beta\beta}^5$ and $Q_{\beta\beta}^{11}$, respectively. Consequently, $\beta\beta$ nuclei with large $Q_{\beta\beta}$ and large nuclear matrix elements are chosen for experimental studies in order to get large transition rates. Since half-lives ($>10^{19}$ yr) for typical $\beta\beta$ decay nuclei are far greater than the age of the universe (10^{10} yr), the $\beta\beta$ isotopes are naturally existing isotopes with finite abundance ratios.

Energy level schemes of typical $\beta\beta$ studies are shown in Fig. 12.4. Among the conditions for the $\beta\beta$ nuclei to be useful in these studies are large $Q_{\beta\beta}$ values and finite natural abundance ratios. Some of them are used as detectors to get large detection efficiencies.

Experimental studies of $\beta\beta$ processes have been carried out by means of indirect geochemical and radiochemical methods and direct-counting methods (see review articles [1–6] and references therein).

The geochemical method measures the isotope concentration of $\beta\beta$ decay products for the geological long period, and the radiochemical method measures the radioactivity of $\beta\beta$ decay products [1–6]. These indirect methods are inclusive measurements of all kinds of $\beta\beta$ decays to ground and excited levels. Finite $\beta\beta$ half-lives for ^{82}Se, ^{96}Zr, ^{128}Te, ^{130}Te, and ^{238}U were measured by the indirect methods. They may be regarded as $2\nu\beta\beta$ half-lives for ground state transitions because $T^{2\nu} \gg T^{0\nu}$ and $T^{2\nu}$(ground state) $\gg$ $T^{2\nu}$(excited state).

The direct-counting method entails the exclusive measurement of $\beta\beta$ rays. The $0\nu\beta\beta$, $0\nu\beta\beta$B, $0\nu\beta\beta$BB, and $2\nu\beta\beta$ are separated from each other by measuring the summed-energy, $E_\beta + E_{\beta'}$ spectra, as shown in Fig. 12.2.

Germanium detectors, CaF_2 scintillators, and $CdWO_4$ scintillators have extensively been used for studying $\beta\beta$ decays of ^{76}Ge, ^{48}Ca, and ^{116}Cd, respectively [1–6]. These detectors contain the $\beta\beta$ source isotopes; thus, the detection efficiencies are nearly 100%. Among these, the Ge detectors are useful for $0\nu\beta\beta$ because of the high energy resolution [9]. Of late, several groups have been using Ge detectors made of enriched ^{76}Ge.

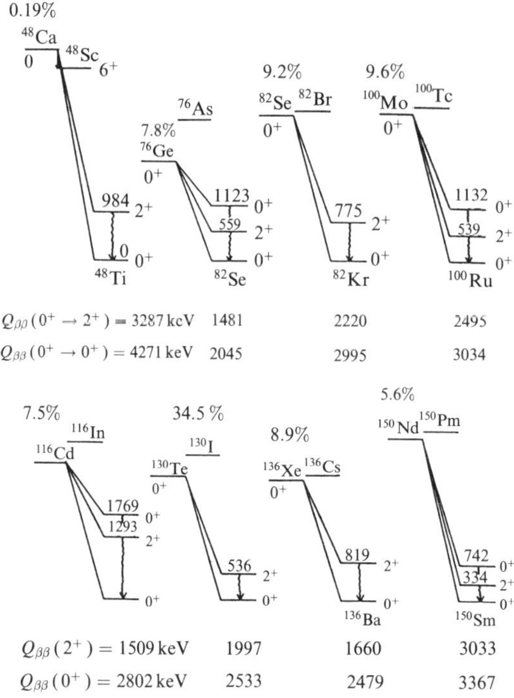

Fig. 12.4   Level schemes of typical $\beta\beta$ decays.

Multi-Si-detector arrays have been used for ^{100}Mo. Here, thin ^{100}Mo foils are inter-leaved between the Si detectors. Thus, two β-rays are measured in coincidence by two adjacent Si detectors.

Thermal detectors at low temperatures have been used for ^{130}Te and others. They are phonon-mediated detectors to measure the $\beta\beta$ energies through the temperature rise in the detectors containing the $\beta\beta$ isotopes. Their energy resolution is considered, in principle, to be quite good. These detectors can be powerful tool for studying $0\nu\beta\beta$. Their practical realization remains a crucial issue [10].

Beta-ray tracking detectors have been used for several $\beta\beta$ nuclei. Observations of β-ray trajectories give energy and angular correlations of the two β-rays, which are crucial for identifying the individual $0\nu\beta\beta$ ($\langle m_\nu \rangle$, $\langle \lambda \rangle$, $\langle \eta \rangle$, in eqn (12.1)) and $2\nu\beta\beta$ modes. Detector systems with drift chambers for β-ray tracking are used for ^{100}Mo and ^{116}Cd and Time Projection Chamber (TPC) for ^{82}Se and ^{136}Xe.

12.2.3   *ELEGANTs for $\beta\beta$ decays of ^{76}Ge, ^{100}Mo, and ^{116}Cd*

High-sensitivity detector systems, namely the ELEctron GAmma-ray Neutrino Tele-scopes (ELEGANTs), have been developed by the Osaka group to study $\beta\beta$ decays and DM. EL III, IV, and V were used at the Kamioka underground laboratory with

2700 m w.e. (water equivalent). EL III consists of a pure Ge detector surrounded by the 4π NaI detector array [11]. Most of the background electron (β-ray) signals are accompanied by γ-rays and/or X-rays, Compton electrons by Compton γ-rays, conversion electrons by X-rays, β-rays by γ-rays, and so on. True $\beta\beta$ rays feeding the ground states, however, are not accompanied by any rays, and those feeding the excited states are accompanied by deexciting γ-rays. Therefore, γ-ray and/or X-ray signals from the NaI detector array are used to eliminate the background events and to select the true $\beta\beta$ events.

EL III was used in the mid-1980s for studying $\beta\beta$ decays of ^{76}Ge. The lower limit on the $0\nu\beta\beta$ half-life for the 0^+ ground state was obtained as $T^{0\nu} > 0.7 \cdot 10^{23}$ yr. It corresponds to the upper limit of 2 eV on the Majorana neutrino mass. The lower limit on the $0\nu\beta\beta$ half-life for the 2^+ excited state was obtained as $T^{0\nu} > 0.6 \cdot 10^{23}$ yr. Measurements of ^{76}Ge $\beta\beta$ decays with natural Ge detectors have been made by several groups in the 1980s. Presently, measurements with enriched Ge detectors are under progress to investigate the Majorana neutrino mass in the region of ≤ 1 eV, as will be discussed later.

EL IV consists of a set of 11 Si detector disks [12]. It was used for studying the $2\nu\beta\beta$ decay of ^{100}Mo.

EL V is a detector complex for studying the $\beta\beta$ decays with external $\beta\beta$ sources. It consists of two drift chambers for β-ray trajectories, 16 modules of plastic scintillators for β-ray energies and times, and 20 modules of NaI detectors for X- and γ-rays [13]. It is shown schematically in Fig. 12.5.

EL V was used for measuring the $\beta\beta$ decays of ^{100}Mo and ^{116}Cd. The first ^{100}Mo run with the 104 gr ^{100}Mo foils gave, for the first time, a finite $2\nu\beta\beta$ half-life of $1.15 \cdot 10^{19}$ yr for the $0^+ \to 0^+$ decay of ^{100}Mo [13].

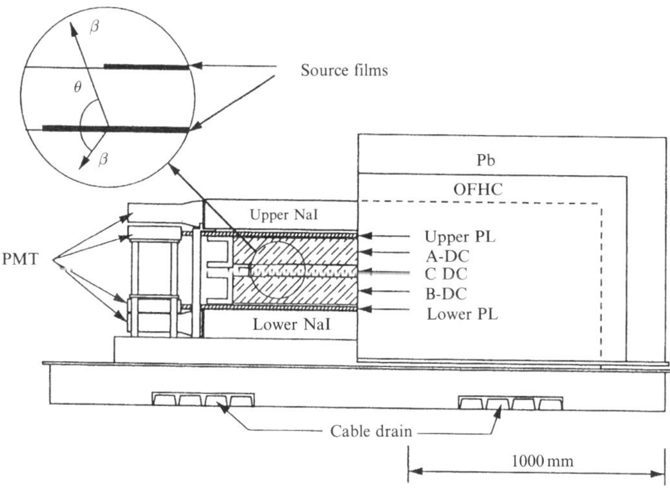

Fig. 12.5 ELEGANT V used for studying the $\beta\beta$ decay of ^{100}Mo and ^{116}Cd. A-, B-, and C-DC are, respectively, the upper, lower, and central drift chambers for tracking β- and α-rays; PLs are plastic scintillators [13].

320   H. EJIRI

The ^{116}Cd run with the 91 gr ^{116}Cd foils gave, again for the first time, a finite $2\nu\beta\beta$ half-life of $2.6 \cdot 10^{19}$ yr for the $0^+ \to 0^+$ decay of ^{116}Cd [14]. The $2\nu\beta\beta$ matrix elements are 0.09 for ^{100}Mo and 0.069 for ^{116}Cd. They are indeed one-tenth of the single-particle values.

The second ^{100}Mo run was carried out with 179 gr ^{100}Mo foils. Here, the background rate was reduced by introducing a new central drift chamber [15]. The result gave the most stringent limits on half-lives for the $0\nu\beta\beta$ and $0\nu\beta\beta$B processes. The upper limits for the $\langle m_\nu \rangle$, $\langle \lambda \rangle$, and $\langle \eta \rangle$ space are shown in Fig. 12.6 [16].

The limits obtained on the $0^+ \to 0^+ 0\nu\beta\beta$ and $0\nu\beta\beta$B processes for ^{100}Mo are as follows:

$$
\begin{aligned}
0\nu\beta\beta : \quad & T_{1/2} > 5.2 \cdot 10^{22} \quad && \langle m_\nu \rangle < 2.2\,\text{eV}, \\
0\nu\beta\beta : \quad & T_{1/2} > 3.9 \cdot 10^{22} \quad && \langle \lambda \rangle < 4.3 \cdot 10^{-6}, \\
0\nu\beta\beta : \quad & T_{1/2} > 5.1 \cdot 10^{22} \quad && \langle \eta \rangle < 2.5 \cdot 10^{-8}, \\
0\nu\beta\beta\text{B} : \quad & T_{1/2} > 0.54 \cdot 10^{22} \quad && \langle g_B \rangle < 7.3 \cdot 10^{-5}.
\end{aligned}
$$

Here, the limits on $\langle m_\nu \rangle$, $\langle \lambda \rangle$, and $\langle \eta \rangle$ are the on-axis values, and the nuclear matrix elements in Ref. [17] are used.

Limits on supersymmetric parameters are derived from the limit on $0\nu\beta\beta$ [18]. The limits obtained for ^{100}Mo are

$$
\lambda'_{111} \le 6.9 \cdot 10^{-9} m_{\tilde{q}}^2 m_{\tilde{g}}^{1/2}, \tag{12.7}
$$

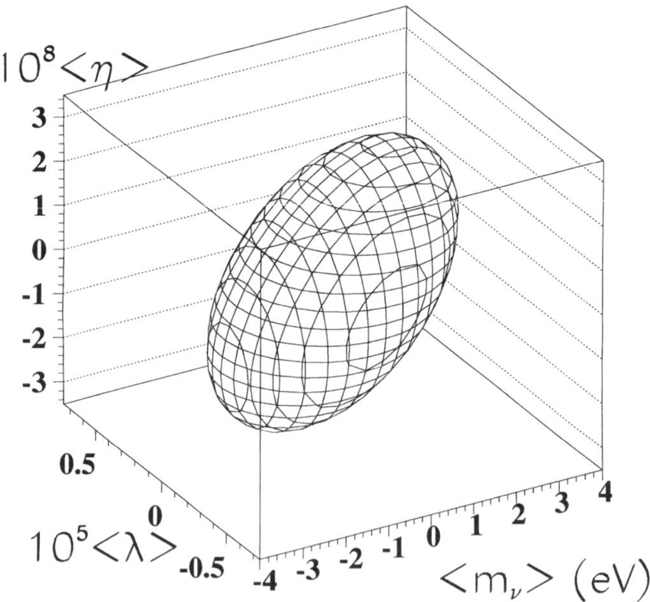

Fig. 12.6   Allowed region for $\langle m_\nu \rangle$, $\langle \lambda \rangle$, and $\langle \eta \rangle$ deduced from T-selection.

where $m_{\tilde{q}}$ and $m_{\tilde{g}}$ are the squark and gluino masses, in units of GeV, respectively, and λ'_{111} is the R-parity-violating (lepton-number non-conserving) coupling. The $0\nu\beta\beta$ limits give limits on the mass of the excited neutrino in the framework of the composite neutrino model [19]. The result obtained from the ^{100}Mo $0\nu\beta\beta$ limit is

$$m(\nu^*) > 1.8 \cdot 10^{-4}\,\text{TeV}. \tag{12.8}$$

EL VI consists of 25 modules of CaF_2 scintillators surrounded by 3π CsI detectors [20]. The CaF_2 scintillator module consists of a central CaF_2 crystal with $45 \times 45 \times 200\,\text{mm}^3$ and two other $45 \times 45 \times 75\,\text{mm}^3$ CaF_2 crystals, placed on either side of the central crystal. The total ^{48}Ca content in the 25 central detectors is 31 gr. Because of the large Q-value of 4.27 MeV, one may study both $0\nu\beta\beta$ and $2\nu\beta\beta$ decays of ^{48}Ca precisely.

EL V and EL VI have been installed at the new underground laboratory of Oto Cosmo Observatory with around 1500 m w.e. It is located about 100 km south of Osaka University. Background levels at the new laboratory are $4 \cdot 10^{-3}\,\text{m}^{-2}\,\text{s}^{-1}$ for cosmic rays, $4 \cdot 10^{-1}\,\text{m}^{-2}\,\text{s}^{-1}$ for neutrons, and 10 Bq m^{-3} for Rn. These are sufficiently small levels.

12.2.4 *Neutrino mass limits and $2\nu\beta\beta$ matrix elements*

Extensive studies of $\beta\beta$ decays have been made at several underground laboratories. Stringent lower limits on $0\nu\beta\beta$ half-lives have been obtained as $6.4 \cdot 10^{24}$ yr (90% CL) and $4.2 \cdot 10^{23}$ yr (90% CL), for ^{76}Ge and ^{136}Xe, respectively [18,21]. These data give upper limits of $\langle m_\nu \rangle < 0.6$–1.5 eV and 1.9–2.8 eV, respectively, depending on the nuclear matrix elements used. A stringent limit on $0\nu\beta\beta$B for ^{136}Xe gives an upper limit of $\langle g_\beta \rangle < 0.8$–$1.3 \cdot 10^{-4}$ [21]. These data, together with the present Osaka data, lead to stringent limits of $\langle m_\nu \rangle < 1$–3 eV, and $\langle g_\beta \rangle < 1$–$2 \cdot 10^{-4}$. Thus, the electron Majorana neutrino is not the major DM component of the universe. Stringent limits on the supersymmetric parameters and on the excited neutrino mass are also deduced from the $0\nu\beta\beta$ limit for ^{76}Ge [18].

It is important to get limits on neutrinos and weak interaction parameters from several high-precision data such as those for ^{76}Ge, ^{100}Mo, and ^{136}Xe, etc., because of the possible uncertainties in the evaluations of the nuclear matrix elements involved.

Nuclear matrix elements $M^{2\nu}$ for $2\nu\beta\beta$ are deduced from the observed $2\nu\beta\beta$ half-lives. Some of the recent data on $2\nu\beta\beta$ [18,22–24], together with the Osaka EL V data [13,14], are shown in Table 12.1. The values obtained for $M^{2\nu}$ are indeed as small as 0.01–0.1 in units of $(m_e c^2)^{-1}$. They are one or two orders of magnitudes smaller than the single-particle values, indicating the destructive effect of the spin–isospin core polarization [25,26].

Let us now discuss the $2\nu\beta\beta$ process for ground state transitions, $0_i^+ \rightarrow 0_f^+$, in even nuclei. Vector-type (Fermi-type) β- and $\beta\beta$ decay strengths are absorbed mostly into isobaric and double-isobaric analogue states, respectively. Thus, the axial-vector-type (GT-type) β-strengths are the major strengths in β- and $\beta\beta$ decays. The $M^{2\nu}$ is written

 H. EJIRI

Table 12.1 Nuclear matrix elements $M^{2\nu}$ for $2\nu\beta\beta$ in medium-heavy nuclei and products of nuclear matrix elements $M_S^\nu M_{S'}^\nu$ with the energy denominator Δ_S for successive single-β decays, through single-particle ground states in intermediate nuclei. $M^{2\nu}$ and Δ_S are in units of $(m_e c^2)^{-1}$ and $m_e c^2$, respectively, and $T_{1/2}^{2\nu}$ is in units of year [28]

Nucleus	$T_{1/2}^{2\nu}$	$M^{2\nu}$	$M_S^\nu M_{S'}^\nu / \Delta_S$
^{76}Ge	$1.77^{+0.13}_{-0.11} \cdot 10^{21}$	$6.5^{+0.3}_{-0.2} \cdot 10^{-2}$	$4.1 \cdot 10^{-2}$
^{82}Se	$1.08^{+0.26}_{-0.06} \cdot 10^{20}$	$4.6^{+0.1}_{-0.5} \cdot 10^{-2}$	$3.0 \cdot 10^{-2}$
^{96}Zr*	$3.9 \pm 0.9 \cdot 10^{19}$	$3.7^{+0.5}_{-0.4} \cdot 10^{-2}$	$3.4 \cdot 10^{-2}$
^{100}Mo	$1.15^{+0.3}_{-0.2} \cdot 10^{19}$	$9.6^{+1.0}_{-1.0} \cdot 10^{-2}$	$1.0 \cdot 10^{-1}$
^{116}Cd	$2.6^{+0.9}_{-0.5} \cdot 10^{19}$	$6.9^{+0.8}_{-0.9} \cdot 10^{-2}$	$1.2 \cdot 10^{-1}$
^{128}Te*	$7.7 \pm 0.4 \cdot 10^{24}$	$1.24^{+0.03}_{-0.03} \cdot 10^{-2}$	$1.0 \cdot 10^{-2}$
^{130}Te*	$2.7 \pm 0.1 \cdot 10^{21}$	$8.8^{+0.2}_{-0.2} \cdot 10^{-3}$	$1.3 \cdot 10^{-2}$

Note: Refs: ^{76}Ge [18], ^{82}Se [22], ^{96}Zr* [23], ^{100}Mo [13], ^{116}Cd [14], ^{128}Te*, ^{130}Te* [24].
*Geochemical method

approximately as follows:

$$M^{2\nu} = \sum \frac{M_i^{2\nu} M_{i'}^{2\nu}}{\Delta_i}, \tag{12.9}$$

where $M_i^{2\nu} = \langle 1_i | \tau\sigma | 0_i \rangle$ and $M_{i'}^{2\nu} = \langle 0_f | \tau\sigma | 1_i^+ \rangle$ are the single-β matrix elements through the ith 1^+ state in the intermediate nucleus, and τ and σ are the isospin and spin operators, respectively.

Theoretical calculations of $M^{2\nu}$ are not straightforward since they are sensitive to the spin–isospin ($\tau\sigma$) responses of the intermediate nucleus for β^- and β^+ operators, and to the spin–isospin ($\tau\sigma$) interaction parameters. Many groups have studied $M^{2\nu}$ and $M^{0\nu}$ in terms of shell models, Quasi-particle Random Phase Approximations (QRPAs), operator expansion methods, and others (see [1–6] and references therein). The QRPA calculations [27] show that $M^{2\nu}$ are sensitive to the particle–particle interaction.

The observed values for $M^{2\nu}$ are compared with the (first) term in eqn (12.9) through the single particle-hole 1^+ state $|S\rangle$ at the intermediate nucleus as shown in Fig. 12.7. It is interesting that $M^{2\nu}$ is given approximately by the product of the successive single-β matrix elements M_S^ν and $M_{S'}^\nu$ through the single particle-hole 1^+ state at the low excitation region [28]. Single-β matrix elements are smaller, by a factor of $(g_A/g_A^0) \sim 0.3$, than the single-particle values [25,26]; thus, the $\beta\beta$ matrix elements are smaller by a factor of $(g_A/g_A^0)^2 \sim 0.1$. Here, g_A^0 is the axial-vector weak coupling and g_A is the effective (renormalized) value incorporating the $\tau\sigma$ core-polarization [26].

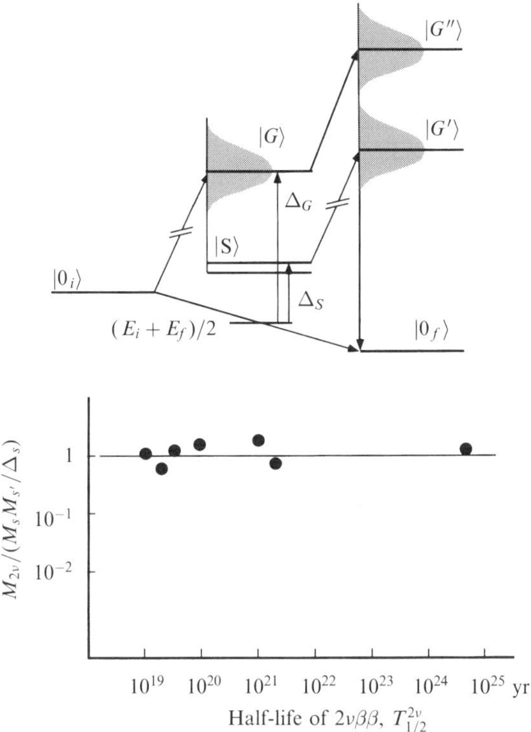

Fig. 12.7 Top: Level scheme of β- and $\beta\beta$ decays and $2v\beta\beta$ matrix elements. $|0_i\rangle$ and $|0_f\rangle$ and $|S\rangle$ are initial 0^+, final 0^+ and intermediate single particle-hole 1^+ states, respectively. $|G\rangle$ and $|G'\rangle$ are the GT giant resonances based on $|0_i\rangle$ and $|S\rangle$, respectively, while $|G''\rangle$ is the GT giant resonance based on $|G\rangle$. Bottom: The ratio of $\beta\beta$ matrix elements M^{2v} and products of single-β matrix elements M^v_S and $M^v_{S'}$ with the energy denominator Δ_S for successive single-β decays through the low-lying single particle-hole 1^+ state $|S\rangle$, in the intermediate nuclei versus the half-lives of $2v\beta\beta$ decays.

12.3 Nuclear spin–isospin responses for $\beta\beta$-v and solar-v

12.3.1 *Nuclear spin–isospin responses for axial weak processes*

Charged axial weak processes involve nuclear spin–isospin ($\tau\sigma$) responses. They are investigated by charge-exchange (τ) spin-flip (σ) nuclear reactions, as shown in Fig. 12.8. The $\tau\sigma$ responses associated with $0v\beta\beta$, $2v\beta\beta$, and solar-v include mainly $\langle\tau\sigma Y_\lambda\rangle$ with $\lambda = 0, 1, \ldots, 6$, $\langle\tau\sigma Y_\lambda\rangle$ with $\lambda = 0$, and $\langle\tau\sigma Y_\lambda\rangle$ with $\lambda = 0, 1$, respectively.

The $\langle\tau\sigma Y_\lambda\rangle$ with $\lambda = 0, 1, \ldots$ have been studied well by (^{3}He, t) reactions at RCNP, Osaka. The RCNP ring cyclotron with $K = 0.4$ GeV provides protons with energy up to 0.4 GeV and ^{3}He up to 0.54 GeV, which are used to excite preferentially the $\tau\sigma$ modes in nuclei.

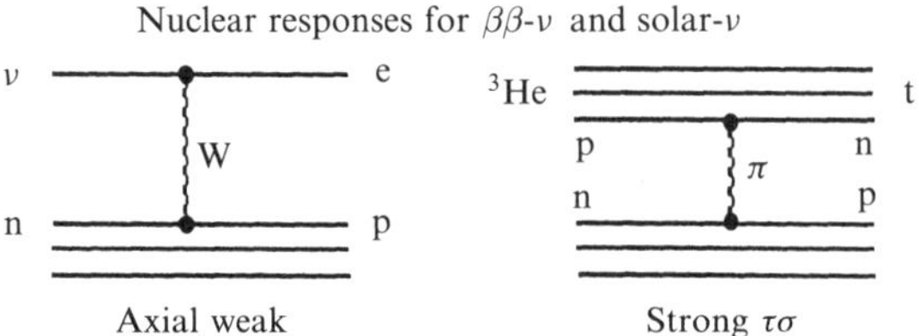

Fig. 12.8 Nuclear spin–isospin responses for axial charged weak processes and those for spin-flip charge-exchange strong processes.

12.3.2 *Spin–isospin responses for* ^{100}Mo *and* $2\nu\beta\beta$ *matrix elements*

Spin–isospin responses for ^{100}Mo were studied in order to investigate axial weak responses relevant to the $\beta\beta$ decay of ^{100}Mo. The ^{100}Mo $(^3$He, t$)^{100}$Tc process was measured by means of the high-resolution spectrometer GRAND RAIDEN. The 0.45 GeV ^{3}He beam obtained from the RCNP cyclotron was used. The $\tau\sigma$ strengths are concentrated, as shown in Fig. 12.9, into three states (resonances), the single particle-hole 1^+ state ($|S\rangle$) in the low excitation region, the Isobaric Analogue State ($|$IAS$\rangle$), and the GT giant resonance state $|G\rangle$. These are common features in medium nuclei.

The $2\nu\beta\beta$ matrix element for the ground state transition, $|0_i\rangle \rightarrow |0_f\rangle$, is discussed, for simplicity, in terms of the two GT(1^+) states, $|S\rangle$ and $|G\rangle$ [28]. Note that the Fermi (τ) transition does not contribute to the $0_i \rightarrow 0_f$ transition in medium nuclei because the Fermi strength is entirely absorbed by the double IAS. The states $|S\rangle$ and $|G\rangle$ are written by introducing an interaction between them as follows:

$$|S\rangle = |S\rangle_0 - \varepsilon|G\rangle_0, \tag{12.10a}$$

$$|G\rangle = |G\rangle_0 + \varepsilon|S\rangle_0, \tag{12.10b}$$

where $|S\rangle_0$ and $|G\rangle_0$ are unperturbed states. Because of the destructive interference between $|S\rangle_0$ and $|G\rangle_0$, the single-β matrix element M^ν_S is reduced by a factor $(g_A/g^0_A) \sim 0.3$ in comparison with the single-particle value. Similarly, the ground state $|0_f\rangle$ in the final nucleus is written as

$$|0_f\rangle = |S'S\rangle - \varepsilon|G'S\rangle, \tag{12.11}$$

where $|S'S\rangle$ and $|G'S\rangle$ are the single particle-hole 1^+ state ($|S'\rangle_0$) and the GT giant resonance ($|G'\rangle_0$) state, respectively, which are built on the intermediate state $|S\rangle$. Then the $2\nu\beta\beta$ matrix element is written as

$$M^{2\nu} = \frac{M^\nu_S M^\nu_{S'}}{\Delta_S} + \frac{M^\nu_G M^\nu_{G'}}{\Delta_G}, \tag{12.12}$$

where the first and the second terms stand for the products of the matrix elements for the successive single-β decays through $|S\rangle$ and $|G\rangle$ in the intermediate nucleus, respectively. The second term in eqn (12.10) vanishes because of the cancellation between the contributions from the admixture of $|S\rangle$ in $|G\rangle$, and of $|G'\rangle$ in $|0_f\rangle$. The former is relevant for the $\tau\sigma$ strength to $|0_f\rangle$, while the latter for the $\tau\sigma$ strength from $|0_i\rangle$. Thus,

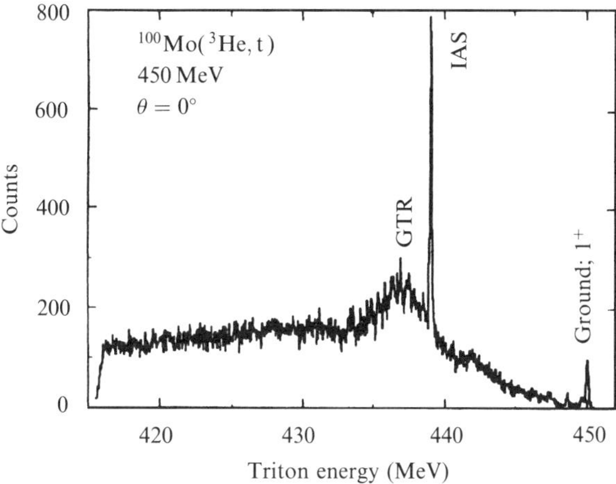

Fig. 12.9 Energy spectrum for ^{100}Mo (^{3}He, t) ^{100}Tc. IAS and GTR are the isobaric analogue state and GT giant resonances, respectively [29].

the vanishing of this term is regarded as a cancellation, in the GR region, of β^+ and β^- amplitudes from $|0_f\rangle$ and $|0_i\rangle$, respectively. Actually, the $\tau\sigma$ strength from $|G\rangle$ is absorbed mainly into the double GT resonance (the GT giant resonance based on $|G\rangle$), leaving little strength to $|0_f\rangle$. Consequently, one gets $M^{2\nu} \approx M_S^\nu M_{S'}^\nu / \Delta_S$, just what is obtained empirically [28].

The GT matrix element $M_{S'}^\nu$ for the 1^+ single-particle state (ground state) of ^{100}Tc(1^+) was obtained from the ^{100}Mo (^{3}He, t) ^{100}Tc reactions [29]. The obtained matrix element M_S^ν, together with $M_{S'}^\nu$ derived from the β-decay rate of ^{100}Tc(1^+) $\rightarrow$ ^{100}Ru(0^+), yields the $2\nu\beta\beta$ matrix elements through the single-particle states of ^{100}Tc(1^+). The value for $M_\nu^2 = M_S^\nu M_{S'}^\nu / \Delta_S$ is derived as 0.1, which agrees with the observed value of $M^{2\nu} = 0.09$ [13].

12.3.3 *Spin–isospin responses for solar-ν*

The solar-ν is mainly the low-energy pp neutrino. It is studied by the inverse β-decay of the ^{71}Ga (ν, β^-) ^{71}Ge process. Here, $\tau\sigma$ responses (solar-ν responses) for ^{71}Ge are crucial for extracting the solar-ν flux from the production of ^{71}Ge. The $\tau\sigma$ responses for ^{71}Ga have been studied by measuring the ^{71}Ga (^{3}He, t) ^{71}Ge reaction [30]. The total GT strength of 3.07 ± 0.11 was found for ^{71}Ge up to the particle threshold energy of 7.4 MeV [30]. It gives the solar-ν absorption rate of 130.3 ± 1.6 (stat) ± 16.8 (syst). Here, the solar-ν flux of 70.8 SNU is assumed for the ground state of ^{71}Ge, where the GT strength is given by the β-decay ft value.

In fact, the solar-ν absorption is investigated by measuring the daughter nuclei of ^{71}Ge produced by the ^{71}Ga ($\nu, e\gamma$) ^{71}Ge process. The corresponding reaction is studied by measuring the ^{71}Ga (^{3}He, tγ) ^{71}Ge process. It is interesting to note that the levels above the neutron threshold energy by a couple of hundred keV undergo γ-decays, producing the ^{71}Ge nuclei [30].

12.4 Spin-coupled dark matter studied by nuclear scatterings

A major component of the mass of the universe is considered to be the invisible DM. Weakly Interacting Massive Particles (WIMPs) are very likely DM candidates. Lightest SUSY Particles (LSPs), such as the neutralinos, are possible WIMPs. Indirect methods of searching for DM investigate products of DM annihilations, while the direct methods investigate elastic and inelastic scattering of DM from nuclei, as shown in Fig. 12.10. Details of DM are given in review articles (see [31] and references therein).

Vector-coupled DM with spin-independent interactions have been studied by measuring nuclear recoil due to elastic (coherent) scattering from Ge and Si nuclei in semiconductor detectors. Axial- (spin-) coupled DM with spin-dependent interactions have not been studied well because the scattering cross sections are very small. Since natural Ge and Si detectors contain mostly even–even spinless isotopes, they are not used for measuring spin-coupled DM. Thus, detectors containing odd-A nuclei with spin $J \neq 0$ are used. Germanium detectors enriched by ^{73}Ge, NaI, and CaF$_2$ scintillators, bolometers, and others are used for studying the spin-coupled DM at several laboratories.

The Osaka group has studied spin-coupled DM by means of the NaI detectors of EL V. The NaI detector with 100% isotope abundance ratios of ^{23}Na$((3/2)^+)$ and ^{127}I$((5/2)^+)$ is found to be very useful because of its large volume (0.76 ton) and low noise level (~ 4 keV) [32]. In particular, the inelastic scattering from the $(5/2)^+$ ground state to the $(7/2)^+$ excited state at $E_x = 58$ keV in ^{127}I is very useful for two reasons [33]: one that the spin response is known from the spin matrix elements of the inverse M1 γ-transition, and the other that the summed-energy signal of the nuclear recoil (E_R) and the γ-decay E_x is much larger than $E_R \approx 0$. The recoil energy for heavy DM is, in most cases, as small as $E_R \leq 2$–6 keV. Consequently, light outputs of these recoil nuclei are hardly observed even with low-noise detectors.

On the other hand, the deexciting γ-energy is of the order of $E_x = 10$–100 keV, which is well above the noise levels. The differential cross section for the inelastic scattering

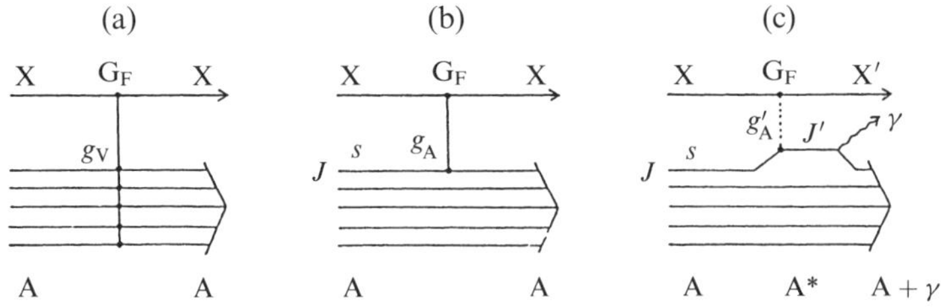

Fig. 12.10 Scattering of DM (X) from a nucleus A. (a) Elastic scattering of vector-coupled DM. (b) Elastic scattering of spin-coupled DM. (c) Inelastic scattering of spin-coupled DM [33].

of the spin-coupled DM from a nucleus of mass M_N is expressed as

$$\frac{\mathrm{d}^2\sigma}{\mathrm{d}p_i\,\mathrm{d}\Omega} = N_k \left(\frac{M_N M_D}{M_N + M_D}\right) \times G_F^2 g_A^2 F^2(q) \frac{P_f}{P_i} f(P_i) \frac{1}{4\pi}, \tag{12.13}$$

where P_f/P_i is the phase space factor, with P_i and P_f the momenta of the incident and outgoing (scattered) DM, Ω is the solid angle for the scattered DM, $f(P_i)$ is the DM momentum distribution and N_k is the constant for the k-type DM. The values of N_k and the coupling terms $G_F g_A$ are given for various types of DM in Table 12.2 [31]. The nuclear form factor is given by $F(q)$, q being the momentum transfer.

The spin matrix element is expressed as

$$\lambda_N = \langle A^*|S|A\rangle, \tag{12.14}$$

where $|A\rangle$ and $|A^*\rangle$ are the ground and excited states and S is the spin operator. The spin matrix elements in nuclei are sensitive to the spin–isospin correlation. Thus, their accurate evaluations are difficult. In the case of the spin-stretched transition to the excited state with spin $J' = J \pm 1$, the spin matrix element is related to the M1 γ matrix element $\langle A|M1|A^*\rangle$ as

$$\langle A^*|S|A\rangle = \left(\frac{2J'+1}{2J+1}\right)^{1/2} \frac{1}{g_M} \langle A|M1|A^*\rangle, \tag{12.15}$$

where g_M is the M1 γ-coupling constant given by $(e\hbar/2M)(3/4\pi)^{1/2}(g_s - g_l)/2$. The value for $\langle A|M1|A^*\rangle$ is obtained from the M1 γ deexcitation (transition) rate.

The upper limits on the elastic and inelastic scatterings lead to upper limits on the axial DM densities. The results are shown in Figs. 12.11 and 12.12. The most stringent limit on the spin-coupled DM was obtained in the heavy-mass region [33].

Search for the spin-coupled DM using CaF_2 (Eu) detectors of ELVI [20] is now going on. Here, the CaF_2 pure crystals used for studying ^{48}Ca $\beta\beta$ decays are replaced by

Table 12.2 The constants N_k and the coupling term for the k-type DM particle. T_q^3: the quark isospin, Δ_q: the spin carried by quark q·Q_q: the charge of quark. $m_{\tilde{q}}$: the mass of squark $\tilde{q}$. $\tan\beta = v_1/v_2$: the ratio of two Higgs vacuum expectation values [31,33]

WIMPs	N_k	Coupling term $g_A G_F$
Dirac neutrino	$2/\pi$	$\lambda_N(\Sigma_q T_q^3 \Delta_q)G_F$
Majorana neutrino	$8/\pi$	$\lambda_N(\Sigma_q T_q^3 \Delta_q)G_F$
Higgsino	$(8/\pi)\cos^2(2\beta)$	$\lambda_N(\Sigma_q T_q^3 \Delta_q)G_F$
Photino	$4/\pi$	$\lambda_N[\Sigma_q(eQ_q/m_{\tilde{q}})\Delta_q]$

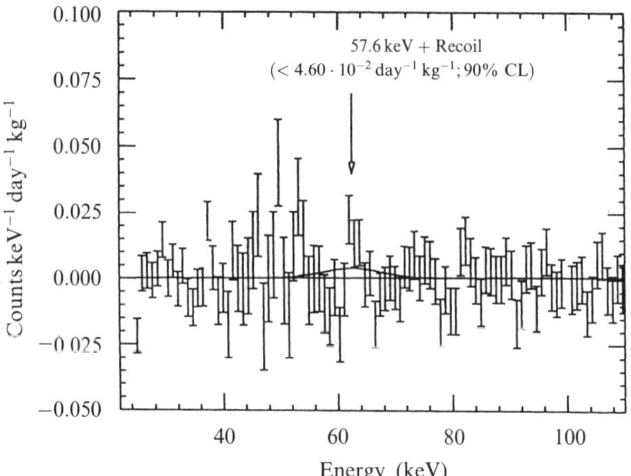

Fig. 12.11 Energy spectrum for the inelastic DM scattering from ^{127}I in EL V with NaI detectors in the energy region of the first excited state in ^{127}I [33].

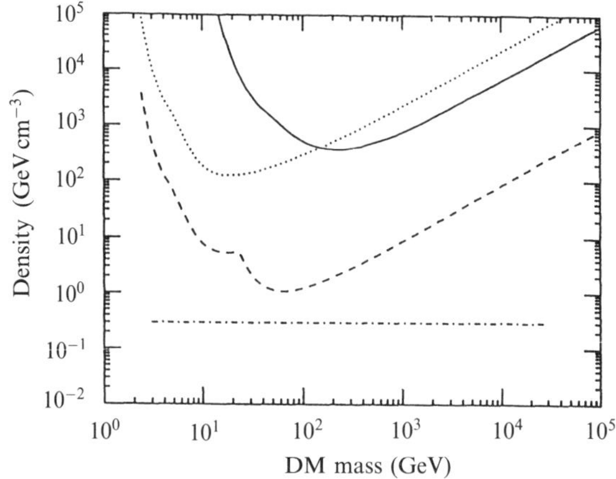

Fig. 12.12 Upper limits on the DM densities derived from the EL V NaI detectors. Dashed line: vector-coupled Dirac ν derived from elastic scatterings. Dotted line: axial-vector-coupled Majorana ν derived from elastic scatterings with the single-particle matrix element. Solid line: axial-coupled Majorana ν derived from inelastic scatterings [33].

25 modules of $45 \times 45 \times 45\ \mathrm{mm}^3$ CaF$_2$ (Eu) crystals in order to get large light output for DM. The ^{19}F isotope with $(1/2)^+$ is useful because of the large spin matrix element. It is expected to be used to search for spin-coupled DM in the region of $\Omega = 1$. A schematic view of EL VI is shown in Fig. 12.13.

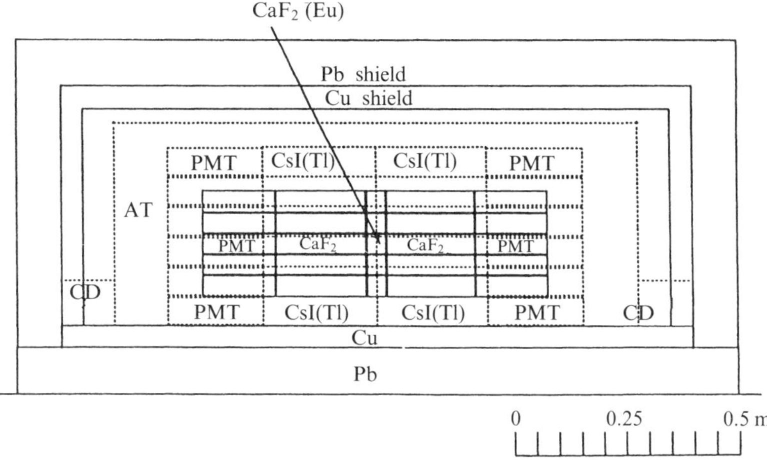

Fig. 12.13   Schematic view of EL VI with CaF_2 (Eu) to search for spin-coupled DMs.

12.5   Nuclear stabilities and nuclear decays

12.5.1   *Nuclear decays and nuclear stabilities studied by exotic nuclear transitions*

Nucleons bound in a nucleus with respect to strong, electromagnetic, and weak decays
are stable. The stability of nucleons in a nucleus is based on the following three facts:
(i) the baryon number is conserved so that nucleons (baryons) cannot decay into non-
nucleon (baryonic) particles; (ii) nucleon as a fermion with the Fermi statistics cannot
decay into a state occupied by other nucleons, by the Pauli principle; and (iii) a nucleon
is the lightest baryon in free space as well as in a nucleus.

High-precision studies of nucleon instabilities in a nucleus provide one with evidence
for violations of these laws. The possible processes associated with the nucleon/nuclear
instabilities are shown in Fig. 12.14.

A search for non-Paulian nuclear transitions associated with nucleon instability and
with the admixture (δ^2) of non-fermion statistics was made by observing the energy
spectra of decay particles from stable Na and I nuclei in the large NaI detector [34].
The upper limit on the probability of the non-Paulian transition with an energy release
greater than 18 MeV is obtained as $\tilde{\Gamma}/\hbar \leq 1.8 \cdot 10^{-33}$ s^{-1}, which corresponds to a lower
limit on the mean lifetime, $\tilde{\tau} \geq 1.7 \cdot 10^{25}$ yr, for the nucleon instability (decay). It gives
an upper limit $\delta^2 \leq 1.4 \cdot 10^{-53}$ for the relative strength of the non-Paulian transition.

12.5.2   *Exotic particles and transitions*

Light exotic particles (X) are searched for by investigating exotic transitions from normal particles N to exotic ones, X, if mass of X is smaller than that of N. If N and X are
bound in a nucleus, one may study exotic nuclear transitions $A(N) \rightarrow B^*(X)$ by measuring deexcitation of $B^*(X) \rightarrow C(X) + x + y + \cdots$, where x, y, ... are decay particles
(n, p, ...) and γ-rays.

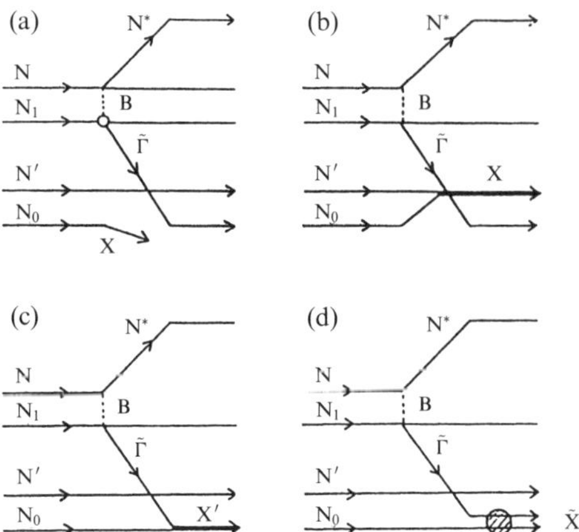

Fig. 12.14 Exotic transitions ($\tilde{\Gamma}$) associated with the nucleon instabilities and with the non-Paulian transition. The mediating boson B may be either a meson, ($\pi^{\pm}, \ldots$), a weak boson ($W^{\pm}, \ldots$), or a photon (γ). N_0, N_1, N, and N' stand for normal states occupied by nucleons, while X and $\tilde{X}$ are exotic states with the baryon number 2 in cases (b) and (c) and the non-Paulian state in the case (d).

One possible candidate for X is the H particle, which is a dihyperon with spin parity $J^{\pi} = 0^+$, isospin $I = 0$, and strangeness $S = -2$ [35]. It is a flavour singlet six-quark state. If the six quarks are so tightly bound that the mass of H is smaller than the two-nucleon mass, exotic transitions $2N \rightarrow H + x + \cdots$ and $A(2N) \rightarrow A^*(H) \rightarrow A(H) + x + y + \cdots$ are possible. These are double weak transition processes because of the strangeness change $\Delta S = 2$. Possible transition processes are shown in Fig. 12.15 [36].

High-sensitivity detectors used for studying rare $\beta\beta$ decays are used to search for these double weak nuclear transitions and for other rare/exotic nuclear processes.

Search for the light H particles has been made by investigating exotic transitions d $\rightarrow$ $H + \beta^+ + \nu$, ^{72,70}Ge $\rightarrow$ ^{72,70}Ge $+ H + \gamma$, and ^{127}I $\rightarrow$ ^{127}I*(H) $\rightarrow$ ^{125}Te $+ H + \beta^+ + \nu$. Lower limits on the half-lives for these exotic transitions were obtained as 10^{20}–10^{24} yr. These half-lives are much greater than the predicted value for light H. Thus, an H particle lighter than the two-nucleon mass of 1875.1 MeV is excluded [36]. The present argument can be applied for any light exotic particle X populated by exotic nuclear transitions of $N \rightarrow X$ with half-lives of the order of or longer than 10^{20}–10^{24} yr.

12.5.3 *Nucleon decays and deexcitation of nucleon holes*

Nucleon decays of $N \rightarrow x + y + \cdots$ have been studied extensively by several groups to search for possible violations of the baryon number conservation ($\Delta B = 0$). Actually, grand unified theories beyond the standard model predict nucleon decays with $\Delta B = 1$. They measure the decay products $x, y, \ldots$, which are mostly charged particles and/or γ-rays, in the high-energy region of 0.3–1 GeV. Nucleon decays can also be studied by investigating the nuclear deexcitation followed by the nucleon hole produced by

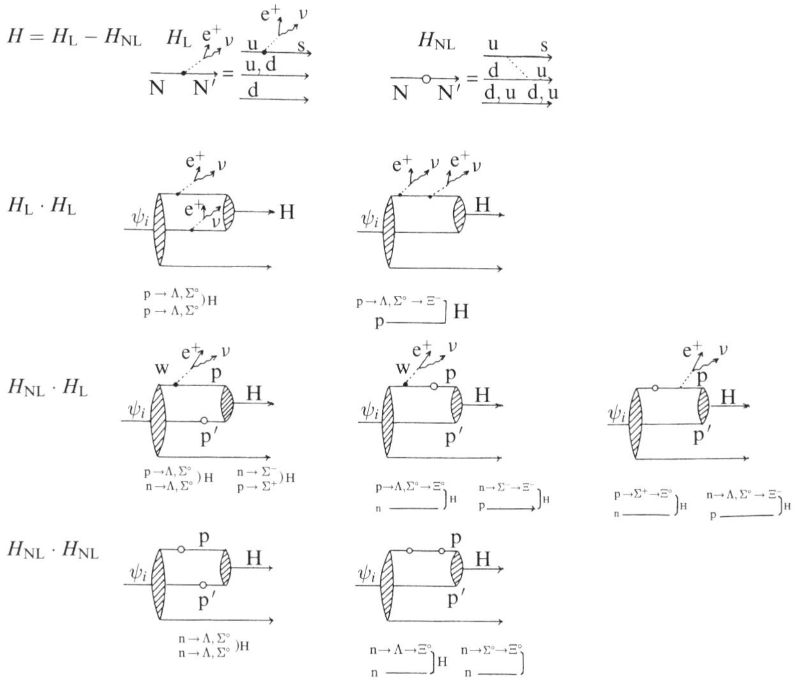

Fig. 12.15 Double weak decay processes for two nucleons feeding the H particle. H_L and H_NL are the leptonic and non-leptonic Hamiltonians, respectively (see text), and the corresponding vertices are indicated by the closed and open circles, respectively [36].

the nucleon decay in a nucleus [37]. This method is very useful for decay modes with unknown and/or neutral decay products since their direct measurements are very difficult and are almost impossible in the case of neutrinos.

Baryon-number-violating nucleon decays were studied by searching for X- and γ-rays associated with radioactive residual nuclei produced by single- and multi-nucleon decays in ^{127}I [38]. Large-volume NaI detectors of the ELEGANT $\beta\beta$ detector were used for this study. New lower limits of $\tau(\mathrm{n}) > 4.7 \cdot 10^{24}$ yr, and $\tau(\mathrm{p}) > 3.0 \cdot 10^{24}$ yr were obtained on the mode-independent mean lives for the neutron and proton decays in ^{127}I, respectively. Lower limits of $\tau(\mathrm{nn}) > 2.1 \cdot 10^{24}$ yr, $\tau(\mathrm{nnn}) > 1.8 \cdot 10^{23}$ yr, and $\tau(\mathrm{nnnn}) > 1.4 \cdot 10^{23}$ yr were deduced for the first time on the mode-independent mean lives for di-, tri-, and tetra-neutron decays in ^{127}I, respectively.

A search was made for anomalous γ-rays associated with the deep S hole in ^{16}O in the large underground water Cherenkov detector [39]. We found no significant excess beyond known backgrounds. The result was used to set a limit on invisible nucleon decay through $\mathrm{n} \to \nu_\mu \nu_\mu \bar{\nu}_\mu$, $\mathrm{n} \to \nu_e \nu_e \bar{\nu}_e$, and $\mathrm{n} \to \nu_\tau \nu_\tau \bar{\nu}_\tau$. A lower limit on the partial lifetime $\tau_\mathrm{B}/\mathrm{Br}(\gamma) = 1.8 \cdot 10^{31}$ yr (90% CL), where $\mathrm{Br}(\gamma) = \Gamma_\gamma / \Gamma_\mathrm{tot}$ is the branching fraction of the γ-decay with respect to the particle emission of the deexcitation of ^{15}O*, was obtained. The lifetime limit becomes $\tau_B = 4.9 \cdot 10^{26}$ yr (90% CL) for a conservative estimate of $Br(\gamma) = 2.7 \cdot 10^{-5}$ assuming a nuclear single-particle model. The result was also used for studying the forbidden transitions in the ^{16}O nucleus. The limit for the relative strength of the forbidden transition to the normal transition is $<2.3 \cdot 10^{-57}$ (90% CL).

12.5.4 *Charge non-conservations and electron decays*

Charge Non-Conservations (CNCs) and electron decays are studied by investigating exotic K X-rays and exotic γ-rays [40]. On the basis of the standard model of the electron, the K X-ray emission is forbidden in stable atoms, where the K-shell is filled by two electrons. Searches for such forbidden K X-rays, however, have recently attracted great interest as the most sensitive test for the stability of electrons in atoms, and for the Charge Conservation (CC) law. The possible process relevant to exotic K X-ray emission is the X-ray transition to the K-electron hole produced by the spontaneous disappearance (decay) of the K-electron (e_K^-). In the framework of the standard model, electron is the lightest charged particle and the CC law is valid. Consequently, the disappearance (decay) of the electron must be associated with the CNC, a process beyond the standard theory. The K-electron decay and the CNC can be studied by investigating the following three modes:

$$e3\nu \text{ mode}: \qquad e_K^- \to \nu_e + \nu_i + \bar{\nu}_i, \tag{12.16}$$

$$\text{ll mode}: \qquad e^-(\nu) + A+ \to e^-(\nu) + B, \tag{12.17}$$

$$e\nu \text{ mode}: \ e\nu NN', \quad e_K^- + A + \nu + A^*; \tag{12.18}$$

$$e\nu NN', \quad e_K^- + A \to \nu + A, \tag{12.19}$$

where A and B are atomic nuclei, and A^* is an excited state of A. The Feynman diagrams are shown in Fig. 12.16. The gauge bosons ($B^{0\pm}$) mediating the CNC process are the weak bosons (Z^0, $W^\pm$) in case of the weak process, and the photon (γ) in case of the electromagnetic process. The CNC process must involve either $e \leftrightarrow \nu$ or $u \leftrightarrow d$ conversions, which violate the CC law in either the lepton or the quark sector, respectively. The CC law is considered to be valid under the framework of the gauge invariance of the QED field theory with massless gauge boson, namely the photon. Experimentally, the CC law has not been checked with an accuracy comparable to other conservation laws.

Experimental studies of the $e3\nu$ mode (eqn (12.16)) have been made by searching for K X-rays [40,41] and those of the ll mode (eqn (12.17)) by searching for the forbidden β-decays (Fig. 12.16, llnp) [42]. Recently, the $e\nu$ mode has been studied by investigating γ-rays following nuclear CNC excitation [43,44], as shown in Fig. 12.16 ($e\nu NN'$). Small-volume Ge detectors have been used for most of the K X-ray studies of the Ge isotopes with atomic number $Z = 32$ and neutron number $N = 38$–42. The $e\nu NN'$ mode (eqn (12.18)) was studied by investigating the CNC process $^{127}I + e_K \to {}^{127}I^* + \nu$, $^{127}I^*$ being the first excited state of ^{127}I. The large NaI detector array of EL V was used

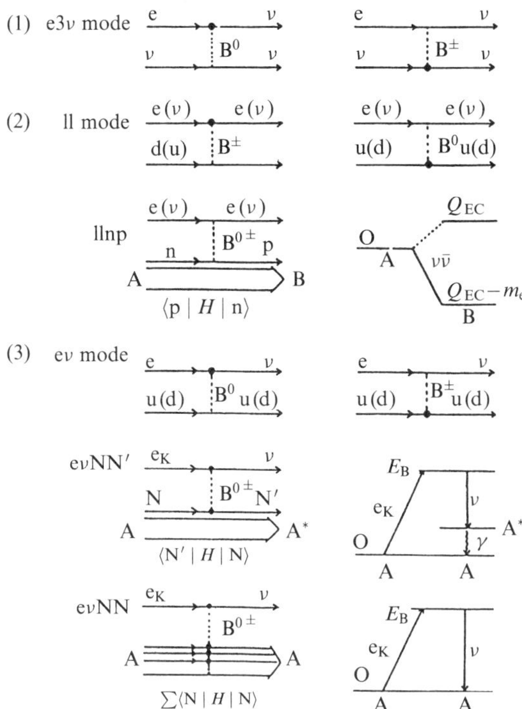

Fig. 12.16 The CNC diagrams for (1) the e3ν mode of e$^-$ $\to$ $\nu_e + \nu_i + \bar{\nu}_i$, (2) ll mode of A $\to$ B $+$ e$^-(\nu)$ $+$ e$^+(\bar{\nu})$, and (3) eν mode of e$^-$ $+$ A $\to$ ν $+$ A (B or A*). llnp, eνNN$'$ and eνNN stand for the CNC process relevant to the β-decay (n$\leftrightarrow$p), the inelastic process (A$\leftrightarrow$A*) and to the elastic process (A$\leftrightarrow$A), respectively. The bosons mediating these processes are given by B$^{0\pm}$. They are the weak bosons (Z, W$^\pm$) in the weak process and the photon (γ) in the electromagnetic process. $Q_{\rm EC}$, $m_{\rm c}$ and $E_{\rm B}$ are the electron capture Q-value, the electron rest mass energy, and the K-electron binding energy, respectively.

to search for the exotic γ-decay from ^{127}I*. The lower limit on the mean lifetime of the process was obtained as $\tau > 0.58 \cdot 10^{23}$ yr. Stringent limits for the CNC coupling strengths have also been derived [44].

The e3ν mode (eqn (12.16)) was studied by investigating the exotic K X-ray emission from neutral iodine atoms of the large NaI detector array [41]. The stringent upper limit on the exotic K X-ray emission was obtained as $\tau > 1.2 \cdot 10^{23}$ yr. It leads to the stringent upper limits of $\varepsilon_{\rm w}^2 < 3 \cdot 10^{-30}$, $\varepsilon_\gamma < 1.5 \cdot 10^{-45}$, and $\delta_{\rm NP}^2 < 1.1 \cdot 10^{-46}$ for the relative strengths of the CNC weak process, the CNC electromagnetic process, and the non-Paulian K X-ray emission, respectively [41]. The non-Paulian K X-ray limit is related to the limit on the finite electron size as $\gamma_0 < 1.1 \cdot 10^{-17}$ cm, which corresponds to the energy scale of $E > 2\,$TeV.

12.6   Concluding remarks

1. Nuclei, as ordered systems of nucleons in eigenstates of E, J^π, T, and so on, are used to study the fundamental properties of particles and interactions.

2. Double beta decays in nuclei are very useful probes for studying the properties of neutrinos and weak interactions.

3. Neutrinoless $\beta\beta$ decays ($0\nu\beta\beta$) are sensitive to the finite Majorana mass $\langle m_\nu \rangle$ of the electron neutrino, the right-handed weak interaction $\langle$RHC$\rangle$, the majoron–ν coupling $\langle g_B \rangle$, the R-parity-violating coupling with SUSY particles, and others. These are all processes beyond the standard theory. The limits on the $0\nu\beta\beta$ of ^{100}Mo studied by EL V and those on other nuclei set upper limits of $\langle m_\nu \rangle < 1\text{–}2\,\text{eV}$, $\langle$RHC$\rangle < 10^{-6}$, $\langle g_B \rangle < 1\text{–}2 \cdot 10^{-4}$, and so on. Further studies on $0\nu\beta\beta$ and $2\nu\beta\beta$ are expected [4–6,13–21,45].

4. Finite half-lives for two-neutrino $\beta\beta$ decays ($2\nu\beta\beta$) of ^{100}Mo and ^{116}Cd were obtained. The $2\nu\beta\beta$ matrix elements are found to be $M^{2\nu} = 0.09$ for ^{100}Mo and $M^{2\nu} = 0.07$ for ^{116}Cd. These values are indeed one-tenth of the single-particle values. The observed values for $M^{2\nu}$ can be written as $M^{2\nu} \approx M_S^\nu M_{S'}^\nu / \Delta_S$, where M_S^ν and $M_{S'}^\nu$ are the matrix elements of successive single-β decays through the single particle-hole 1^+ state in the intermediate nucleus, and Δ_S is the energy denominator.

5. Nuclear spin–isospin ($\sigma\tau$) responses for axial charged-current weak processes were studied by investigating the spin-flip charge-exchange nuclear reactions.

6. The $\tau\sigma$ strength for ^{100}Mo is found to be concentrated in two states of the single particle-hole 1^+ state $|S\rangle$ and the GT giant resonance state $|G\rangle$. Analysis of $M^{2\nu}$ in terms of the coupled states $|S\rangle$ and $|G\rangle$ of the intermediate nucleus is found to explain the empirical relations $M^{2\nu} \approx M_S^\nu M_{S'}^\nu / \Delta_S$.

7. Spin-coupled DM is studied by investigating the inelastic scattering from ^{127}I in the EL V NaI detector and the elastic scattering from ^{19}F in the EL VI CaF$_2$ detector. The search for spin-coupled DM by investigating inelastic scatterings from odd-A nuclei is shown to be very powerful because of the large energy signal and the reliable nuclear responses derived from the γ-decay.

8. Studies of rare nuclear processes are extended to investigate conservation laws and exotic transitions. Nuclear/nucleon stabilities in nuclei along with the Pauli principle for composite particles are studied by measuring exotic γ- and nucleon decays of stable nuclei. Most stringent limits were derived for various modes relevant to nuclear/nucleon instabilities.

9. Nucleon decays and baryon number non-conservations are studied by measuring nuclear deexcitations of the nucleon hole produced after the nucleon decay in a nucleus. These methods give most stringent limits on nucleon decays into invisible particles (neutrinos, etc.) and on multi-nucleon decays.

10. Electron decays and CNCs are studied by measuring exotic K X-rays following $e_K \rightarrow 3\nu$ and exotic γ-rays following $e_K \rightarrow \nu\gamma$. Stringent limits are given on these processes.

11. The new underground laboratory of Oto Cosmo Observatory with the ELEGANT series of high-sensitivity detectors is expected to open up a new and rich field of

the lepton nuclear physics. Here, the accelerator laboratory with strong probes is expected to play an important role by providing nuclear responses relevant to weak leptonic processes.

References

1. W. C. Haxton and G. J. Stephenson Jr., *Prog. Part. Nucl. Phys.* **12** (1984) 409.
2. M. Doi, T. Kotani, and E. Takasugi, *Prog. Theor. Phys.* **83** (Suppl.) (1981) 1.
3. D. O. Caldwell, *Int. J. Mod. Phys.* **A4** (1989) 1851.
4. H. Ejiri, *Nucl. Phys.* **A522** (1991) 305c; H. Ejiri in *Phys. astrophys. neutrinos*, M. Fukugita and A. Suzuki, eds, Springer-Verlag (1994), p. 500; H. Ejiri, *Int. J. Mod. Phys.* **E6**(1) (1997) 1.
5. A. Morales, LFNAE-94-004 (1994).
6. M. Moe and P. Vogel, *Ann. Review. Nucl. Part. Science* **44** (1994) 247; H. Ejiri *et al.*, eds, *Proc. WEIN'95*, Osaka, June 1995, World Scientific (1995); K. Enqrist *et al.*, eds, *Proc. ν'96*, Helsinki, June 1996, World Scientific (1996).
7. F. J. Gilman and S. H. Rhie, *Phys. Rev.* **D32** (1985) 324; R. N. Mohapatra, *Phys. Rev.* **D34** (1986) 909, 3457; M. Hirsh *et al.*, *Phys. Rev.* **D53** (1996) 1392.
8. R. H. Mohapatra and E. Takasugi, *Phys. Lett.* **B211** (1988) 192.
9. E. Fiorini *et al.*, *Nuovo Cimento* **13A** (1973) 747.
10. E. Fiorini and T. D. Ninikoski, *Nucl. Instr. Meth.* **224** (1984) 83; A. Giuliani *et al.*, *Proc. WEIN'95*, Osaka, H. Ejiri, T. Kishimoto, and T. Sato, eds, World Scientific (1995), p. 583.
11. H. Ejiri *et al.*, *Nucl. Phys.* **A448** (1986) 271; *J. Phys. G. Nucl. Phys.* **13** (1987) 839; K. Kamikubota *et al.*, *Nucl. Instr. Meth.* **A245** (1986) 379.
12. H. Ejiri, *Proc. Yamada conf. nuclear weak processes and nuclear structures*, Osaka 1986, Morita *et al.*, eds, World Scientific (1989), p. 190; T. Watanabe, Ph. D. Thesis, Osaka University, 1988.
13. H. Ejiri *et al.*, *Nucl. Instr. Meth.* **A302** (1991) 304; H. Ejiri *et al.*, *Phys. Lett.* **258B** (1991) 17; *J. Phys. G. Nucl. Phys.* **17** (1991) S155; J. Tanaka and H. Ejiri, *Phys. Rev.* **D48** (1993) 5412; N. Kudomi *et al.*, *Phys. Rev.* **C46** (1992) R2132.
14. H. Ejiri *et al.*, *J. Phys. Soc. Japan* **64** (1995) 339; K. Kume *et al.*, *Nucl. Phys.* **A577** (1994) 405c.
15. K. Nagata, H. Ejiri, K. Hayashi, T. Shima, and J. Tanaka, *Nucl. Instr. Meth.* **A362** (1994) 261.
16. N. Kudomi *et al.*, *Proc. WEIN'95*, H. Ejiri, T. Kishimoto, and T. Sato, eds, World Scientific (1995), p. 204; H. Ejiri *et al.*, *Nucl. Phys.* **A611** (1996) 85.
17. T. Tomoda, *Rep. Prog. Part. Phys.* **54** (1991) 53.
18. H. V. Klapdor-Kleingrothaus, *Proc. WEIN'95*, H. Ejiri, T. Kishimoto, and T. Sato, eds, World Scientific (1995); M. Hirsh and H. V. Klapdor-Kleingrothaus, *ibid.*, p. 217.
19. E. Takasugi, *Prog. Theor. Phys.* **95** (1995).
20. R. Hazama *et al.*, *Proc. WEIN'95*, H. Ejiri, T. Kishimoto, and T. Sato, eds, World Scientific (1995), p. 635.

21. K. Lou, F. Boehm *et al.*, *Proc. WEIN'95*, H. Ejiri, T. Kishimoto, and T. Sato, eds, World Scientific (1995), p. 192.
22. S. R. Elliot *et al.*, *Phys. Rev.* **C46** (1992) 1535.
23. A. Kawashima *et al.*, *Phys. Rev.* **C47** (1993) 2452.
24. T. Bernatowicz *et al.*, *Phys. Rev.* **C47** (1993) 806; *Phys. Rev. Lett.* **69** (1992) 2341.
25. K. Ikeda, S. Fujii, and J. I. Fujita, *Phys. Lett.* **3** (1963) 271.
26. H. Ejiri and J. I. Fujita, *Phys. Rep.* **38C** (1978) 85.
27. P. Vogel and M. R. Zirnbauer, *Phys. Lett.* **57** (1986) 3148, and also references in review articles [1–6].
28. H. Ejiri and H. Toki, *J. Phys. Soc. Japan* **65** (1996) 7.
29. H. Akimune, H. Ejiri *et al.*, *Phys. Lett.* **B394** (1997) 23.
30. M. Fujiwara *et al.*, *Nucl. Phys.* **A599** (1996) 223c; H. Ejiri *et al.*, private communication (1995).
31. V. Trimble, *Ann. Rev. Astron. Astrophys.* **56** (1987) 425; P. F. Smith and J. D. Lewin, *Phys. Rep.* **187** (1990) 203; D. O. Caldwell, *Proc. part. nucl. astrophys. cosmology in the next millenium*, Snowmass (1994).
32. K. Fushimi *et al.*, *Phys. Rev.* **C47** (1993) R425.
33. H. Ejiri *et al.*, *Phys. Lett.* **B317** (1993) 14; K. Fushimi and H. Ejiri, private communication (1995).
34. H. Ejiri and H. Toki, *Phys. Lett.* **B306** (1993) 218.
35. R. L. Jaffe, *Phys. Lett.* **38** (1977) 24.
36. H. Ejiri *et al.*, *Phys. Lett.* **B228** (1989) 24.
37. H. Ejiri, *Phys. Rev.* **C48** (1993) 1442.
38. R. Hazama, H. Ejiri, K. Fushimi, and H. Ohsumi, *Phys. Rev.* **C49** (1994) 2407.
39. Y. Suzuki *et al.*, *Phys. Lett.* **B311** (1993) 357.
40. A. W. Sunyar and M. Goldhaber, *Phys. Rev.* **120** (1960) 871; E. Bellotti *et al.*, *Phys. Lett.* **B124** (1983) 435; E. T. Avigon II *et al.*, *Phys. Rev.* **D34** (1986) 97.
41. H. Ejiri *et al.*, *Phys. Lett.* **B282** (1992) 281.
42. B. E. Normal and A. G. Seamster, *Phys. Rev. Lett.* **43** (1979) 1226; A. Roy *et al.*, *Phys. Rev.* **D28** (1983) 1770.
43. S. Holjevic *et al.*, *Phys. Rev.* **C35** (1987) 502.
44. H. Ejiri *et al.*, *Phys. Rev.* **C44** (1987) 502.
45. Ph. Hubert and NEMO collaboration, *Proc. WEIN'95*, H. Ejiri., T. Kishimoto, and T. Sato, eds, World Scientific (1995), p. 184.

Index

abelian gauge
  fixing 304, 307–8
  MA gauge 304, 311
  SU(N) invariance condition 296, 309
α-cluster model 150–89
  α-clustering in ^{94}Mo region 172–6
  α-clustering in ^{212}Po region 176–81
  and its dilemma 157–8
  experimental studies of α-cluster
     structure 151–2
  extension to ^{44}Ti region 157–72
  formalism of cluster models 182–5
  history 150
  revival 150–1
  structure of light nuclei 153–7
α-decay 152
α-particles
  neutron decay of GQR induced by 109–12
  transfer reactions 132, 152
α-width 176–81
analysing power 19, 20–1, 44–5, 47–8
Anomalous Large Angle Scattering
    (ALAS) 155–7, 158
anomalous renormalization factor 142
asymmetry parameter 267, 269–71
  polarization of hypernuclei 274–83
atomic physics 226–49
  channelling in a bent crystal 229–30
  charge-state-dependent stopping power of ions
     in thin carbon foils 240–2
  convoy electrons 230–4
  direct measurement of the binding energy
     effect 246
  electron capture and loss of light
     projectiles 234–9
  L_γ X-ray emission in heavy-ion bombardment
     of Bi 227–9
  stopping power of carbon for H_2^+ and H_3^+ ions
     242–4
  stopping power of metallic elements for light
     projectiles 245–6
AVF cyclotron 2
axial weak processes 323–4

Backward Angle Anomaly (BAA) *see* Anomalous
    Large Angle Scattering
baryon–baryon interactions 271–4

baryon-number violating nucleon decays 331–2
baryon structure 10–11
bent crystal: channelling in 229–30
beryllium: structure of ^{8}Be 150–1, 153
beta decays 202–6, 332–3
  double *see* double beta decays
binding energy effect 246
bismuth (Bi)
  decay from spin–isospin resonance in
     ^{208}Bi 67–9
  L_γ X-ray emission in heavy-ion bombardment
     of 227–9
boron: snowballs in ^{12}B 198–207
bound states: unification with scattering states
    154–7, 158–60, 179–81
branching ratios 264–5, 282–3
breakup processes 4, 126–49
  CDCC method 130–1, 134–6
  elastic and inelastic 126
  elastic scattering and breakup reactions of
     composite projectiles 138–46
    ^{12}C $\rightarrow$ 3α breaking effect on the excitation
     of the 3α-cluster state 145–6
    Coulomb breakup of ^{6}Li and
     Coulomb–nuclear interference 144–5
    deuteron breakup 138–40
    energy dependence of dynamical
     effects 142–4
    ^{6}Li and ^{7}Li at low energies 141–2
    light–heavy ions 140–6
  excitation of cluster states in heavier
     nuclei 132–3
  inelastic scattering 132
  microscopic cluster model 134
  microscopic coupled-channel formalism 133–8
  projectile breakup effects on elastic scattering
     129–30
  projectile-target folding model
     interactions 136–8
  transfer reactions 131–2
  types of 127–9
'bubbles' 190, 196
bulk capture effect 229–30

cadmium: $\beta\beta$ decay of ^{116}Cd 319–21
calcium 167–72
  α-cluster structure of ^{42}Ca 170–2

calcium (*contd.*)
  higher-nodal positive parity band in
    ^{40}Ca 168–70
  $K = 0^-$ band in ^{40}Ca 167–8
calcium fluoride (CaF$_2$) detector array 15
carbon
  α-cluster model for ^{12}C 153
  breakup of ^{12}C to 3α 145–6
  DBPAs in ^{12}C 216–21
  foils and cyclotron beams 231–4
    charge-state-dependent stopping power of
      ions in 240–2
    stopping power for H$_2^+$ and H$_3^+$ ions 242–4
    hypernuclear polarization 257–9, 260–2,
      275–80
centrifugal potential 106
channelling in a bent crystal 229–30
charge distribution 29–31
charge-exchange reactions 54–5, 107–9, 323–4
  see also spin–isospin giant resonances
Charge Non-Conservations (CNCs) 332–3
chiral symmetry 292–3
  recovery at finite temperature 299–300
  spontaneous breaking 297–9
Cluster Folding (CF) model 136
cluster structures 4
  breakup processes and 126–49
    excitation of cluster states in heavier nuclei
      132–3, 145–6
    microscopic cluster model 134
  see also α-cluster model; 'snowballs'
coexistence model 157
colour confinement 9–10
  deconfinement phase transition 300–2
  modelling 293–4
  static confining potential 295–7
  see also quark nuclear physics
composite projectiles, light 126–49
Compton scattering 3
confinement–deconfinement phase
    transition 300–2
Continuum-Discretized Coupled-Channel
    (CDCC) method 130–1, 134–6
convoy electrons 230–4
  by foil-excited He ions 233–4
  secondary electrons 234
  target material and thickness
    dependence 231–3
copper: ^{58}Cu 118–22
Coulomb breakup 138, 144–5
Coulomb-nuclear interference 144–5
Coulomb potential 106
  finite-size 211, 212, 213
Coupled-Channel Born Approximation
    (CCBA) 131
coupled-channel formalism 36, 37
  microscopic CC theory 133–8

coupling form factors 136, 144–5
cross sections
  DBPAs 212–13, 219–21
  elastic scattering 19, 20–1, 44–5, 47–8
cumulative-sum plot 94, 95

dark matter (DM) 15–16, 326–9
deconfinement phase transition 300–2
deeply bound pionic atoms (DBPAs) 8, 210–25
  (d, ^{3}He) reactions for 221–3, 223–4
  formation of 211–16
    (n, d) reactions 214–16
    (n, p) reactions 212–14
  search for using (p, ^{2}He) reactions 216–21
  structure of 210–11
deexcitation of nucleon holes 14, 330–2
deformed nuclei: inelastic scattering from 33–43
Deformed Optical Potential (DOP) 34–8
density
  density dependence of interactions 31–2
  distributions 46–8
depolarization processes 257–9
deuterons
  breakup effects on the transfer reactions 131
  breakup and elastic scattering 138–40
    CDCC calculations 140
    spin-singlet ($S = 0$) states 140
    spin-triplet ($S = 1$) states 138–40
  (d, ^{3}He) reactions for DBPAs 221–3, 223–4
Dirac equation 43
direct breakup process 127–9
direct transfer 200–2
dispersion relation 194
Distorted-Wave Born Approximation
    (DWBA) 131, 152, 163–7
Distorted Wave Impulse Approximation (DWIA)
    theory 252
double beta decays 6–7, 12–14, 314–23
  ELEGANTs for 318–21
  experiments 317–18
  GT transitions 64–6
  neutrino mass limits and $2\nu\beta\beta$ matrix
    elements 321–3
double-closedness 154
Double-Folding (DF) model 137, 173–4
double weak decay processes 330, 331
dual gauge field Lagrangian 310
Dual Ginzburg–Landau (DGL) theory 9–10,
    293–302
  colour confinement modelling 293–4
  deconfinement phase transition 300–2
  Lagrangian 294–5
  recovery of chiral symmetry 299–300
  spontaneous chiral symmetry breaking 297–9
  static confining potential 295–7
dual Higgs mechanism 9–10
dual Meissner effect 293, 295

dynamical polarization potential 130, 141–2
dynamical spin-dependent interactions 130,
    142–4

effective number approach 215
elastic breakup 126
elastic scattering 4, 18–49
    and breakup reactions of composite projectiles
        see breakup processes
    comparison with microscopic theory 27–33
    effects of medium 46–8
    at intermediate energies 43–8
    microscopic optical potential 23–7
    projectile breakup effects on 129–30
    proton elastic scattering around 65 MeV 19–23
    spin-coupled dark matter 326–9
electric mode giant resonances 77, 78, 90–104
    highly excited region 101–4
    isovector electric modes 103–4, 105
    $L = 4$ and other multipolarities 98–101
    LEOR 90–7
electron capture 230–4
    and loss of light projectiles 234–9
electron decays 202–6, 332–3
electron loss capture (ELC) 230–4
electrostriction 196
ELEGANTs (ELEctron GAmma-ray Neutrino
        Telescopes) 3, 13–14, 318–21
energy
    characteristic incident energies for GR studies
        89–90
    decay of GRs 108–9
    and elastic scattering 32–3
        intermediate energies 43–8
        low energy protons 19–23
    energy dependence of breakup of light-heavy
        ions 142–4
    G-T strengths at high excitation energy 55–7
    loss for cyclotron beams in carbon 240–2
Energy-Weighted Sum Rule (EWSR) 80
equilibrium stage 107–9
escape width 81, 107–8
excitation
    of cluster states in heavier nuclei 132–3, 145–6
    excitations in liquid helium 194, 197–8
    giant resonances 101–24
    excitation and decay of GRs 107–9
    in highly excited region 101–4
    magnetic GRs 112–22
    particle decay from highly excited states 104–6
exotic nuclear beams 7
exotic particles 329–30
exotic transitions 122, 329–30

flavour 8–9, 250
foil thickness
    convoy electrons 231–4

foil-excited He ions 233–4
    target material and thickness dependence
        231–3
    energy losses of cyclotron beams 241–2
folding model 30–1, 36–8
    Double-Folding model 137, 173–4
    projectile breakup processes 136–8
fp-shell region 151, 157–72, 178, 181
FRagment Separator (FRS) 221–2
fragmentations 199–202
freezing-out of nuclear polarization 202–7
fusion excitation function 158–60

G-matrix theory 23–4
Gallium: ^{71}Ga as a neutrino detector 61–3
(γ, K^+) reaction
    and its characteristics 259–64
    new possibilities 287
γ-decay 63, 68–9, 81, 105, 109
    isovector electric modes 103–4, 105
γ-rays, exotic 332–3
γ-vibrations 38–42, 43
Gamow–Teller (GT) resonances 4–5, 50–3, 79
    charge-exchange reactions 54–5
    double beta decay 64–6
    G-T strengths at high excitation energy 55–7
    G-T transitions to the β^+-side 57–61
    isospin structure 112–13
        $T_0 = 0$ nucleus 113–15
        $T_0 \geq 0$ nuclei 117–22
    particle decays 67–9
gauge bosons 1
Generator Coordinate Method (GCM) 182–3
germanium: double beta decays of ^{76}Ge 319
Giant Dipole Resonance (GDR) 50, 52, 79, 82–4
Giant Monopole Resonance (GMR) 52, 66, 79,
        87–9, 101
    highly excited region 101, 103, 104
Giant Quadrupole Resonance (GQR) 52, 66, 79,
        84–5, 86, 87, 101
    highly excited region 101, 102, 103
    neutron decay of GQR induced by α-particles
        109–12
giant resonances 77–125
    characteristic incident energies 89–90
    fine structure of electric GRs 90–101
        discrete levels 90
        $L = 4$ and other multipolarities 98–101
        low-energy octupole resonance 90–7
    formation and decay 104–12
        excitation and decay 107–9
        neutron decay of GQR induced by
            α-particles 109–12
        particle decay from highly excited
            states 104–6
    in highly excited region 101–4
        high-energy giant octupole resonance 101–3

giant resonances (*contd.*)
  in highly excited region (*contd.*)
    isovector electric modes 103–4
    observed as a bump 101
  isoscalar giant quadrupole resonance 84–5
  isoscalar monopole resonance 87–9
  isoscalar octupole resonance 85–7
  isospin structure of magnetic GRs 112–22
    M1 and Gamow–Teller resonances 112–13
    $T_0 = 0$ nucleus 113–17
    $T_0 \geq 1$ nuclei 117–22
  isovector giant dipole resonance 82–4
  spin–isospin GRs *see* spin–isospin giant
    resonances
  sum rule and general features 79–81
  width 81–2
glueballs 11

H particle 330, 331
hadron–nucleon systems 1–2, 7–9
  hyperon interactions and flavour nuclear
    physics 8–9
  nucleons, mesons and isobars in NN
    channels 7–8
heavy nuclei: α-clustering 176–81
hedgehog configuration 302, 308–9
helium
  helium ion beams
    convoy electrons 231–4
    energy loss in carbon foils 240–2
    K-shell electron capture 235–9
  hypernuclear weak decays 269–71
    study of $^5_\Lambda$He 280–4
  impurity ions in superfluid *see* snowballs
Hexadecapole Resonance (HDR) 103
  low energy 98–101
High-Energy Octupole Resonance
    (HEOR) 101–3
higher nodal band 155–7, 168–70
highly excited states 104–6
  giant resonances and 77–125
hybrid quark model 272–4
hydrogen
  secondary electrons produced by fast H^+ and
    H_2^+ ions 234
  stopping power of carbon for H_2^+ and H_3^+ ions
    242–4
hypernuclei 250–91
  basic aspects of hypernuclear weak
    decays 264–71
    asymmetry in the pionic weak decays of
      $^5_\Lambda$He 269–71
    calculation of the mesonic decay rate 265–7
    mesonic decays of p-shell
      hypernuclei 267–9

non-mesonic weak decay and baryon–baryon
    interactions 271–4
  production of polarized hypernuclei 251–64
    effects of depolarization processes before the
      weak decay 257–9
    the (γ, K^+) reaction and its characteristics
      259–64
    (π^+, K^+) and (K^-, π^-) reactions 252–7
  weak decays of polarized hypernuclei 274–84
    study of the ^{12}C(π^+, K^+) reaction 275–80
    study of $^5_\Lambda$He 280–4
  see also strangeness
hyperon interactions 8–9

Ikeda diagram 151
Ikeda sum rule 53
impurity ions in liquid helium *see* 'snowballs'
inelastic breakup 126, 132
inelastic scattering 4, 107
  breakup effects 129, 132
  from deformed nuclei 33–43
    polar cap model 42–3
  spin-coupled dark matter 326–9
inner-shell ionization 227–9
instabilities, nuclear 14–15, 329–33
instantons 10, 302–3
  multi-instantons and monopole condensation
    303–4
  multi-instantons and the Wilson loop 304
Isobaric Analogue State (IAS) 4, 50–2, 53, 64,
    65, 79
  particle decay from spin–isospin
    resonances 67–9
isobaric mirrors 58–61
isobars 7–8, 56–7
isoscalar giant dipole resonance (ISGDR) 79, 84
isoscalar giant monopole resonance (ISGMR) 79,
    87–9
isoscalar giant octupole resonance (ISGOR) 79,
    85–7
isoscalar giant quadrupole resonance
    (ISGQR) 79, 84–5
isoscalar (IS) modes 77, 78
isospin structure of magnetic giant resonances
    112–22, 123
  M1 and Gamow–Teller resonances 112–13
  $T_0 = 0$ nucleus 113–17
  $T_0 \geq 1$ nuclei 117–22
isospin symmetry 250
isovector giant dipole resonance (IVGDR) 79,
    82–4
isovector (IV) modes 77, 78
  electric 103–4, 105
isovector spin monopole resonance (IVSM) 52

JLM optical potential 29

K-shell electron capture 235–9
K X-rays 246
  exotic 332–3
kaons
  (γ, K^+) reaction 259–64, 287
  (K^-, π^-) reaction 251–7
  (π^+, K^+) reaction 251–7, 275–84
Klein–Gordon equation 210–11
knock-out (KO) process 107

L_γ X-ray emission 227–9
lambda hyperon (Λ) 8–9
  hypernuclear weak decays 264–71
    non-mesonic weak decay and baryon–baryon
        interactions 271–4
    strangeness production by the weak
        interaction 284–7
    weak decays of polarized hypernuclei 274–84
λ-point 192–3
Landau's critical velocity 197
Large Acceptance Spectrograph (LAS) 217–18
laser electron photons 3, 11, 304–6
lattice QCD 298–9
lead
  $\alpha + {}^{208}$Pb system 179–81
  DBPAs 215–23
lepton nuclear physics 12–16, 313–36
  neutrinos and weak interactions 12–14, 314–23
    double beta decay experiments 317–18
    double beta decays 314–17
    ELEGANTs for double beta decays 318–21
    neutrino mass limits and double beta matrix
        elements 321–3
  nuclear spin–isospin responses 14–15, 323–6
  nuclear stabilities and nuclear decays 14–15,
      329–33
    charge non-conservations and electron
        decays 332–3
    exotic particles and transitions 329–30
    nucleon decays and deexcitation of nucleon
        holes 330–2
  spin-coupled dark matter 15, 16, 326–9
light exotic particles 329–30
light nuclei: α-clustering 153–7
light projectiles see projectiles
liquid helium see 'snowballs'
lithium 242
  elastic scattering and breakup of lithium ions
      141–6
  ^{6}Li hypernucleus decay to polarized
      ${}^5_\Lambda$He 280–4
Local-Potential Cluster Model (LPCM) 184–5
Low-Energy Hexadecapole Resonance (LEHR)
    98–101
Low-Energy Octupole Resonance (LEOR) 87,
    88, 90–7, 100, 102

M1 resonances 54–5, 56, 58–61
  isospin structure 112–13
    $T_0 = 0$ nucleus 113–17
    $T_0 \geq 1$ nuclei 117–18
Madison conventions 19, 22
magnetic giant resonances 77, 78
  isospin structure 112–22, 123
    M1 and Gamow–Teller resonances 112–13
    $T_0 = 0$ nucleus 113–17
    $T_0 \geq 1$ nuclei 117–22
magnetic monopole 310
Majorana neutrino exchange 12, 314, 315, 316
Majoron–neutrino coupling 12, 315, 316
mass number, target 28–32, 42–3
Massey criterion 226
Maximally Abelian (MA) gauge 10, 304, 311
Maxwell equations 294, 310
Mean Square Radius (MSR) of the optical
    potential 27–33
medium, nuclear see nuclear medium
Meissner effect 293
  dual Meissner effect 293, 295
meson-exchange models 272–4
mesonic decay 264–71, 275–84
  asymmetry in the pionic decays 269–71
  calculational framework of the decay
      rate 265–7
  p-shell nuclei 267–9
  weak decays of polarized hypernuclei 274–5,
      281–3
metallic elements 245–6
microscopic cluster model 134
microscopic coupled-channel formalism 133–8
microscopic optical potential 18–19, 23–7
  comparison with phenomenological optical
      potential 27–33
mobility of charge carriers 195–6
molybdenum
  α-clustering in ^{94}Mo region 172–6
  double beta decay of ^{100}Mo 319–21
    spin–isospin responses for ^{100}Mo 324–5
  structure of ^{94}Mo 174–6
momentum bins, method of 134–6
monopole
  Maxwell equations with magnetic
      monopole 310
  QCD 10, 11, 293–4
    appearance of the QCD monopole 308–9
    monopole condensation 294–5, 298
    multi-instantons and 303–4, 305
multi-instantons 302–4, 305
multipole moments 36–43
multistep transitions/processes 128–9, 132

negative-parity band 154–5, 160–8
neon: α-cluster structure of ^{20}Ne 154–7
neutrino capture cross section 53

neutrinoless double beta decay 12–14, 64,
      314–17
neutrinos 5–7
   double beta decays 12–14, 314–23
   gallium detector (^{71}Ga) 61–3
   mass limits 321–3
   solar 6, 14–15, 61–3, 325–6
neutron decay 105–6, 109
   of GQR induced by α-particles 109–12
neutron halo nuclei 129–30
neutrons
   density 46–8
   formation of DBPAs
      neutron–deuteron reactions 214–16
      neutron–proton reactions 212–14
      neutron–proton charge-exchange
            reactions 57–8
nickel: ^{58}Ni(^{3}He, t) ^{58}Cu reaction 118–22
nitrogen: decay from spin–isospin resonance in
      ^{13}N 69–70
non-mesonic decay 264–5
   and baryon–baryon interactions 271–4
   weak decays of polarized hypernuclei 274–5,
      275–80
non-Paulian transition 329, 330
non-resonant breakup process 127–9
nuclear clustering *see* α-cluster model; cluster
      structures
nuclear decays 329–33
   charge non-conservations and electron decays
         332–3
   and deexcitation of nucleon holes 330–2
nuclear instabilities 14–15, 329–33
nuclear medium 3–4
   effects in elastic scattering 46–8
nucleon–hadron many-body systems 1–2, 7–9
   hyperon interactions and flavour nuclear
         physics 8–9
   nucleons, mesons and isobars 7–8
   sensitivity and quality in studying 2
nucleon holes, deexcitation of 14, 330–2
nucleon many-body systems 3–7
   exotic nuclear beams and spin polarizations 7
   neutrino nuclear responses 5–7
   nucleon and nuclear interactions in nuclear
         medium 3–4
   spin–isospin waves and interactions in
         nuclei 4–5
nucleon–nucleon (NN) interactions
   in nuclear medium 3–4
   nucleons, mesons and isobars in NN
         channels 7–8

One-Pion-Exchange (OPE) model 272–4
optical potential 18–19, 19–23
   comparison of real parts of phenomenological
         and microscopic potentials 27–33

DBPAs 210–11, 212, 213
   deformed 34–8
   microscopic 18–19, 23–7
   phenomenological 19–23, 24, 27
Orthogonality-Condition Model (OCM) 184
Oto Cosmo observatory 2–3
oxygen 242
   hypernuclear polarization of ^{16}O 262–4
   structure of ^{16}O 154

p-shell hypernuclei 267–9
parity doublets 154–7
   higher-nodal positive-parity band 168–70
   $K = 0^-$ band in ^{40}Ca 167–8
   $K = 0^-$ band in ^{44}Ti 160–7
      observation 162–7
      prediction 160–2
particle decay
   excitation and decay 107–9
   of giant resonances 66–70, 81–2, 104–12, 123
      from highly excited states 104–6
      see also neutron decay; proton decay
Pauli-blocking effect 23–4
Perturbed Stationary State (PSS) theory 246
phenomenological optical potential 19–23, 24, 27
   compared with microscopic optical
         potential 27–33
phonons 194
pions 8
   deeply bound pionic atoms *see* deeply bound
         pionic atoms
   (K^-, π^-) reaction 251–7
   mesonic decay of hypernuclei 264–71, 275–84
      asymmetry in weak decays of $^5_\Lambda$He 269–71
      decay rate 265–7
      p-shell hypernuclei 267–9
   (π^+, K^+) reaction 251–7, 275–84
Plane Wave Born Approximation
         (PWBA) 212–14
polar cap model 42–3
polarization 7, 190–1
   exotic nuclear beams and spin polarizations 7
   monitoring beam polarization 19, 22
   snowball experiments 198–208
      measurement of freezing-out of polarization
            202–7
      nuclear polarization from heavy-ion
            reactions 199–202
polarized hypernuclei *see* hypernuclei
polarized ion sources 18
polonium: α-clustering in ^{212}Po region 176–81
Polyakov-like gauge 303
pomerons 11
pre-equilibrium (PEQ) excitation 107–8
projectiles
   breakup processes *see* breakup processes

electron capture and loss of light
projectiles 234–9
projectile–target interactions 136–8
stopping power of metallic elements for light
projectiles 245–6
proton decay of giant resonances 67–8, 69–70,
105–6, 109, 121–2
protons
density distributions 46–8
elastic scattering at around 65 MeV 19–33
elastic scattering at intermediate energies 43–8
inelastic scattering from deformed
nuclei 33–43
pion production reactions 8, 216–21
pn → pΛ process 284–7
properties of the reaction 286–7
strangeness production 284–6
pp bremsstrahlung 7
proton–neutron charge-exchange reactions 54
secondary electrons produced by 234
pseudo-states, method of 136

Quantum ChromoDynamics (QCD) 9–10, 11,
250, 292, 294
monopole *see* monopole
see also quark nuclear physics
quark–gluon coupling constant α_s 293
quark–lepton systems 1
quark nuclear physics (QNP) 9–11, 292–312
abelian gauge fixing 307–8
appearance of QCD monopole 308–9
baryon structure 10–11
dual Ginzburg–Landau theory 9–10, 293–302
colour confinement modelling 293–4
deconfinement phase transition 300–2
Lagrangian 294–5
recovery of chiral symmetry 299–300
spontaneous chiral symmetry
breaking 297–9
static confining potential 295–7
experimental facilities 11, 304–6
GeV photon facility 306
SPring-8 light source facility 304–6
instantons 302–3
maximally abelian gauge 311
Maxwell equations with magnetic
monopole 310
multi-instantons and monopole condensation
303–4
multi-instantons and the Wilson loop 304
SU(N) invariance condition 309
quartet model 157
quasi-free scattering (QFS) 3–4, 107
quenching 55–7, 116–17

R-matrix theory 180–1
Radioactive Nuclear Beams (RNBs) 57

RAIDEN magnetic spectrograph 2, 90, 91, 235–6
Random Phase Approximation (RPA) 1
Relativistic Hartree (RH) approximation 44,
46–7
Relativistic Impulse Approximation (RIA) 44–8
Relativistic Love–Franey model 44–5
Research Center for Nuclear Physics (RCNP) at
Osaka University 2–3
ELEGANTs 3, 13–14, 318–21
GRAND RAIDEN 2, 90, 91, 235–6
SPring-8 3, 287, 304–6
resonant breakup process 127–9
Resonating Group Method (RGM) 183
right–handed weak current (RHC) 314
ring cyclotron 2
rotational cluster bands 151–2
rotons 194, 197–8
Rydberg states 233–4

Satchler's theorem 36
scalar densities 46–8
scattering states: unification with bound states
154–7, 158–60, 179–81
Schrödinger equation 36, 43
non–relativistic 43–5
Schwinger–Dyson (SD) equation 297–8
sequential breakup process 127–9
silicon: isospin structure of M1 states in
^{28}Si 115–17
Single–Nucleon Knock–on Exchange (SNKE)
term 137
'snowballs' 7, 190–209
detection methods and production of
polarization 198–202
impurity ions in liquid helium 194–8
measurement of freezing–out of polarization
202–7
sodium iodide (NaI) detector array 15, 16
solar neutrinos 6, 14–15
gallium detector ^{71}Ga 61–3
spin–isospin responses 325–6
solidification of liquid helium 193, 195
spatial spreading 207
spin-coupled dark matter 326–9
spin-dependent interactions 130, 142–4
spin-flip Dipole Resonance (SDR) 50–2, 53, 54,
64, 65, 67–9
spin-isospin giant resonances 4, 5, 50–76
application of the (^{3}He, t) reaction 61–6
double beta decay 64–6
neutrino detector ^{71}Ga 61–3
charge-exchange reactions 54–5
Gamow–Teller strengths at high excitation
energy 55–7
Gamow–Teller transitions to the
β^+-side 57–61
isospin structure 113–22

spin-isospin giant resonances (*contd.*)
  particle decay of giant resonances 66–70
spin–isospin responses 4–5, 323–6
  for axial weak processes 323–4
  for ^{100}Mo and double beta decay matrix
     elements 324–5
  for solar neutrinos 325–6
spin-orbit structures 113–17
spin-singlet states 140
spin-triplet states 138–40
spreading width 81, 107–8
SPring-8 3, 287, 304–6
  GeV photon facility 306
  light source facility 304–6
Standard Solar Model (SSM) 61, 62, 63
static confining potential 295–7
stopping power 240–6
  of carbon for H_2^+ and H_3^+ ions 242–4
  charge-state-dependent 240–2
  of metallic elements for light projectiles 245–6
strange-flavour nuclear physics 8–9, 250–91
  basic aspects of hypernuclear weak decays
     264–71
  non–mesonic weak decay and baryon–baryon
     interactions 271–4
  production of polarized hypernuclei 251–64
  weak decays of polarized hypernuclei 274–84
strangeness 250, 284–7
  new possibilities of the (γ, K^+) reaction at
     SPring-8 287
  production by the weak pn $\rightarrow$ pΛ
     process 284–6
  properties of the np $\rightarrow$ Λp reaction 286–7
  see also hypernuclei
SU(N) invariance 296, 309
sum rule 79–81
superfluid helium *see* 'snowballs'
surface tension 301–2
SUSY model 314–15
symmetry studies 12–16
  nuclear instabilities and solar neutrinos 14–15

target mass number 28–32, 42–3
temperature, finite 299–302
thermodynamical potential 300–1
thickness, foil *see* foil thickness
3α-cluster state: breakup effect on excitation of
    145–6
3α structures 153
threshold rule 132–3, 151
time reversal (T) violation 285–6
titanium: α-cluster model in ^{44}Ti region 157–72
  α-cluster model and its dilemma 157–8
  higher–nodal positive–parity band 168–70

observation of parity doublet band 162–7
prediction of parity doublet band 160–2
unification of bound and scattering states of
    ^{44}Ti 158–60
transfer reactions 129, 131–2, 152
transition strengths 40–3
trapping time 198
tritons: in charge-exchange reactions 54–5
  application of $(^3$He, t$)$ reaction 61–6
  isobaric mirrors 58–61
  isospin structure of GT strength in ^{58}Ni
    $(^3$He, t$)$ ^{58}Cu reaction 118–22
2α structures 153
two-neutrino double beta decay 12–14, 64–6,
    314–17
  neutrino mass limits and 2$\nu\beta\beta$ matrix
     elements 321–3
    spin–isospin responses for 2$\nu\beta\beta$ matrix
      elements 324–5
  see also double beta decay
2p–2h configuration 55–7

unification of bound and scattering states 154–7,
    158–60, 179–81

vector densities 46–8
vector mesons 11
volume integral of the optical potential 27–33
vortices 197

weak decays
  double weak decay processes 330, 331
  hypernuclear 264–71
    experimental studies of polarized
      hypernuclei 274–84
  non-mesonic and baryon–baryon interactions
    271–4
weak processes
  axial 323–4
  pn $\rightarrow$ pΛ process 8, 284–7
weakly interacting massive particles (WIMPs)
    15–16
Wigner supermultiplet scheme 122
Wilson loop 304
Woods–Saxon form for optical potential 22–3,
    24, 27

X–rays
  exotic K X-rays 332–3
  L_γ X–ray emission in heavy-ion bombardment
    of Bi 227–9

zirconium: α-^{90}Zr interaction 172–4